Compendium of Organic Synthetic Methods

Volume 2

IAN T. HARRISON

and

SHUYEN HARRISON

SYNTEX RESEARCH
PALO ALTO, CALIFORNIA

A Wiley-Interscience Publication

JOHN WILEY & SONS, New York • London • Sydney • Toronto

Library of Congress Cataloging in Publication Data:

Harrison, Ian T
  Compendium of organic synthetic methods.

  1. Chemistry, Organic—Synthesis.   I. Harrison,
Shuyen, joint author.   II. Title

QD262.H32   547′.2   71-162800
ISBN 0-471-35551-8

Printed in the United States of America

10 9 8 7 6 5 4 3 2

PREFACE

Compendium of Organic Synthetic Methods, Volume 2 presents the early 1971 to early 1974 crop of published functional group transformations, plus the gleanings from previous years. It is in part a supplement to *Volume 1* and in addition contains a new chapter on the preparation of difunctional compounds. Previous reviewers and abstractors have tended to avoid difunctional compounds, having found the data particularly well concealed in the literature. Addition of this new chapter increases the complexity of the classification schemes, requiring more instruction in the use of the book. However, the system used is still simple enough that those who wish to just jump in will do well with no more guidance than that offered by the indexes.

The task of abstracting the approx. 90 journals covered by *Volume 1* and *2* of the *Compendium* becomes ever more difficult as chemical publications continue to appear in increasing numbers. Within a few more years it will not be possible for a team of two to continue as sole abstractors, writers, artists, typists and proofreaders for the *Compendium*. Perhaps a fully organized abstracting service, responsive to the needs of the synthetic chemist, will someday be set up. It would help, of course, if chemists would learn to write orderly, concise papers that can actually be read rather than deciphered.

Ian T. Harrison
Shuyen Harrison

Palo Alto, California
May 1974

CONTENTS

ABBREVIATIONS

Ac	acetyl
9-BBN	9-borabicyclo[3.3.1]nonane
Bu	butyl
DCC	dicyclohexylcarbodiimide
DDQ	2,3-dichloro-5,6-dicyanobenzoquinone
DMA	dimethylacetamide
DMF	dimethylformamide
Et	ethyl
HMPA	hexamethylphosphoramide (hexamethylphosphoric triamide)
Me	methyl
Ms	methanesulfonyl
NBA	*N*-bromoacetamide
NBS	*N*-bromosuccinimide
NCS	*N*-chlorosuccinimide
Ni	Raney nickel
NIS	*N*-iodosuccinimide
Ph	phenyl
Pr	propyl
Pyr	pyridine
THF	tetrahydrofuran
THP	tetrahydropyranyl
Ts	*p*-toluenesulfonyl

INDEX, MONOFUNCTIONAL COMPOUNDS

Sections—heavy type
Pages—light type

PREPARATION OF →
FROM ↓

(Each cell gives section number — heavy type — over page number — light type. Blank cells correspond to sections for which no additional examples were found in the literature.)

FROM ↓ \ PREPARATION OF →	Acetylenes	Carboxylic acids, acid halides, anhydrides	Alcohols, phenols	Aldehydes	Alkyls, methylenes, aryls	Amides	Amines	Esters	Ethers, epoxides	Halides, sulfonates, sulfates	Hydrides (RH)	Ketones	Nitriles	Olefins
Acetylenes	1 / 1	16 / 7	31 / 28	46 / 53			91 / 92	106 / 108				166 / 155		196 / 193
Carboxylic acids, acid halides, anhydrides	2 / 1	17 / 8	32 / 29	47 / 53	62 / 70	77 / 82	92 / 92	107 / 108		137 / 136	152 / 146	167 / 156	182 / 185	197 / 194
Alcohols, phenols		18 / 10	33 / 30	48 / 55	63 / 71	78 / 84	93 / 93	108 / 113	123 / 127	138 / 137	153 / 146	168 / 157		198 / 195
Aldehydes	4 / 2	19 / 11	34 / 31	49 / 57	64 / 71	79 / 84	94 / 94	109 / 115	124 / 130		154 / 148	169 / 159	184 / 186	199 / 197
Alkyls, methylenes, aryls		20 / 12		50 / 59	65 / 72							170 / 160		200 / 199
Amides		21 / 13		51 / 59		81 / 85	96 / 95				156 / 149	171 / 161	186 / 187	
Amines			37 / 32	52 / 60		82 / 86	97 / 96	112 / 117		142 / 141	157 / 149	172 / 161	187 / 188	202 / 199
Esters		23 / 13	38 / 32			83 / 87		113 / 117			158 / 150	173 / 162		203 / 200
Ethers, epoxides		24 / 15	39 / 34	54 / 61	69 / 73		99 / 98	114 / 118	129 / 131	144 / 141	159 / 150	174 / 162		204 / 200
Halides, sulfonates, sulfates	10 / 3	25 / 16	40 / 36	55 / 61	70 / 73	85 / 88	100 / 99	115 / 119	130 / 132	145 / 141	160 / 151	175 / 163	190 / 189	205 / 201
Hydrides (RH)		26 / 18	41 / 38	56 / 63	71 / 76		101 / 101	116 / 121	131 / 132	146 / 142		176 / 167	191 / 189	206 / 204
Ketones	12 / 4	27 / 20	42 / 41	57 / 64	72 / 77		102 / 101	117 / 122	132 / 132		162 / 152	177 / 168	192 / 189	207 / 204
Nitriles				58 / 65		88 / 89	103 / 102	118 / 124			163 / 153	178 / 174	193 / 190	
Olefins	14 / 5	29 / 23	44 / 45	59 / 66	74 / 79	89 / 90	104 / 103	119 / 124	134 / 133	149 / 144		179 / 175		209 / 208
Miscellaneous compounds	15 / 5	30 / 24	45 / 46	60 / 66			105 / 104	120 / 125	135 / 135	150 / 145	165 / 154	180 / 178	195 / 191	210 / 211

PROTECTION

	Sect.	Pg.
Acetylenes	15A	5
Carboxylic acids	30A	26
Alcohols, phenols	45A	47
Aldehydes	60A	67
Amides	90A	90
Amines	105A	105
Esters	120A	126
Ketones	180A	181
Olefins	210A	211

Blanks in the table correspond to sections for which no additional examples were found in the literature.

INDEX, DIFUNCTIONAL COMPOUNDS

Sections—heavy type

Pages—light type

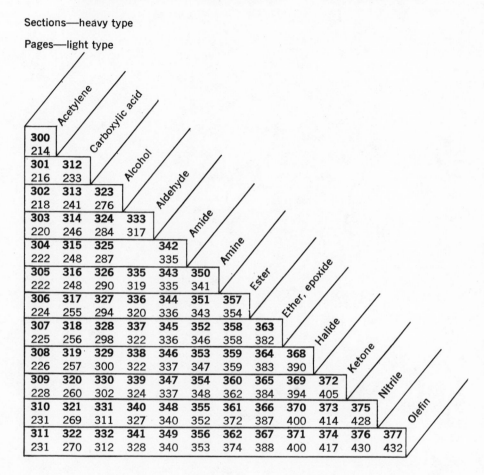

	Acetylene	Carboxylic acid	Alcohol	Aldehyde	Amide	Amine	Ester	Ether, epoxide	Halide	Ketone	Nitrile	Olefin
300												
214												
301	**312**											
216	233											
302	**313**	**323**										
218	241	276										
303	**314**	**324**	**333**									
220	246	284	317									
304	**315**	**325**		**342**								
222	248	287		335								
305	**316**	**326**	**335**	**343**	**350**							
222	248	290	319	335	341							
306	**317**	**327**	**336**	**344**	**351**	**357**						
224	255	294	320	336	343	354						
307	**318**	**328**	**337**	**345**	**352**	**358**	**363**					
225	256	298	322	336	346	358	382					
308	**319**	**329**	**338**	**346**	**353**	**359**	**364**	**368**				
226	257	300	322	337	347	359	383	390				
309	**320**	**330**	**339**	**347**	**354**	**360**	**365**	**369**	**372**			
228	260	302	324	337	348	362	384	394	405			
310	**321**	**331**	**340**	**348**	**355**	**361**	**366**	**370**	**373**	**375**		
231	269	311	327	340	352	372	387	400	414	428		
311	**322**	**332**	**341**	**349**	**356**	**362**	**367**	**371**	**374**	**376**	**377**	
231	270	312	328	340	353	374	388	400	417	430	432	

Blanks in the table correspond to sections for which no examples were found in the literature.

HOW TO USE THE COMPENDIUM

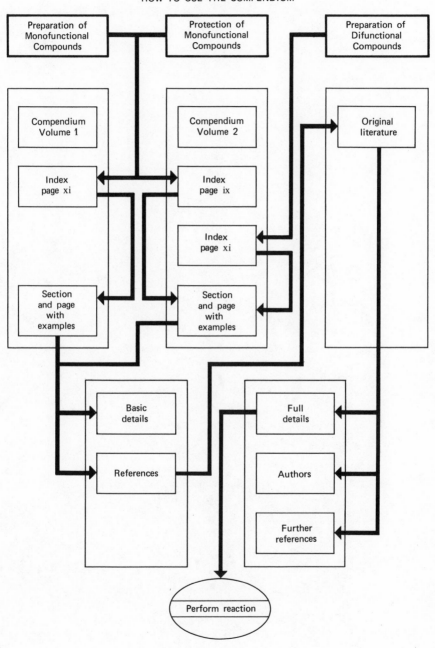

INTRODUCTION

Relationship Between Volume 1 and Volume 2. *Compendium of Organic Synthetic Methods, Volume 2,* presents more than 1000 examples of published methods for the preparation of monofunctional compounds, updating the 3000 or so in Volume 1. In addition Volume 2 presents a new chapter with about 1000 examples of the preparation of difunctional compounds. Methods for the protection of carboxylic acids, alcohols, phenols, aldehydes, amines, and ketones were included in Volume 1. In Volume 2 protective methods for acetylenes, amides, esters, and olefins are also covered.

The same systems of section and chapter numbering are used in the two volumes.

Classification and Organization of Reactions Forming Monofunctional Compounds. Examples of published chemical transformations are classified according to the reacting functional group of the starting material and the functional group formed. Those reactions that give products with the same functional group form a chapter. The reactions in each chapter are further classified into sections on the basis of the functional group of the starting material. Within each section reactions are listed in a somewhat arbitrary order, although an effort has been made to put chain-lengthening processes before degradations.

The classification is unaffected by allylic, vinylic, or acetylenic unsaturation, which appears in both starting material and product, or increases or decreases in the length of carbon chains; for example, the reactions t-BuOH $\rightarrow$ t-BuCOOH, $PhCH_2OH \rightarrow PhCOOH$ and $PhCH=CHCH_2OH \rightarrow PhCH=CHCOOH$ would all be considered as preparations of carboxylic acids from alcohols.

The terms hydrides, alkyls, and aryls classify compounds containing reacting hydrogens, alkyl groups, and aryl groups, respectively; for example, $RCH_2\text{-}H \rightarrow RCH_2COOH$ (carboxylic acids from hydrides), RMe $\rightarrow$ RCOOH (carboxylic acids from alkyls), RPh $\rightarrow$ RCOOH (carboxylic acids from aryls). Note the distinction between $R_2CO \rightarrow R_2CH_2$ (methylenes from ketones) and $RCOR' \rightarrow RH$ (hydrides from ketones).

The following examples illustrate the application of the classification scheme to some potentially confusing cases:

$RCH=CHCOOH \rightarrow RCH=CH_2$	(hydrides from carboxylic acids)
$RCH=CH_2 \rightarrow RCH=CHCOOH$	(carboxylic acids from hydrides)
$ArH \rightarrow ArCOOH$	(carboxylic acids from hydrides)

$$ArH \rightarrow ArOAc$$ (esters from hydrides)
$$RCHO \rightarrow RH$$ (hydrides from aldehydes)
$$RCH = CHCHO \rightarrow RCH = CH_2$$ (hydrides from aldehydes)
$$RCHO \rightarrow RCH_3$$ (alkyls from aldehydes)
$$R_2CH_2 \rightarrow R_2CO$$ (ketones from methylenes)
$$RCH_2COR \rightarrow R_2CHCOR$$ (ketones from ketones)
$$RCH = CH_2 \rightarrow RCH_2CH_3$$ (alkyls from olefins)
$$RBr + RC \equiv CH \rightarrow RC \equiv CR$$ (acetylenes from halides; also acetylenes from acetylenes)
$$ROH + RCOOH \rightarrow RCOOR$$ (esters from alcohols; also esters from carboxylic acids)

Yields quoted are overall; they are reduced to allow for incomplete conversion and impurities in the product.

Reactions not described in the given references, but required to complete a sequence, are indicated by a dashed arrow.

Reactions are included even when full experimental details are lacking in the given reference. In some cases the quoted reaction is a minor part of a paper or may have been investigated from a purely mechanistic aspect. When several references are given, the first refers to the reaction illustrated; others give further examples, related reactions, or reviews.

How to Use the Book to Locate Examples of the Preparation or Protection of Monofunctional Compounds. Examples of the preparation of one functional group from another are located via the monofunctional index on p. ix, which lists the corresponding section and page. Thus Section 1 contains examples of the preparation of acetylenes from other acetylenes; Section 2, acetylenes from carboxylic acids; and so forth.

Sections that contain examples of the reactions of a functional group are found in the horizontal rows of the index. Thus Section 1 gives examples of the reactions of acetylenes that form other acetylenes; Section 16, reactions of acetylenes that form carboxylic acids; and Section 31, reactions of acetylenes that form alcohols.

Examples of alkylation, dealkylation, homologation, isomerization, transposition are found in Sections 1, 17, 33, and so on, which lie close to a diagonal of the index. These sections correspond to such topics as the preparation of acetylenes from acetylenes, carboxylic acids from carboxylic acids, and alcohols and phenols from alcohols and phenols.

Examples of name reactions can be found by first considering the nature of the starting material and product. The Wittig reaction, for instance, is in Section 199 on olefins from aldehydes and Section 207 on olefins from ketones.

Examples of the protection of acetylenes, carboxylic acids, alcohols, phenols, aldehydes, amides, amines, esters, ketones, and olefins are also indexed on p. ix.

The pairs of functional groups alcohol, ester; carboxylic acid, ester; amine, amide; carboxylic acid, amide can be interconverted by quite trivial reactions. When a member of these groups is the desired product or starting material, the other member should, of course, also be consulted in the text.

A few reactions already presented in Volume 1 are given again in Volume 2 when significant new publications have appeared. In such cases the starting material and product are shown in a contracted form; for example, ROH instead of $PhCH_2CH_2OH$.

The original literature must be used to determine the generality of reactions. A reaction given in this book for a primary aliphatic substrate may also be applicable to tertiary or aromatic compounds.

The references usually yield a further set of references to previous work. Subsequent publications can be found by consulting the Science Citation Index.

Classification and Organization of Reactions forming Difunctional Compounds. This new chapter considers all possible difunctional compounds formed from the groups acetylene, carboxylic acid, alcohol, aldehyde, amide, amine, ester, ether, epoxide, halide, ketone, nitrile, and olefin. Reactions that form difunctional compounds are classified into sections on the basis of the two functional groups of the product. The relative positions of the groups do not affect the classification. Thus preparations of 1,2-aminoalcohols, 1,3-aminoalcohols and 1,4-aminoalcohols are included in a single section. It is recommended that the following illustrative examples of the classification of difunctional compounds be scrutinized closely.

Difunctional Product	*Section Title*
$RC \equiv C\text{-}C \equiv CR$	Acetylene—Acetylene
$RCH(OH)COOH$	Carboxylic Acid—Alcohol
$RCH(COOH)CH_2COOMe$	Carboxylic Acid—Ester
$RCH(OAc)COOH$	Carboxylic Acid—Ester
$RCH = CHOMe$	Ether—Olefin
$RCH(OMe)_2$	Ether—Ether
$RCHF_2$	Halide—Halide
$RCH(Br)CH_2F$	Halide—Halide
$RCH(OAc)CH_2OH$	Alcohol—Ester
$RCH(OH)COOMe$	Alcohol—Ester
$RCOCOOEt$	Ester—Ketone
$RCOCH_2OAc$	Ester—Ketone
$RCH = CHCH_2COOMe$	Ester—Olefin
$RCH = CHOAc$	Ester—Olefin
$RCH(Br)COOEt$	Ester—Halide
$RCH(Br)CH_2OAc$	Ester—Halide
$RCH = CHCH_2CH = CH_2$	Olefin—Olefin

How to Use the Book to Locate Examples of the Preparation of Difunctional Compounds. The difunctional index on p. xi gives the section and page corresponding to each difunctional product. Thus Section 327 (Alcohol—Ester) contains examples of the preparation of hydroxyesters; Section 323 (Alcohol —Alcohol) contains examples of the preparation of diols.

Some preparations of olefinic and acetylenic compounds from olefinic and acetylenic starting materials can, in principle, be classified in either the monofunctional or difunctional sections; for example, $RCH=CHBr \rightarrow RCH= CHCOOH$, Carboxylic acids from Halides (monofunctional sections) or Carboxylic acid—Olefin (difunctional sections). In such cases both sections should be consulted.

Reactions applicable to both aldehyde and ketone starting materials are in many cases illustrated by an example that uses only one of them.

Many literature preparations of difunctional compounds are extensions of the methods applicable to monofunctional compounds. Thus the reaction $RCl \rightarrow ROH$ can clearly be extended to the preparation of diols by using the corresponding dichloro compound as a starting material. Such methods are not fully covered in the difunctional sections.

The user should bear in mind that the pairs of functional groups alcohol, ester; carboxylic acid, ester; amine, amide; carboxylic acid, amide can be interconverted by quite trivial reactions. Compounds of the type $RCH(OAc)CH_2OAc$ (Ester—Ester) would thus be of interest to anyone preparing the diol $RCH(OH)CH_2OH$ (Alcohol—Alcohol).

Chapter 1 PREPARATION OF ACETYLENES

Section 1 <u>Acetylenes from Acetylenes</u>

$$t\text{-}BuC{\equiv}CH \xrightarrow[\text{2 } I_2 \quad Et_2O]{\text{1 } Ph_3B \quad BuLi \quad THF} t\text{-}BuC{\equiv}CPh \qquad 83\%$$

JACS (1973) <u>95</u> 3080

Section 2 <u>Acetylenes from Carboxylic Acids</u>

$$\underset{(CH_2)_4COOH}{\overset{COOMe}{|}} \xrightarrow[\text{electrolysis} \quad MeOH]{C_8H_{17}C{\equiv}C(CH_2)_3COOH} \underset{(CH_2)_7C{\equiv}CC_8H_{17}}{\overset{COOMe}{|}} \qquad 24\%$$

JCS (1955) 2218

Section 3 <u>Acetylenes from Alcohols</u>

No additional examples

Section 4   Acetylenes from Aldehydes

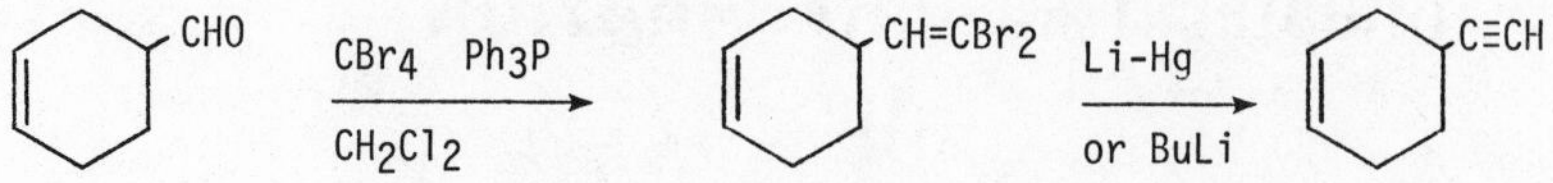

Tetr Lett (1972) 3769

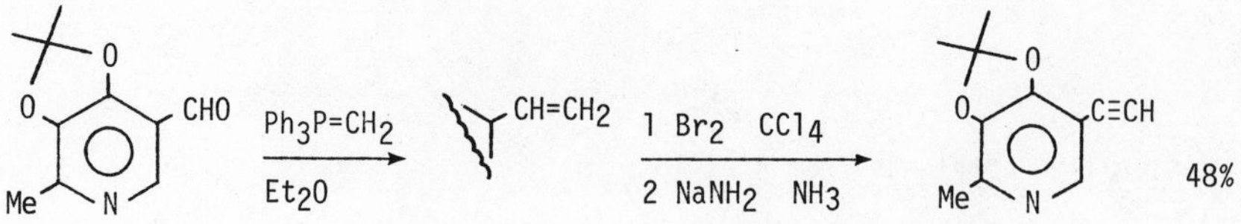

48%

J Med Chem (1972) 15 1262

Section 5   Acetylenes from Alkyls, Methylenes and Aryls

No examples

Section 6   Acetylenes from Amides

No examples

Section 7   Acetylenes from Amines

No additional examples

Section 8   Acetylenes from Esters

No additional examples

Section 9   Acetylenes from Ethers

No examples

Section 10   Acetylenes from Halides

Examples of the conversion of dibromides into acetylenes are included
in section 14 (Acetylenes from Olefins)

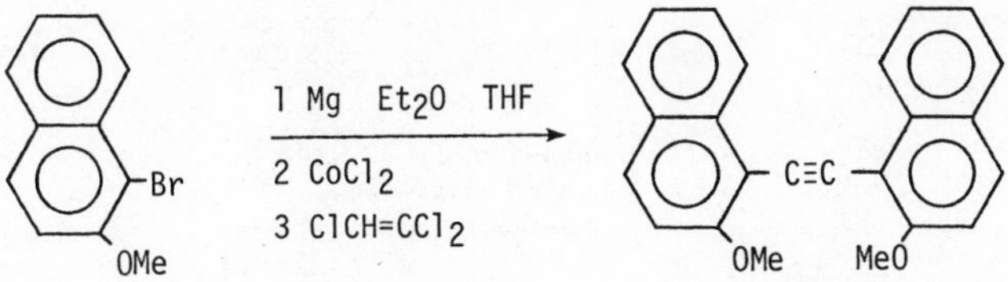

1 Mg Et$_2$O THF

2 CoCl$_2$

3 ClCH=CCl$_2$

55%

Synthesis (1972) 38

1 Li THF

2 CuBr Et$_2$O

3 IC≡CSiMe$_3$

4 Base

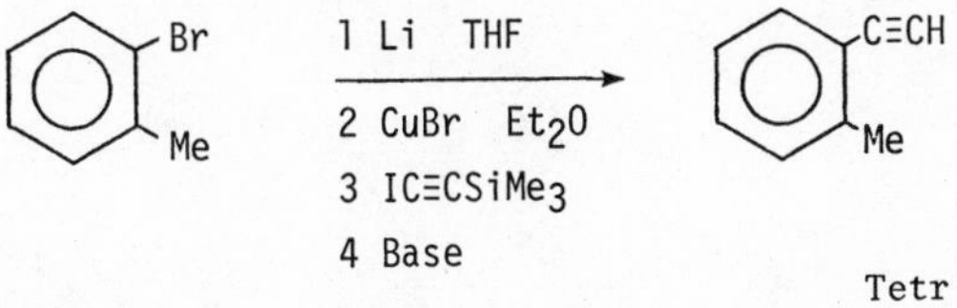

<48%

Tetr Lett (1972) 5209

RBr $\xrightarrow[\text{NH}_3]{\text{NaC≡CH}}$ RC≡CH

Org Synth (1963) Coll Vol 4 117
Angew (1959) 71 245

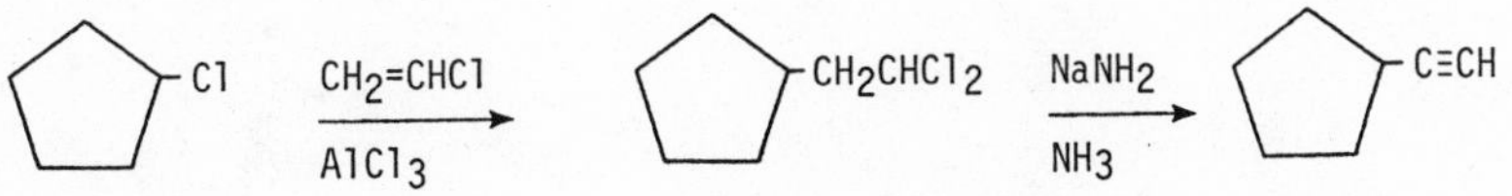

CH$_2$=CHCl / AlCl$_3$

-CH$_2$CHCl$_2$

NaNH$_2$ / NH$_3$

-C≡CH   38%

Rec Trav Chim (1965) 84 31

Section 11   Acetylenes from Hydrides

No examples

For examples of the reaction RC≡CH ⟶ RC≡C-C≡CR' see section 300
(Acetylene - Acetylene)

Section 12   Acetylenes from Ketones

$PhCOCH_2Me$ $\xrightarrow{\text{1 NH}_2\text{CONHNH}_2\;\;\text{2 SeO}_2\;\;\text{HOAc}\;\;\text{3 }\Delta}$ $PhC≡CMe$ 49%

1 $NH_2CONHNH_2$

2 SeO_2 HOAc

3 Δ

Angew (1970) 82 484
Internat Ed 9 464
Chem Comm (1971) 1059

$(CH_2)_9$ $\begin{array}{c}CH_2\\CO\\CH_2\end{array}$ $\xrightarrow[\text{2 Et}_3\text{N}]{\text{1 Br}_2}$ $(CH_2)_9$ $\begin{array}{c}C\\||\;CO\\C\end{array}$ $\xrightarrow{210°}$ $(CH_2)_9$ $\begin{array}{c}C\\|||\\C\end{array}$ ~7%

JACS (1965) 87 1326

$\begin{pmatrix}COCH_2\\(CH_2)_{13}\end{pmatrix}$ $\xrightarrow[\text{dioxane}]{\text{SeO}_2\;\;\text{H}_2\text{O}}$ $\begin{pmatrix}COCO\\(CH_2)_{13}\end{pmatrix}$ $\xrightarrow[\substack{\text{2 h}\nu\;\;\text{NaOH}\;\;\text{MeOH}\\\text{H}_2\text{O}}]{\text{1 NH}_2\text{NHTs}\;\;\text{HOAc}}$ $\begin{pmatrix}C≡C\\(CH_2)_{13}\end{pmatrix}$ 23%

Synthesis (1971) 215

Section 13   Acetylenes from Nitriles

No examples

Section 14   Acetylenes from Olefins

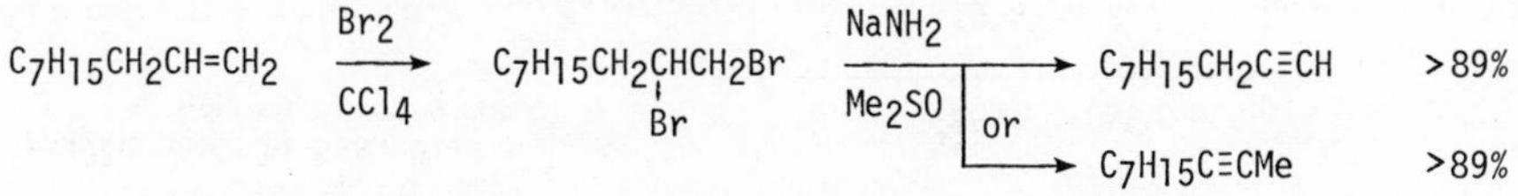

$$C_7H_{15}CH_2CH=CH_2 \xrightarrow[CCl_4]{Br_2} C_7H_{15}CH_2\underset{Br}{\underset{|}{C}H}CH_2Br \xrightarrow[Me_2SO]{NaNH_2}$$

or

$$\rightarrow C_7H_{15}CH_2C\equiv CH \qquad >89\%$$

$$\rightarrow C_7H_{15}C\equiv CMe \qquad >89\%$$

Tetrahedron (1970) <u>26</u> 2127 2637

Section 15   Acetylenes from Miscellaneous Compounds

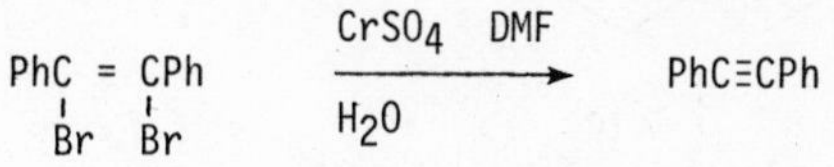

$$\underset{Br \quad Br}{PhC = CPh} \xrightarrow[H_2O]{CrSO_4 \quad DMF} PhC\equiv CPh \qquad 96\%$$

JACS (1964) <u>86</u> 4603

$$\xrightarrow[HOAc \quad CH_2Cl_2]{TsNHNH_2 \quad KOAc}$$

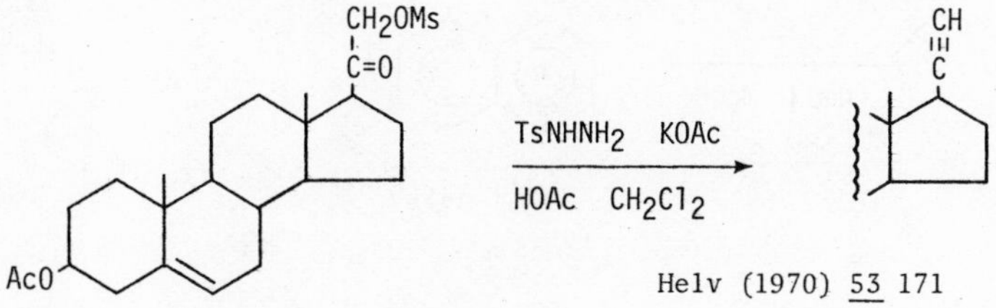

50%

Helv (1970) <u>53</u> 171

Section 15A   Protection of Acetylenes

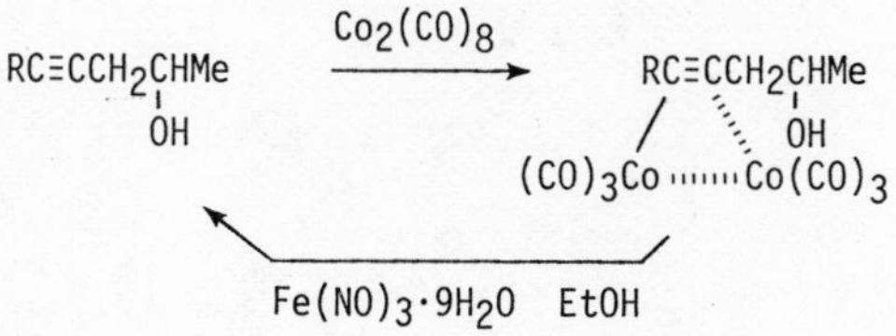

$$\underset{OH}{\underset{|}{RC\equiv CCH_2CHMe}} \xrightarrow{Co_2(CO)_8}$$

(Stable to B_2H_6, diimide and acid)

$Fe(NO)_3\cdot 9H_2O$ EtOH

Tetr Lett (1971) 3475

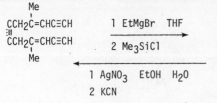

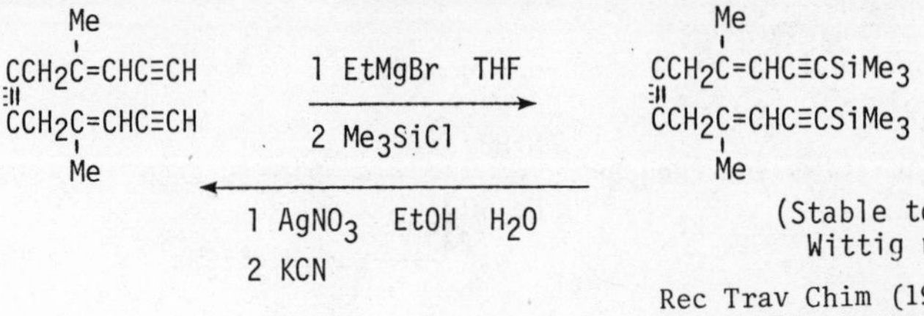

$$\begin{array}{c} \text{Me} \\ \text{CCH}_2\overset{|}{\text{C}}\text{=CHC}\equiv\text{CH} \\ \overset{||}{\text{CCH}_2}\text{C=CHC}\equiv\text{CH} \\ \overset{|}{\text{Me}} \end{array}$$

1 EtMgBr   THF
─────────────→
2 Me$_3$SiCl

←─────────────

1 AgNO$_3$ EtOH   H$_2$O

2 KCN

$$\begin{array}{c} \text{Me} \\ \text{CCH}_2\overset{|}{\text{C}}\text{=CHC}\equiv\text{CSiMe}_3 \\ \overset{||}{\text{CCH}_2}\text{C=CHC}\equiv\text{CSiMe}_3 \\ \overset{|}{\text{Me}} \end{array}$$

(Stable to acid, NaH
       Wittig reagents and H$_2$/Pd)

Rec Trav Chim (1967) 86 1138
JCS Perkin I (1972) 361
JACS (1968) 90 5618
Tetrahedron (1972) 28 4591 4601 5221

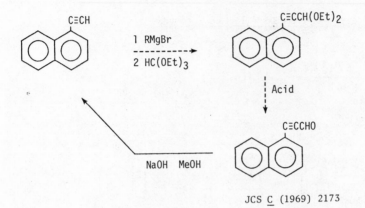

1 RMgBr
- - - - - - - - →
2 HC(OEt)$_3$

Acid

NaOH   MeOH

JCS C (1969) 2173

Chapter 2

PREPARATION OF CARBOXYLIC ACIDS ACID HALIDES AND ANHYDRIDES

Section 16   Carboxylic Acids from Acetylenes

$C_8H_{17}C{\equiv}CH$ $\xrightarrow{\begin{array}{l} 1 \ LiNH_2 \quad NH_3 \\ 2 \ Br(CH_2)_5COOH \\ 3 \ H_2 \quad Pd{-}C \quad EtOH \end{array}}$ $C_8H_{17}(CH_2)_7COOH$ 92%

JCS (1963) 775
(1965) 894

$t{-}BuC{\equiv}CH$ $\xrightarrow{\begin{array}{l} RuO_2 \quad NaOCl \\ CCl_4 \quad H_2O \end{array}}$ $t{-}BuCOOH$ 60%

Tetr Lett (1971) 2941

$C_6H_{13}C{\equiv}CH$ $\xrightarrow{\begin{array}{l} Tl(NO_3)_3 \quad HClO_4 \\ MeOCH_2CH_2OMe \quad H_2O \end{array}}$ $C_6H_{13}COOH$ 80%

JACS (1973) 95 1296

7

Also via: Section

Section 17   <u>Carboxylic Acids, Acid Halides and Anhydrides</u>

              <u>from Carboxylic Acids</u>

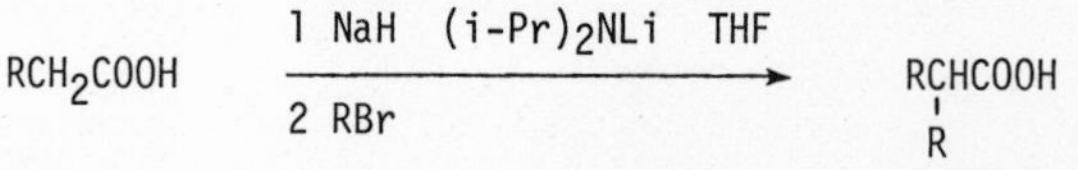

$$RCH_2COOH \quad \xrightarrow[\text{2 RBr}]{\text{1 NaH (i-Pr)}_2\text{NLi   THF}} \quad \underset{\overset{|}{R}}{RCHCOOH}$$

                                    Org Synth (1970) <u>50</u> 58
                                    JOC (1972) <u>37</u> 451

Carboxylic Acids may be alkylated and homologated via ketoacid, ketoester
and olefinic acid intermediates. See section 320 (Carboxylic Acid -
Ketone), section 360 (Ester - Ketone) and section 322 (Carboxylic Acid -
Olefin)

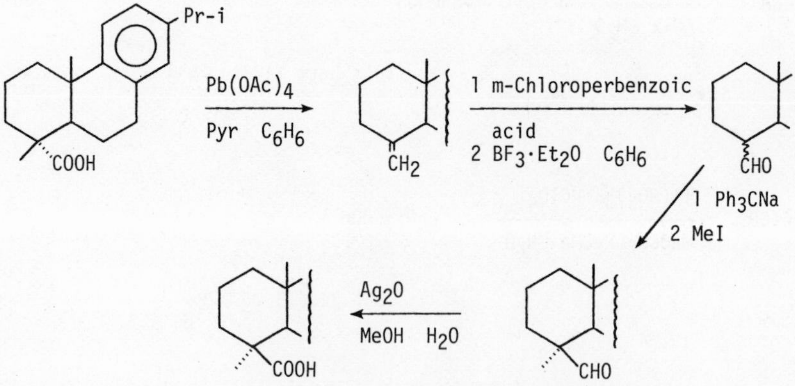

Chem Comm (1971) 10

$$C_{14}H_{29}CH_2COOH \xrightarrow[\substack{2 \ PhH \\ AlCl_3}]{1 \ SOCl_2} C_{14}H_{29}CH_2COPh \xrightarrow[\substack{HCl \ dioxane \\ 2 \ TsCl \ NaOH \\ 3 \ KOH \ PrOH}]{1 \ i\text{-}PrCH_2CH_2ONO} C_{14}H_{29}COOH \quad 63\%$$

JACS (1953) 75 2347
JOC (1961) 26 3507

$$C_7H_{15}COOH \xrightarrow[\substack{BF_3 \cdot Et_2O \ CH_2Cl_2}]{MeC{\equiv}CH \ Hg(OAc)_2} C_7H_{15}COOC{=}CH_2 \xrightarrow{HF} C_7H_{15}COF \quad 35\%$$
$$\underset{Me}{|}$$

JOC (1969) 34 2486

$$C_{17}H_{35}COOH \xrightarrow{PCl_5 \ C_6H_6} C_{17}H_{35}COCl \quad 70\%$$

JACS (1945) 67 2239

$$RCOOK \xrightarrow[CH_2Cl_2]{SOCl_2 \ Pyr} RCOCl$$

Can J Chem (1968) 46 2549
JACS (1963) 85 643

$$PhCOOH \xrightarrow{(Me_2N)_3P \ CCl_4} (PhCO)_2O \quad 90\%$$

Bull Soc Chim Fr (1971) 3034

$$C_{17}H_{35}COOH \xrightarrow[C_6H_6 \ Et_2O]{Resin\text{-}CH_2N{=}C{=}NPr\text{-}i} (C_{17}H_{35}CO)_2O \quad 65\%$$

Tetr Lett (1972) 3281

Section 18   Carboxylic Acids from Alcohols

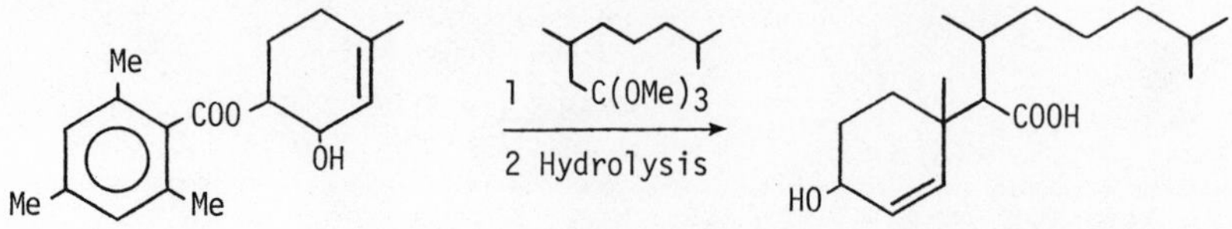

$$\xrightarrow[\text{2 Hydrolysis}]{1 \quad \text{C(OMe)}_3}$$

Chem Comm (1970) 1513 1512

PrOH $\xrightarrow[\text{Ph}_3\text{P} \quad \text{THF}]{\overset{\displaystyle \text{CH}_2\text{COOEt}}{\overset{\displaystyle \text{CN}}{|}} \quad \overset{\displaystyle \text{NCOOEt}}{\overset{\displaystyle \text{NCOOEt}}{\overset{\displaystyle \|}{}}}}$ $\overset{\displaystyle \text{CN}}{\overset{\displaystyle |}{\text{PrCHCOOEt}}}$ $\xrightarrow{\text{Hydrolysis}}$ PrCH$_2$COOH   < 52%

Tetr Lett (1972) 1279

$C_6H_{13}CH_2OH$ $\xrightarrow[\text{C}_6\text{H}_6]{\text{KMnO}_4 \quad \text{dicyclohexyl-18-crown-6}}$ $C_6H_{13}COOH$ 70%

JACS (1972) 94 4024

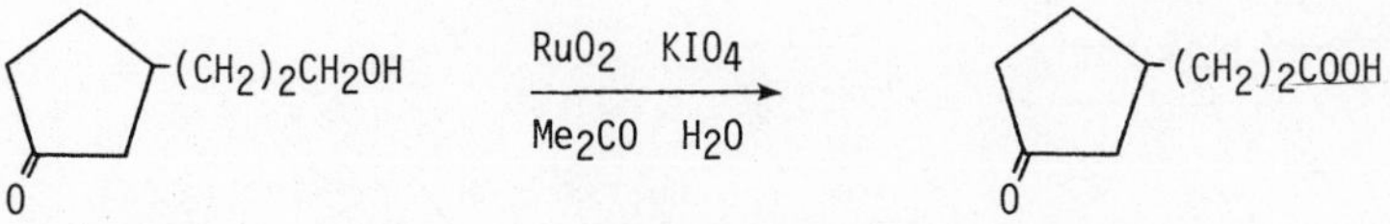

$$\xrightarrow[\text{Me}_2\text{CO} \quad \text{H}_2\text{O}]{\text{RuO}_2 \quad \text{KIO}_4}$$

JOC (1972) 37 1947

RCH$_2$OH $\xrightarrow{\text{KOH} \quad 245\text{-}255°}$ RCOOH

JACS (1948) 70 3485

$$\underset{\underset{\text{Et}}{|}}{\overset{\overset{\text{Me}}{|}}{\text{t-BuCHCHOH}}} \xrightarrow{\text{Br}_2 \quad \text{NaOH} \quad \text{H}_2\text{O}} \underset{\underset{\text{Et}}{|}}{\text{t-BuCHCOOH}} \qquad 38\%$$

JACS (1950) <u>72</u> 3701

Section 19   <u>Carboxylic Acids and Anhydrides from Aldehydes</u>

$$\text{C}_6\text{H}_{13}\text{CHO} \xrightarrow[\text{dibenzoyl peroxide}]{\text{CH}_2\text{=CH(CH}_2)_8\text{COOEt}} \text{C}_6\text{H}_{13}\text{CO(CH}_2)_{10}\text{COOEt} \xrightarrow[\substack{\text{diethylene} \\ \text{glycol}}]{\text{N}_2\text{H}_4 \quad \text{KOH}} \text{C}_6\text{H}_{13}\text{(CH}_2)_{11}\text{COOH}$$

Rec Trav Chim (1953) <u>72</u> 84

$$\text{PrCHO} \xrightarrow[\text{2 Acid}]{\text{1 Me}_3\text{SiCLi}} \text{PrCH=C} \overset{\text{S}}{\underset{\text{S}}{\Big\langle}} \xrightarrow{\text{HgCl}_2} \text{PrCH}_2\text{COOH}$$

Chem Comm (1972) 526
Ber (1973) <u>106</u> 2277

$$\text{C}_5\text{H}_{11}\text{CHO} \xrightarrow{\text{m-Chloroperbenzoic acid} \quad \text{THF}} \text{C}_5\text{H}_{11}\text{COOH}$$

JACS (1967) <u>89</u> 291

$$\text{PhCH=CHCHO} \dashrightarrow \text{PhCH=CHCH} \overset{\text{S}}{\underset{\text{S}}{\Big\langle}} \xrightarrow[\text{2 MeSSMe}]{\text{1 BuLi}} \text{PhCH=CHCSMe} \overset{\text{S}}{\underset{\text{S}}{\Big\langle}} \xrightarrow[\text{Me}_2\text{CO} \quad \text{H}_2\text{O}]{\text{HgCl}_2 \quad \text{HgO}}$$

PhCH=CHCOOH   < 86%

JOC (1972) <u>37</u> 2757

PhCHO $\xrightarrow[\text{C}_6\text{H}_6]{\text{KMnO}_4 \quad \text{dicyclohexyl-18-crown-6}}$ PhCOOH

JACS (1972) 94 4024

Related methods: Carboxylic Acids from Ketones (Section 27)

		Section
Also via:	Esters	109
	Amides	79
	Ketoacids	320
	Ketoesters	360
	Olefinic acids	322
	Olefinic esters	362
	Olefinic amides	349
	Acetylenic acids	301

PhCHO $\xrightarrow{\text{PhCOO}_2\text{Bu-t} \quad \text{CuBr}}$ (PhCO)$_2$O   70%

JOC (1960) 25 899

Section 20   Carboxylic Acids from Alkyls
°°°°°°°°°°°°°°°°°°°°°°°°°°°°°°°°°°°°

KMnO$_4$ dicyclohexyl-18-crown-6

JACS (1972) 94 4024

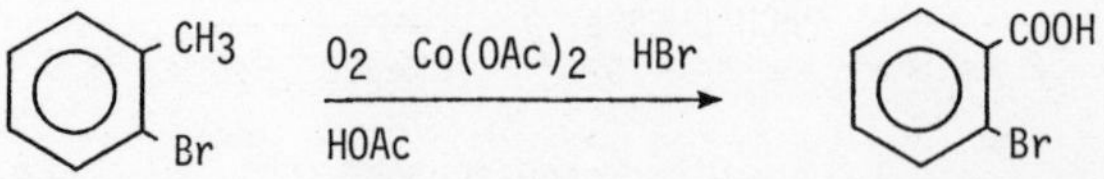

O$_2$ Co(OAc)$_2$ HBr

HOAc   91%

Can J Chem (1965) 43 1306
Chem Comm (1971) 1166
JOC (1972) 37 2564

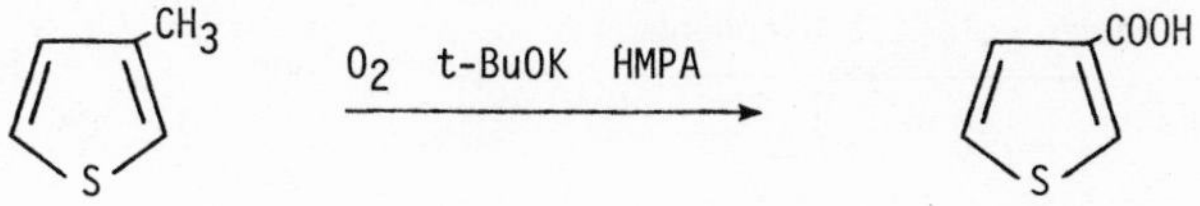

O_2 t-BuOK   HMPA   80%

JOC (1965) <u>30</u> 3520
JACS (1965) <u>87</u> 2523

Section 21   Carboxylic Acids from Amides

Et$_3$CCONH$_2$ $\xrightarrow{\text{NaNO}_2 \quad \text{H}_2\text{SO}_4 \quad \text{H}_2\text{O}}$ Et$_3$CCOOH   >80%

M. S. Newman, Steric Effects in
Organic Chemistry (J. Wiley 1956)
228 footnote 73

Section 22   Carboxylic Acids, Acid Halides and Anhydrides from Amines

No additional examples

Section 23   Carboxylic Acids from Esters

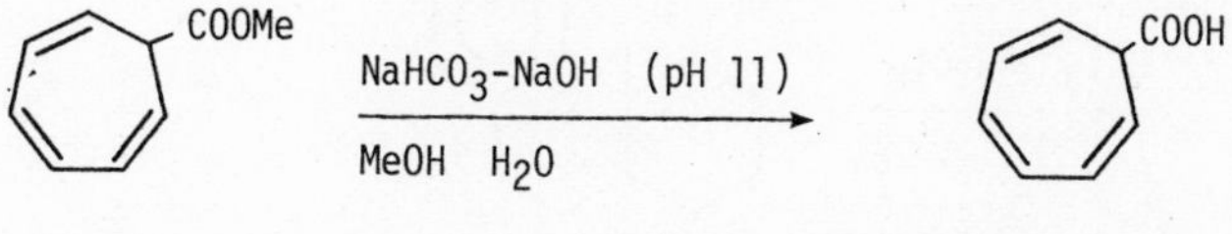

COOMe $\xrightarrow[\text{MeOH} \quad \text{H}_2\text{O}]{\text{NaHCO}_3\text{-NaOH} \quad (\text{pH 11})}$ COOH   92%

Ber (1972) <u>105</u> 1778

$(i\text{-}Pr)_3CCOOMe$ $\xrightarrow[\text{xylene}]{\text{1,5-Diazabicyclo[4.3.0]nonene}}$ $(i\text{-}Pr)_3CCOOH$ 94%

Tetr Lett (1972) 3987
JOC (1973) 38 1223

PhCOOMe $\xrightarrow{\text{LiI   NaCN   DMF}}$ PhCOOH

Synth Comm (1972) 2 389

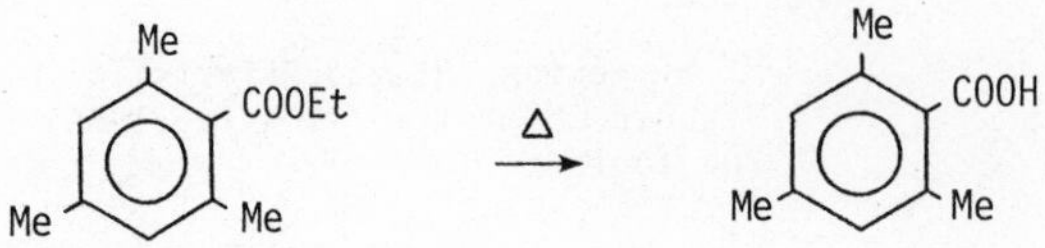

91%

J Am Oil Chem Soc (1956) 33 317
JOC (1964) 29 1252

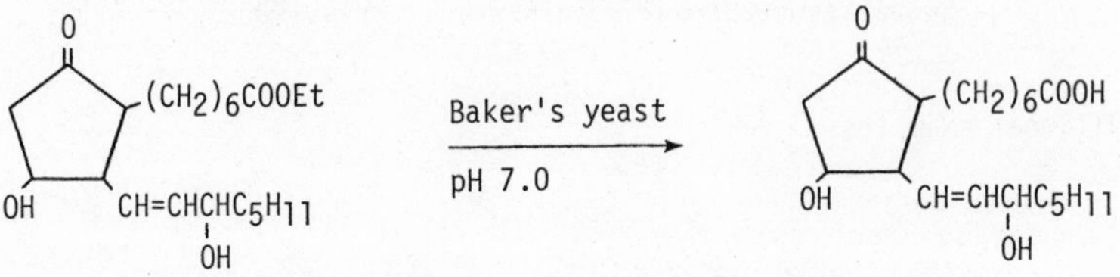

JACS (1972) 94 3643

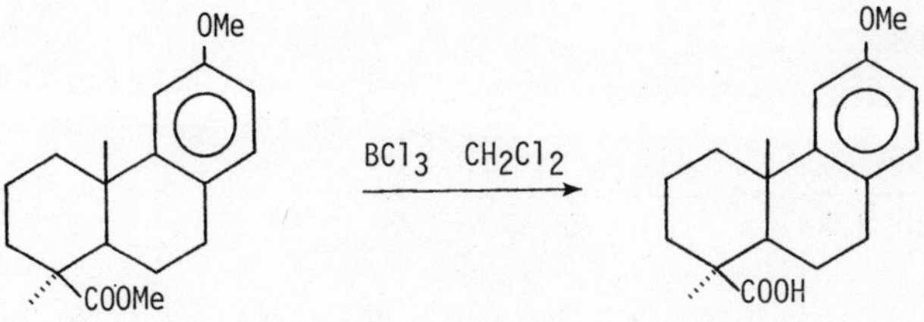

90%

Chem Comm (1971) 667

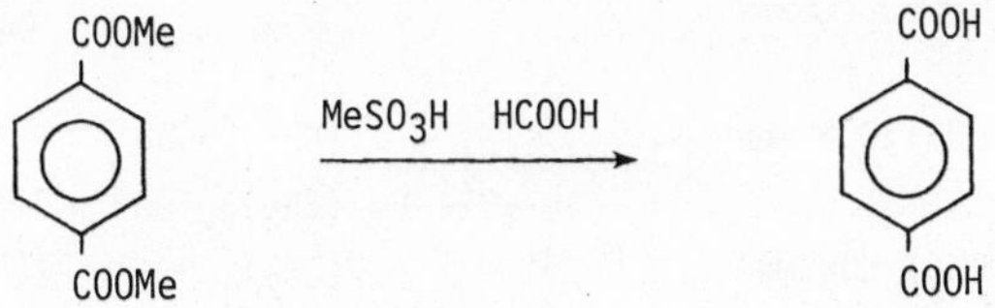

COOMe

$\xrightarrow{\text{MeSO}_3\text{H} \quad \text{HCOOH}}$

COOH   99%

COOMe   COOH

Chem Ind (1964) 193

$C_{17}H_{35}COOMe \xrightarrow[\text{}]{\text{Ph}_3\text{P·HBr} \quad 170\text{-}180°} C_{17}H_{35}COOH$ 62%

Annalen (1967) 709 105

$RCH_2COOEt \xrightarrow[\text{2} \quad 220\text{-}230°]{\text{1} \quad \text{PhMgBr} \quad \text{Et}_2\text{O}} RCH=CPh_2 \xrightarrow[\text{H}_2\text{O}]{\text{CrO}_3 \quad \text{HOAc}} RCOOH$

JCS (1957) 1622
     (1953) 1785

Section 24   Carboxylic Acids from Ethers

COOH
|
$(CH_2)_9CH_2OEt$ $\xrightarrow{\text{KOH} \quad 360°}$ COOH
|
$(CH_2)_9COOH$

JCS C (1971) 1840

Section 25   <u>Carboxylic Acids from Halides</u>

$$\begin{array}{c} Br \\ | \\ (CH_2)_9 \\ | \\ Br \end{array} \xrightarrow[\substack{2 \quad \text{(glutarimide ring with CO-NMe-CO)}}]{1 \;\; Mg \quad Et_2O} \begin{array}{c} CO(CH_2)_3CONHMe \\ | \\ (CH_2)_9 \\ | \\ CO(CH_2)_3CONHMe \end{array} \xrightarrow[\substack{2 \;\; N_2H_4 \\ Na\text{-ethylene} \\ glycol}]{1 \;\; H_2SO_4 \quad H_2O} \begin{array}{c} (CH_2)_4COOH \\ | \\ (CH_2)_9 \\ | \\ (CH_2)_4COOH \end{array} \quad \sim 12\%$$

Chem Listy (1958) <u>52</u> 1926
(Chem Abs <u>53</u> 3055)

$$\xrightarrow[\substack{2 \;\; Me_2C=C(COOEt)_2}]{1 \;\; Mg \quad Et_2O} \quad \text{CH(COOEt)}_2 \xrightarrow[\substack{2 \;\; \Delta}]{1 \;\; NaOH} \quad CH_2COOH \quad 35\%$$

JOC (1971) <u>36</u> 3260
      (1972) <u>37</u> 825

$$BuI \xrightarrow[\substack{2 \;\; Acid \; alumina}]{1 \;\; LiCH_2C\equiv CN} \quad BuCH_2CH_2CON \xrightarrow[\substack{ethylene \\ glycol}]{KOH} \quad BuCH_2CH_2COOH \quad \sim 20\%$$

JOC (1970) <u>35</u> 3405

$$\begin{array}{c} RCl \\ (3ry) \end{array} \xrightarrow{CH_2=CCl_2 \quad H_2SO_4 \quad BF_3} \quad RCH_2COOH$$

Ber (1967) <u>100</u> 978

$$BuBr \xrightarrow[\substack{NH_3}]{NaC\equiv COEt} \quad BuC\equiv COEt \xrightarrow[\substack{2 \;\; NaOH \quad EtOH}]{1 \;\; HgO \quad H_2SO_4 \quad EtOH} \quad BuCH_2COOH \quad 17\%$$

JCS (1954) 1860
Advances in Org Chem (1960) <u>2</u> 117

$$PhCH_2Cl \xrightarrow{\quad NaCH_2COONa \quad decalin \quad} PhCH_2CH_2COOH \qquad 34\%$$

Annalen (1966) <u>691</u> 61

$$RBr \xrightarrow[EtOH]{CH_2(COOEt)_2 \quad EtONa} RCH(COOEt)_2 \xrightarrow[2 \ HCl \quad H_2O]{1 \ KOH \quad EtOH} RCH_2COOH$$

JOC (1971) <u>36</u> 3944

$$PhCH=CHCH_2Cl \xrightarrow[EtONa \quad EtOH]{CH_2(COOEt)_2} PhCH=CHCH_2CH(COOEt)_2 \xrightarrow[\substack{EtONa \\ 2 \ Ac_2O \\ 3 \ KOH}]{1 \ C_5H_{11}ONO} PhCH=CHCH_2COOH$$

JCS (1950) 926

$$C_{12}H_{25}Br \xrightarrow[2 \ O_2 \quad THF]{1 \ Na_2Fe(CO)_4} C_{12}H_{25}COOH \qquad 84\%$$

JACS (1973) <u>95</u> 249

$$PhF \xrightarrow{\quad Lithium \ dihydronaphthylide \quad} PhLi \xrightarrow{CO_2} PhCOOH$$

Chem Comm (1972) 752

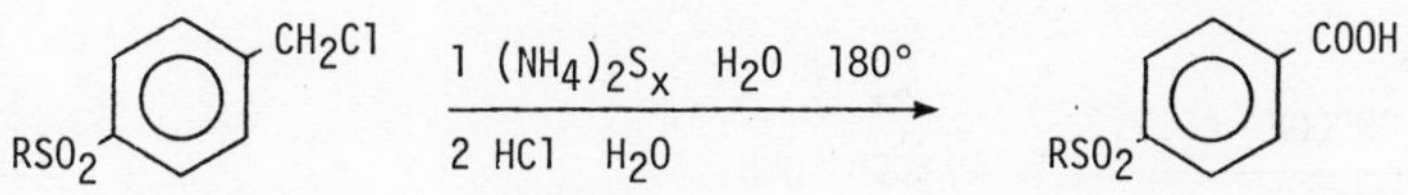

$$48\%$$

Ber (1971) <u>104</u> 1697

Also via:		Section
	Esters	115
	Amides	85
	Ketoacids	320
	Ketoesters	360
	Olefinic acids	322
	Olefinic esters	362
	Acetylenic acids	301
	Acetylenic esters	306

Section 26 Carboxylic Acids from Hydrides

Reactions in which a hydrogen is replaced by carboxyl or a carboxyl containing chain are included in this section. For reactions in which alkyl groups are oxidised to carboxyl see section 20 (Carboxylic Acids from Alkyls)

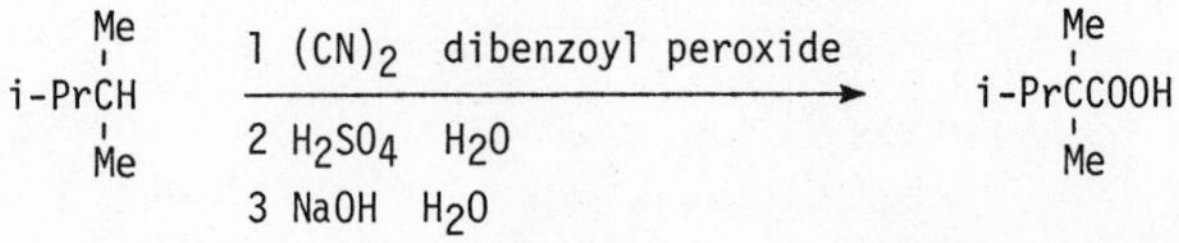
Me
i-PrCH
Me
1 (CN)2 dibenzoyl peroxide
2 H2SO4 H2O
3 NaOH H2O
Me
i-PrCCOOH
Me

JACS (1969) 91 3028

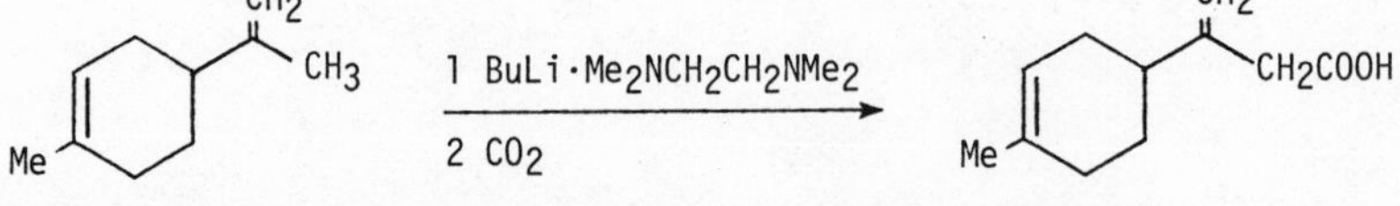
CH2
CH3
Me
1 BuLi·Me2NCH2CH2NMe2
2 CO2
CH2
CH2COOH
Me

JACS (1972) 94 4298

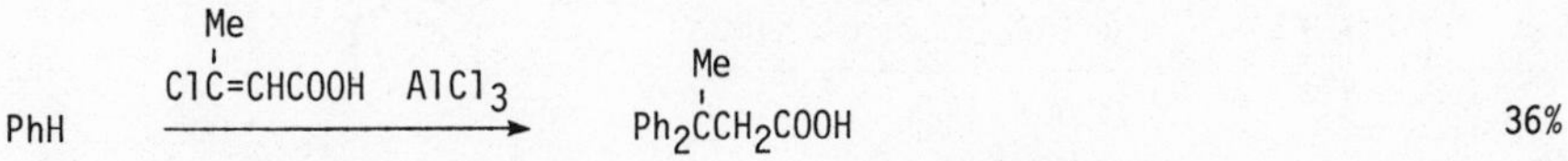
Me
ClC=CHCOOH AlCl3
PhH
Me
Ph2CCH2COOH
36%

JACS (1943) 65 59

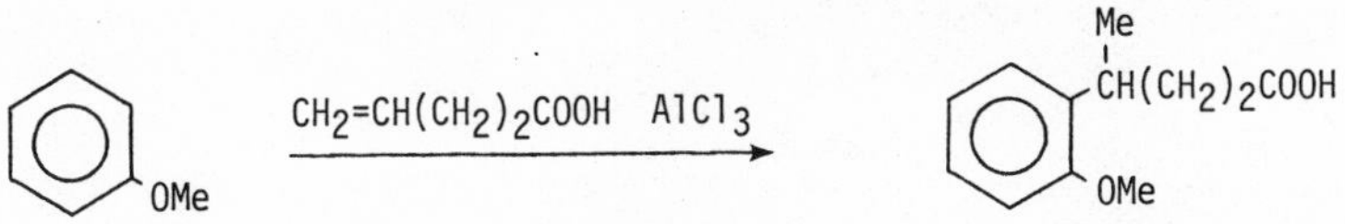
OMe
$CH_2=CH(CH_2)_2COOH$ $AlCl_3$
Me
$CH(CH_2)_2COOH$
OMe

23%

Tetrahedron (1967) 23 2481

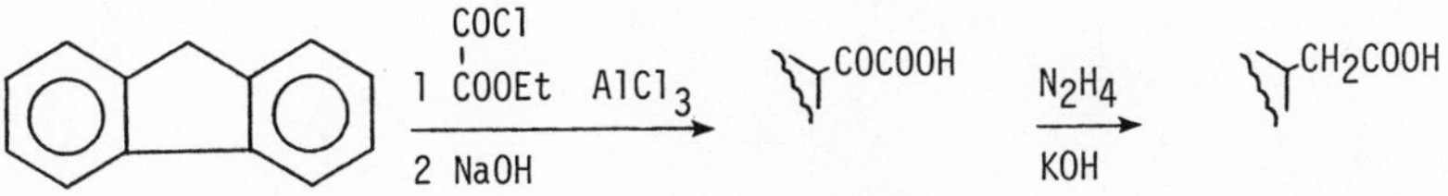
COCl
1 COOEt $AlCl_3$
2 NaOH
COCOOH
N_2H_4
KOH
CH_2COOH

75%

J Med Chem (1972) 15 1029

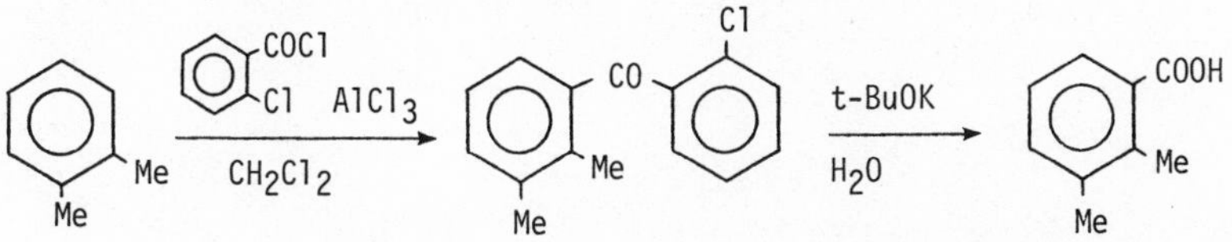
Me
Me
COCl
Cl $AlCl_3$
CH_2Cl_2
CO
Cl
Me
Me
t-BuOK
H_2O
COOH
Me
Me

58%

Tetr Lett (1971) 3825

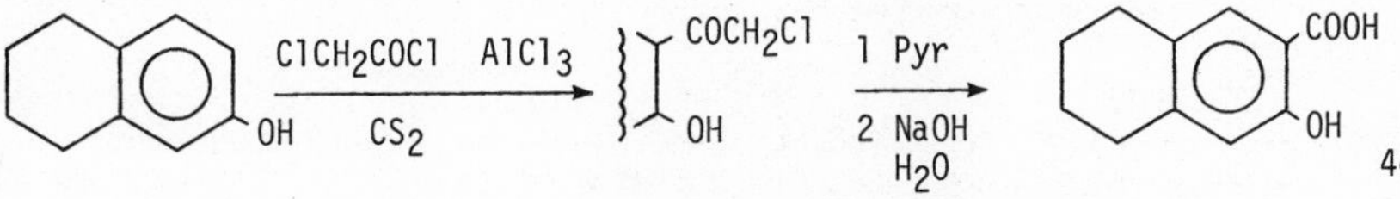
OH
$ClCH_2COCl$ $AlCl_3$
CS_2
$COCH_2Cl$
OH
1 Pyr
2 NaOH
H_2O
COOH
OH

44%

Tetrahedron (1970) 26 201

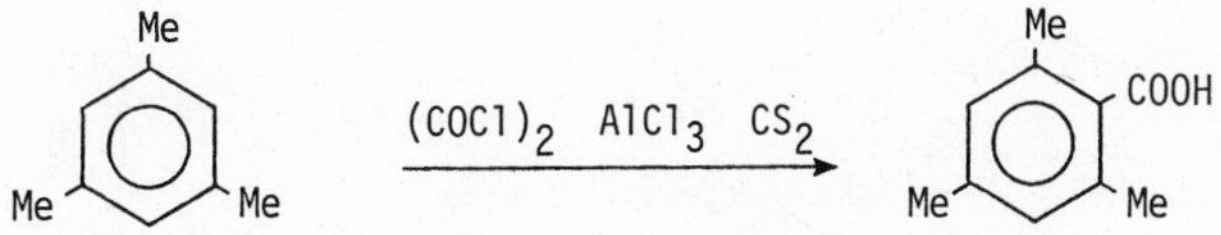
Me
Me
Me
$(COCl)_2$ $AlCl_3$ CS_2
Me
COOH
Me
Me

65-76%

Org Synth (1964) 44 69

PhH $\xrightarrow[\text{AlCl}_3]{\text{ClCOSEt}}$ PhCOSEt $\xrightarrow{\text{Hydrolysis}}$ PhCOOH   ~49%

Annalen (1972) <u>761</u> 77

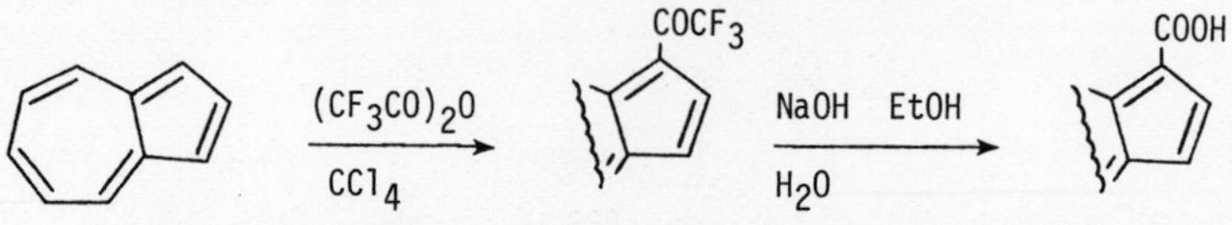

JOC (1962) <u>27</u> 3578   81%

PhH $\xrightarrow{\text{TlCl}_4 \cdot 4\text{H}_2\text{O} \quad \text{CCl}_4}$ PhCOOH   10%

JCS Perkin I (1972) 2268

Also via: Esters (Section 116)

Section 27   <u>Carboxylic Acids from Ketones</u>

Me$_2$CO $\xrightarrow[\substack{\text{ZnCl}_2 \quad \text{Ac}_2\text{O} \\ 2 \text{ MeMgI} \quad \text{CuCl} \\ \text{Et}_2\text{O}}]{1 \text{ CH}_2(\text{COOEt})_2}$ Me$_3$CCH(COOEt)$_2$ $\xrightarrow[2 \Delta]{1 \text{ Hydrolysis}}$ Me$_3$CCH$_2$COOH

Org Synth (1970) <u>50</u> 38

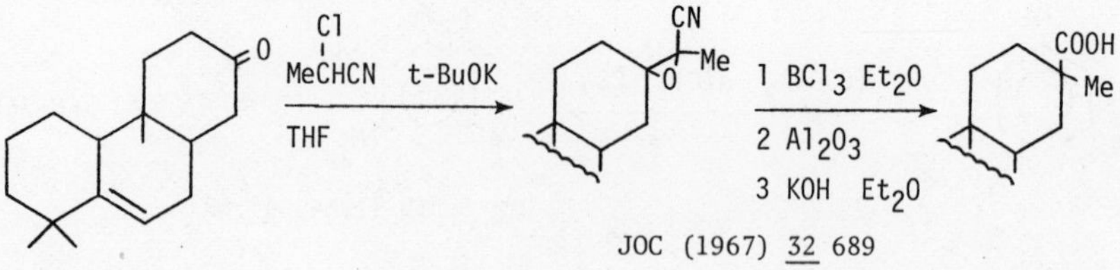

JOC (1967) <u>32</u> 689   14%

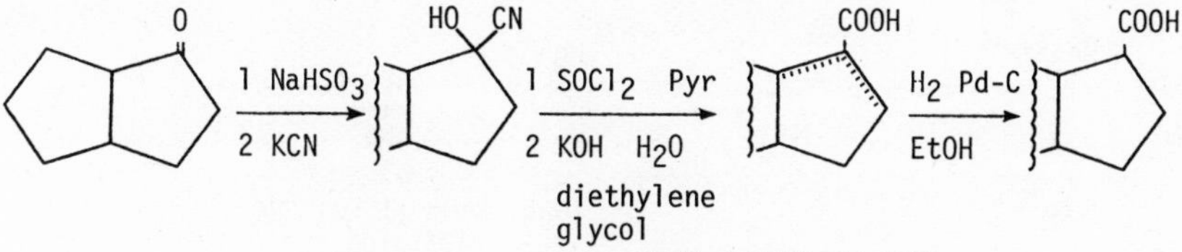

JOC (1970) 35 2376

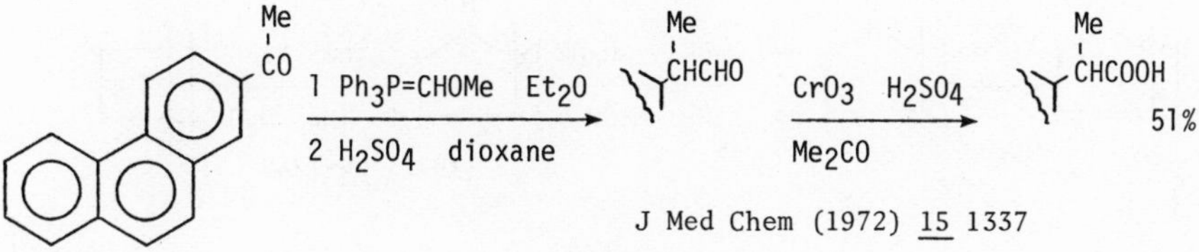

J Med Chem (1972) 15 1337

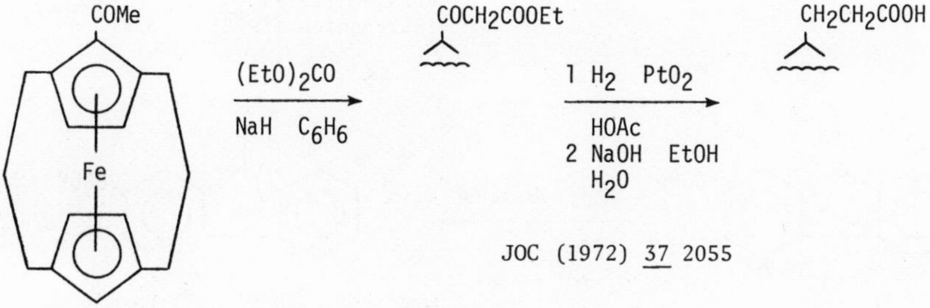

JOC (1972) 37 2055

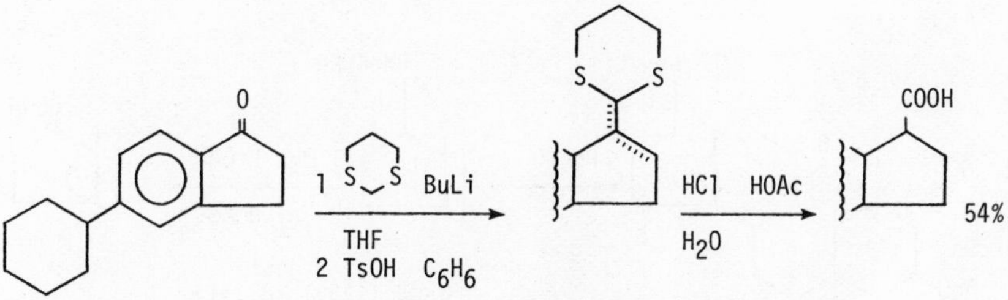

J Med Chem (1972) 15 1297

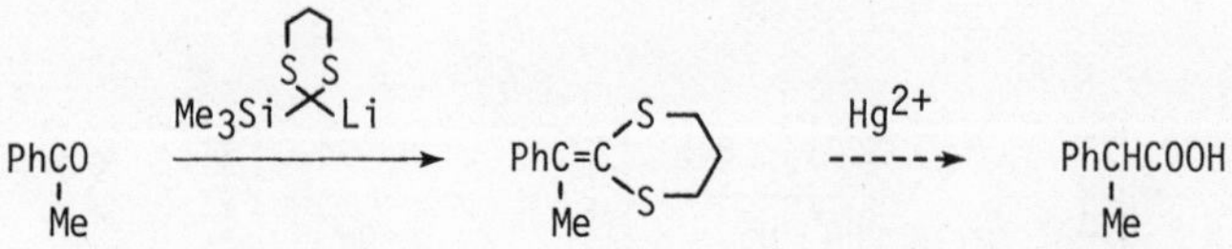

PhCO—Me →(Me₃Si—(S-S)—Li)→ PhC=C(S-S)—Me --→(Hg²⁺) PhCHCOOH—Me

$$PhCO\text{–}Me \xrightarrow{Me_3Si\text{-}C(S\text{-}S)\text{-}Li} PhC{=}C(S\text{-}S)\text{-}Me \xdashrightarrow{Hg^{2+}} PhCH(Me)COOH$$

Chem Comm (1972) 526
Ber (1973) 106 2277

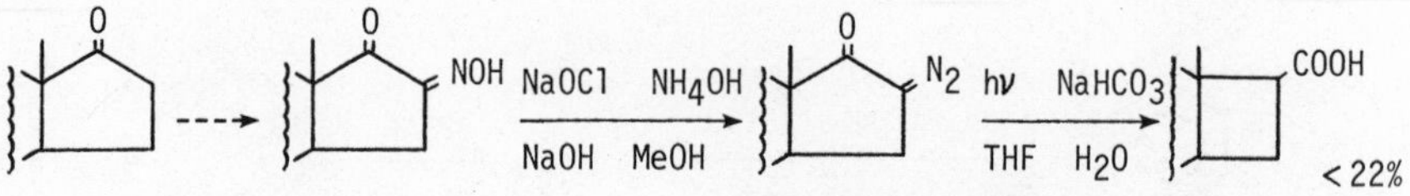

(cyclopentanone) ---→ (=NOH oxime) →(NaOCl NaOH / NH₄OH MeOH)→ (=N₂ diazo) →(hν NaHCO₃ / THF H₂O)→ (cyclobutane COOH) < 22%

JACS (1972) 94 1282

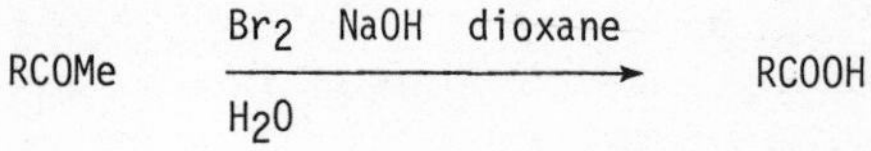

RCOMe →(Br₂ NaOH dioxane / H₂O)→ RCOOH

JOC (1972) 37 25
Annalen (1971) 747 14
Tetrahedron (1972) 28 3381

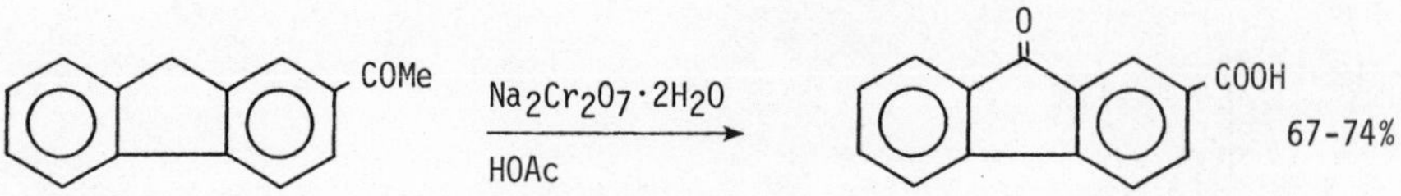

(fluorene COMe) →(Na₂Cr₂O₇·2H₂O / HOAc)→ (fluorenone COOH) 67-74%

Org Synth (1960) Coll Vol 3 420

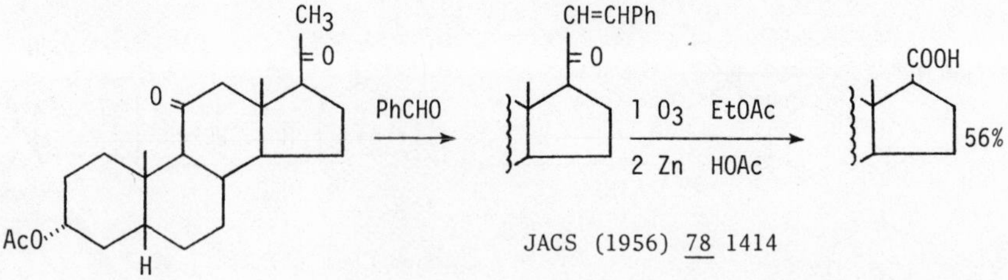

(steroid with CH₃—C=O, AcO, H) →(PhCHO)→ (CH=CHPh, C=O) →(1 O₃ EtOAc / 2 Zn HOAc)→ (COOH) 56%

JACS (1956) 78 1414

$$RCOPh \xrightarrow[\text{150°}]{\text{t-BuOK \quad H}_2\text{O \quad anisole}} RCOOH$$

Ber (1971) <u>104</u> 2637

$$C_{14}H_{29}CH_2COPh \xrightarrow[\text{HCl \quad dioxane \quad H}_2\text{O}]{\text{i-PrCH}_2\text{CH}_2\text{ONO}} C_{14}H_{29}\overset{\text{NOH}}{\underset{\parallel}{C}}COPh \xrightarrow[\text{2 KOH \quad PrOH}]{\text{1 TsCl \quad NaOH \quad H}_2\text{O}} C_{14}H_{29}COOH$$

JACS (1953) <u>75</u> 2347

Also via:

	Section
Esters	117
Ketoacids	320
Ketoesters	360
Ketoamides	347
Olefinic acids	322
Olefinic esters	362
Olefinic amides	349

Section 28   <u>Carboxylic Acids from Nitriles</u>
ooooooooooooooooooooooooooooooooo

No additional examples

Section 29   <u>Carboxylic Acids from Olefins</u>
ooooooooooooooooooooooooooooooooo

$$RCH=CH_2 \xrightarrow[\text{di-t-butyl peroxide}]{\text{CH}_3\text{COOH}} R(CH_2)_3COOH$$

Synthesis (1970) 99
Chem Comm (1973) 694

$$R_2C=CH_2 \xrightarrow{\text{CH}_2=\text{CCl}_2 \quad \text{BF}_3 \quad \text{H}_2\text{SO}_4} R_2\underset{\text{Me}}{C}CH_2COOH$$

Ber (1967) <u>100</u> 978

$C_6H_{13}CH=CH_2$ $\xrightarrow{\begin{array}{c} 1 \ CH(NO_2)_3 \ \ Hg(NO_3)_2 \ \ HClO_4 \\ \hline MeOCH_2CH_2OMe \ \ H_2O \\ 2 \ NaBH_4 \ \ NaOH \ \ THF \\ 3 \ KMnO_4 \ \ NaOH \ \ H_2O \end{array}}$ $\underset{\underset{Me}{|}}{C_6H_{13}CHCOOH}$ 8%

Acta Chem Scand (1970) 24 550

$\xrightarrow{CO \ \ Cu_2O \ \ H_2SO_4}$ $\overset{Me}{\underset{COOH}{}}$ 63%

JOC (1973) 38 2016

$R_2C=CH_2$ $\xrightarrow{\begin{array}{c} 1 \ NaBH_4 \ \ AlCl_3 \ \ diglyme \\ \hline 2 \ H_2O_2 \ \ NaOH \ \ H_2O \end{array}}$ R_2CHCH_2OH $\xrightarrow{\begin{array}{c} CrO_3 \ \ H_2SO_4 \\ \hline HOAc \ \ H_2O \end{array}}$ $R_2CHCOOH$ 53%

JACS (1960) 82 2498

$PhCH=CHPh$ $\xrightarrow{\begin{array}{c} KMnO_4 \ \ dicyclohexyl-18-crown-6 \\ \hline C_6H_6 \end{array}}$ $PhCOOH$ 100%

JACS (1972) 94 4024

Also via: Section
                    Esters   119
          Olefinic esters   362

Section 30   Carboxylic Acids from Miscellaneous Compounds
oooooooooooooooooooooooooooooooooooooooooooooooo

$MeCH=CHCOOH$ $\xrightarrow{EtMgBr \ \ Et_2O}$ $\underset{\underset{Et}{|}}{MeCHCH_2COOH}$ 40%

JACS (1953) 75 6342

PhCH=CHCOOH $\xrightarrow[\text{2 MeI}]{\text{1 Li HMPA}}$ PhCH$_2$CHCOOH   63%
                                                                                Me

Compt Rend (1969) C 268 640

R$_2$C=CCOOH $\xrightarrow{\hspace{3cm}}$ R$_2$CHCHCOOH
        R   R

                            Li-NH$_3$ JOC (1971) 36 1151
                            Na-Hg   J Med Chem (1965) 8 598
                    NiCl$_2$-KCN-NaBH$_4$ Tetr Lett (1968) 1821

PhC≡CCOOH $\xrightarrow[\text{Me}_2\text{SO}_4 \quad \text{HOAc} \quad \text{H}_2\text{O}]{\text{N}_2\text{H}_4 \quad \text{NaIO}_4 \quad \text{CuSO}_4}$ PhCH$_2$CH$_2$COOH   85%

Chem Comm (1971) 1245

1 NaOH EtOH
2 Li NH$_3$ Et$_2$O

JOC (1971) 36 1151

C$_8$H$_{17}$CH=CH(CH$_2$)$_7$COOH $\xrightarrow{\text{KOH   360}°}$ C$_8$H$_{17}$(CH$_2$)$_7$COOH   80-85%

Tetrahedron (1960) 8 221

C$_6$H$_{13}$CH=CHCOOH $\xrightarrow{\text{KOH   360}°}$ C$_6$H$_{13}$COOH   ~80%

Tetrahedron (1960) 8 221

Section 30A   Protection of Carboxylic Acids
               °°°°°°°°°°°°°°°°°°°°°°°°°°°°°°°°°°°°

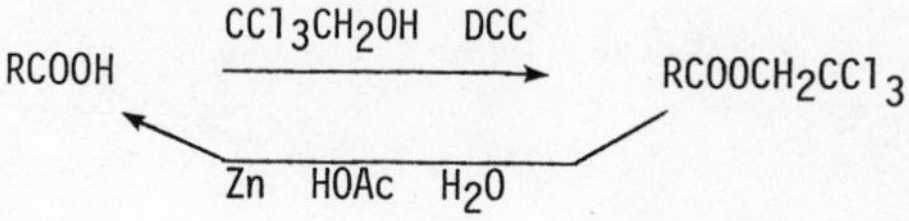

RCOOH $\xrightarrow{\text{CCl}_3\text{CH}_2\text{OH DCC}}$ RCOOCH$_2$CCl$_3$

$\xleftarrow{\text{Zn HOAc H}_2\text{O}}$

JOC (1970) 35 2430
       (1971) 36 1259
J Med Chem (1971) 14 420 426

Selective removal of haloethoxy groups by electrolysis
                              JACS (1972) 94 5139

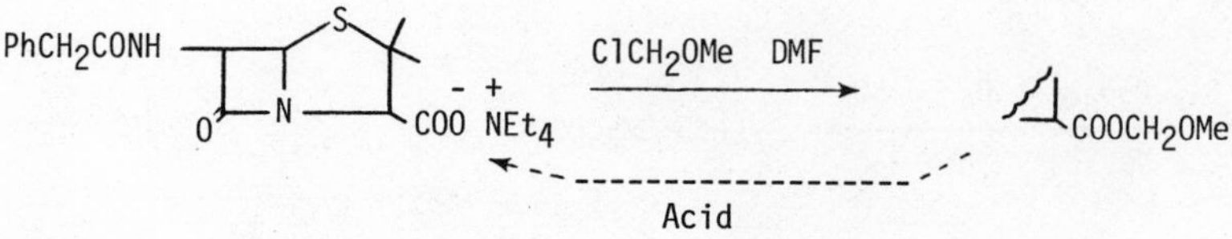

PhCH$_2$CONH ... COO$^-$ NEt$_4^+$ $\xrightarrow{\text{ClCH}_2\text{OMe DMF}}$ ... COOCH$_2$OMe

$\xleftarrow{}$

Acid

JCS (1965) 2127

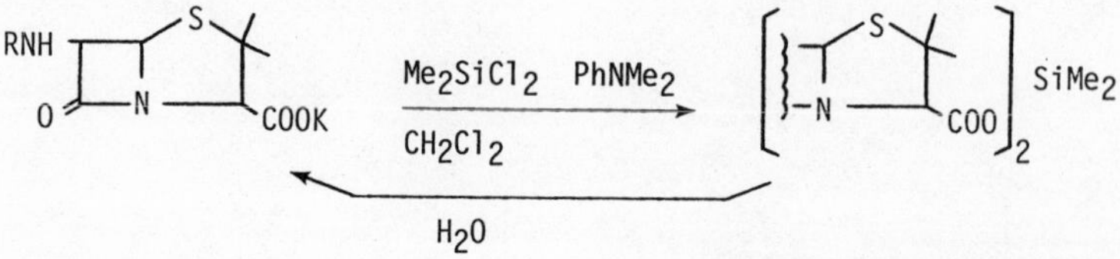

RNH ... COOK $\xrightarrow[\text{CH}_2\text{Cl}_2]{\text{Me}_2\text{SiCl}_2 \ \text{PhNMe}_2}$ $\left[\text{-N} ... \text{COO} \right]_2$ SiMe$_2$

$\xleftarrow{\text{H}_2\text{O}}$

Rec Trav Chim (1970) 89 1081

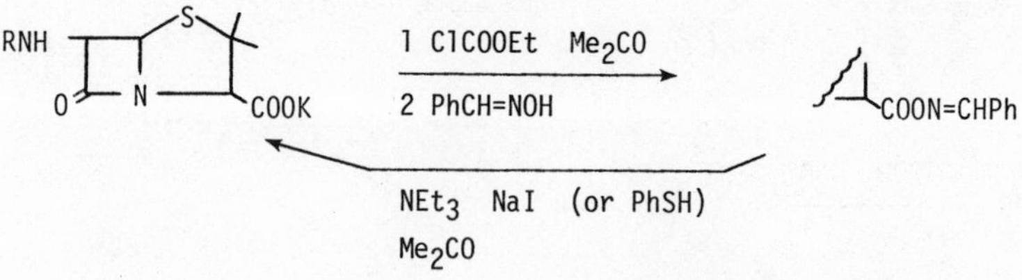

RNH ... COOK $\xrightarrow[\text{2 PhCH=NOH}]{\text{1 ClCOOEt Me}_2\text{CO}}$ ... COON=CHPh

$\xleftarrow[\text{Me}_2\text{CO}]{\text{NEt}_3 \ \text{NaI \ (or PhSH)}}$

JCS C (1971) 1917

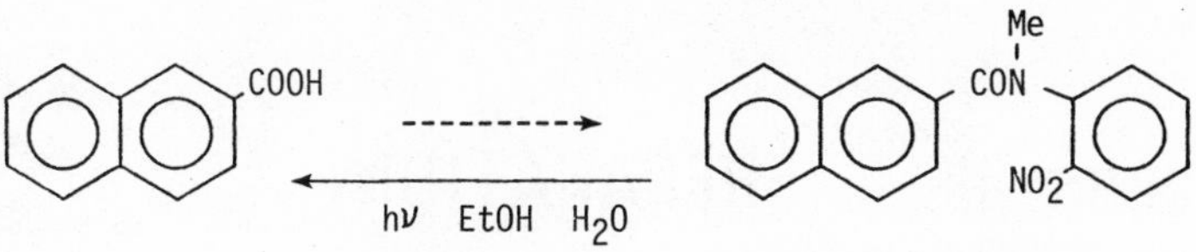

Tetr Lett (1973) 2205

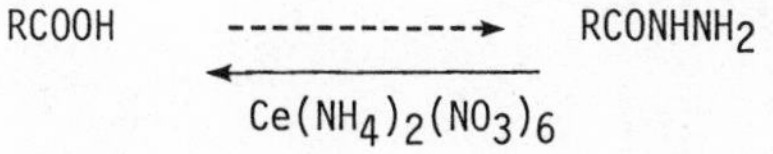

RCOOH $\dashrightarrow$ RCONHNH$_2$

$\xleftarrow{\quad\quad}$

Ce(NH$_4$)$_2$(NO$_3$)$_6$ Synthesis (1972) 562

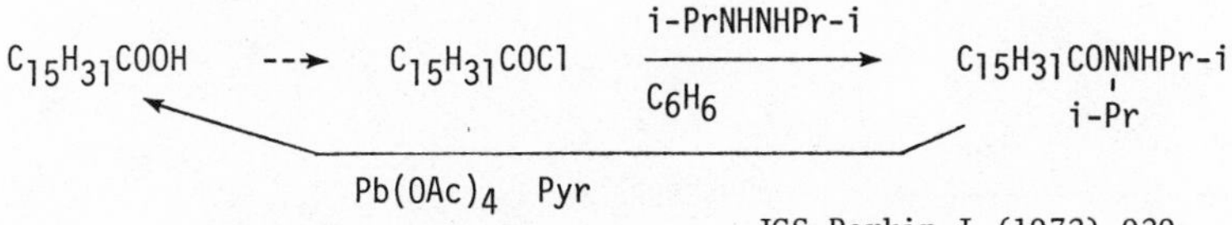

$$C_{15}H_{31}COOH \dashrightarrow C_{15}H_{31}COCl \xrightarrow[C_6H_6]{\text{i-PrNHNHPr-i}} C_{15}H_{31}CONNHPr-i$$

Pb(OAc)$_4$ Pyr   i-Pr

JCS Perkin I (1972) 929

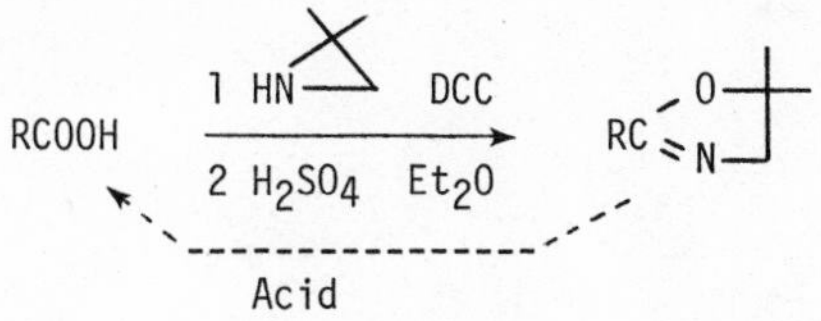

RCOOH $\xrightarrow[\text{2 H}_2\text{SO}_4 \quad \text{Et}_2\text{O}]{\text{1 HN} \quad \text{DCC}}$ RC $\underset{N}{\overset{O}{\lessgtr}}$ (Stable to RMgX and LiAlH$_4$)

Acid   Tetr Lett (1972) 3031

Other reactions useful for the protection of carboxylic acids are included in section 107 (Esters from Carboxylic Acids and Acid Halides) and section 23 (Carboxylic Acids from Esters)

Chapter 3   PREPARATION OF ALCOHOLS AND PHENOLS

Section 31   <u>Alcohols from Acetylenes</u>

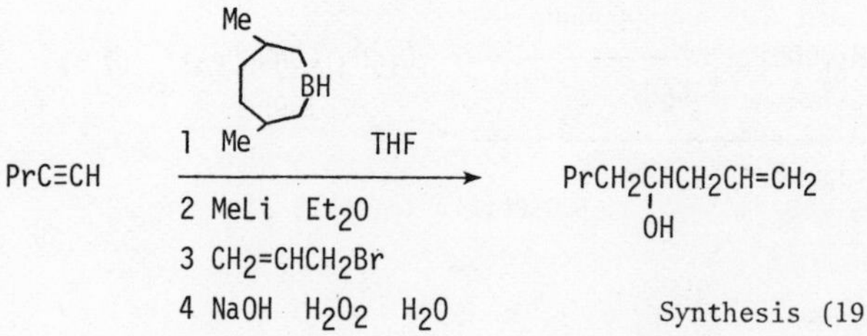

PrC≡CH

$$\xrightarrow[\substack{2\ \text{MeLi}\ \ \text{Et}_2\text{O} \\ 3\ \text{CH}_2=\text{CHCH}_2\text{Br} \\ 4\ \text{NaOH}\ \ \text{H}_2\text{O}_2\ \ \text{H}_2\text{O}}]{\substack{\text{Me} \\ \text{BH} \\ \text{Me} \\ 1\ \qquad\qquad \text{THF}}}$$

PrCH$_2$CHCH$_2$CH=CH$_2$ 84%
$\quad\quad\ |$
$\quad\quad$ OH

Synthesis (1973) 37

EtCHC≡CH
 |
Me

$$\xrightarrow[\substack{2\ \text{HC}\equiv\text{CCH}_2\text{OH} \\ \text{CuCl}_2\ \ \text{NH}_2\text{OH} \\ \text{EtNH}_2}]{1\ \text{NaOBr}}$$

EtCH(C≡C)$_2$CH$_2$OH
   |
   Me

$$\xrightarrow[\text{EtOH}]{\text{H}_2\ \ \text{PtO}_2}$$

EtCH(CH$_2$)$_4$CH$_2$OH   55%
   |
   Me

JOC (1971) <u>36</u> 2902

Also via: Section

        Acetylenic alcohols   302
          Olefinic alcohols   332

28

Section 32 Alcohols from Carboxylic Acids

$$\mathrm{BuCOOH} \xrightarrow{C_5H_{11}MgBr\quad Et_2O\quad C_6H_6} \mathrm{BuC(OH)(C_5H_{11})_2} \qquad 40\text{-}60\%$$

JACS (1946) 68 1382

$$\mathrm{Ph(CH_2)_2COOH} \xrightarrow[(CF_3CO)_2O]{CF_3CH_2OH} \mathrm{Ph(CH_2)_2COOCH_2CF_3} \xrightarrow[MeOCH_2CH_2OMe]{NaBH_4} \mathrm{Ph(CH_2)_2CH_2OH}$$

JOC (1970) 35 1505

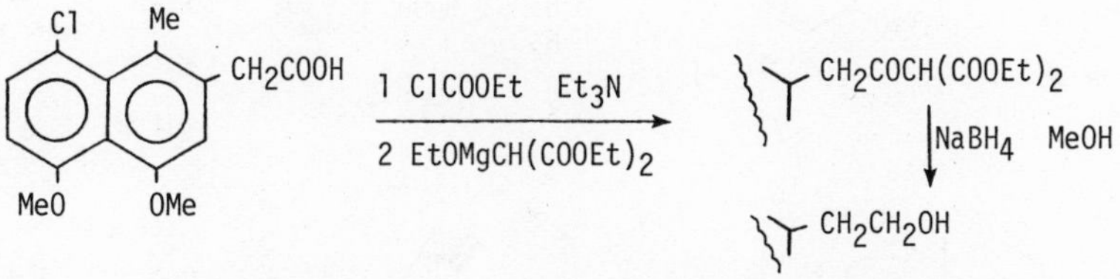

Cl
Me
CH2COOH
MeO
OMe
1 ClCOOEt Et3N
2 EtOMgCH(COOEt)2
CH2COCH(COOEt)2
NaBH4 MeOH
CH2CH2OH

66%

Ber (1965) 98 3040

$$\mathrm{RCOOH} \xrightarrow{B_2H_6\quad THF} \mathrm{RCH_2OH}$$

JOC (1973) 38 2786

$$\mathrm{RCOOH} \xrightarrow[EtOH\quad H_2O]{Electrolysis\quad H_2SO_4} \mathrm{RCH_2OH}$$

Synthesis (1971) 285

Also via: Esters (Section 38)

Section 33   Alcohols from Alcohols and Phenols
○○○○○○○○○○○○○○○○○○○○○○○○○○○○○○○○○○○○○○

$C_6H_{13}CH_2CH_2OH$ $\xrightarrow[295°]{C_8H_{17}ONa \quad \text{copper bronze}}$ $C_6H_{13}\underset{\underset{CH_2CH_2C_6H_{13}}{|}}{C}HCH_2OH$ 52%

JOC (1950) 15 54
JACS (1954) 76 52

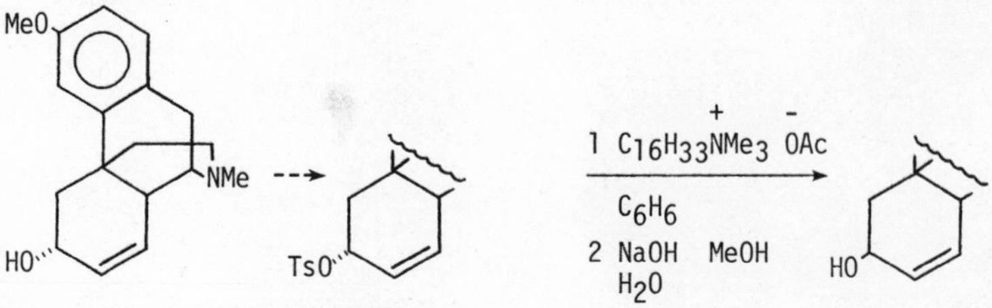

1 $C_{16}H_{33}\overset{+}{N}Me_3 \; \overset{-}{O}Ac$
$\xrightarrow{\qquad\qquad}$
C_6H_6
2 NaOH MeOH
H_2O

JCS C (1971) 3047

$\begin{array}{c}NCOOEt\\ \| \\ NCOOEt \quad Ph_3P\\ \hline HCOOH \quad THF\end{array}$

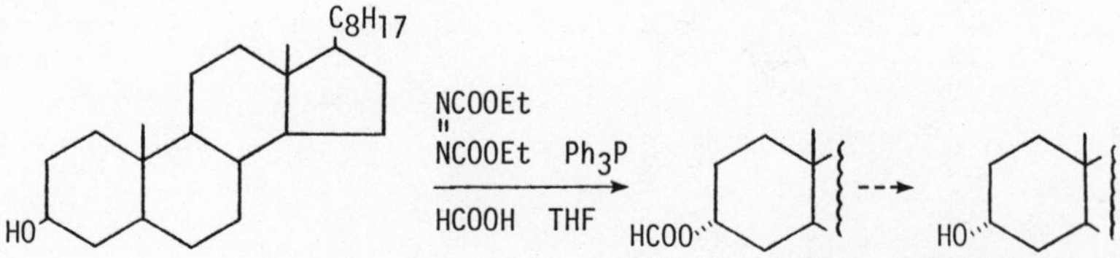

97%

Tetr Lett (1973) 1619

$PrCH_2OH$ --→ $PrCH_2OOH$ $\xrightarrow[CH_2Cl_2]{H_2SO_4 \quad H_2O}$ $PrOH$ < 60%

JOC (1970) 35 3080

$PhOH$ --→ $PhOPO(OEt)_2$ $\xrightarrow[K \quad NH_3]{MeCOCH_2K}$ $PhCH_2\underset{\underset{OH}{|}}{C}HMe$ < 42%

JACS (1972) 94 683

Section 34   Alcohols from Aldehydes
              oooooooooooooooooooooooo

$$RCHO \xrightarrow{\text{MeBr Li THF}} \underset{\overset{|}{Me}}{RCHOH}$$ (One-step procedure)

JCS Perkin I (1972) 1655

$$PhCHO \xrightarrow[\text{2 Ni}]{\text{1 MeSO} \langle \bigcirc \rangle \text{-Me Et}_2\text{NLi THF}} \underset{\overset{|}{Me}}{PhCHOH}$$ 50%

Tetr Lett (1972) 4605

$$C_5H_{11}CHO \xrightarrow[\underset{Me}{\overset{Me}{\overset{|}{i\text{-PrCBH}_2}}} \quad THF]{} C_5H_{11}CH_2OH$$

JOC (1972) 37 2942

Related methods: Alcohols from Ketones (Section 42)

Also via:
                                    Section
              Acetylenic alcohols   302
              Olefinic alcohols   332

Section 35   Alcohols and Phenols from Alkyls, Methylenes and Aryls
              ooooooooooooooooooooooooooooooooooooooooooooooooooooooooo

No examples of the reaction RR' ⟶ ROH (R'=alkyl, aryl etc.) occur in the
literature. For reactions of the type RH ⟶ ROH (R=alkyl or aryl) see
section 41 (Alcohols and Phenols from Hydrides)

Section 36   Alcohols and Phenols from Amides
              ooooooooooooooooooooooooooooooooo

No additional examples

Section 37   Alcohols and Phenols from Amines

$ArNH_2$ $\xrightarrow{\text{1 NaNO}_2 \text{ HCl H}_2\text{O}}_{\text{2 HPF}_6}$ $ArN_2^+ PF_6^-$ $\xrightarrow{\text{1 Me}_2\text{NCONMe}_2}_{\text{2 NaOH}}$ $ArOH$

JOC (1963) <u>28</u> 568

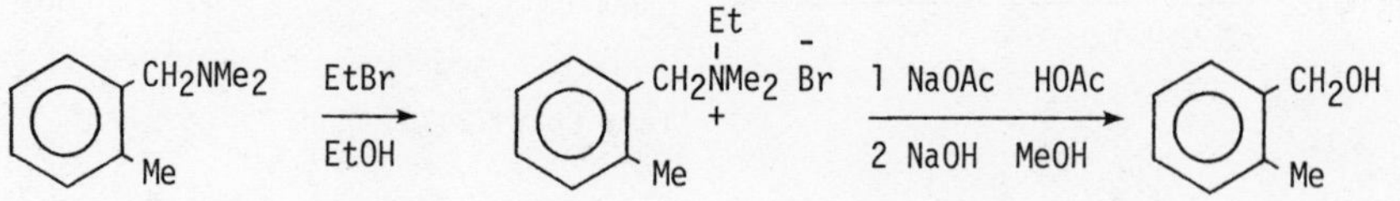

Org Synth (1963) Coll Vol 4 582   ~80%

Also via: Esters (Section 112)

Section 38   Alcohols from Esters

$PhCH_2COOEt$ $\xrightarrow{\text{Na   NH}_3}$ $PhCH_2CH_2OH$

Compt Rend (1913) <u>156</u> 1020

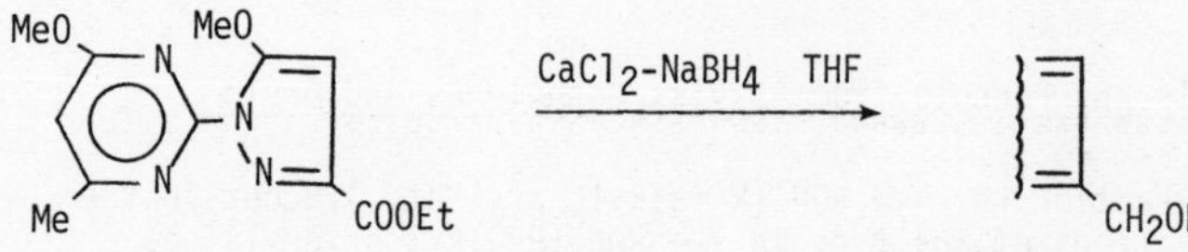

76%

Chem Pharm Bull (1970) <u>18</u> 1908

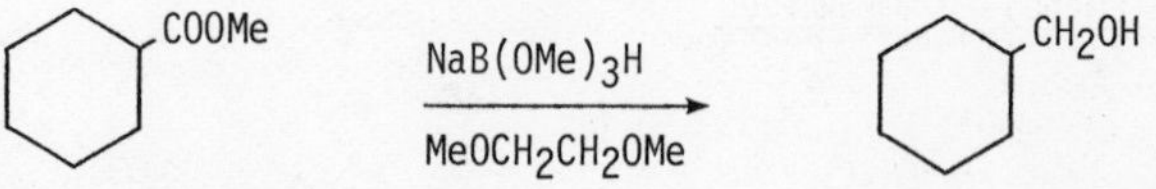

Can J Chem (1969) <u>47</u> 2099

PrC=CHCOOEt $\xrightarrow{\text{AlH}_3 \quad \text{THF}}$ PrC=CHCH$_2$OH   74%
| |
Me   Me

Tetr Lett (1973) 1277

C$_8$H$_{17}$COOEt $\xrightarrow[\text{C}_6\text{H}_6]{\text{NaAl(OCH}_2\text{CH}_2\text{OMe)}_2\text{H}_2}$ C$_8$H$_{17}$CH$_2$OH   92%

Coll Czech (1969) $\underline{34}$ 1025

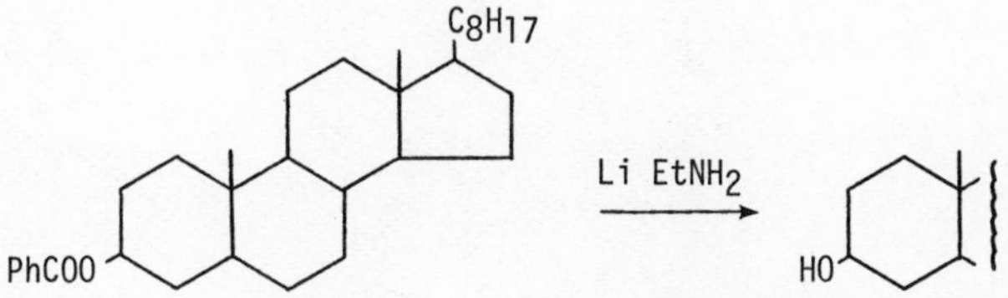

$\xrightarrow{\text{Li} \quad \text{EtNH}_2}$ 62%

Synthesis (1972) 391

HCOOC$_8$H$_{17}$ $\xrightarrow{\text{Pd-C} \quad 200°}$ C$_8$H$_{17}$OH   JOC (1970) $\underline{35}$ 1694   94%
                                                    (1973) $\underline{38}$ 3954

ROAc $\xrightarrow{\text{HCl} \quad \text{MeOH}}$ ROH

Acta Chem Scand (1952) $\underline{6}$ 1127

Related methods: Section
                    Carboxylic Acids from Esters   23
                         Protection of Alcohols   45A

Section 39   Alcohols and Phenols from Ethers and Epoxides
ooooooooooooooooooooooooooooooooooooooooooooooooooo

PPh$_3$·HBr   170-180°

74%

Annalen (1967) 709 105

ArOMe $\xrightarrow{\text{EtSNa   DMF}}$ ArOH

Tetr Lett (1972) 241

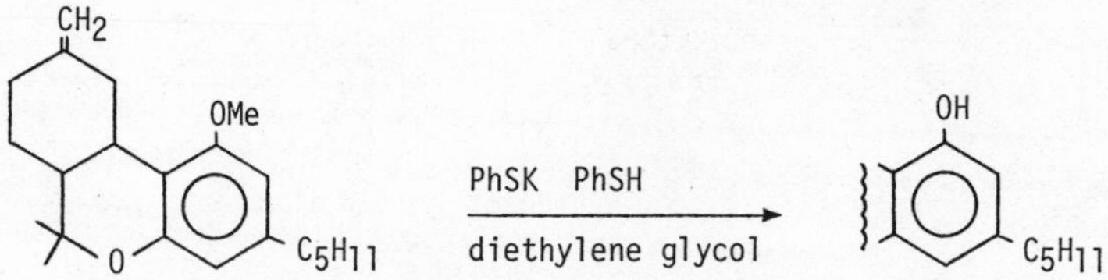

PhSK   PhSH

diethylene glycol

(Applicable to compounds with labile double bonds)

JOC (1971) 36 721

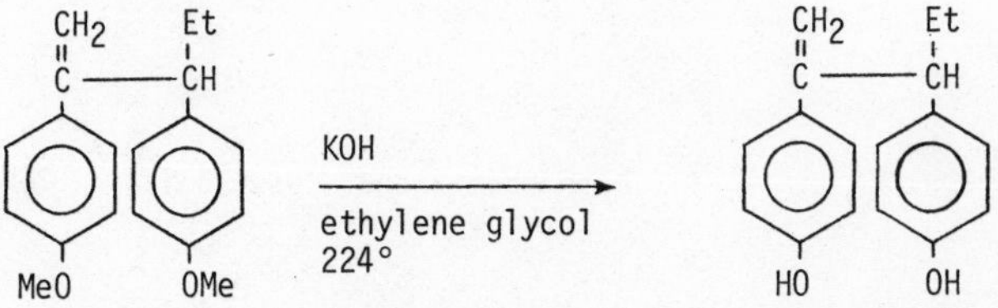

KOH

ethylene glycol
224°

<55%

JACS (1943) 65 11

$\begin{array}{l}\text{COOH}\\ |\\ (\text{CH}_2)_9\text{CH}_2\text{OEt}\end{array}$ $\xrightarrow{\text{KOH   200°}}$ $\begin{array}{l}\text{COOH}\\ |\\ (\text{CH}_2)_9\text{CH}_2\text{OH}\end{array}$ 26%

JCS C (1971) 1840

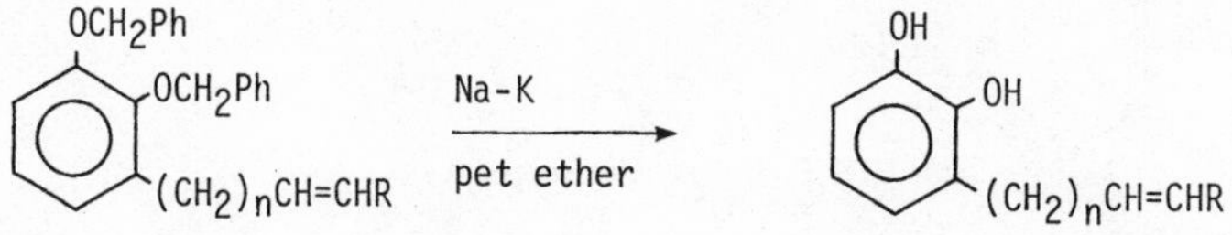

OCH$_2$Ph

OCH$_2$Ph

(CH$_2$)$_n$CH=CHR → Na-K / pet ether →

OH

OH

(CH$_2$)$_n$CH=CHR   75%

JOC (1965) 30 1610

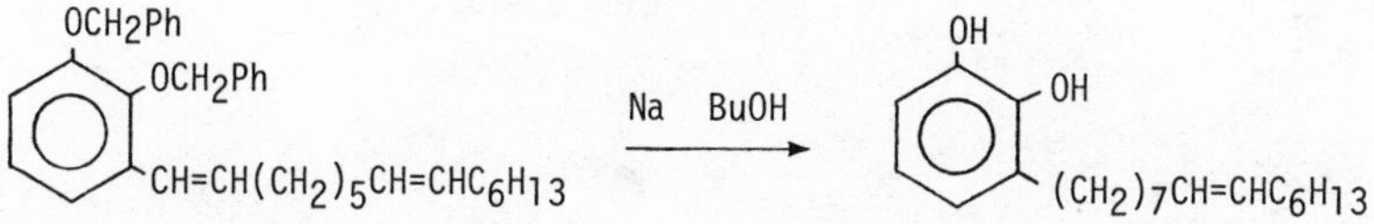

OCH$_2$Ph

OCH$_2$Ph

CH=CH(CH$_2$)$_5$CH=CHC$_6$H$_{13}$ → Na BuOH →

OH

OH

(CH$_2$)$_7$CH=CHC$_6$H$_{13}$

JOC (1959) 24 980

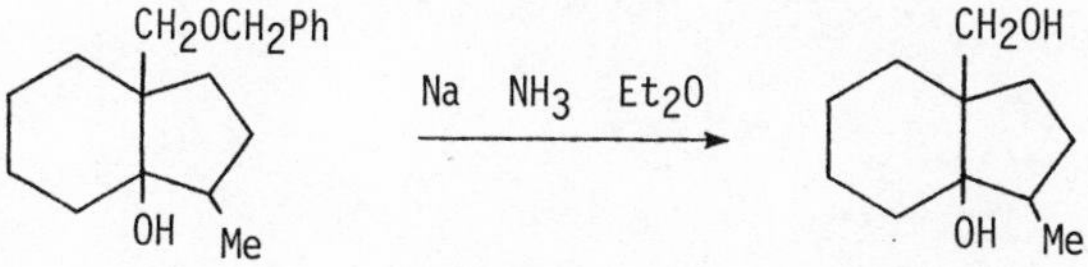

CH$_2$OCH$_2$Ph

OH Me → Na NH$_3$ Et$_2$O →

CH$_2$OH

OH Me   70%

JOC (1971) 36 2035

Related methods: Protection of Alcohols and Phenols (section 45A)

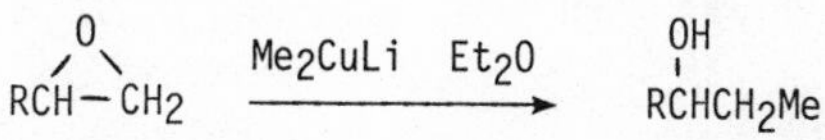

O

RCH—CH$_2$ → Me$_2$CuLi Et$_2$O →

OH

RCHCH$_2$Me

JACS (1970) 92 3813
JOC (1973) 38 4263 4346

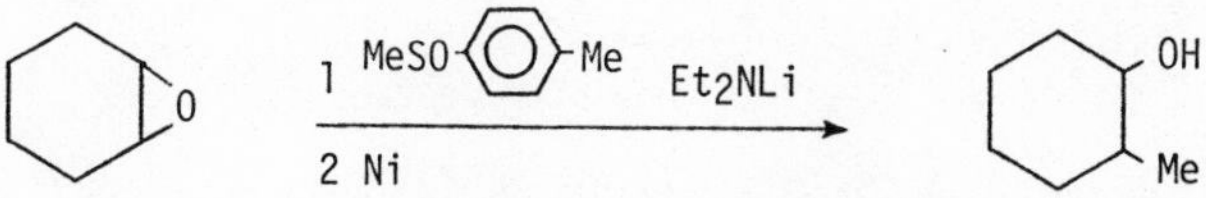

O → 1 MeSO⟨◯⟩Me Et$_2$NLi / 2 Ni →

OH

Me   ~44%

Tetr Lett (1972) 4605

MeCH—CHMe (O) → Na NH₃ → MeCHCH₂Me (OH)   55%

JOC (1971) 36 330

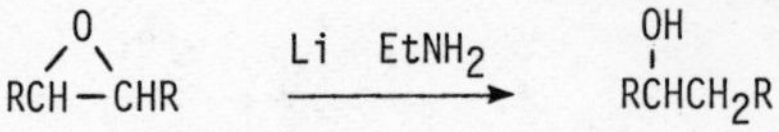

RCH—CHR (O) → Li EtNH₂ → RCHCH₂R (OH)

Synthesis (1972) 391

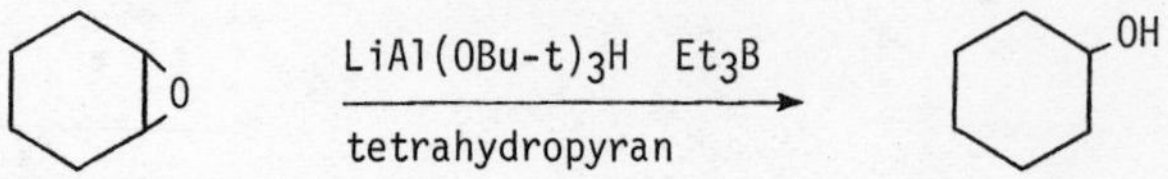

epoxide → LiAl(OBu-t)₃H Et₃B / tetrahydropyran → cyclohexanol-OH   100%

JACS (1972) 94 1750

Section 40   Alcohols and Phenols from Halides and Sulfonates

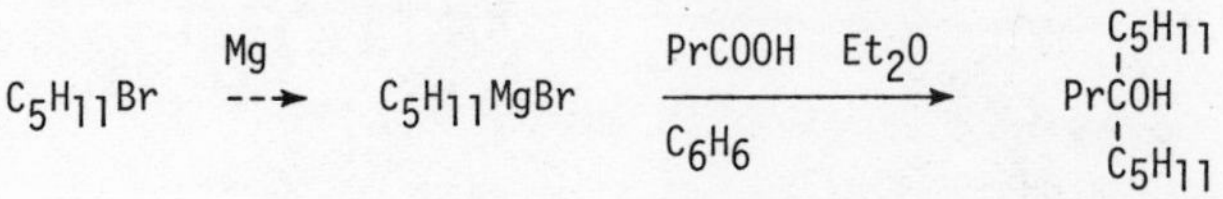

C₅H₁₁Br → Mg → C₅H₁₁MgBr → PrCOOH Et₂O / C₆H₆ → PrCOH (C₅H₁₁)(C₅H₁₁)   40-60%

JACS (1946) 68 1382

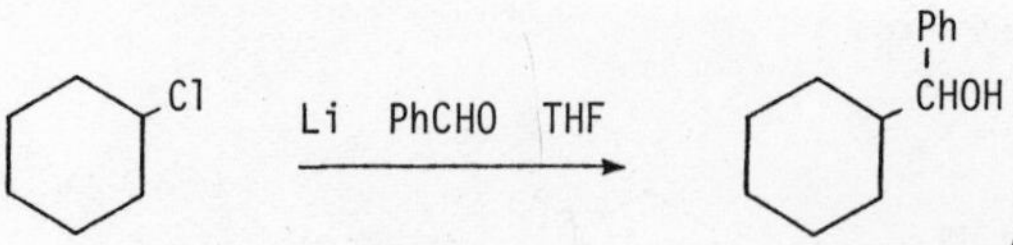

cyclohexyl-Cl → Li PhCHO THF → cyclohexyl-CHOH(Ph)   60%

(One-step procedure)

JCS Perkin I (1972) 1655

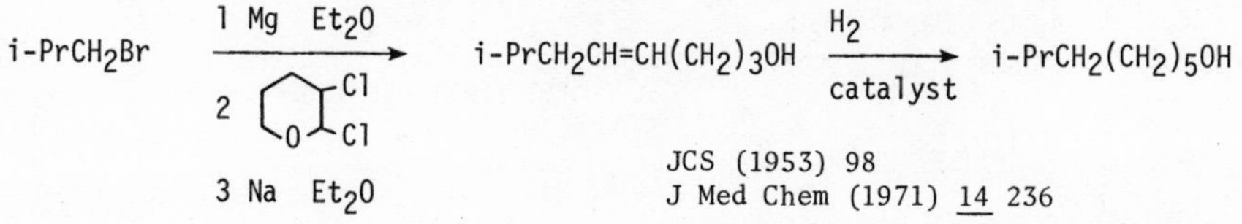

i-PrCH2Br
1 Mg Et2O
2
Cl
Cl
O
3 Na Et2O
i-PrCH2CH=CH(CH2)3OH
H2
catalyst
i-PrCH2(CH2)5OH

JCS (1953) 98
J Med Chem (1971) 14 236

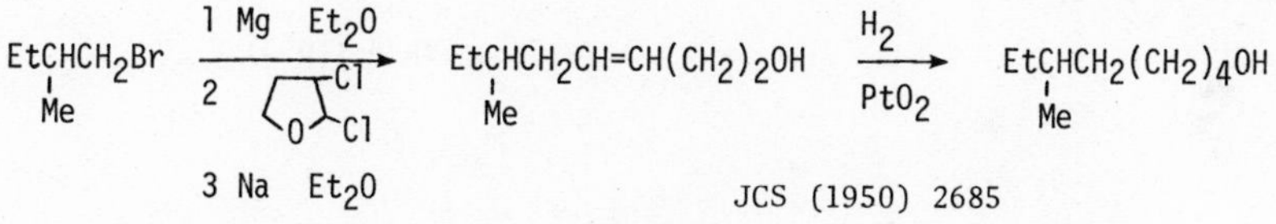

EtCHCH2Br
Me
1 Mg Et2O
2
Cl
Cl
O
3 Na Et2O
EtCHCH2CH=CH(CH2)2OH
Me
H2
PtO2
EtCHCH2(CH2)4OH
Me

36%

JCS (1950) 2685

$C_7H_{15}Br$ —(1 Mg Et_2O; 2 CO_2; 3 $NaAl(OCH_2CH_2OMe)_2H_2$)→ $C_7H_{15}CH_2OH$ 85%

Coll Czech (1971) 36 2394

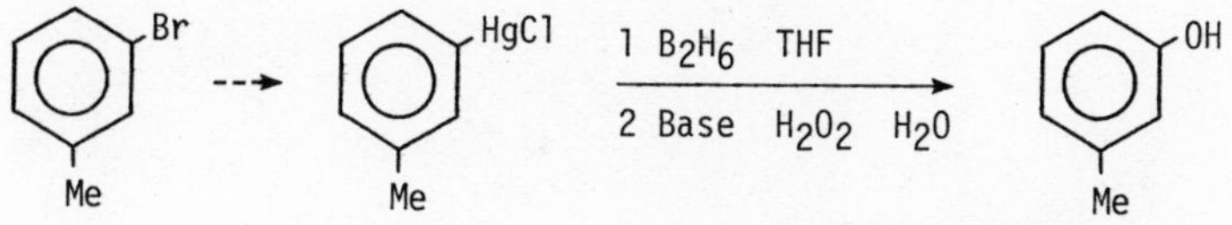

Br
Me
HgCl
Me
1 B2H6 THF
2 Base H2O2 H2O
OH
Me

70%

Chem Comm (1971) 1475

Br
1 Li Et2O
2 PhNO2
OH

51%

JACS (1969) 91 3544

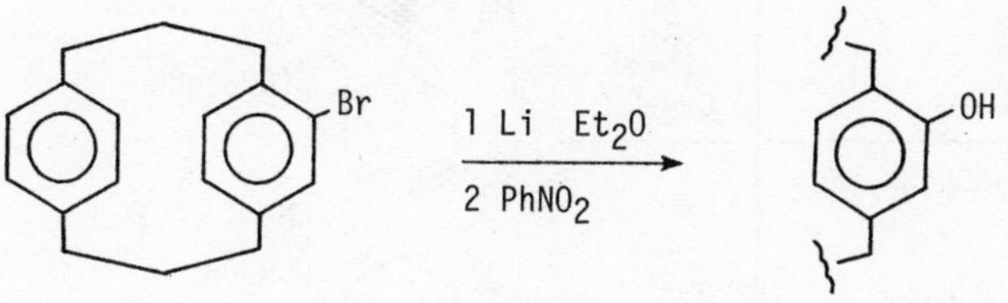

$$ArCH_2Cl \xrightarrow{\text{AgNO}_3 \quad \text{Me}_2\text{CO} \quad \text{H}_2\text{O}} ArCH_2OH \qquad 79\%$$

JOC (1971) <u>36</u> 1765

$$C_5H_{11}CH\text{-}CHCH_2CH\text{-}CH(CH_2)_7Br \xrightarrow[\text{PhBr} \quad \text{EtOH}]{\text{CF}_3\text{COONa}} C_5H_{11}CH\text{-}CHCH_2CH\text{-}CH(CH_2)_7OH \qquad 39\%$$
$$\overset{|}{Br}\ \overset{|}{Br}\quad \overset{|}{Br}\ \overset{|}{Br} \qquad\qquad\qquad \overset{|}{Br}\ \overset{|}{Br}\quad \overset{|}{Br}\ \overset{|}{Br}$$

Acta Chem Scand (1953) <u>7</u> 1001

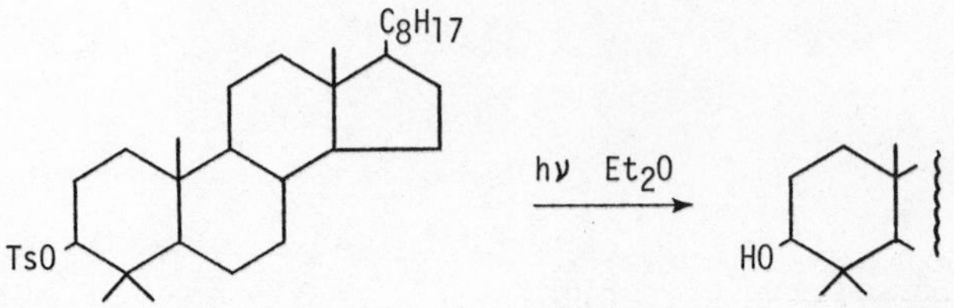

$$\xrightarrow{h\nu \quad \text{Et}_2\text{O}} \qquad 78\%$$

Tetr Lett (1971) 4555 4559
Bull Chem Soc Jap (1968) <u>41</u> 3025

Also via: Section
             Olefinic alcohols   332
             Acetylenic alcohols   302

Section 41   <u>Alcohols and Phenols from Hydrides</u>
○○

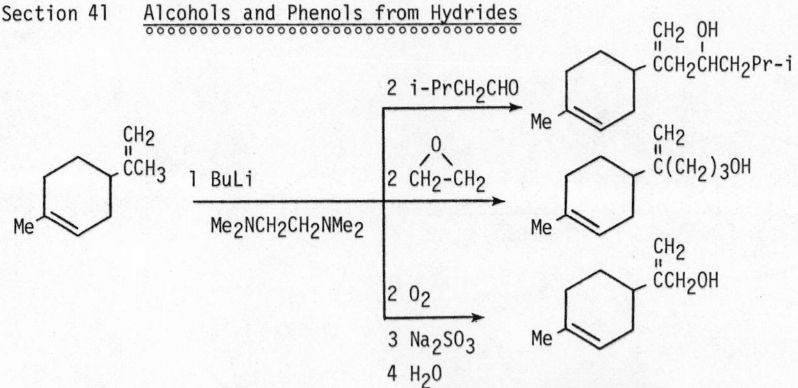

JACS (1972) <u>94</u> 4298
JOC (1972) <u>37</u> 3543

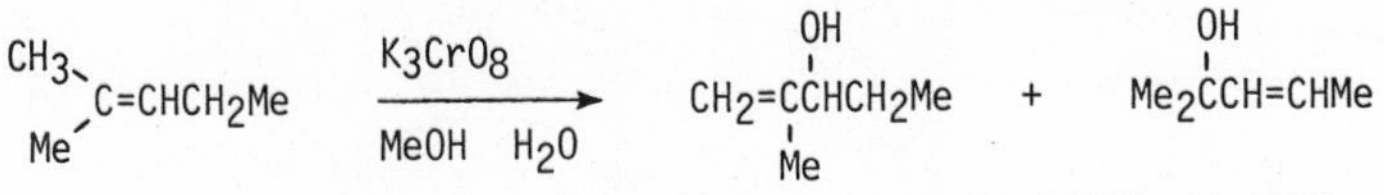

$$\underset{Me}{\overset{CH_3}{\diagdown}}C=CHCH_2Me \quad \xrightarrow[\text{MeOH} \quad H_2O]{K_3CrO_8} \quad \underset{Me}{\overset{OH}{\underset{|}{CH_2=CCHCH_2Me}}} \quad + \quad \overset{OH}{\overset{|}{Me_2CCH=CHMe}}$$

JACS (1972) **94** 4348

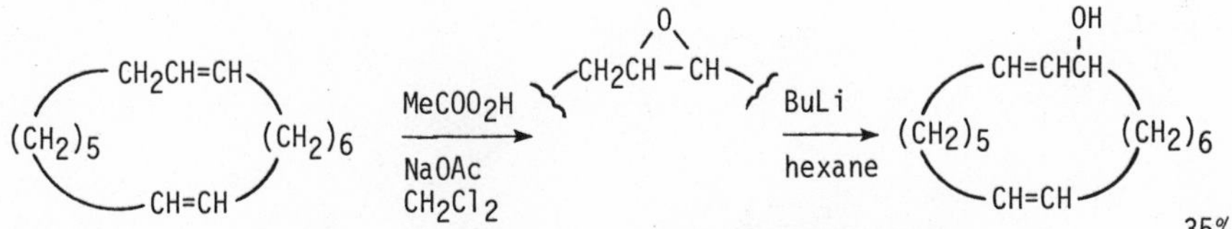

JOC (1971) **36** 3266
Synthesis (1972) 194

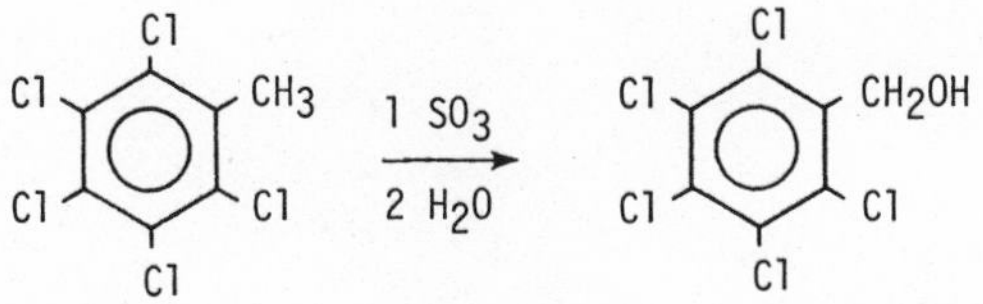

35%

JACS (1971) **93** 3538

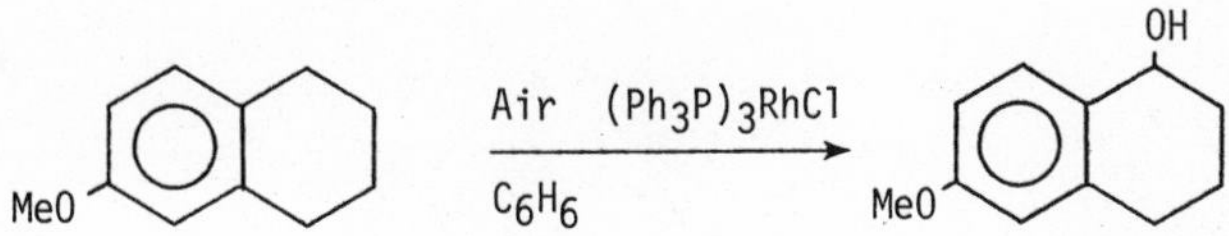

~91%

Tetr Lett (1968) 2917

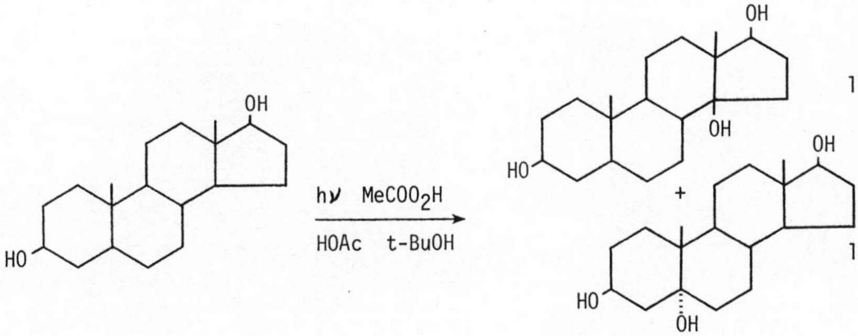

48%

JACS (1972) **94** 6228

10%

10%

Chem Comm (1970) 1390

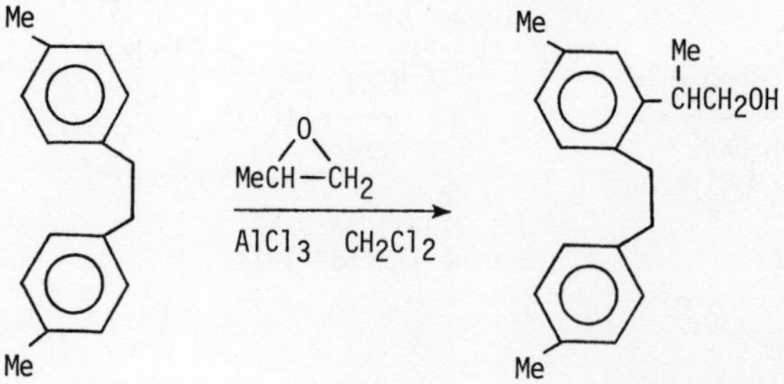

JACS (1972) 94 4241

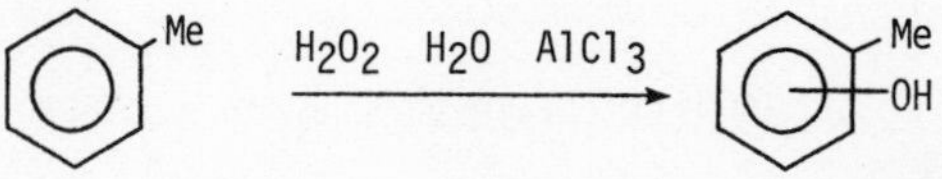

JOC (1971) 36 3184

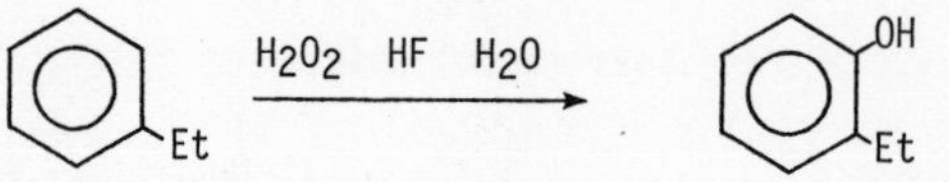

15%

JOC (1970) 35 4028

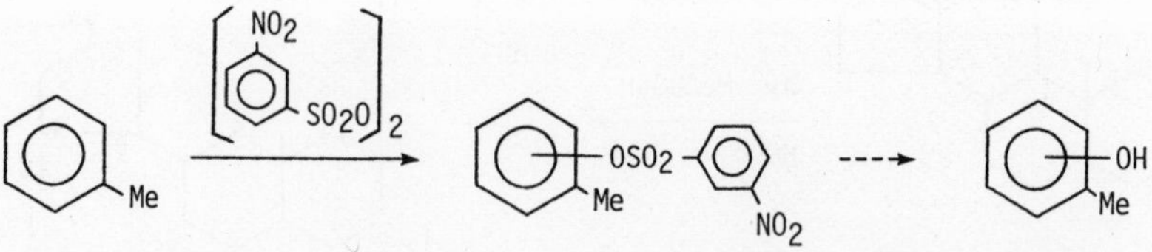

<71%

JOC (1966) 31 153

Also via: Esters (Section 116)

Section 42   Alcohols from Ketones
Conversion of Ketones to alcohols with longer carbon chains.page 41-42
Reduction of Ketones to alcohols 42-45

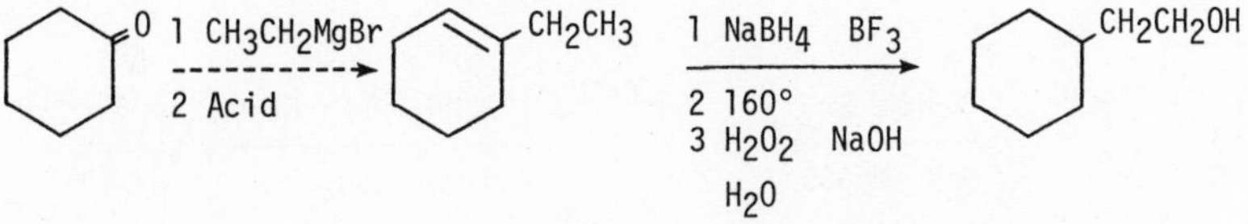

JACS (1967) <u>89</u> 561

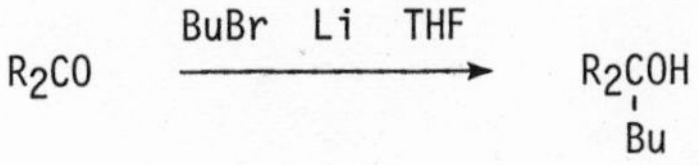

$$R_2CO \xrightarrow{\text{BuBr \quad Li \quad THF}} R_2\underset{\underset{Bu}{|}}{C}OH$$

(One-step procedure)

JCS Perkin I (1972) 1655

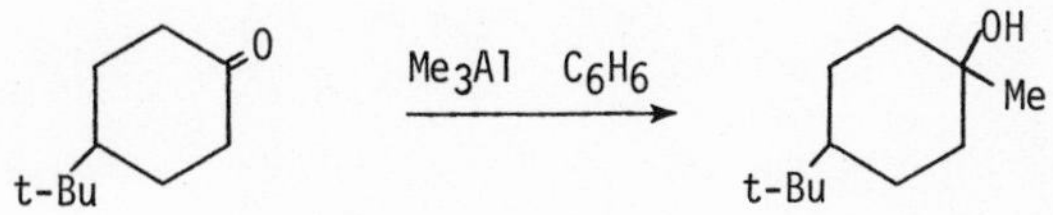

87%

Chem Comm (1971) 351

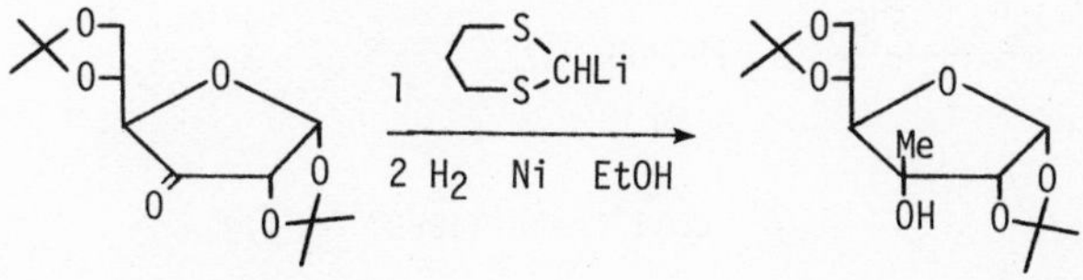

53%

Ber (1972) <u>105</u> 1978

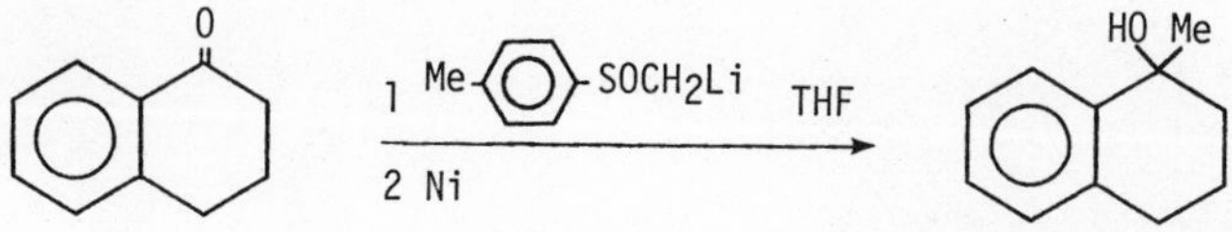

~68%

Tetr Lett (1972) 4605

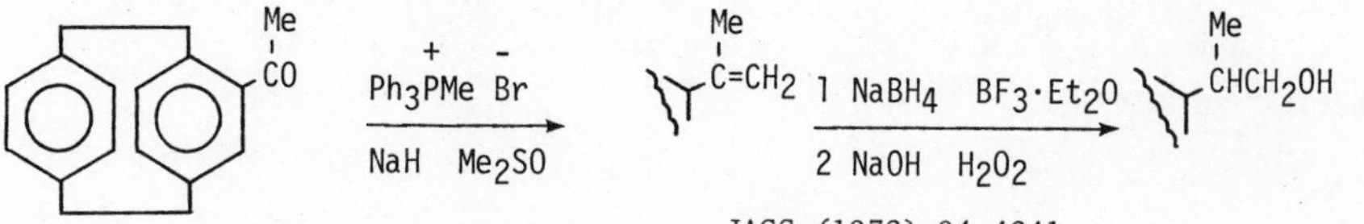

JACS (1972) 94 4241

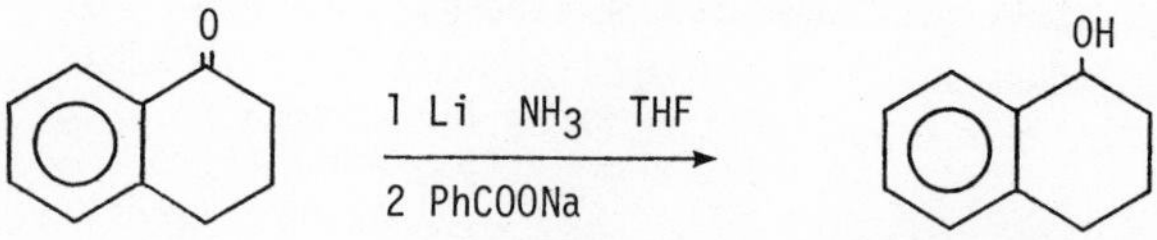

Tetr Lett (1971) 1853

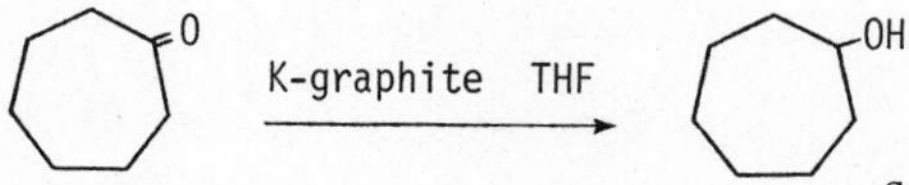

92%

Can J Chem (1972) 50 3058

PhCO   NaAl(OCH$_2$CH$_2$OMe)$_2$H$_2$ C$_6$H$_6$ PhCHOH   90%
| ————————————————————————————————→ |
Me   Me

Coll Czech (1969) 34 118

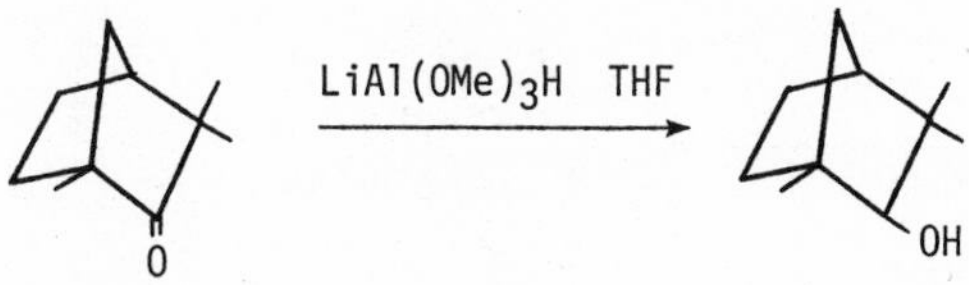
LiAl(OMe)3H THF
O
OH

80%

Synthesis (1972) 217

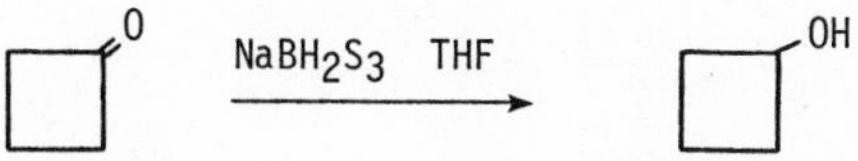
O
NaBH2S3 THF
OH

90%

Can J Chem (1970) 48 2366
Synthesis (1972) 526

$$\mathrm{PhCOMe} \xrightarrow[\text{pH 4}]{\mathrm{NaBH_3CN}\ \ \mathrm{MeOH}} \mathrm{PhCH(OH)Me}$$

93%

JACS (1971) 93 2897

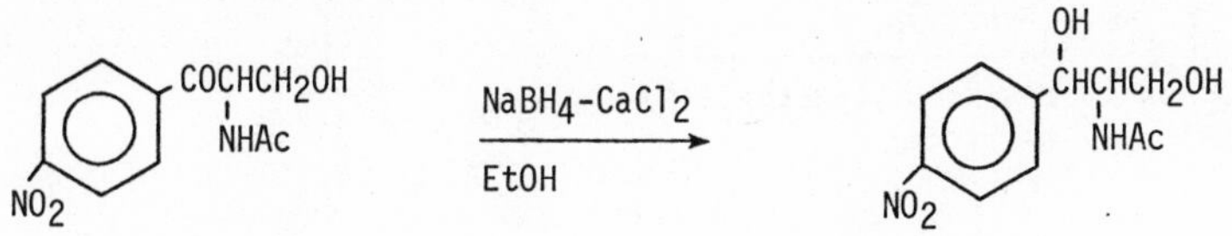
COCHCH2OH
NHAc
NO2
NaBH4-CaCl2
EtOH
OH
CHCHCH2OH
NHAc
NO2

70%

Ber (1960) 93 387

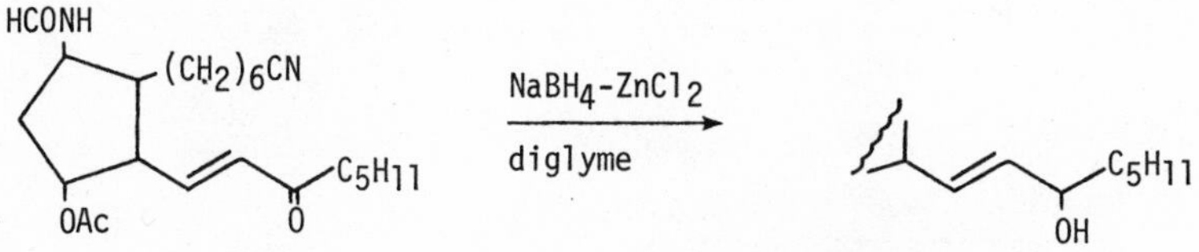
HCONH
(CH2)6CN
C5H11
OAc
O
NaBH4-ZnCl2
diglyme
C5H11
OH

JACS (1968) 90 3245
(1960) 82 6074

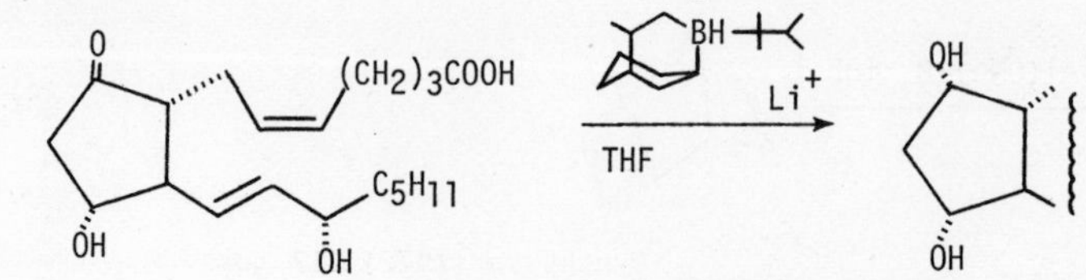

JACS (1971) <u>93</u> 7319

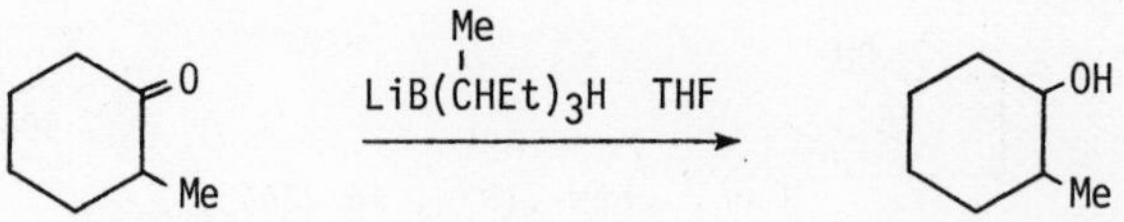

JACS (1972) <u>94</u> 7159

100%

(99% cis)

$R_2CO \xrightarrow{\text{R'}_2SnH_2 \text{ or } \text{R'}_3SnH} R_2CHOH$

Synthesis (1970) 499

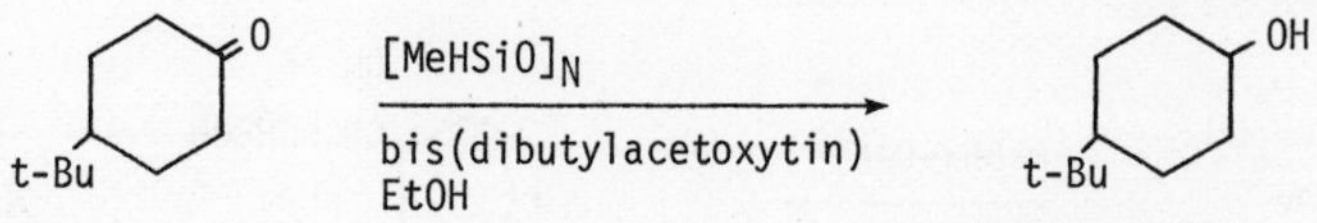

$\xrightarrow[\substack{\text{bis(dibutylacetoxytin)} \\ \text{EtOH}}]{[MeHSiO]_N}$

65%

JOC (1973) <u>38</u> 162

$R_2CO \xrightarrow[\substack{\text{i-PrOH} \quad H_2O}]{IrCl_4 \quad (MeO)_3P \quad HCl} R_2CHOH$

Org Synth (1970) <u>50</u> 13

$C_9H_{19}\underset{\underset{Me}{|}}{CO} \xrightarrow[\substack{\text{EtOH} \quad H_2O}]{\overset{\overset{NH_2}{|}}{NH_2}=CSO_2^- \quad NaOH} C_9H_{19}\underset{\underset{Me}{|}}{CHOH}$

74%

Tetr Lett (1972) 343
JCS Perkin I (1973) 2633

Related methods: Alcohols from Aldehydes (section 34)

		Section
Also via:	Olefinic alcohols	332
	Acetylenic alcohols	302

Section 43   Alcohols and Phenols from Nitriles
ooooooooooooooooooooooooooooooooooooooooo

No examples

Section 44   Alcohols from Olefins
ooooooooooooooooooooooooo

For the preparation of diols from olefins see section 323 (Alcohol-Alcohol)

$C_6H_{13}CH=CH_2$ $\xrightarrow{\begin{array}{l}1\ B_2H_6\quad THF\\ \hline 2\ NaCN\\ 3\ (CF_3CO)_2O\\ 4\ H_2O_2\quad NaOH\quad H_2O\end{array}}$ $(C_6H_{13}CH_2CH_2)_3COH$ 75%

Chem Comm (1971) 1048
JACS (1971) 93 2070
JOC (1973) 38 3968

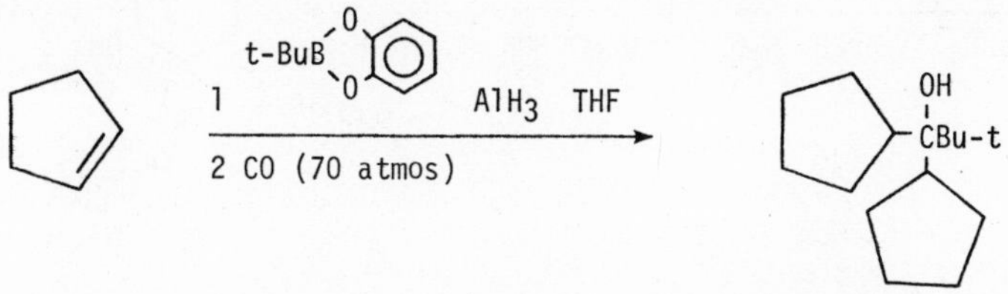

Synthesis (1972) 197   60%

$EtCH=CH_2$ $\xrightarrow{\begin{array}{l}1\ B_2H_6\quad THF\\ \hline 2\ Br_2\quad h\nu\quad CH_2Cl_2\quad H_2O\\ 3\ H_2O_2\quad NaOH\quad H_2O\end{array}}$ $(EtCH_2CH_2)_2COH$ 76%
$\qquad\qquad\qquad\qquad\qquad\qquad\qquad EtCH_2$

JACS (1971) 93 1025
Synthesis (1972) 699

Me$_2$C=CHMe

1 B$_2$H$_6$ THF

2 MeOH

3 Br$_2$ hν CH$_2$Cl$_2$ H$_2$O

4 H$_2$O$_2$ NaOH   EtOH   H$_2$O

Me
|
Me$_2$CHCOH
Me$_2$CHCHMe   82%

Synthesis (1972) 303 304 699

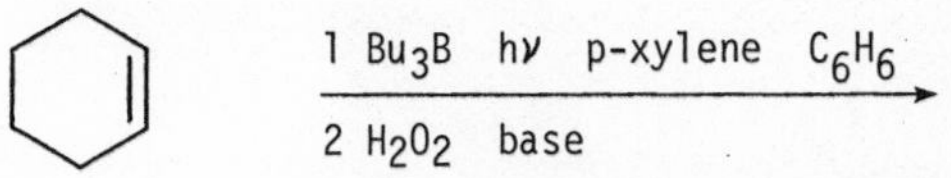

1 Bu$_3$B   hν p-xylene   C$_6$H$_6$

2 H$_2$O$_2$ base

80%

Tetrahedron (1973) 29 2365

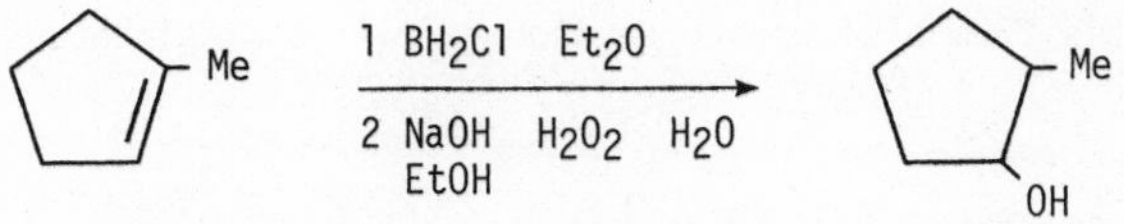

1 BH$_2$Cl   Et$_2$O

2 NaOH   H$_2$O$_2$ H$_2$O
  EtOH

JOC (1973) 38 182

1 NOCl

2 NaBH$_4$ i-PrOH   H$_2$O

NaBH$_4$

NaOH   H$_2$O

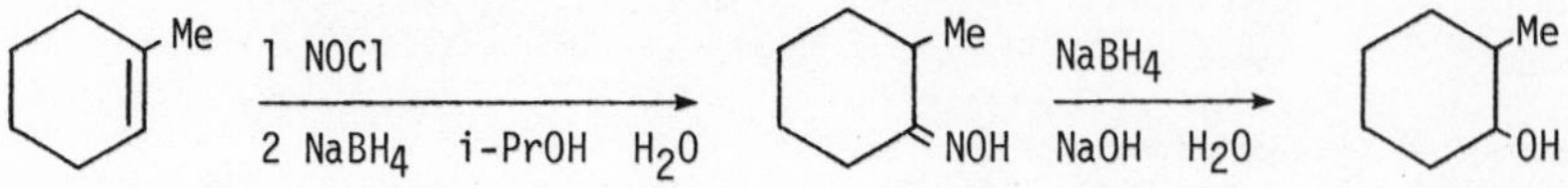

74%

Aust J Chem (1971) 24 1089

Section 45   Alcohols from Miscellaneous Compounds

PhCH=CHCH$_2$OH

H$_2$ nickel boride   EtOH

Ph(CH$_2$)$_3$OH   100%

JOC (1971) 36 2018

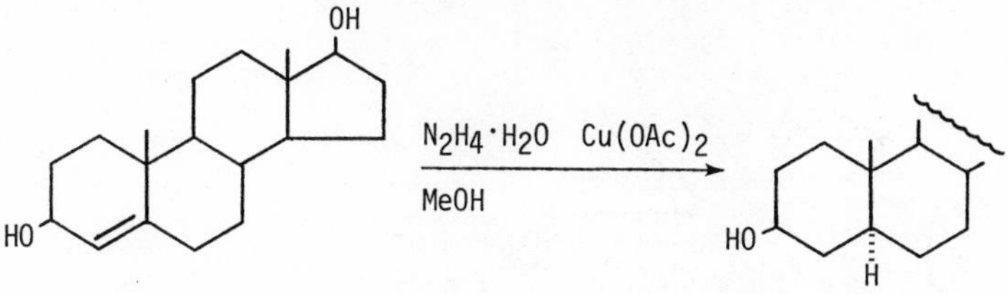

OH
N2H4·H2O Cu(OAc)2
MeOH
HO
HO
H

JOC (1971) 36 2730

SO3H
Me
NaOH
280-300°
OH
Me

Ber (1925) 58B 1211
JACS (1949) 71 3889

$$Ph_2S \xrightarrow{MeCOCH_2K \quad K \quad NH_3} PhCH_2\underset{\displaystyle Me}{CHOH}$$

71%

JACS (1972) 94 683

Et
O
CH2
Electrolysis Me4N+ OTs-
dioxane MeOH
OH
Et
Me

47%

JACS (1971) 93 5284

Section 45A Protection of Alcohols and Phenols

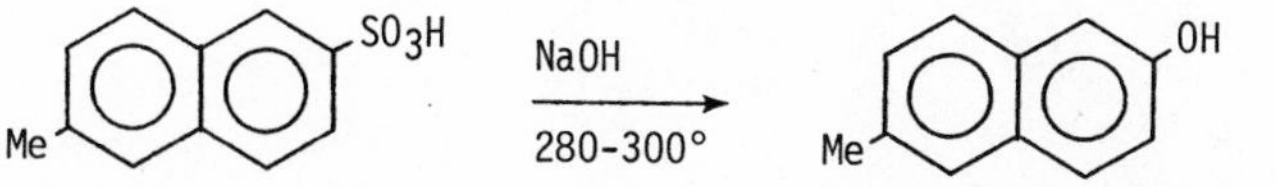

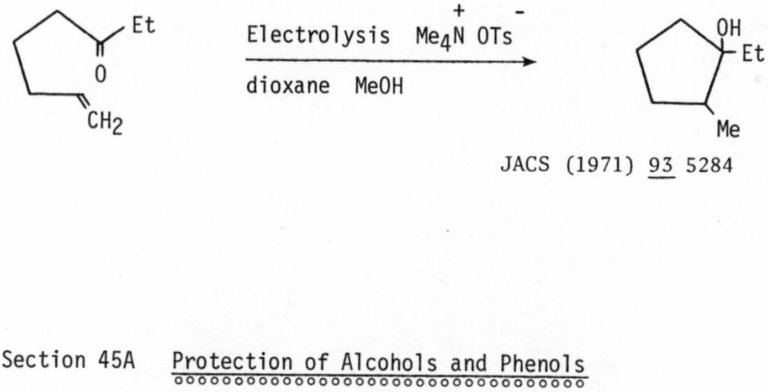

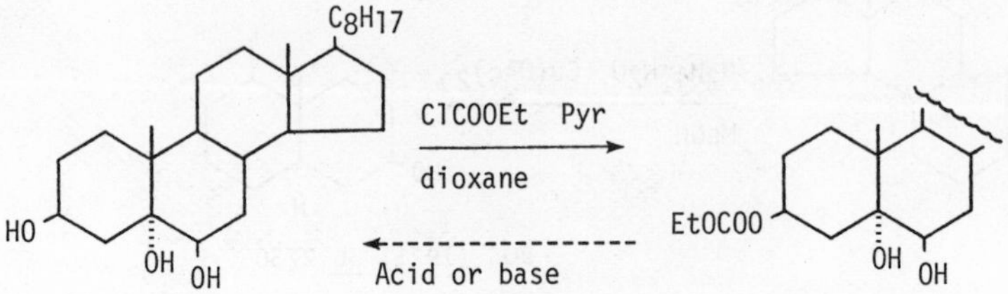

C8H17
ClCOOEt Pyr
dioxane
Acid or base
HO
OH
OH
EtOCOO
OH
OH

JACS (1952) 74 3309

ROH
ClCOOCH2CCl3 Pyr
Electrolysis
ROCOOCH2CCl3

JACS (1972) 94 5139

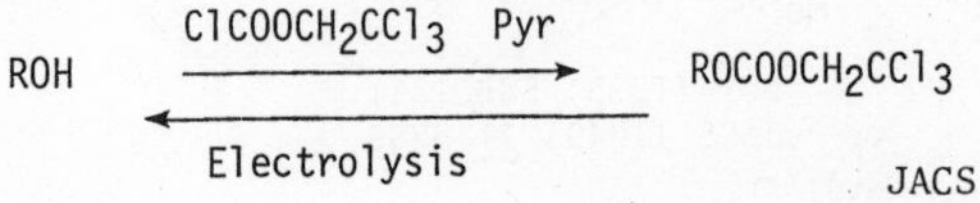

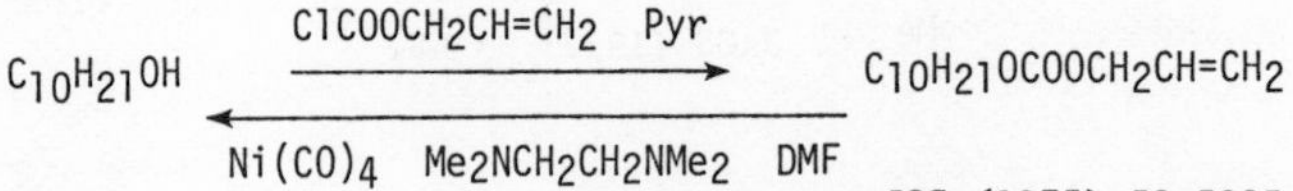

C10H21OH
ClCOOCH2CH=CH2 Pyr
Ni(CO)4 Me2NCH2CH2NMe2 DMF
C10H21OCOOCH2CH=CH2

JOC (1973) 38 3223 3575

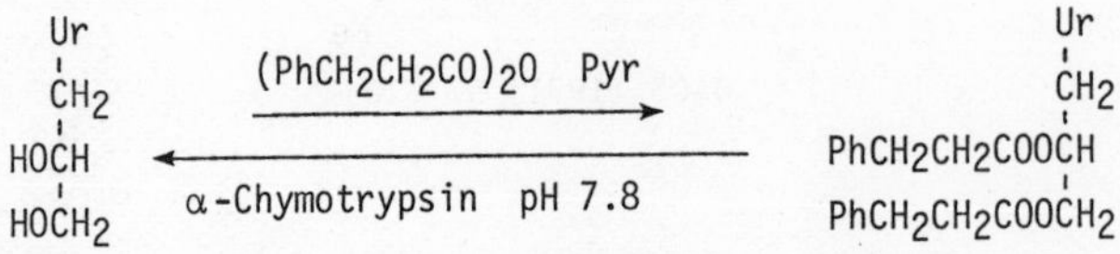

Ur
CH2
HOCH
HOCH2
(PhCH2CH2CO)2O Pyr
α-Chymotrypsin pH 7.8
Ur
CH2
PhCH2CH2COOCH
PhCH2CH2COOCH2

JOC (1973) 38 977

C8H17
COCl
NO2
Pyr
Zn NH4Cl THF H2O
HO
COO
NO2
(Applicable to phenols)

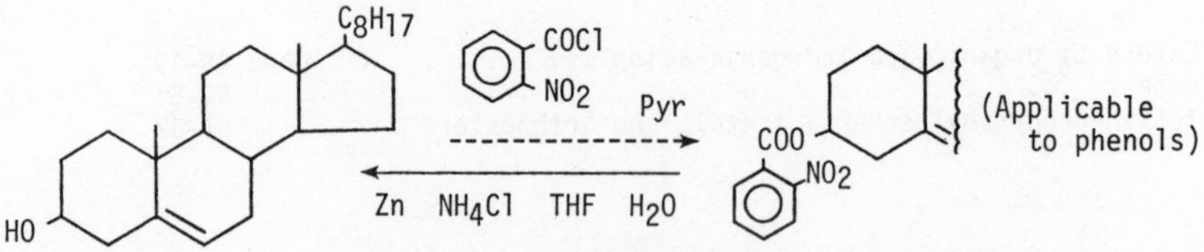

JCS Perkin I (1973) 599

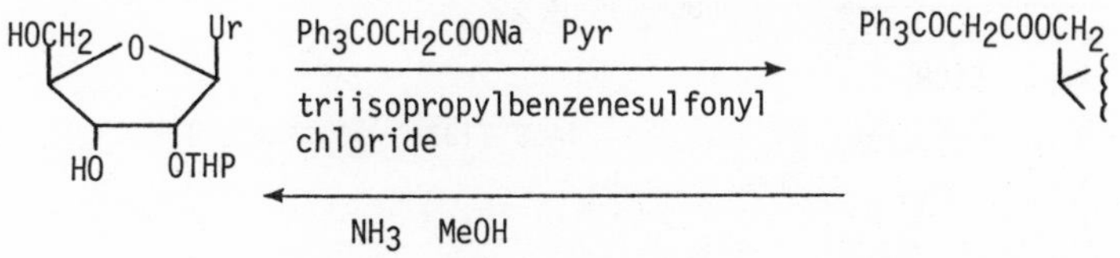
HOCH2
O
Ur
HO
OTHP
Ph3COCH2COONa Pyr
triisopropylbenzenesulfonyl chloride
NH3 MeOH
Ph3COCH2COOCH2

Can J Chem (1972) 50 1283

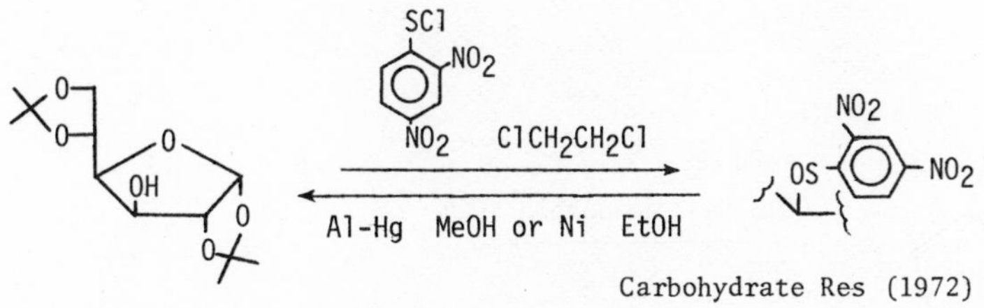
SCl
NO2
NO2
ClCH2CH2Cl
Al-Hg MeOH or Ni EtOH
O
OH
NO2
OS
NO2

Carbohydrate Res (1972) 21 301

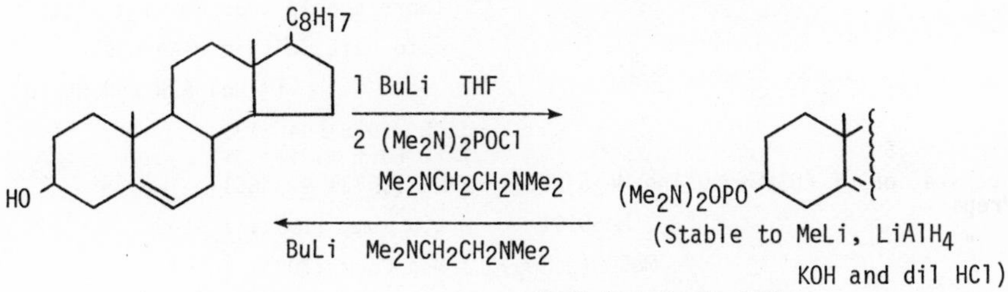
C8H17
HO
1 BuLi THF
2 (Me2N)2POCl
Me2NCH2CH2NMe2
BuLi Me2NCH2CH2NMe2
(Me2N)2OPO
(Stable to MeLi, LiAlH4 KOH and dil HCl)

JACS (1972) 94 5098

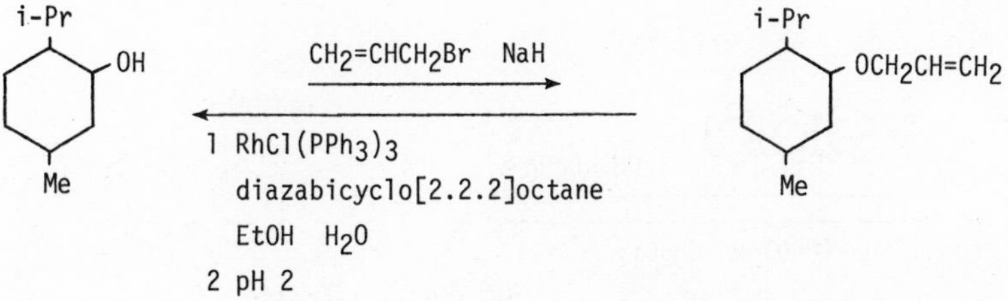
i-Pr
OH
Me
CH2=CHCH2Br NaH
1 RhCl(PPh3)3
diazabicyclo[2.2.2]octane
EtOH H2O
2 pH 2
i-Pr
OCH2CH=CH2
Me

JOC (1973) 38 3224

ROH $\xrightarrow[\text{H}_2 \quad \text{Pd-C} \quad \text{EtOH}]{\text{NaH} \quad \text{PhCH}_2\text{Br} \quad \text{C}_6\text{H}_6}$ ROCH$_2$Ph

JACS (1971) <u>93</u> 1746

Cleavage of benzyl ethers with Ph$_3$C$^+$ BF$_3^-$ JCS Perkin I (1972) 542

Further examples of the cleavage of benzyl ethers are included in Section 39 (Alcohols and Phenols from Ethers and Epoxides)

PhCH$_2$OCH$_2$ OH
→ (t-Bu, ClSiMe$_2$ imidazole, THF)
← (Bu$_4$N$^+$ F$^-$ THF or HOAc H$_2$O)
PhCH$_2$OCH$_2$ OSiMe$_2$ / t-Bu

(More stable than OSiMe$_3$. Stable to base, Wittig reagents, (i-Bu)$_2$AlH and H$_2$/Pd)

Preparation of ROSiMe$_3$ using Me$_3$SiNEt$_2$
JACS (1972) <u>94</u> 6190
Tetr Lett (1973) 317
JACS (1972) <u>94</u> 3651

"   "   "   "   Me$_3$SiCl   Synth Comm (1971) <u>1</u> 81

"   "   "   "   MeCON(SiMe$_3$)$_2$ J Med Chem (1973) <u>16</u> 54

"   " ROSi(Me$_2$)Pr-i using i-Pr(Me$_2$)SiCl JACS (1971) <u>93</u> 7319

Cleavage of silyl ethers with H$_2$/Pd   JACS (1972) <u>94</u> 6190 footnote 4

C$_{16}$H$_{33}$OH $\xrightarrow[\text{O}_3\text{-(PhO)}_3\text{P} \quad \text{CH}_2\text{Cl}_2]{\text{TsOH} \quad \text{C}_6\text{H}_6}$

OMe →

C$_{16}$H$_{33}$O

JOC (1971) <u>36</u> 4134

$$\text{ROH} \xrightleftharpoons[\text{H}_2 \quad \text{Pd-C} \quad \text{EtOH}]{\text{Ph}_3\text{CCl} \quad \text{Pyr} \quad \text{C}_6\text{H}_6} \text{ROCPh}_3$$

JOC (1972) <u>37</u> 2877

Cleavage of trityl ethers with Na/NH₃ Tetr Lett (1972) 2349

Preparation of substituted trityl ethers JOC (1972) <u>37</u> 956

Polymer-bound trityl ethers Annalen (1972) <u>766</u> 6

$$\text{C}_{16}\text{H}_{33}\text{OH} \xrightleftharpoons[\text{N}_2\text{H}_4 \quad \text{NaOH} \quad \text{triethylene glycol}]{\text{TsOH} \quad \text{C}_6\text{H}_6}$$

Tetrahedron (1972) <u>28</u> 419

$$\text{ROH} \xrightleftharpoons[\text{acid}]{\text{catalyst}}$$

Catalyst:	TsOH	Synthesis (1973) 169
		Steroids (1964) <u>4</u> 229
	Conc HCl	JCS Perkin I (1973) 720
	POCl₃	JCS (1950) 3646
	BF₃·Et₂O	Synthesis (1972) 81
	Picric acid	Tetr Lett (1972) 3333
	Ion exch resin (Acid)	JCS (1956) 4665

Cleavage of tetrahydropyranyl ethers with Ph₃C⁺ BF₄⁻ Chem Comm (1971) 1109

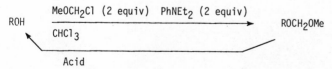

$$\text{ROH} \xrightleftharpoons[\text{Acid}]{\begin{array}{c}\text{MeOCH}_2\text{Cl (2 equiv)} \quad \text{PhNEt}_2 \text{ (2 equiv)} \\ \text{CHCl}_3 \end{array}} \text{ROCH}_2\text{OMe}$$

V. Fletcher (Columbia Univ),
unpublished

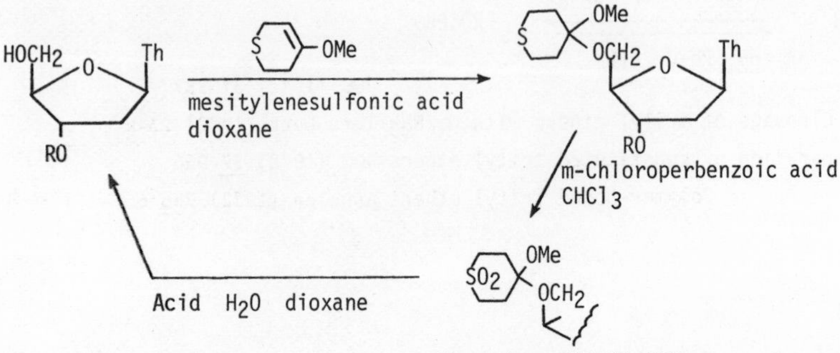

HOCH₂ ─ O ─ Th
RO

mesitylenesulfonic acid
dioxane

⟶

OMe / OCH₂ ─ O ─ Th
RO

m-Chloroperbenzoic acid
CHCl₃

Acid   H₂O   dioxane

SO₂ ─ OMe / OCH₂

Chem Comm (1972) 766

OR / OH
Me

EtOCH=CH₂   TsOH   Et₂O
⟶
HOAc   dioxane   H₂O

OR / OCHMe / OEt
Me

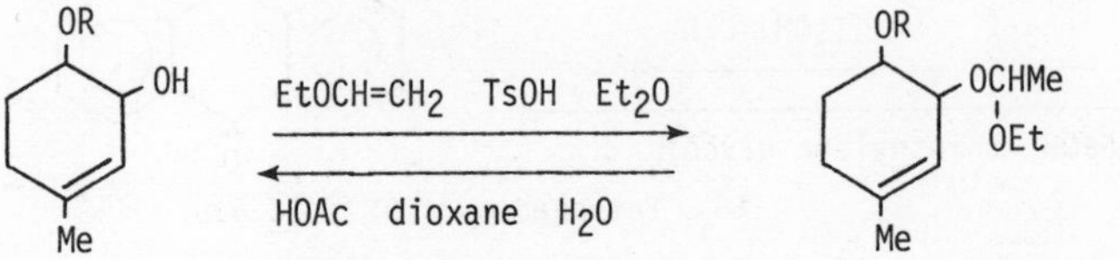

(Stable to LiAlH₄ and base)

JCS C (1971) 2944

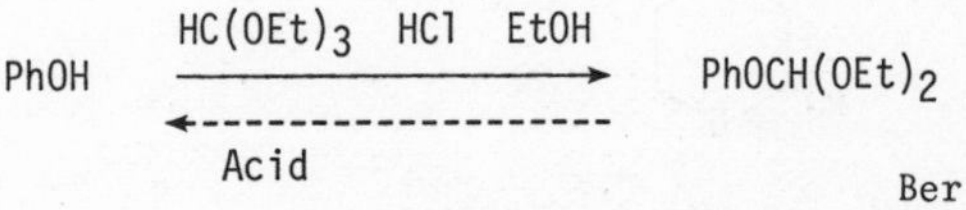

PhOH

HC(OEt)₃   HCl   EtOH
⟶
Acid
←- - - - - - - - - - - - - -

PhOCH(OEt)₂

Ber (1970) 103 643

Related methods: Section
        Alcohols and Phenols from Ethers and Epoxides   39
                Ethers from Alcohols and Phenols   123

Chapter 4   PREPARATION
OF
ALDEHYDES

Section 46   <u>Aldehydes from Acetylenes</u>

$$EtC{\equiv}CH \xrightarrow[\text{toluene}]{CO \quad H_2 \quad HRh(CO)(Ph_3P)_3} \quad \underset{\overset{|}{CHO}}{EtCHMe} \qquad < 75\%$$

Tetr Lett (1972) 3455

$$C_6H_{13}C{\equiv}CH \xrightarrow[\substack{2 \text{ m-Chloroperbenzoic acid} \\ 3 \text{ } H_2SO_4 \text{ MeOH} \\ 4 \text{ HOAc } H_2O}]{1 \text{ Et}_3\text{SiH } H_2PtCl_6} \quad C_6H_{13}CH_2CHO \qquad \sim 60\%$$

JACS (1971) <u>93</u> 2080

Section 47   <u>Aldehydes from Carboxylic Acids and Acid Halides</u>

$$C_5H_{11}COOH \xrightarrow[\text{}]{\overset{\displaystyle \overset{Me}{|}}{\underset{\displaystyle \underset{Me}{|}}{\text{i-PrCBH}_2}} \quad \text{THF}} \quad C_5H_{11}CHO \qquad 98\%$$

JOC (1972) <u>37</u> 2942

PhCH$_2$COCl $\xrightarrow[\text{2 HOAc EtOH H}_2\text{O}]{\text{1 Na}_2\text{Fe(CO)}_4}$ PhCH$_2$CHO   ~59%

JACS (1972) 94 2516
Bull Chem Soc Jap (1971) 44 2569
Tetr Lett (1973) 3535

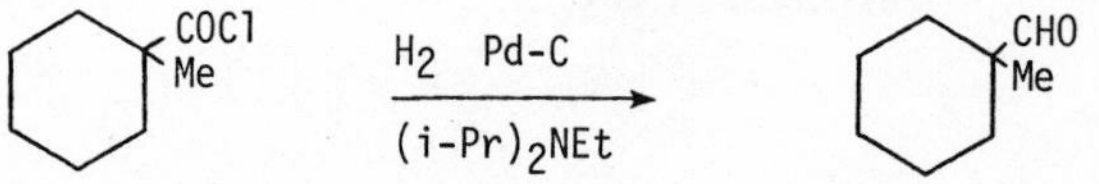

$\xrightarrow[\text{(i-Pr)}_2\text{NEt}]{\text{H}_2\quad\text{Pd-C}}$

78%

Rec Trav Chim (1971) 90 1323

C$_7$H$_{15}$COCl $\xrightarrow{\text{Et}_3\text{SiH}\quad\text{Pd-C}}$ C$_7$H$_{15}$CHO   51%

JOC (1969) 34 1977
Chem Comm (1970) 1703

RCOCl $\xrightarrow{\text{LiAl(OBu-t)}_3\text{H}\quad\text{diglyme}}$ RCHO

Synthesis (1972) 217

PhCH=CHCOCl $\xrightarrow[\substack{\text{2 NaBH}_4\quad\text{MeOH}\\ \text{3 NaOH}}]{\text{1 (EtO)}_3\text{P}}$ PhCH=CHCHO   100%

Ber (1970) 103 2984

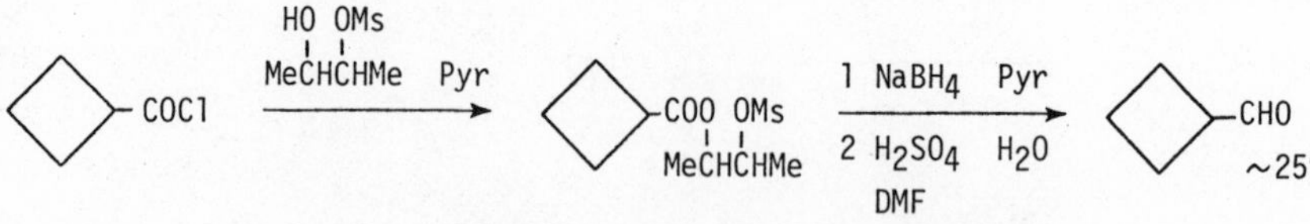

Org Synth (1971) 51 11

$PrCH_2COOH$ $\dashrightarrow$ $\underset{Br}{PrCHCOOH}$ $\xrightarrow[\text{xylene}]{\text{Pyridine N-oxide}}$ $PrCHO$ $<67\%$

JOC (1966) <u>31</u> 3058

$PhCH_2COOH$ $\xrightarrow{Na_2S_2O_8 \quad AgNO_3 \quad H_2O}$ $PhCHO$ 48%

JCS (1960) 1332

Section 48   <u>Aldehydes from Alcohols and Phenols</u>
oooooooooooooooooooooooooooooooooooooooooo

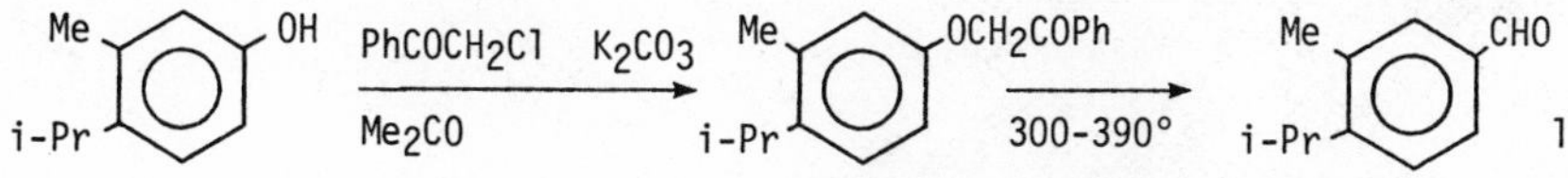

PhCOCH2Cl K2CO3 / Me2CO ; 300-390°

Bull Soc Chim Fr (1960) 278
JOC (1961) <u>26</u> 4308   17%

$CH_2=CHCH_2OH$ $Hg(OAc)_2$ / $BF_3 \cdot Et_2O$ C_6H_6 ; 540-545° (Vapor phase)

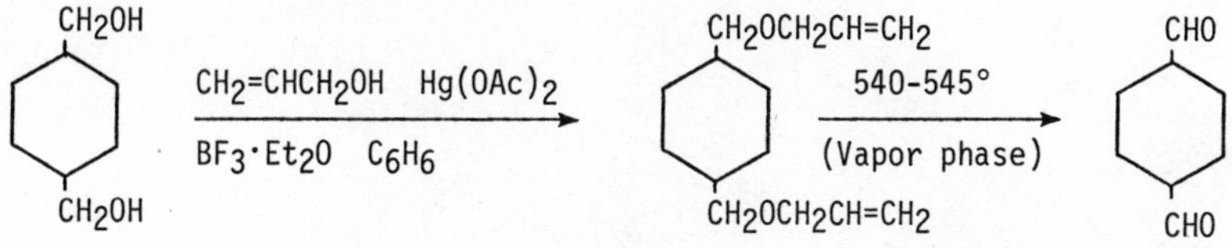

JCS <u>C</u> (1971) 2730

$C_7H_{15}CH_2OH$ $\dashrightarrow$ $C_7H_{15}CH_2OCPh_3$ $\xrightarrow[CH_2Cl_2]{Ph_3C^+ \ AsF_6^-}$ $C_7H_{15}CHO$ $<56\%$

JOC (1973) <u>38</u> 625

$$C_7H_{15}CH_2OH \xrightarrow[\text{Et}_3\text{N} \quad \text{toluene}]{\begin{array}{c} \text{CH}_2\text{CO} \\ | \qquad \overset{+}{>} \text{N-SMe}_2 \overset{-}{\text{Cl}} \\ \text{CH}_2\text{CO} \end{array}} C_7H_{15}CHO \qquad 96\%$$

JACS (1972) <u>94</u> 7586

$$\text{CH}_2\text{OH} \xrightarrow[\text{2 Et}_3\text{N}]{\text{1 Me}_2\text{SO·Cl}_2 \quad \text{CH}_2\text{Cl}_2} \text{CHO} \qquad >95\%$$

Tetr Lett (1973) 919

$$C_6H_{13}CH_2OH \xrightarrow[\text{H}_3\text{PO}_4 \quad \text{Me}_2\text{SO} \quad \text{C}_6\text{H}_6]{\text{Resin-CH}_2\text{N=C=NPr-i}} C_6H_{13}CHO$$

Tetr Lett (1972) 3285

$$C_5H_{11}CH_2OH \xrightarrow[\text{Pyr} \quad \text{C}_6\text{H}_6]{\text{NO}_2\text{-}\bigcirc\text{-SO}_2\text{Cl}} C_5H_{11}CH_2OSO_2 \bigcirc NO_2 \xrightarrow[]{\text{Me}_2\text{SO} \quad \text{NaHCO}_3} C_5H_{11}CHO \quad 33\%$$

Synthesis (1971) 655

$$RCH_2OH \xrightarrow{\text{Argentic picolinate} \quad \text{H}_2\text{O}} RCHO$$

Tetrahedron (1973) <u>29</u> 751

$$C_{15}H_{31}CH_2OH \xrightarrow[\text{toluene}]{\text{CrO}_3\text{-graphite}} C_{15}H_{31}CHO \qquad 95\%$$

Can J Chem (1972) <u>50</u> 3058

$C_7H_{15}CH_2OH$ $\xrightarrow[\text{(vapor phase)}]{Cu_2O \quad 250\text{-}300°}$ $C_7H_{15}CHO$ 98%

Tetr Lett (1972) 257

$\triangleright\!\!-CH_2OH$ $\xrightarrow{H_2O_2 \quad FeSO_4 \quad H_2O}$ $\triangleright\!\!-CHO$ 37%

Can J Chem (1965) 43 2924

Related methods: Ketones from Alcohols and Phenols (Section 168)

Section 49   Aldehydes from Aldehydes
              ○○○○○○○○○○○○○○○○○○○○○○○○○○○

$C_{10}H_{21}CH_2CHO$ $\xrightarrow[\text{KOH \quad EtOH}]{Fe(CO)_5 \quad HCHO}$ $C_{10}H_{21}\underset{\underset{Me}{|}}{C}HCHO$ 55%

Tetr Lett (1973) 2491

Me_2CHCHO $\xrightarrow[C_6H_6 \quad H_2O]{PhCH_2Cl \quad NaOH \quad Bu_4\overset{+}{N}\ \overset{-}{I}}$ $Me_2\underset{\underset{PhCH_2}{|}}{C}CHO$ 75%

Tetr Lett (1973) 1273

$PhCHO$ $\xrightarrow[\substack{THF \quad MeOH \\ 2\ LiAlH_4}]{1\ MeSOCH_2SMe \quad PhCH_2\overset{+}{N}Me_3\ \overset{-}{O}H}$ $PhCH_2CH(SMe)_2$ $--\!\!\rightarrow$ $PhCH_2CHO$

Tetr Lett (1972) 1383

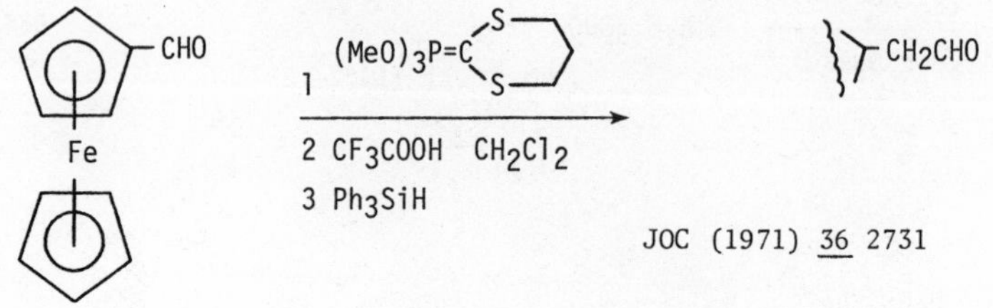

JOC (1971) 36 2731

$$\text{EtCHO} \xrightarrow[\text{THF}]{\text{BrCH}_2\text{COOEt} \quad (\text{Me}_3\text{Si})_2\text{NLi}} \text{EtCHCHCOOEt} \xrightarrow[2 \, \Delta]{1 \text{ Base}} \text{EtCH}_2\text{CHO}$$

Tetr Lett (1972) 3761

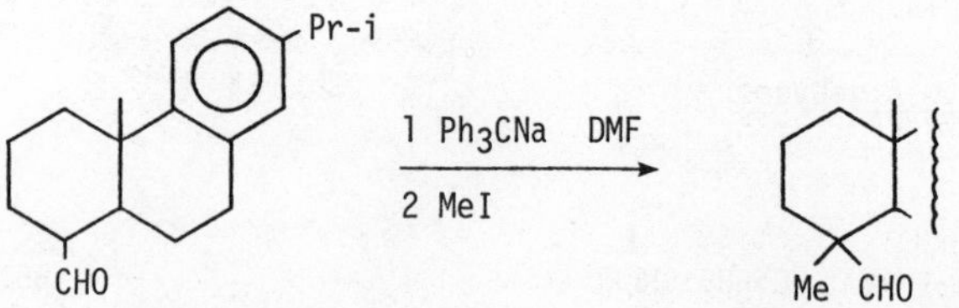

Chem Comm (1971) 10

66%

$$\text{Et}_2\text{CHCHO} \dashrightarrow \text{Et}_2\text{C=CHOAc} \xrightarrow[2 \text{ MeI}]{1 \text{ Bu}_3\text{SnOMe} \quad \text{MeCOOMe}} \underset{\text{Me}}{\text{Et}_2\text{CCHO}}$$

<82%

Compt Rend (1970) C 270 100

$$\text{C}_5\text{H}_{11}\text{CH}_2\text{CHO} \xrightarrow[\substack{\text{DMF} \\ 2 \text{ CH}_2\text{I}_2 \quad \text{Zn-Ag}}]{1 \text{ ClSiMe}_3 \quad \text{Et}_3\text{N}} \text{C}_5\text{H}_{11}\text{CH-CHOSiMe}_3 \xrightarrow[\text{H}_2\text{O}]{\text{NaOH} \quad \text{EtOH}} \underset{\text{Me}}{\text{C}_5\text{H}_{11}\text{CHCHO}} \quad >53\%$$

Tetr Lett (1973) 2767

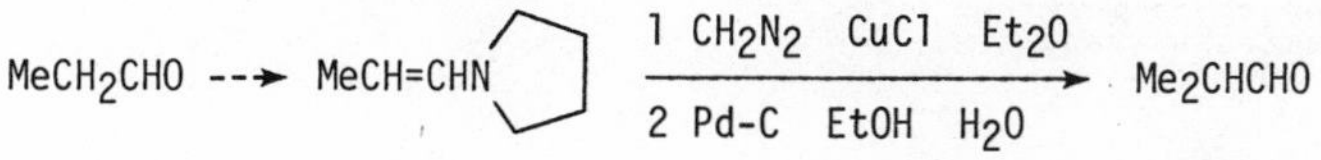

MeCH$_2$CHO --► MeCH=CHN⟨⟩ $\xrightarrow[\text{2 Pd-C EtOH H}_2\text{O}]{\text{1 CH}_2\text{N}_2\text{ CuCl Et}_2\text{O}}$ Me$_2$CHCHO   < 28%

JOC (1973) <u>38</u> 304

Related methods: Aldehydes from Ketones (Section 57)

Also via: Olefinic aldehydes (Section 341)

Section 50   <u>Aldehydes from Alkyls</u>

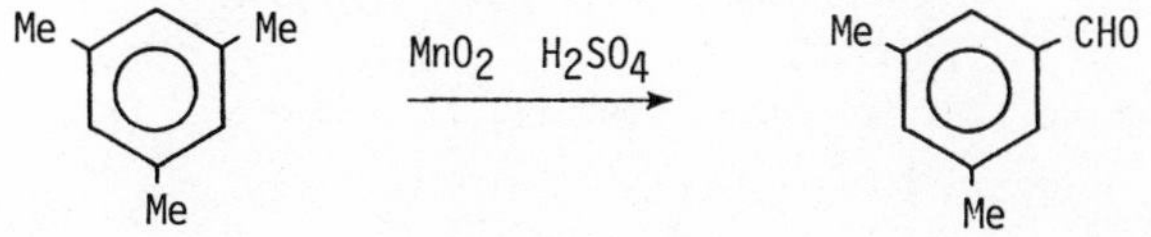

Me⟨⟩Me (Me below) $\xrightarrow{\text{MnO}_2\text{ H}_2\text{SO}_4}$ Me⟨⟩CHO (Me below)

JACS (1946) <u>68</u> 1085

Section 51   <u>Aldehydes from Amides</u>

RCONMe$_2$ $\xrightarrow{\text{LiAl(OEt)}_3\text{H Et}_2\text{O}}$ RCHO

Synthesis (1972) 217

Section 52   Aldehydes from Amines

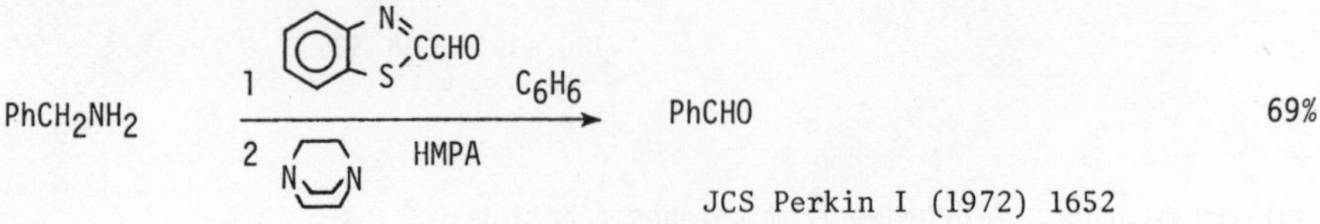

$PhCH_2NH_2$

$\xrightarrow{\text{1 } \quad C_6H_6 \\ \text{2 } \quad HMPA}$

PhCHO   69%

JCS Perkin I (1972) 1652

$PrCH_2NH_2$ $\xrightarrow[H_2O]{[(\pi\text{-}C_5H_5)_2Mo(SMe_2)Br]^+ PF_6^-}$ PrCHO

Chem Comm (1971) 1274

$C_5H_{11}CH_2NH_2$ $\xrightarrow{h\nu \quad cyclohexane}$ $C_5H_{11}CH=NCH_2C_5H_{11}$ $\dashrightarrow$ $C_5H_{11}CHO$

Chem Comm (1971) 1430

$PrCH_2NH_2$ $\xrightarrow{PhCOCH_2Br}$ $PrCH_2NHCH_2COPh$ $\xrightarrow{h\nu \quad MeOH \quad H_2O}$ PrCHO   <57%

JOC (1972) 37 1254

Section 53   Aldehydes from Esters

No additional examples

Section 54   Aldehydes from Ethers and Epoxides

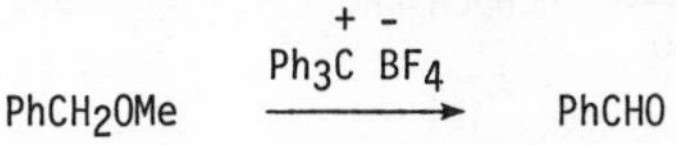

PhCH$_2$OMe $\xrightarrow{\overset{+ \quad -}{Ph_3C \; BF_4}}$ PhCHO   75%

Chem Comm (1971) 1109

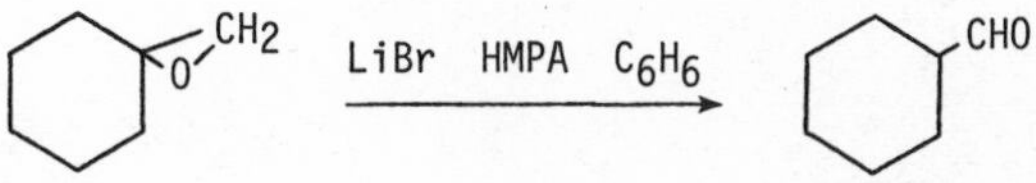

JACS (1971) 93 1693

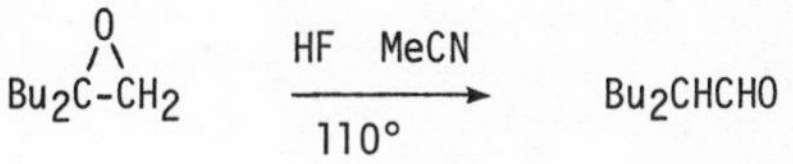

Bu$_2$$\overset{\overset{O}{\diagup\!\backslash}}{C}$-CH$_2$ $\xrightarrow[110°]{HF \quad MeCN}$ Bu$_2$CHCHO   74%

Z Chem (1967) 7 229

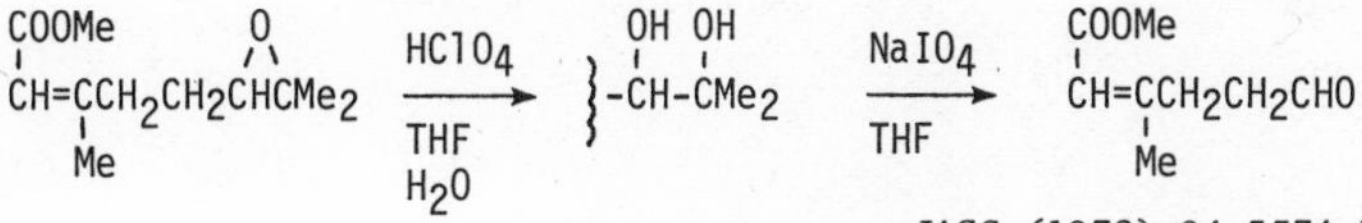

JACS (1972) 94 5374 5379

Section 55   Aldehydes from Halides

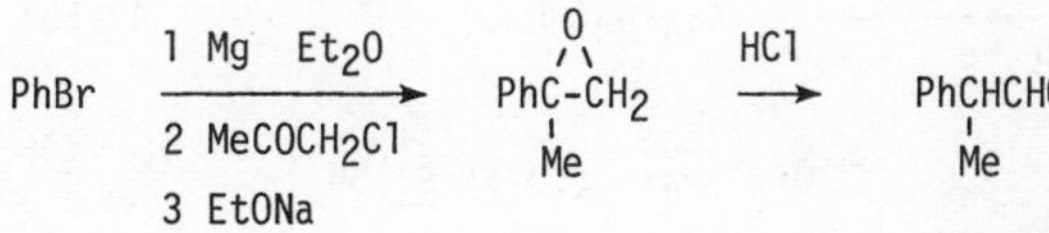

PhBr $\xrightarrow[\substack{2 \; MeCOCH_2Cl \\ 3 \; EtONa}]{1 \; Mg \quad Et_2O}$ Ph$\overset{\overset{O}{\diagup\!\backslash}}{C}$-CH$_2$ $\xrightarrow{HCl}$ PhCHCHO
                                                                                             | |
                                                                                            Me   Me

JACS (1939) 61 2383
(1946) 68 2339

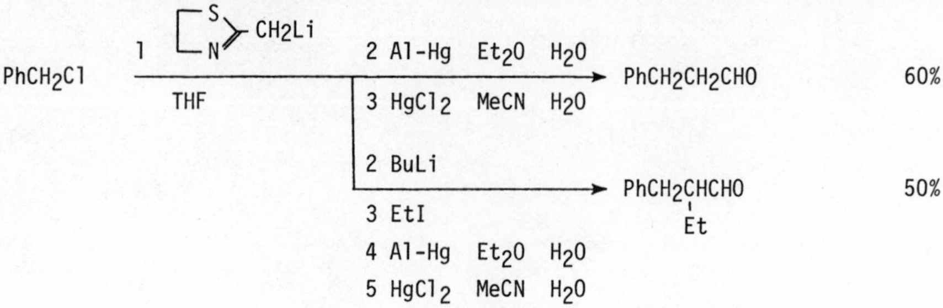

PhCH₂Cl → (1. [thiazoline-CH₂Li structure], THF) → (2 Al-Hg Et₂O H₂O; 3 HgCl₂ MeCN H₂O) → PhCH₂CH₂CHO   60%

→ (2 BuLi; 3 EtI; 4 Al-Hg Et₂O H₂O; 5 HgCl₂ MeCN H₂O) → PhCH₂CHCHO with Et   50%

Tetr Lett (1972) 3929

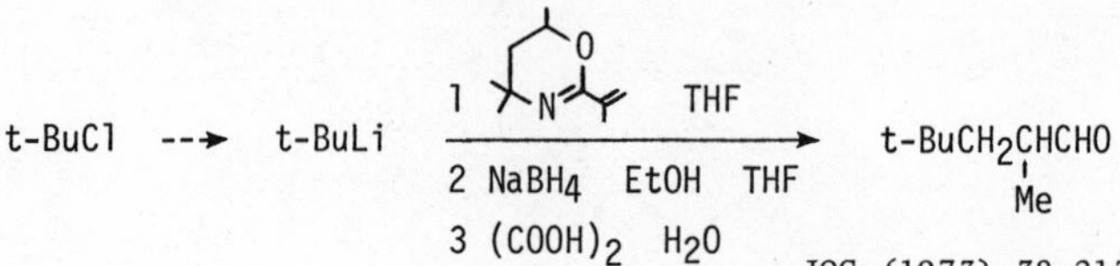

t-BuCl ⟶ t-BuLi → (1. [oxazoline structure] THF; 2 NaBH₄ EtOH THF; 3 (COOH)₂ H₂O) → t-BuCH₂CHCHO with Me

JOC (1973) 38 2136

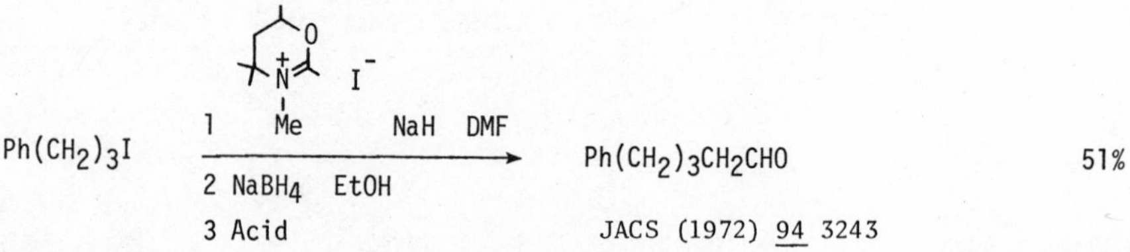

Ph(CH₂)₃I → (1. [N-Me oxazoline] I⁻ NaH DMF; 2 NaBH₄ EtOH; 3 Acid) → Ph(CH₂)₃CH₂CHO   51%

JACS (1972) 94 3243

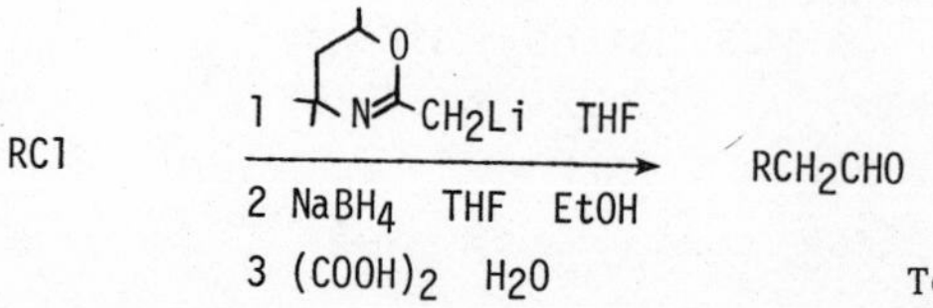

RCl → (1. [oxazoline-CH₂Li] THF; 2 NaBH₄ THF EtOH; 3 (COOH)₂ H₂O) → RCH₂CHO

Tetrahedron (1971) 27 5979
JOC (1973) 38 36

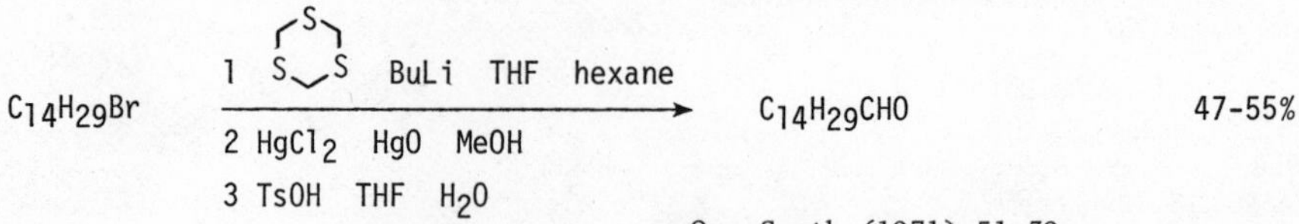

C₁₄H₂₉Br → (1. [1,3,5-trithiane] BuLi THF hexane; 2 HgCl₂ HgO MeOH; 3 TsOH THF H₂O) → C₁₄H₂₉CHO   47-55%

Org Synth (1971) 51 39

$$BuI \xrightarrow[\text{NaH\quad THF}]{\text{MeSOCH}_2\text{SMe}} \underset{\underset{\text{SMe}}{|}}{BuCHSOMe} \xrightarrow[\text{THF}]{\text{HCl\quad}\cdot\text{H}_2\text{O}} BuCHO \qquad 66\%$$

Tetr Lett (1971) 3151
(1973) 3267

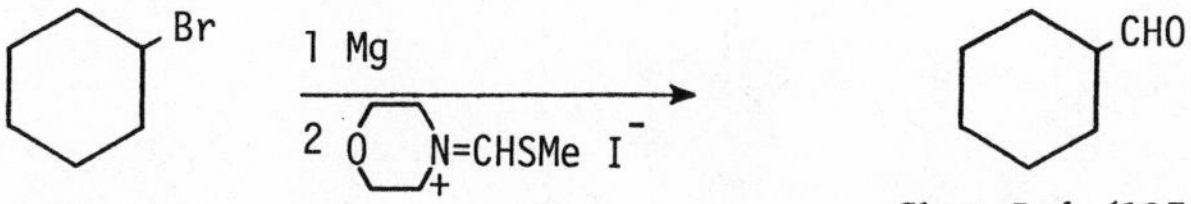

Chem Ind (1972) 380　　　75%

$$RBr \xrightarrow[\text{hexane}]{\text{Li\quad Et}_2\text{O}} RLi \xrightarrow[]{\overset{\displaystyle\text{Me}}{\underset{\displaystyle\text{Me}}{|}}\atop t\text{-BuCNC}} RCHO$$

Org Synth (1971) <u>51</u> 31

$$C_7H_{15}Br \xrightarrow[\text{THF}]{\overset{\displaystyle\text{Me}}{|}\atop \text{HCONPh\quad Li}} C_7H_{15}CHO \qquad \text{(One-step procedure)} \qquad 73\%$$

Synthesis (1973) 160

Also via:　Olefinic aldehydes (Section 341)

Section 56　　<u>Aldehydes from Hydrides</u>
　　　　　　　ooooooooooooooooooooooooo

$$ArH \xrightarrow[]{\text{Cl}_2\text{CHOMe\quad TiCl}_4\quad \text{CH}_2\text{Cl}_2} ArCHO$$

Org Synth (1967) <u>47</u> 1

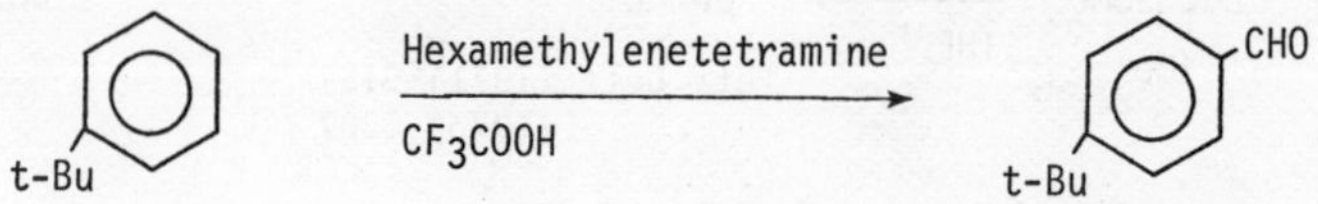

Hexamethylenetetramine
$\xrightarrow{\hspace{2cm}}$
CF$_3$COOH

75%

JOC (1972) **37** 3972

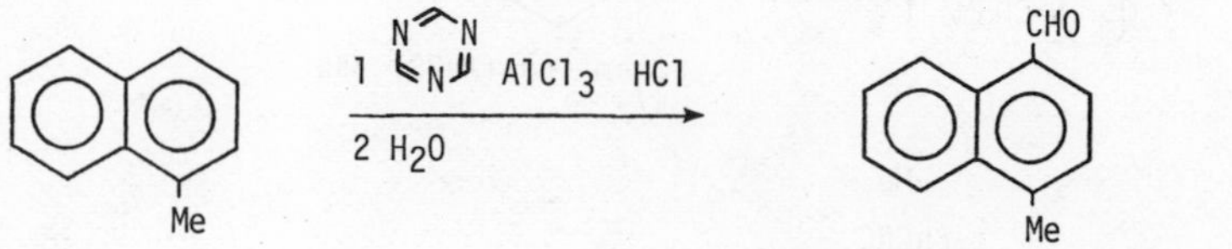

1 (triazine) AlCl$_3$ HCl

2 H$_2$O

$\xrightarrow{\hspace{2cm}}$

Arch Pharm (1971) **304** 362

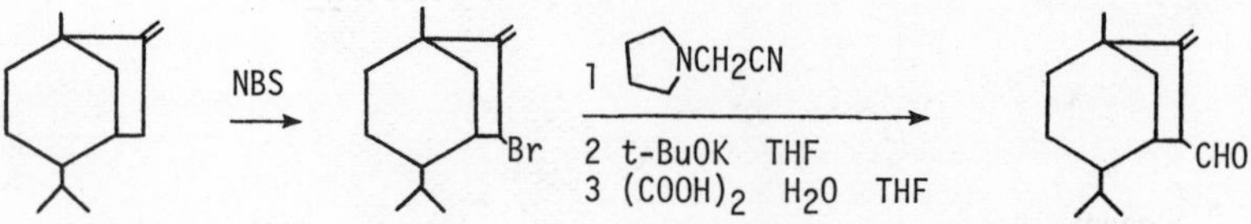

NBS
$\xrightarrow{\hspace{1cm}}$
Br

1 NCH$_2$CN

2 t-BuOK THF

3 (COOH)$_2$ H$_2$O THF

$\xrightarrow{\hspace{2cm}}$ CHO

JOC (1973) **38** 2915

PhC≡CH $\xrightarrow[\text{ZnI}_2]{\text{HC(OEt)}_3}$ PhC≡CCH(OEt)$_2$ $\dashrightarrow$ PhC≡CCHO   ∼75%

Org Synth (1963) Coll Vol 4 801

Section 57   Aldehydes from Ketones

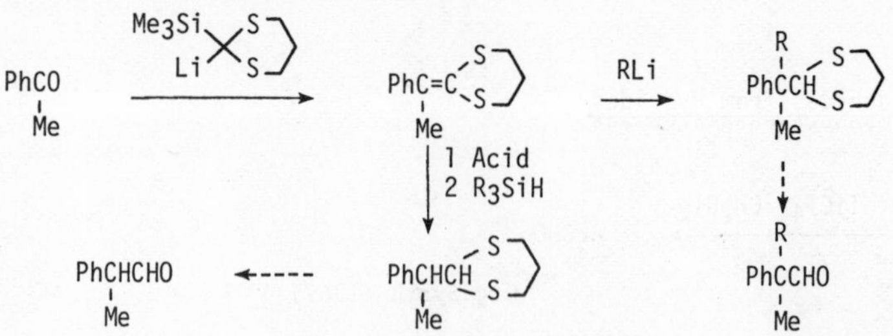

PhCO
|
Me

$\xrightarrow[\text{Li}]{\text{Me}_3\text{Si}}$ (S,S dithiane)

PhC=C
|
Me
(S S)

$\xrightarrow{\text{RLi}}$

R
|
PhCCH
|
Me
(S S)

1 Acid
2 R$_3$SiH

$\downarrow$

R
|
PhCCHO
|
Me

PhCHCHO
|
Me
$\dashleftarrow$
PhCHCH
|
Me
(S S)

Chem Comm (1972) 526

$$Ph_2CO \xrightarrow[\text{2 } H_2SO_4 \quad H_2O]{\overset{\text{Li}}{\text{1 } Ph_2P(O)\overset{|}{C}HNMe_2 \quad Et_2O}} Ph_2CHCHO$$

44%

JACS (1971) 93 4027

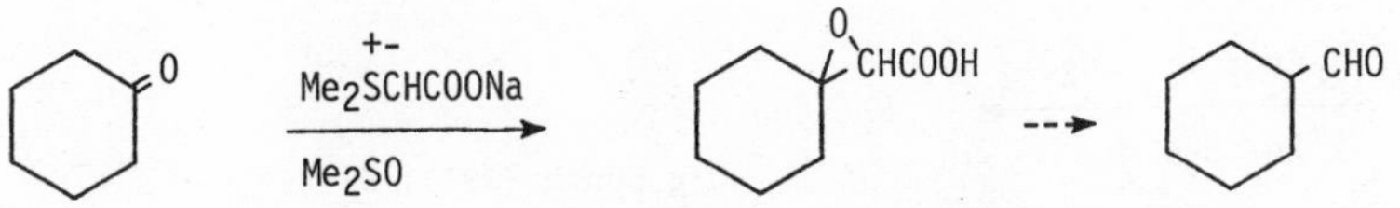

$$\text{(cyclohexanone)} \xrightarrow[\text{Me}_2\text{SO}]{\overset{+-}{Me_2SCHCOONa}} \text{(epoxide CHCOOH)} \dashrightarrow \text{(CHO)}$$

< 60%

JOC (1970) 35 1600

$$\text{(cyclohexanone)} \xrightarrow[\text{(Me}_3Si)_2NLi \quad THF]{BrCH_2COOEt} \text{(epoxide CHCOOH)} \dashrightarrow \text{(CHO)}$$

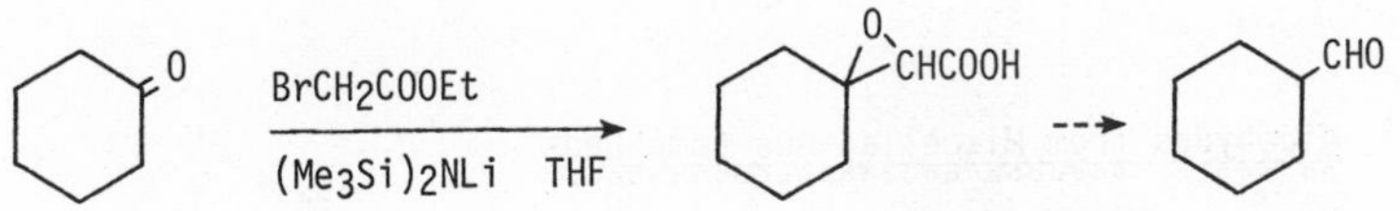

< 83%

Tetr Lett (1972) 3761

Further methods for the preparation of glycidic esters are included in section 358 (Ester — Ether, Epoxide)

Also via: Olefinic aldehydes (Section 341)

Section 58   Aldehydes from Nitriles

$$ArCN \xrightarrow{Ni-Al \quad HCOOH \quad H_2O} ArCHO$$

Org Synth (1971) 51 20

$$RCN \xrightarrow{LiAl(OEt)_3H \quad Et_2O} RCHO$$

Synthesis (1972) 217

Section 59   Aldehydes from Olefins

$EtCH=CH_2$ $\xrightarrow{B_2H_6}$ $(EtCH_2CH_2)_3B$ $\xrightarrow{N_2CHCHO}$ $Et(CH_2)_3CHO$ ~77%

Can J Chem (1970) 48 868

$R_2C=CH_2$ $\xrightarrow[\text{2 Zn } H_2O]{\text{1 } CrO_2Cl_2 \text{ } CH_2Cl_2}$ R_2CHCHO

Org Synth (1971) 51 4

Section 60   Aldehydes from Miscellaneous Compounds

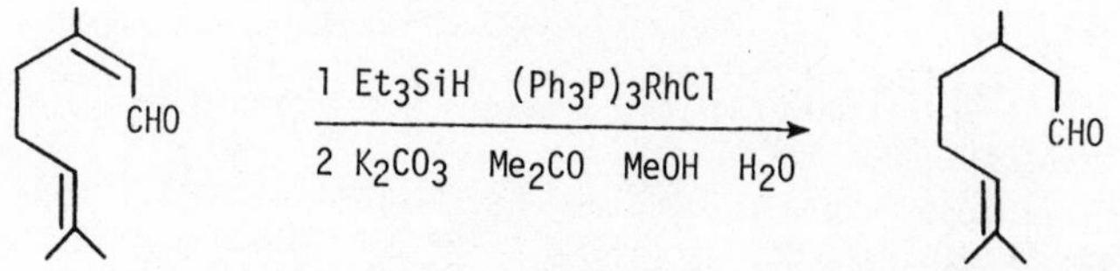

$\xrightarrow[\text{2 } K_2CO_3 \text{ } Me_2CO \text{ } MeOH \text{ } H_2O]{\text{1 } Et_3SiH \text{ } (Ph_3P)_3RhCl}$

Tetr Lett (1972) 5035

$EtCH_2NO_2$ $\xrightarrow[\text{2 } H_2SO_4 \text{ } H_2O]{\text{1 Zn HOAc}}$ $EtCHO$ 43%

JACS (1939) 61 3194

MeONa, O_3 JOC (1974) 39 259

$C_6H_{13}\underset{NH_2}{CHCOOH}$ $\xrightarrow[H_2O]{\text{Argentic picolinate}}$ $C_6H_{13}CHO$ 75%

JCS C (1970) 815

$$C_6H_{13}\underset{\text{OH}}{CH}CO(CH_2)_8COOH \xrightarrow{Pb(OAc)_4 \quad HOAc \quad H_2O} C_6H_{13}CHO$$

JCS (1957) 1622

Section 60A Protection of Aldehydes

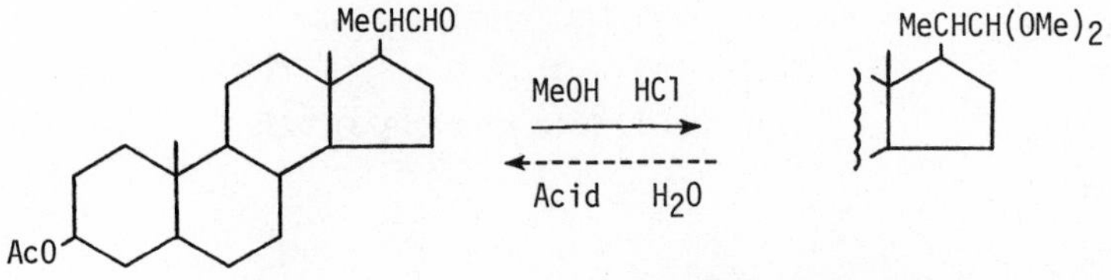

MeCHCHO
MeCHCH(OMe)2
MeOH HCl
Acid H2O
AcO

JCS (1953) 3864

Me
C=CH2
CH2CH2CHO
1 SnCl4 DCC C6H6
2 HOCH2CH2OH (COOH)2
Acid H2O
CH2CH2CH

Synth Comm (1973) 3 125

CH2OH
CH2OH
CH2Br TsOH C6H6
C11H23CHO
Zn . MeOH
C11H23CH
CH2Br

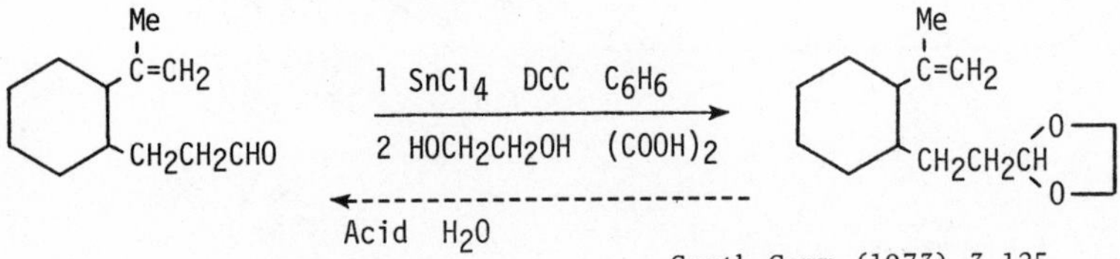

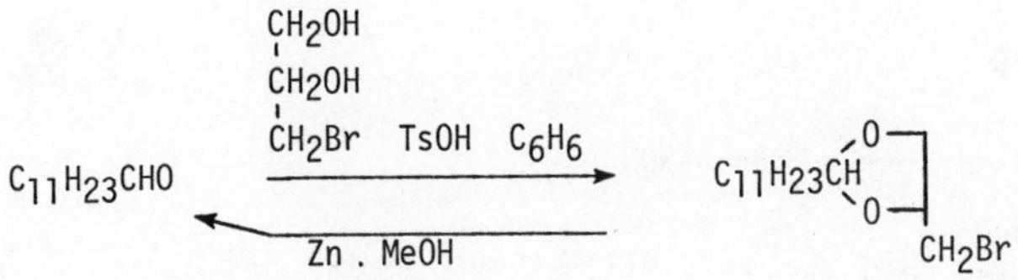

(Stable to NH_3, $NaBH_4$, peracids, MeLi and CrO_3-H_2SO_4)

JOC (1973) 38 834

Trichloroethyl acetals JOC (1973) 38 554

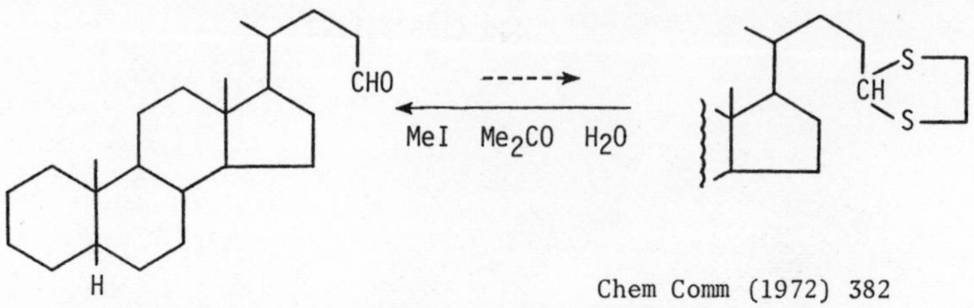

$$\xleftarrow[\text{MeI \quad Me}_2\text{CO \quad H}_2\text{O}]{\text{- - - - - ->}}$$

Thioacetal cleavage with NBS
chloramine-T
HgO, $BF_3 \cdot Et_2O$
$CuCl_2$, CuO
I_2, Me_2SO

Chem Comm (1972) 382
Synthesis (1969) 17
Tetr Lett (1971) 3445 3449
JOC (1971) 36 366
Bull Chem Soc Jap (1972) 45 3724
Tetr Lett (1973) 3735

$$\xrightarrow[\text{HgO \quad CaSO}_4]{\text{PhCH}_2\text{OH \quad HgCl}_2}$$

H_2 Pd-C   MeOH

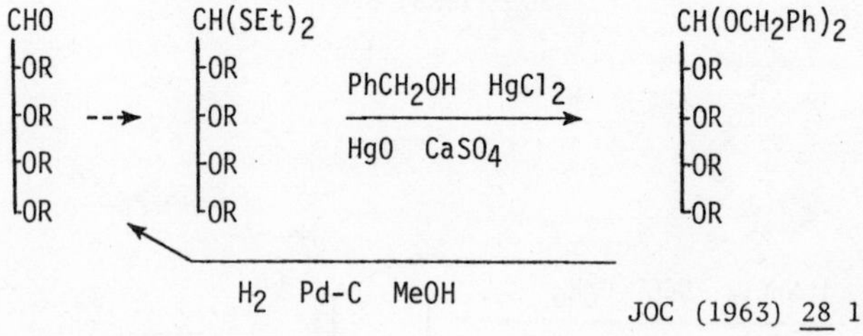

JOC (1963) 28 1395

PhCHO $\xrightarrow[\text{- - - - - - - - - - - - -}]{\text{(EtS)}_3\text{B \quad Pet ether}}$ $PhCH(SEt)_2$ (Neutral conditions)

Can J Chem (1965) 43 307
(1969) 47 859

PhCHO $\xrightleftharpoons[\text{TiCl}_3 \quad \text{NH}_4\text{OAc \quad HOAc}]{\text{NH}_2\text{OH}}$ PhCH=NOH
dioxane H_2O

Tetr Lett (1971) 195

PhCHO $\xrightleftharpoons[\text{- - - - - - - - -}]{\text{Ph}_3\text{CONH}_2}$ $PhCH=NOCPh_3$

JOC (1971) 36 3835

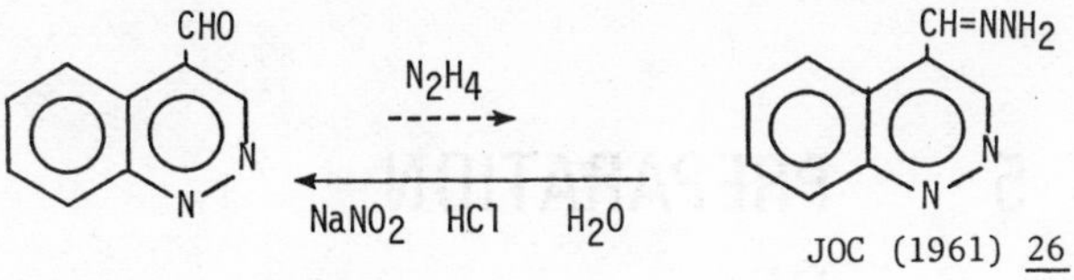

JOC (1961) 26 4465

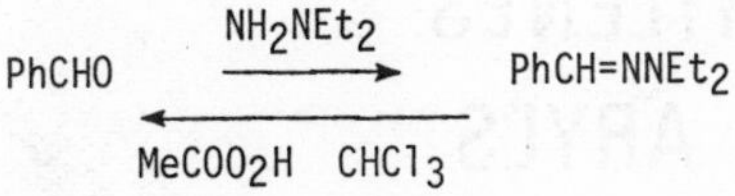

PhCHO $\xrightarrow[\text{MeCOO}_2\text{H} \quad \text{CHCl}_3]{\text{NH}_2\text{NEt}_2}$ PhCH=NNEt$_2$

Ber (1961) 94 712

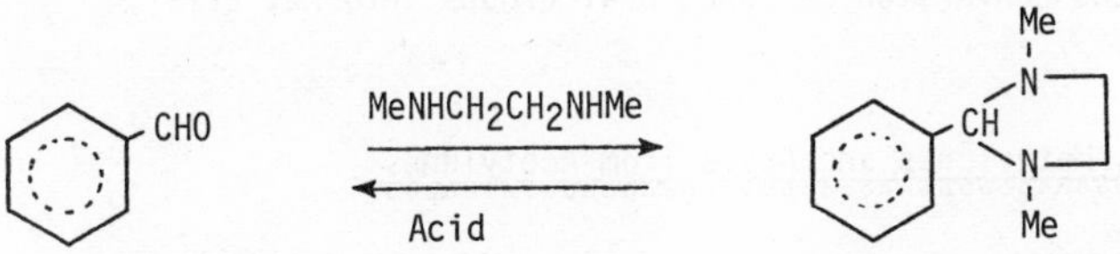

(Stable to Na-NH$_3$)

Aust J Chem (1955) 8 512

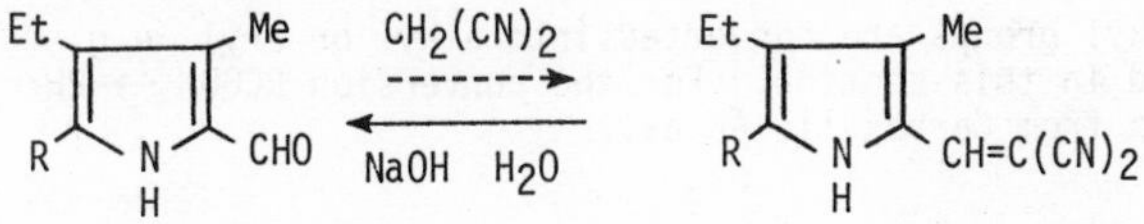

Angew (1960) 72 651

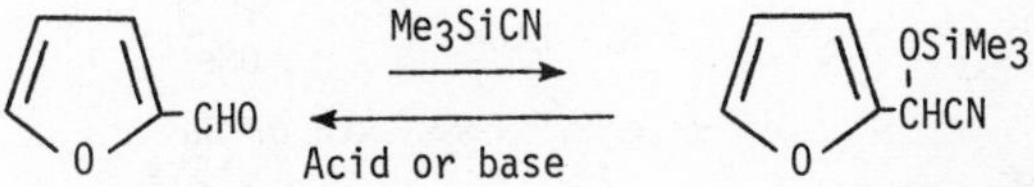

Chem Comm (1973) 55
Tetr Lett (1973) 4929

Related methods: Protection of Ketones   Section
                                          180A
                 Enol ethers   367
                 Acetals   363

Chapter 5   PREPARATION
OF
ALKYLS
METHYLENES
AND ARYLS

This chapter lists the conversion of functional groups into Me, Et···, CH_2, Ph etc.

Section 61   Alkyls, Methylenes and Aryls from Acetylenes
oooooooooooooooooooooooooooooooooooooooooooooooo

No additional examples

Section 62   Alkyls and Aryls from Carboxylic Acids
oooooooooooooooooooooooooooooooooooooooooo

Reactions in which carboxyl groups are converted into alkyl or aryl, e.g. $RCOOH \rightarrow RR'$ are included in this section. For the conversion $RCOOH \rightarrow RH$ see section 152 (Hydrides from Carboxylic Acids)

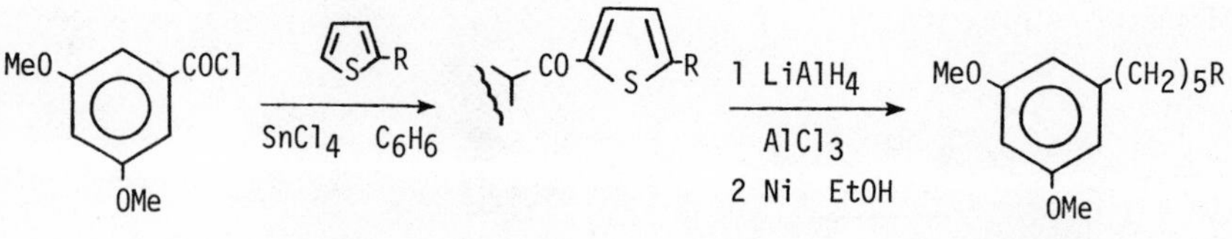

R=H or Bu

Aust J Chem (1971) <u>24</u> 2655

70

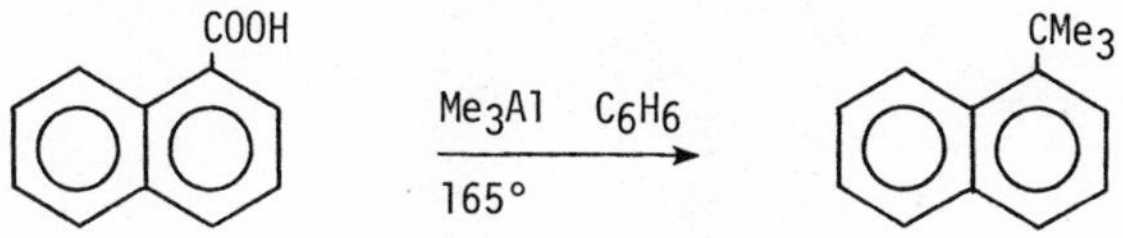

75%

Chem Comm (1972) 595

PhCOOH --→ PhCONHNHCOPh $\xrightarrow{\text{AgO}\quad C_6H_6}$ Ph-Ph   < 37%

JOC (1972) 37 2748

Section 63   Alkyls from Alcohols

Reactions in which hydroxyl groups are replaced by alkyl e.g. ROH →RMe,
are included in this section. For the conversion ROH →RH see section 153
(Hydrides from Alcohols and Phenols)

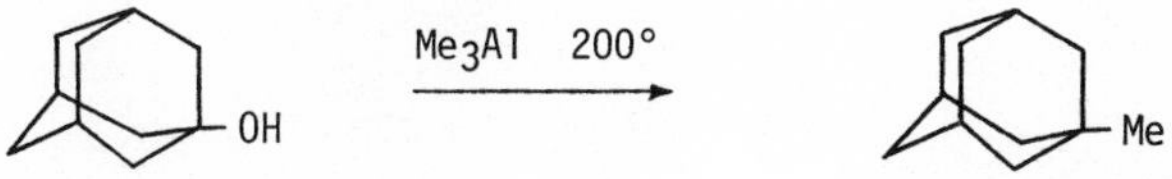

83%

Chem Comm (1972) 595

Section 64   Alkyls from Aldehydes

Reactions which convert the group CHO into CH_3 or larger alkyl groups are
included in this section. For the conversion RCHO → RH see section 154
(Hydrides from Aldehydes)

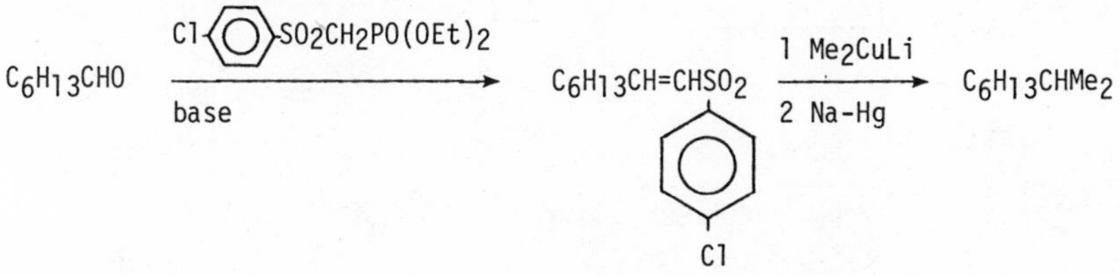

82%

Tetr Lett (1973) 935
          (1972) 293

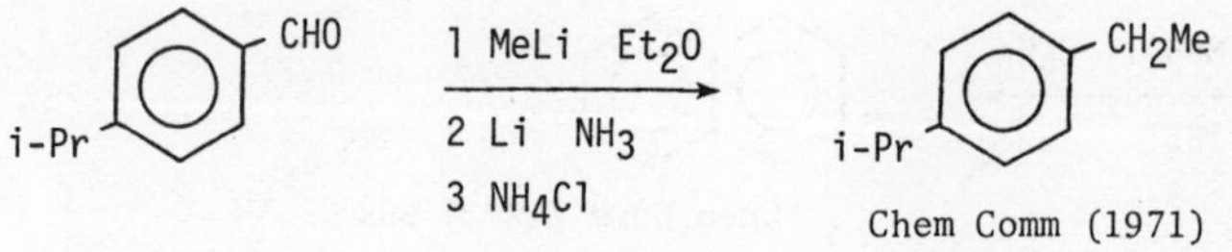

i-Pr—C6H4—CHO
$\xrightarrow{\begin{array}{l}1 \text{ MeLi } Et_2O \\ 2 \text{ Li } NH_3 \\ 3 \text{ } NH_4Cl\end{array}}$
i-Pr—C6H4—CH2Me

96%

Chem Comm (1971) 1242

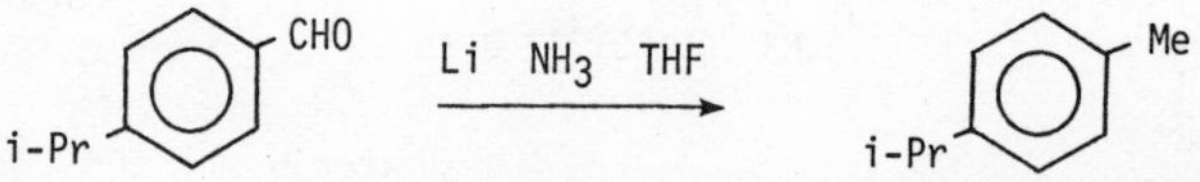

i-Pr—C6H4—CHO
$\xrightarrow{\text{Li } NH_3 \text{ THF}}$
i-Pr—C6H4—Me

90%

JOC (1972) 37 760

Me2N—C6H4—CHO
$\xrightarrow{B_2H_6 \text{ } BF_3}$
Me2N—C6H4—Me

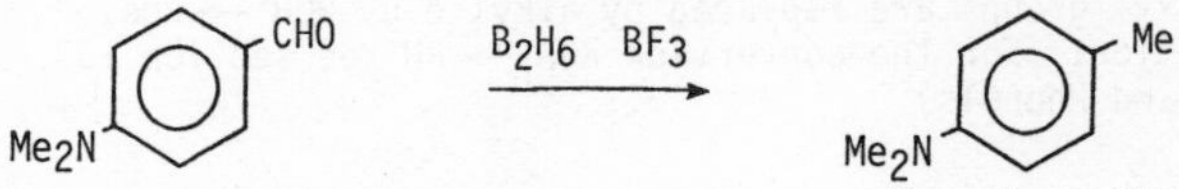

Tetr Lett (1967) 1849

RCHO
$\xrightarrow{\begin{array}{l}1 \text{ TsNHNH}_2 \text{ TsOH DMF} \\ \text{sulfolane} \\ 2 \text{ NaBH}_3CN\end{array}}$
RMe

JACS (1971) 93 1793

Related methods: Alkyls and Methylenes from Ketones (Section 72)

Section 65   Alkyls from Alkyls
ooooooooooooooooooo

Et, Et, Et-benzene
$\xrightarrow{\begin{array}{l}1 \text{ H}_2SO_4 \\ 2 \text{ H}_2O \text{ 140-150°}\end{array}}$
Et, Et, Et, Et-benzene

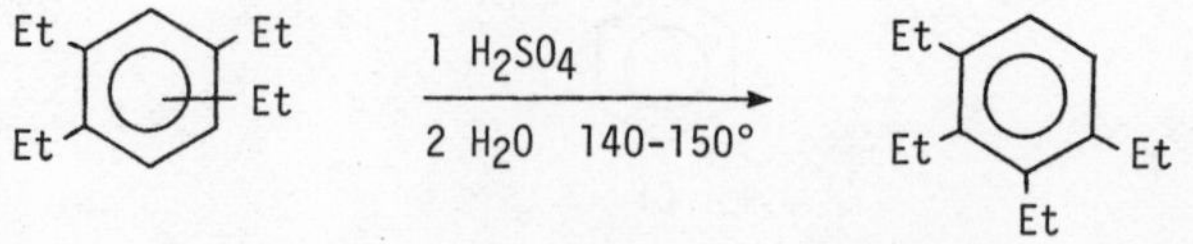

91%

Org React (1942) 1 370

Section 66   Alkyls, Methylenes and Aryls from Amides

No additional examples

Section 67   Alkyls, Methylenes and Aryls from Amines

No additional examples

Section 68   Alkyls, Methylenes and Aryls from Esters

No additional examples

Section 69   Alkyls from Ethers

The conversion ROR →RR' (R'=alkyl) is included in this section. For
the hydrogenolysis of ethers (ROR →RH) see section 159 (Hydrides from
Ethers)

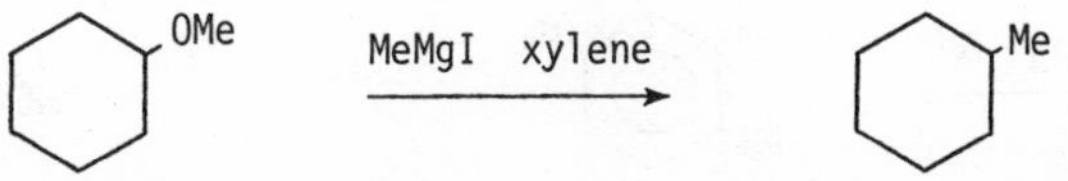

16%

Gazz (1970) 100 939

Section 70   Alkyls and Aryls from Halides

The replacement of halogen by alkyl or aryl groups is included in this
section. For the conversion RX → RH (X=halo) see section 160 (Hydrides
from Halides and Sulfonates)

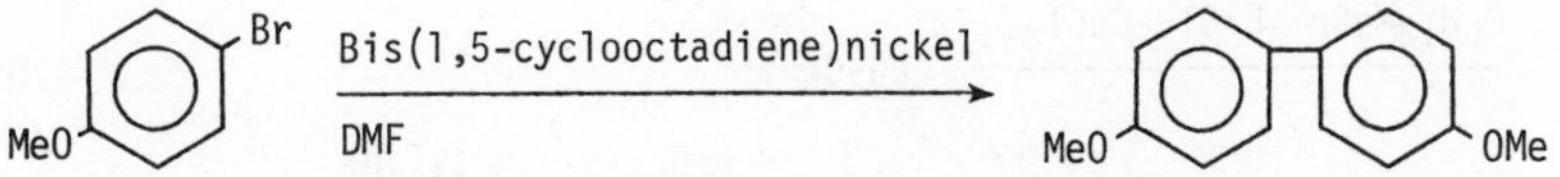

83%

JACS (1971) 93 5908

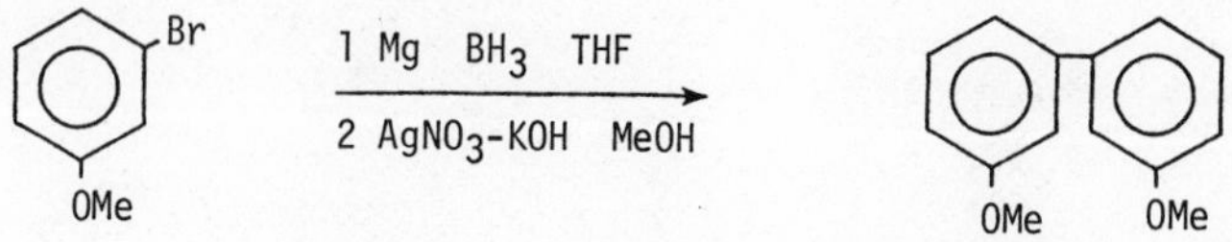

$$\begin{array}{c} \xrightarrow{\text{1 Mg BH}_3\text{ THF}} \\ \text{2 AgNO}_3\text{-KOH MeOH} \end{array}$$

52%

Tetr Lett (1972) 2193

PhI $\xrightarrow[\text{DMF}]{\pi\text{-allyl nickel bromide}}$ PhCH$_2$CH=CH$_2$ 82%

JOC (1972) 37 462
Org React (1972) 19 115

PhCH$_2$I $\xrightarrow{\text{CH}_2\text{=CHCH}_2\text{I Et}_3\text{B O}_2}$ PhCH$_2$CH$_2$CH=CH$_2$ 64%

JACS (1971) 93 1508

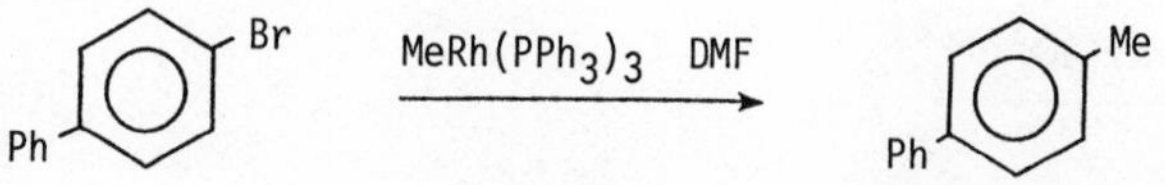

$\xrightarrow{\text{MeRh(PPh}_3)_3\text{ DMF}}$

65%

Tetr Lett (1973) 2967

C$_{16}$H$_{33}$I $\xrightarrow{\text{Zn-Ag H}_2\text{O}}$ C$_{16}$H$_{33}$-C$_{16}$H$_{33}$ 70%

Coll Czech (1964) 29 597

BuBr $\xrightarrow{\text{BuMgBr Ag THF}}$ Bu-Bu   79%

$\xrightarrow{\text{C}_6\text{H}_{13}\text{MgBr LiCl CuCl}_2}$ Bu-C$_6$H$_{13}$ 78%

Synthesis (1971) 303

$C_8H_{17}Br$ $\xrightarrow[\text{2 AgNO}_3\text{-KOH MeOH}]{\text{1 Mg BH}_3\text{ THF}}$ $C_8H_{17}\text{-}C_8H_{17}$ 90%

Tetr Lett (1972) 2193
JACS (1938) <u>60</u> 105

$Me_2C=CH(CH_2)_2\overset{\overset{\displaystyle Me}{|}}{C}=CH(CH_2)_2\overset{\overset{\displaystyle Me}{|}}{C}=CHCH_2Br$ $\xrightarrow[\substack{\text{2 PhLi}\\\text{3 (A)}\\\text{4 Li}\\\text{EtNH}_2}]{\text{1 Bu}_3\text{P}}$ $Me_2C=CH(CH_2)_2\overset{\overset{\displaystyle Me}{|}}{C}=CH(CH_2)_2\overset{\overset{\displaystyle Me}{|}}{C}=CHCH_2$
(A)

$Me_2C=CH(CH_2)_2\underset{\underset{\displaystyle Me}{|}}{C}=CH(CH_2)_2\underset{\underset{\displaystyle Me}{|}}{C}=CHCH_2$ 76%

JACS (1970) <u>92</u> 2139

$PhCH_2Br$ $\xrightarrow{\quad\quad}$ $PhCH_2S$ $\xrightarrow[\substack{\text{2 MeI}\\\text{3 Ni}}]{\text{1 BuLi}}$ $PhCH_2Me$ <60%

Tetr Lett (1971) 4359

$\xrightarrow[\text{2 Me}_2\text{SO}_4]{\text{1 BuLi Et}_2\text{O}}$ 58%

JACS (1969) <u>91</u> 3544

$PhCl$ $\xrightarrow{\text{BuMgBr [NiCl}_2\text{(dpe)] Et}_2\text{O}}$ $PhBu$ 76%

JACS (1972) <u>94</u> 4374

Also via: Section
            Acetylenes   10
            Olefins   205

Section 71   <u>Alkyls and Aryls from Hydrides</u>

This section lists examples of the reaction RH → RR' (R,R'=alkyl or aryl). For the reaction C=CH → C=CR (R=alkyl or aryl) see section 209 (Olefins from Olefins)

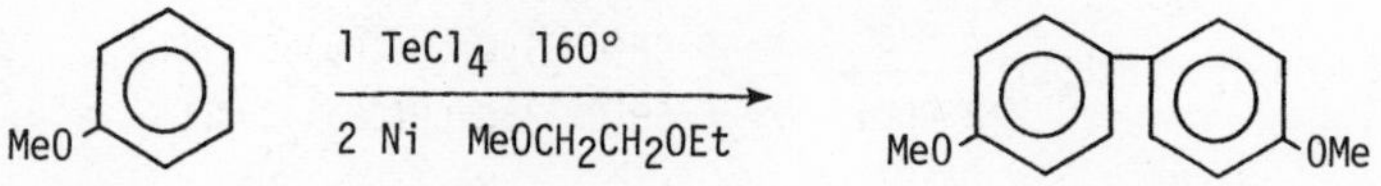

Tetrahedron (1972) <u>28</u> 3323

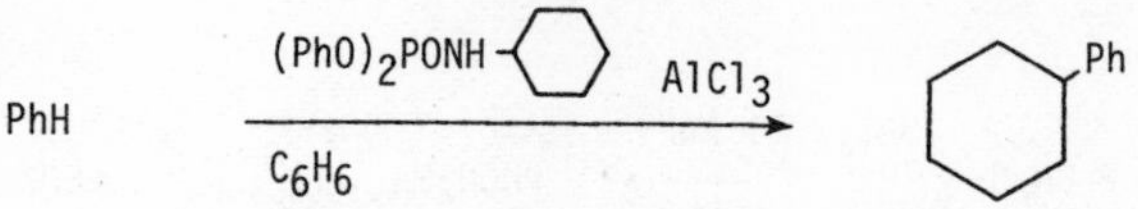

74%

Synthesis (1971) 203

27%

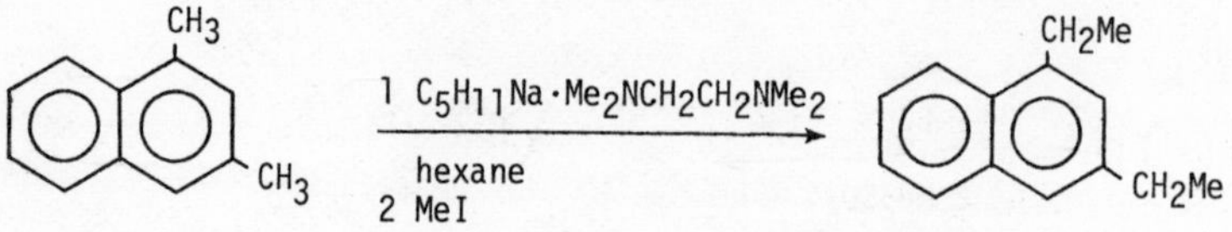

JACS (1972) <u>94</u> 2152

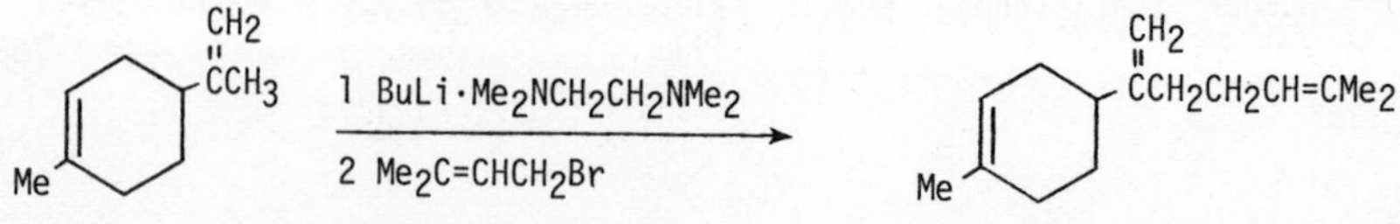

JACS (1972) <u>94</u> 4298

Section 72   Alkyls and Methylenes from Ketones

The conversions $R_2CO \rightarrow RR$, R_2CH_2, R_2CHR, etc. are listed in this section.
For the conversion $R_2CO \rightarrow RH$ see section 162 (Hydrides from Ketones)

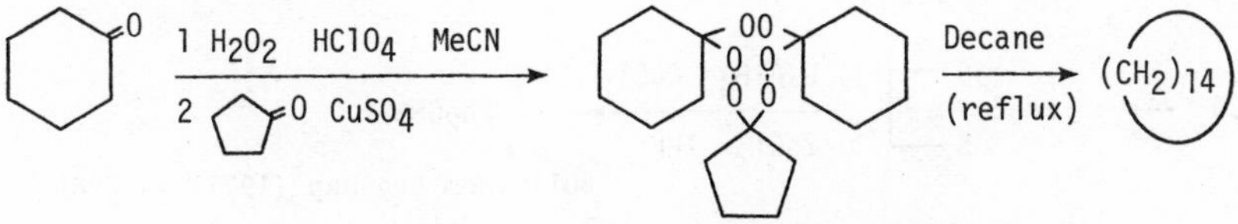

Synthesis (1970) 181
Advances in Org Chem (1972) 8 67

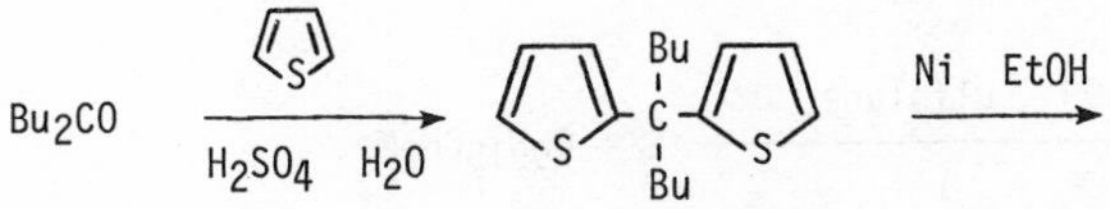

Bull Soc Chim Fr (1966) 2253

Ph_2CO $\xrightarrow[162°]{NaAlH_2(OCH_2CH_2OMe)_2 \ PhPr}$ Ph_2CMe_2 71%

Tetr Lett (1972) 691

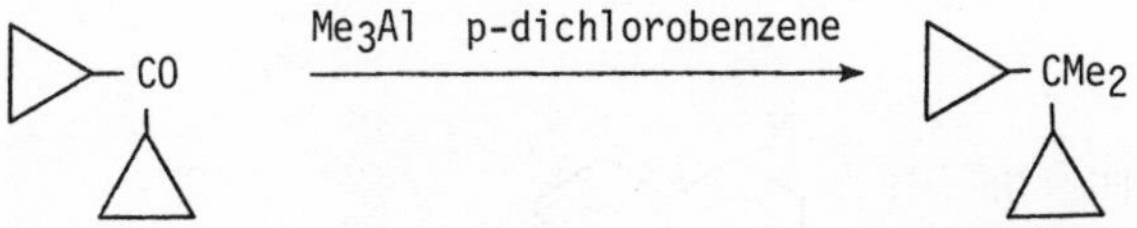

43%

Chem Comm (1972) 595

R_2CO $\xrightarrow[2 \ BuLi \ C_6H_6]{1 \ TsNHNH_2}$ R_2CHBu

JCS C (1971) 2294

Ph_2CO $\xrightarrow{\begin{array}{l}\text{1 MeLi   Et}_2\text{O}\\\text{2 Li   NH}_3\\\text{3 NH}_4\text{Cl}\end{array}}$ Ph_2CHMe 95%

JOC (1973) <u>38</u> 1735 1738

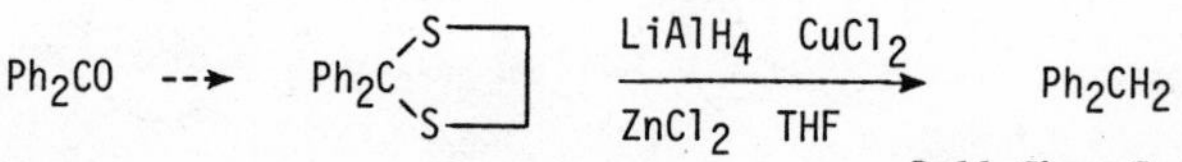

Ph_2CO --→ $Ph_2C\begin{array}{c}S\\S\end{array}$ $\xrightarrow{\begin{array}{l}\text{LiAlH}_4\quad\text{CuCl}_2\\\text{ZnCl}_2\quad\text{THF}\end{array}}$ Ph_2CH_2

Bull Chem Soc Jap (1971) <u>44</u> 2285

$C_9H_{19}COMe$ $\xrightarrow{\begin{array}{l}\text{1 TsNHNH}_2\quad\text{TsOH   sulfolane   DMF}\\\text{2 NaBH}_3\text{CN}\end{array}}$ $C_9H_{19}CH_2Me$

JACS (1971) <u>93</u> 1793

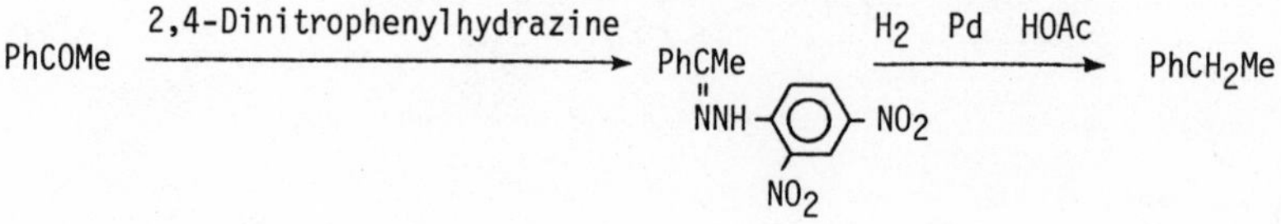

PhCOMe $\xrightarrow{\text{2,4-Dinitrophenylhydrazine}}$ PhCMe $\underset{\overset{\|}{\text{NNH}}}{}$ —⟨⟩— NO_2 (NO_2) $\xrightarrow{\text{H}_2\quad\text{Pd}\quad\text{HOAc}}$ PhCH$_2$Me

JOC (1971) <u>36</u> 737

$\xrightarrow{\begin{array}{l}\text{1 Li   NH}_3\quad\text{THF}\\\text{2 NH}_4\text{Cl}\end{array}}$

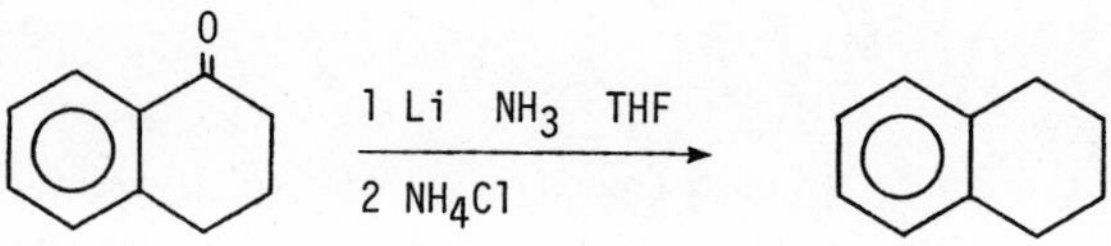

96%

JOC (1971) <u>36</u> 2588

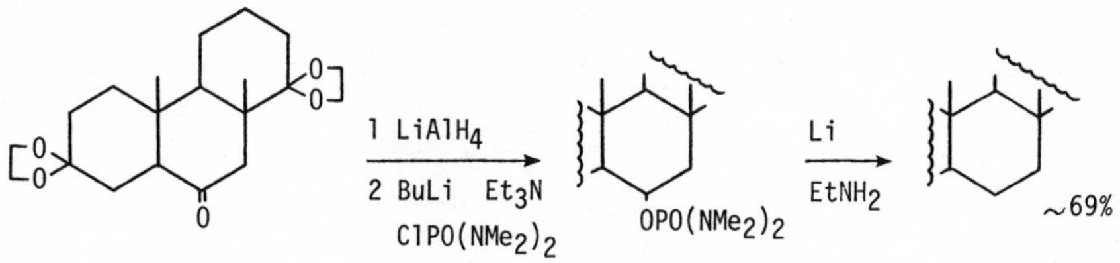

1LiAlH_4

2BuLi Et_3N

$\text{ClPO(NMe}_2)_2$

$\text{OPO(NMe}_2)_2$

Li

EtNH_2

~69%

JACS (1972) 94 5098

Related methods: Alkyls from Aldehydes (Section 64)

Also via: Olefins (Section 207)

Section 73   Alkyls, Methylenes and Aryls from Nitriles

No additional examples

Section 74   Alkyls, Methylenes and Aryls from Olefins

The hydrogenation, alkylation and arylation of olefins forming alkanes
or aryl-substituted alkanes, are included in this section.

BuCH=CH_2

$1 \text{B}_2\text{H}_6$ THF

$2 \text{Ph}_3\text{AsCHPh}$
$+ -$

$\text{Bu(CH}_2)_3\text{Ph}$

JACS (1967) 89 6804

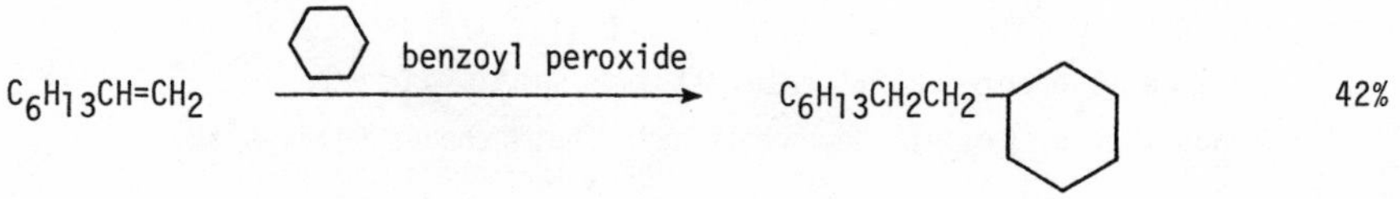

$\text{C}_6\text{H}_{13}\text{CH=CH}_2$

benzoyl peroxide

$\text{C}_6\text{H}_{13}\text{CH}_2\text{CH}_2$

42%

JCS (1965) 1939

Chem Comm (1968) 569

$C_8H_{17}CH=CH(CH_2)_7COOH$ $\xrightarrow[\text{MeOH}]{\text{NaBH}_4 \quad \text{NiCl}_2}$ $C_8H_{17}(CH_2)_9COOH$ 92%

Chem Pharm Bull (1971) 19 817
JOC (1971) 36 2018

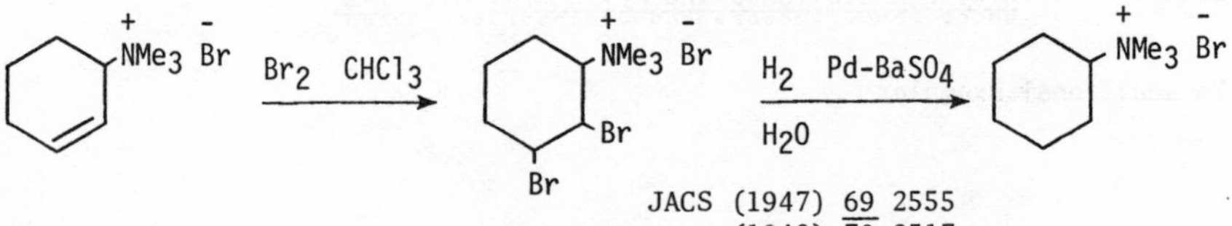

JACS (1947) 69 2555
(1948) 70 2517

$C_6H_{13}CH=CH_2$ $\xrightarrow[\text{EtOH}]{[\text{MeHSiO}]_N \quad \text{Pd-C}}$ $C_6H_{13}CH_2Me$

JOC (1973) 38 162

$EtCH=CMe_2$ $\xrightarrow[\text{cyclohexane}]{\text{H}_2 \quad \text{BuLi-cobalt 2-ethylhexanoate}}$ $EtCH_2CHMe_2$ 92%

JOC (1971) 36 1445
Polymer-supported Rh(I) catalyst JACS (1971) 93 3062
"Hydrogenation of Organic Compounds using Homogeneous Catalysts"
Chem Rev (1973) 73 21

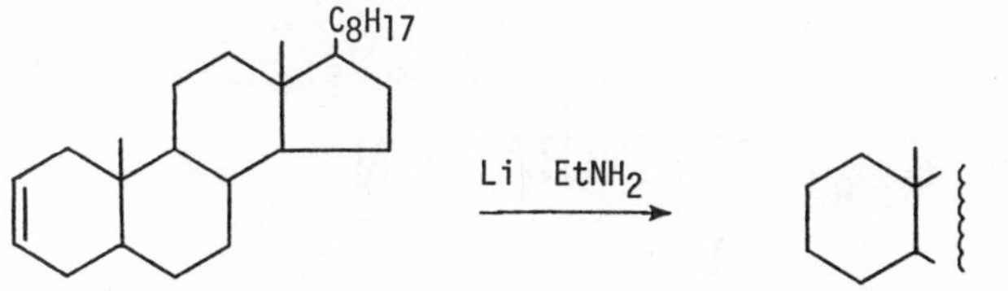

$$\xrightarrow{\text{Li} \quad \text{EtNH}_2}$$

JCS (1957) 1969

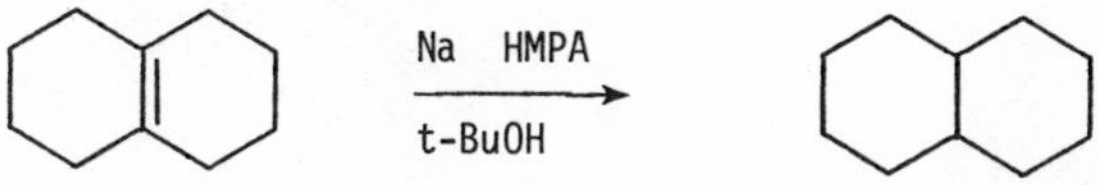

77%

$$\xrightarrow[\text{t-BuOH}]{\text{Na} \quad \text{HMPA}}$$

JOC (1970) 35 3565

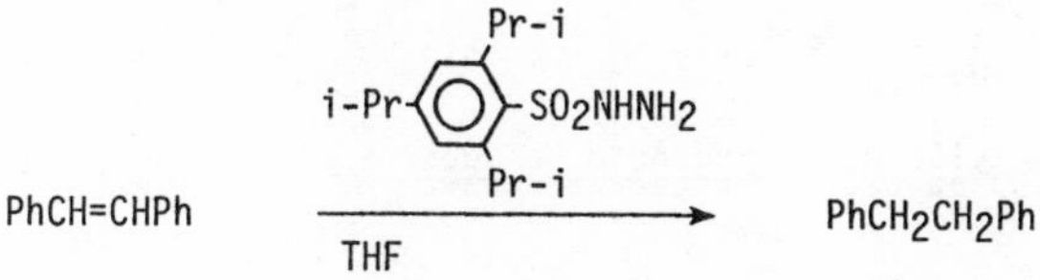

91%

PhCH=CHPh

$$\xrightarrow[\text{THF}]{\text{i-Pr}\overset{\text{Pr-i}}{\underset{\text{Pr-i}}{\bigcirc}}\text{SO}_2\text{NHNH}_2}$$

PhCH$_2$CH$_2$Ph

Chem Comm (1972) 1132

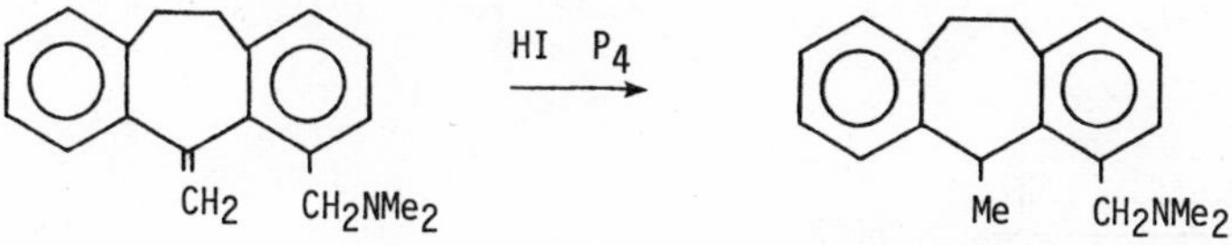

~60%

$$\xrightarrow{\text{HI} \quad \text{P}_4}$$

J Med Chem (1971) 14 982

94%

Section 75   Alkyls, Methylenes and Aryls from Miscellaneous Compounds
oooooooooooooooooooooooooooooooooooooooooooooooooooooooooooooooooo

No additional examples

Chapter 6 PREPARATION OF AMIDES

Section 76 Amides from Acetylenes
°°°°°°°°°°°°°°°°°°°°°°°°°°°

No additional examples

Section 77 Amides from Carboxylic acids and Acid Halides
°°°

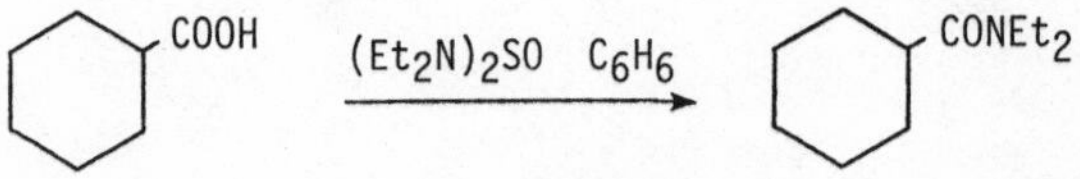

$\text{(Et}_2\text{N)}_2\text{SO} \quad \text{C}_6\text{H}_6$

Il Farmaco Ed Sc (1971) 26 153
(Chem Abs 75 4874)

PhCOOH $\xrightarrow[\text{2 BuNH}_2]{\text{1 Ph}_3\text{P} \quad \text{CCl}_4 \quad \text{THF}}$ PhCONHBu   85%

JOC (1971) 36 1305

PhCOOH $\xrightarrow[\text{2 Et}_2\text{NH}]{\text{1 (Me}_2\text{N)}_3\text{P} \quad \text{CCl}_4 \quad \text{Et}_2\text{O}}$ PhCONEt$_2$

Bull Soc Chim Fr (1971) 3034

$C_5H_{11}COOH$ $\xrightarrow[\text{Ph}_3\text{P}\quad\text{Et}_3\text{N}\quad\text{CuCl}_2\quad\text{CH}_2\text{Cl}_2]{\text{BuNH}_2\quad Cl-\bigcirc-SS-\bigcirc-Cl}$ $C_5H_{11}CONHBu$ 62%

Bull Chem Soc Jap (1971) <u>44</u> 1373

PhCOOH $\xrightarrow[\text{Et}_3\text{N}\quad\text{DMF}]{\text{BuNH}_2\quad\text{NCP(O)(OEt)}_2}$ PhCONHBu   94%

Tetr Lett (1973) 1595

PhCOOH $\xrightarrow[\text{2 } \text{HON}\bigcirc]{\text{1 ClCOOMe}}$ PhCOON$\bigcirc$ $\xrightarrow{\text{BuNH}_2}$ PhCONHBu

Tetr Lett (1971) 2697

RCOOH $\xrightarrow[\text{2 t-BuNH}_2]{\text{1 ClCOOEt}\quad\text{Et}_3\text{N}\quad\text{CH}_2\text{Cl}_2}$ RCONHBu-t

JCS <u>C</u> (1971) 3540

$(t\text{-Bu})_2CHCOCl$ $\xrightarrow{\text{NH}_3\quad\text{NaNH}_2\quad\text{Et}_2\text{O}}$ $(t\text{-Bu})_2CHCONH_2$ 81%

JACS (1960) <u>82</u> 2498

$\begin{array}{c}\text{COOMe}\\ | \\ (CH_2)_4COOH\end{array}$ $\xrightarrow[\text{MeONa}\quad\text{MeOH}]{\begin{array}{c}\text{COOH}\\ | \\ (CH_2)_5NHAc\quad\text{electrolysis}\end{array}}$ $\begin{array}{c}\text{COOMe}\\ | \\ (CH_2)_9NHAc\end{array}$

Z Naturforsch (1947) <u>2b</u> 185

Electrolysis MeCN H_2O

t-BuCOOK $\xrightarrow{\hspace{3cm}}$ t-BuNHAc   40%

Acta Chem Scand (1964) <u>18</u> 1567
Bull Soc Chim Fr (1968) 3657
Angew (1971) <u>83</u> 579
(Internat Ed <u>10</u> 557)

Related methods: Amides from Amines (Section 82)

Also via: Esters (Section 83)

Section 78   <u>Amides from Alcohols</u>

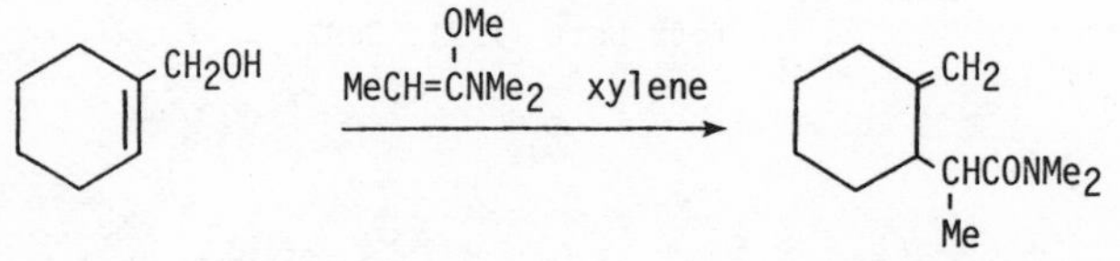

$$JCS \underline{C} (1971) 2950$$
$$Ber (1971) \underline{104} 3689$$

Ph_2CHOH $\dashrightarrow$ Ph_2CHN_3 $\xrightarrow[\hspace{1.5cm}]{\overset{+}{NO} \overset{-}{BF_4}\ \ MeCN}$ $Ph_2CHNHAc$ ~53%

JACS (1970) <u>92</u> 4999

Section 79   <u>Amides from Aldehydes</u>

PhCHO $\xrightarrow{Li[Me_2NCONi(CO)_3]\ \ Et_2O}$ $PhCONMe_2$ < 84%

JOC (1971) <u>36</u> 2721

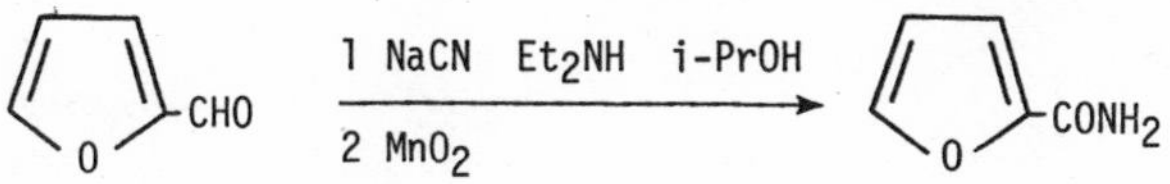

1 NaCN Et$_2$NH i-PrOH

2 MnO$_2$

98%

(Ar and αβ-unsaturated only)   Chem Comm (1971) 733

PhCHO $\xrightarrow{\text{PhNH}_2}$ PhCH=NPh $\xrightarrow{\text{CrO}_2\text{Cl}_2 \quad \text{CCl}_4}$ PhCONHPh   80%

Chem Ind (1970) 1297

Also via: Olefinic amides (Section 349)

Section 80   Amides from Alkyls, Methylenes and Aryls

No additional examples

Section 81   Amides from Amides

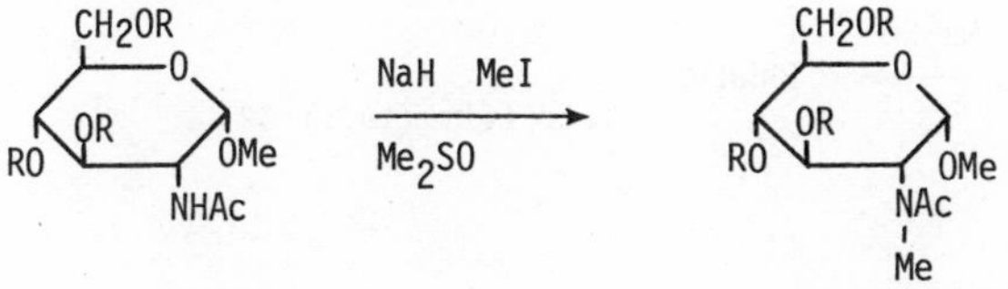

NaH MeI

Me$_2$SO

89%

Carbohydrate Res (1972) 23 251

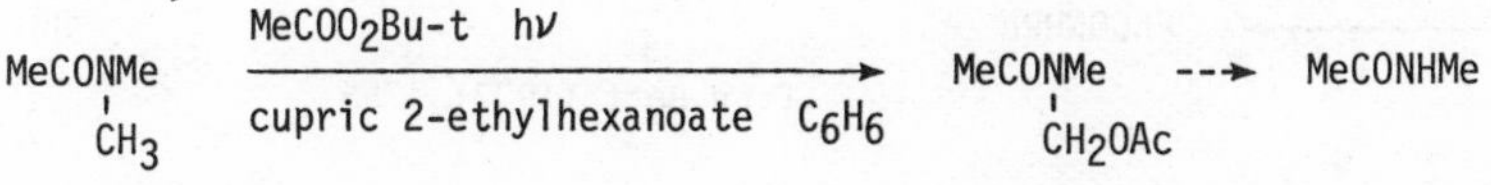

MeCONMe $\xrightarrow[\text{cupric 2-ethylhexanoate C}_6\text{H}_6]{\text{MeCOO}_2\text{Bu-t h}\nu}$ MeCONMe $\dashrightarrow$ MeCONHMe   < 67%

CH$_3$ CH$_2$OAc

Synthesis (1972) 1

Related methods: Amides from Halides (Section 85)

Section 82   Amides from Amines

$$BuNH_2 \xrightarrow[\text{CCl}_4 \quad \text{THF}]{\text{HOAc} \quad \text{Ph}_3\text{P}} BuNHAc \qquad\qquad 91\%$$

JOC (1971) <u>36</u> 1305

$$PhNH_2 \xrightarrow[\text{toluene}]{\text{Resin-CO-O-COPh}} PhCONHPh$$

Tetr Lett (1973) 1627

$$PhNH_2 \xrightarrow{C_{17}H_{35}COOH \quad SiCl_4 \quad Pyr} PhNHCOC_{17}H_{35} \qquad\qquad 70\%$$

JOC (1969) <u>34</u> 2766

Further examples of the reaction $RNH_2 + R'COOH \longrightarrow RNHCOR'$ are included in section 77 (Amides from Carboxylic Acids and Acid Halides)

$$PhNH_2 \xrightarrow{\text{MeCOOEt} \quad \text{NaH} \quad \text{Me}_2\text{SO}} PhNHAc \qquad\qquad 85\%$$

Tetr Lett (1971) 321

$$PhNH_2 \xrightarrow{\text{PhCOSCSNMe}_2} PhCONHPh \qquad\qquad 80\%$$

Tetr Lett (1972) 4785

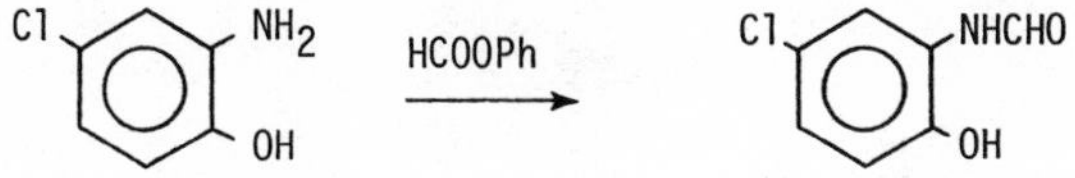

JOC (1971) 36 3238

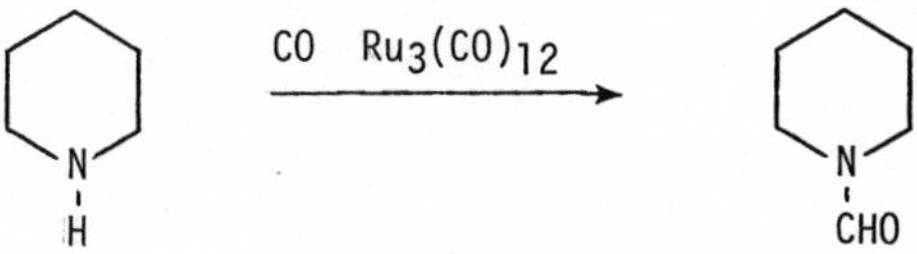

95%

Chem Comm (1971) 1482

45%

Related methods: Protection of Amines (Section 105A)

Section 83   Amides from Esters

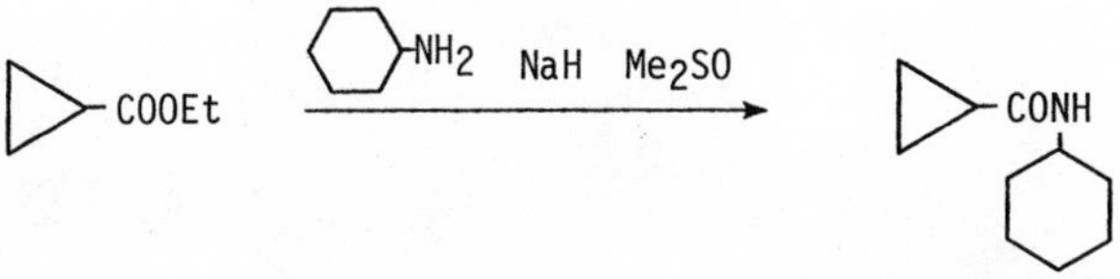

Tetr Lett (1971) 321

68%

$(EtO)_2CHCH_2COOEt$ $\xrightarrow{\text{NH}_3 \quad \text{H}_2\text{O}}$ $(EtO)_2CHCH_2CONH_2$ 58%

J Med Chem (1973) 16 1

Section 84   Amides from Ethers and Epoxides

No additional examples

Section 85   Amides from Halides

$C_5H_{11}Br$ $\xrightarrow[\text{2 } I_2 \quad Et_2NH \quad THF]{\text{1 } Na_2Fe(CO)_4}$ $C_5H_{11}CONEt_2$ 80%

JACS (1973) <u>95</u> 249

$\xrightarrow[Et_2O]{Li[Me_2NCONi(CO)_3]}$ 99%

(Applicable to aromatic and vinyl
halides)

JOC (1971) <u>36</u> 2721

$C_7H_{15}Br$ $\xrightarrow[THF]{Et_2NCOOEt \quad Li}$ $C_7H_{15}CONEt_2$ 70%

(One-step procedure)

Synthesis (1973) 160

$C_{18}H_{37}Cl$ $\xrightarrow{PhNHAc \quad KOH \quad Me_2SO}$ $C_{18}H_{37}\underset{Ph}{N}Ac$ 90%

Synthesis (1971) 266

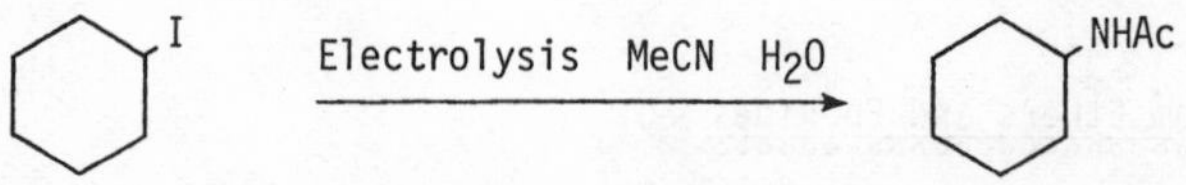

$\xrightarrow{Electrolysis \quad MeCN \quad H_2O}$ 85%

Compt Rend (1970) <u>C</u> <u>271</u> 324

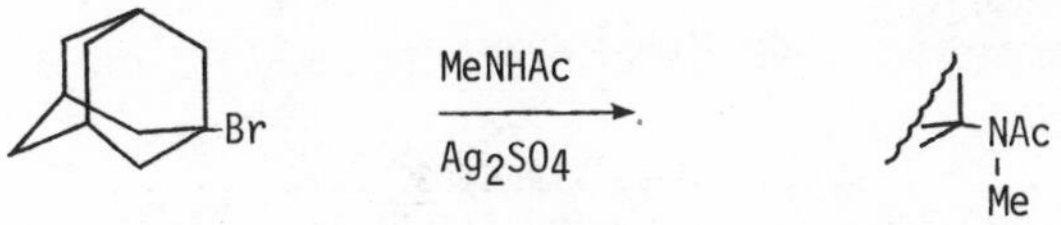

$$\xrightarrow[\text{Ag}_2\text{SO}_4]{\text{MeNHAc}}$$

J Med Chem (1971) <u>14</u> 535

Section 86   Amides from Hydrides

No additional examples

Section 87   Amides from Ketones

No additional examples

Section 88   Amides from Nitriles

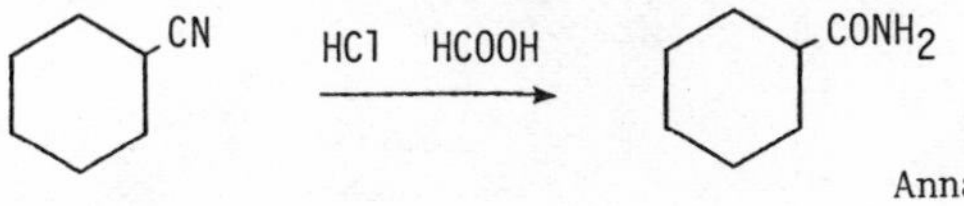

$$\xrightarrow{\text{HCl} \quad \text{HCOOH}}$$

92%

Annalen (1971) <u>749</u> 198

PhCN $\xrightarrow[\text{2 H}_2\text{O}]{\text{1 BBr}_3}$ PhCONH$_2$

JCS <u>C</u> (1969) 739

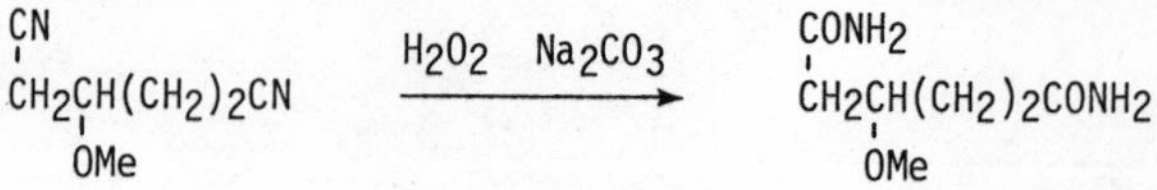

CN
|
CH$_2$CH(CH$_2$)$_2$CN $\xrightarrow{\text{H}_2\text{O}_2 \quad \text{Na}_2\text{CO}_3}$
|
OMe

CONH$_2$
|
CH$_2$CH(CH$_2$)$_2$CONH$_2$
|
OMe

Helv (1956) <u>39</u> 1233
Chem Ind (1961) 987

Also via: Amines (Section 103)

Section 89   <u>Amides from Olefins</u>
              ◦◦◦◦◦◦◦◦◦◦◦◦◦◦◦◦◦◦

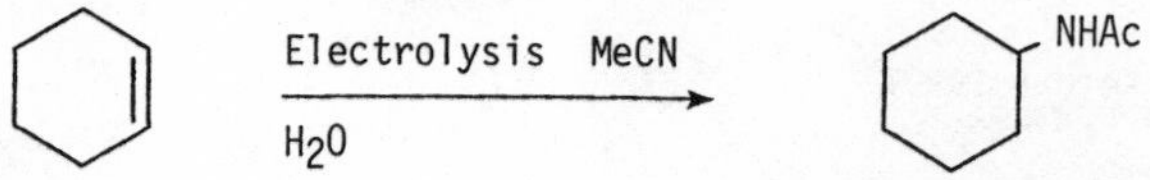

$\xrightarrow[\text{H}_2\text{O}]{\text{Electrolysis MeCN}}$ 80%

Compt Rend (1970) <u>C 271</u> 324

Also via: Amines (Section 104)

Section 90   <u>Amides from Miscellaneous Compounds</u>
              ◦◦◦◦◦◦◦◦◦◦◦◦◦◦◦◦◦◦◦◦◦◦◦◦◦◦◦◦◦◦◦◦◦◦

No additional examples

Section 90A   <u>Protection of Amides</u>
               ◦◦◦◦◦◦◦◦◦◦◦◦◦◦◦◦◦◦◦◦◦

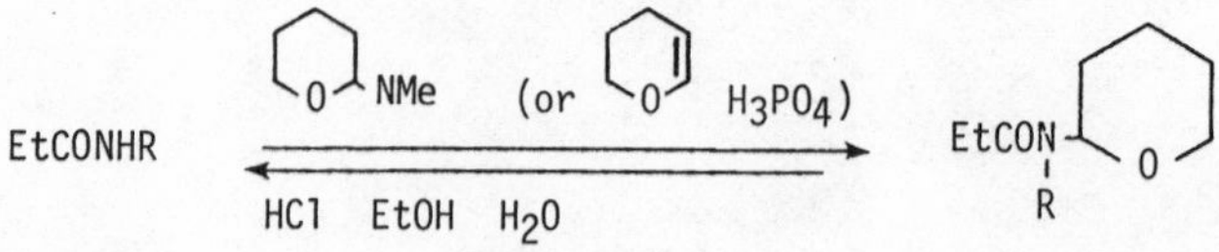

EtCONHR $\xrightleftharpoons[\text{HCl EtOH H}_2\text{O}]{}$ EtCON

Bull Soc Chim Fr (1964) 292

NHAc
R
Ac_2O
Na_2CO_3 H_2O
NAc_2
R

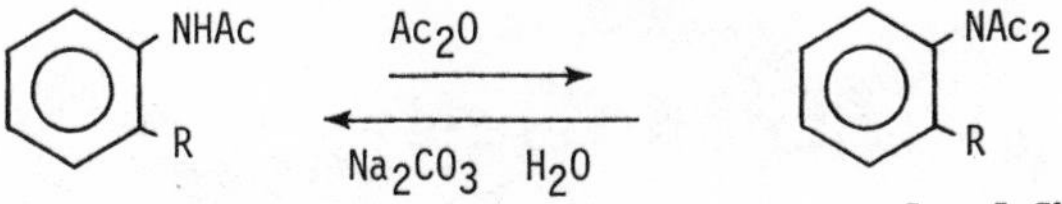

(Stable to NBS)

Can J Chem (1955) 33 1819

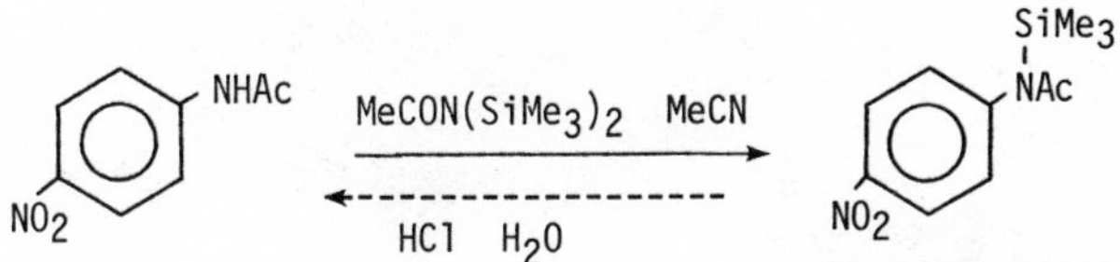

NHAc
NO_2
$MeCON(SiMe_3)_2$ MeCN
HCl H_2O
$SiMe_3$
NAc
NO_2

JACS (1966) 88 3390

$RCH_2CH_2CONH_2$
OH
MeO CH OMe
H_2SO_4
CF_3COOH anisole
RCH_2CH_2CONH
MeO CH OMe

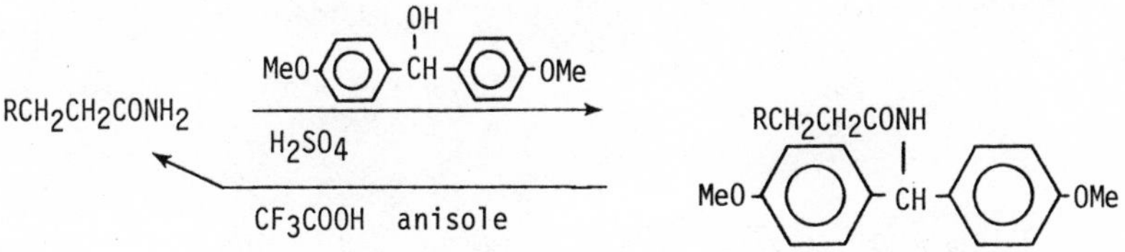

(Stable to H_2/Pd)

Ber (1970) 103 2041

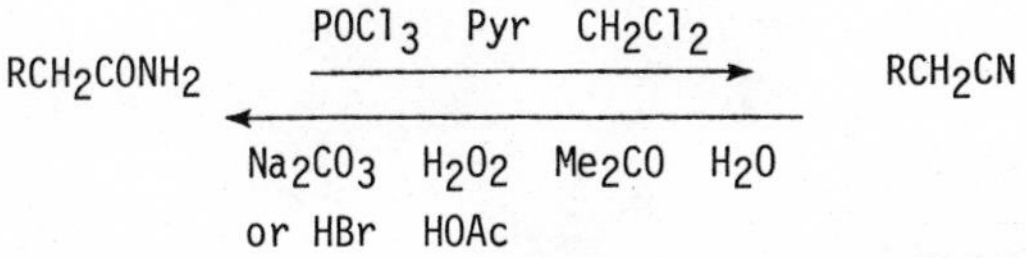

RCH_2CONH_2
$POCl_3$ Pyr CH_2Cl_2
Na_2CO_3 H_2O_2 Me_2CO H_2O
or HBr HOAc
RCH_2CN

Chem Ind (1961) 987

Chapter 7   PREPARATION
OF
AMINES

Section 91   <u>Amines from Acetylenes</u>

$$PhC{\equiv}CC_6H_{13} \xrightarrow[\text{2 } H_2 \text{ Ni NaOH MeOH}]{\text{1 NaNHNH}_2 \text{ N}_2\text{H}_4 \text{ Et}_2\text{O}} PhCH_2\underset{\underset{NH_2}{|}}{C}HC_6H_{13}$$

71%

Ber (1966) <u>99</u> 1843

Section 92   <u>Amines from Carboxylic Acids</u>

$$C_7H_{15}COOH \xrightarrow[\text{2 t-BuOH}]{\text{1 N}_3\text{PO(OPh)}_2 \text{ Et}_3\text{N DMF}} C_7H_{15}NHCOOBu\text{-}t \dashrightarrow C_7H_{15}NH_2$$

~67%

JACS (1972) <u>94</u> 6203

Section 93   <u>Amines from Alcohols and Phenols</u>

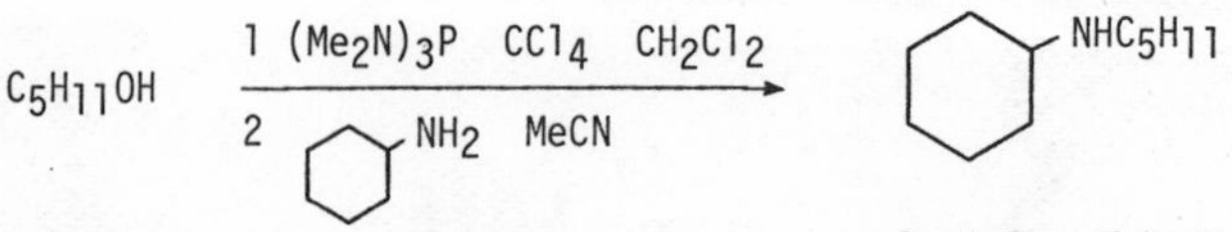

$$C_5H_{11}OH \xrightarrow[\text{2}\quad\text{NH}_2\text{ MeCN}]{\text{1 (Me}_2\text{N)}_3\text{P CCl}_4 \text{ CH}_2\text{Cl}_2}$$

58%

Bull Soc Chim Fr (1971) 4368

PhCHOH $\xrightarrow{\begin{array}{l}\text{1 ClSO}_2\text{NCO hexane}\\ \text{2 NH}_2\text{NHCOOBu-t}\\ \text{3 Pb(OAc)}_4\text{ CH}_2\text{Cl}_2\end{array}}$ PhCHNH$_2$ 52%
  | |
  Me   Me

JACS (1973) <u>95</u> 4083

C$_6$H$_{13}$CHOH $\xrightarrow{\begin{array}{l}\text{1 Phthalimide Ph}_3\text{P }\overset{\text{NCOOEt}}{\overset{\|}{\text{NCOOEt}}}\text{ THF}\\ \text{2 N}_2\text{H}_4\text{ EtOH}\end{array}}$ C$_6$H$_{13}$CHNH$_2$
   | |
   Me   Me

JACS (1972) <u>94</u> 679
Carbohydrate Res (1972) <u>23</u> 460

$\xrightarrow[230\text{-}240°]{(\text{Me}_2\text{N})_3\text{PO}}$

Chem Comm (1971) 1018   68%

$\xrightarrow[\text{NaOH}]{(\text{EtO})_2\text{POCl}}$ $\xrightarrow[\text{NH}_3]{\text{KNH}_2\text{ K}}$ ~78%

JOC (1972) <u>37</u> 3570

$\xrightarrow{\begin{array}{l}\text{1 NaH}\\ \text{2 330-335°}\\ \text{3 NaOH}\end{array}}$ 58%

J Med Chem (1966) <u>9</u> 249
Steroids (1968) <u>11</u> 151

Section 94   Amines from Aldehydes

$$RCHO \xrightarrow[\text{HOAc}]{CH_3NO_2 \quad NH_4OAc} RCH=CHNO_2 \xrightarrow{LiAlH_4 \quad THF} RCH_2CH_2NH_2$$

Ber (1971) 104 2517
JCS (1965) 3645
      (1957) 2926

$$RCHO \xrightarrow[\text{MeOH \quad pH 6-8}]{R'NH_2 \quad NaBH_3CN} RCH_2NHR'$$

JACS (1971) 93 2897

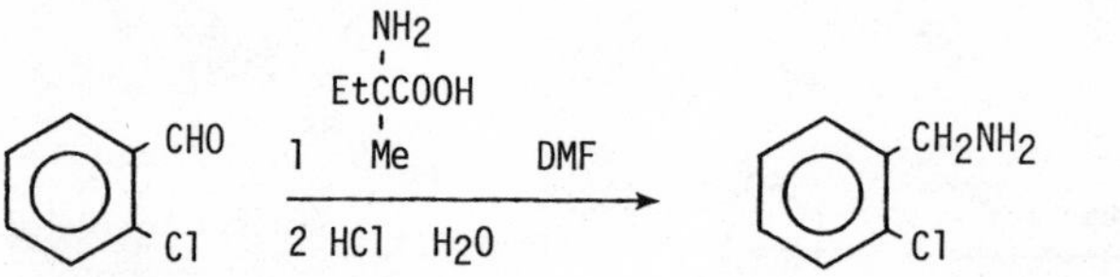

(Ar aldehydes only)

JOC (1971) 36 1710

70%

$$C_6H_{13}CHO \xrightarrow{NH_2OH} C_6H_{13}CH=NOH \xrightarrow[\text{CHCl}_3]{H_2 \quad PtO_2 \quad EtOH} C_6H_{13}CH_2NH_2$$

JOC (1972) 37 335

Related methods: Amines from Ketones (Section 102)

Section 95   Amines from Alkyls, Methylenes and Aryls

No examples

Section 96 Amines from Amides

$C_5H_{11}CONR_2 \xrightarrow{B_2H_6 \quad THF} C_5H_{11}CH_2NR_2$ 87% R=H; 95% R=Me

JOC (1973) 38 912

$PhCONH_2 \xrightarrow{NaBH_2S_3 \quad THF} PhCH_2NH_2$ 46%

Can J Chem (1971) 49 2990

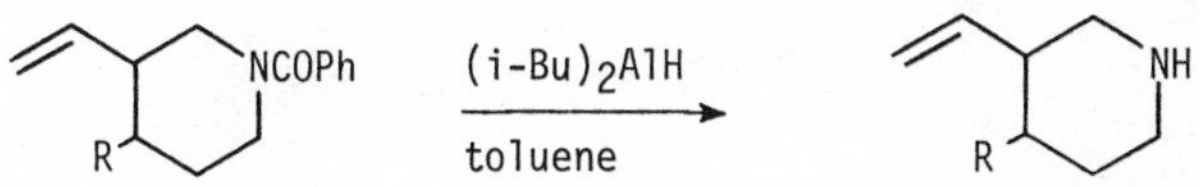

NCOPh
R
(i-Bu)2AlH
toluene
NH
R

JACS (1970) 92 204

Ac
NMe
Li EtNH2
NHMe

Tetr Lett (1971) 1935

PhCH2CONH
S
N
O
COON=CHPh
1 PCl5
N-methylmorpholine
2 MeOH
NH2
N
O

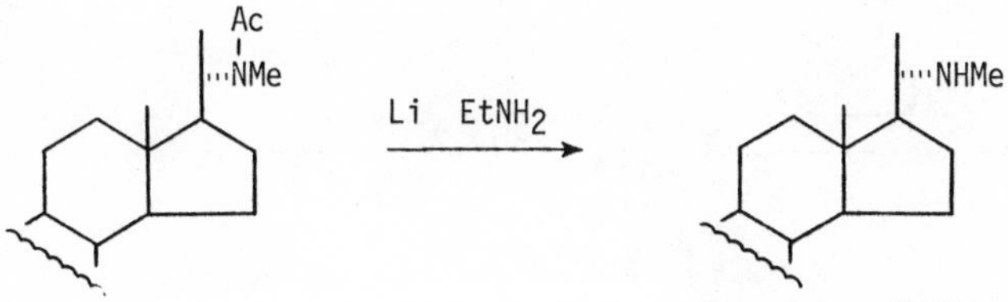

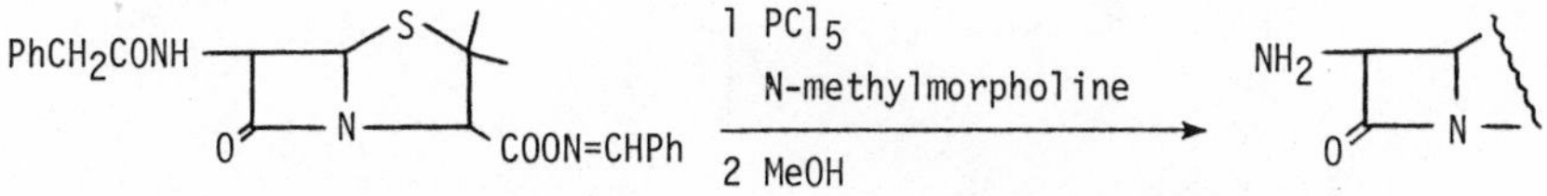

JCS C (1971) 1917
Rec Trav Chim (1970) 89 1081
JOC (1971) 36 1259

Section 97   Amines from Amines

Me
|
PhCHNH$_2$ $\xrightarrow{\text{BuLi}}$ Me
|
PhCNH$_2$ 25%
|
Bu   Tetr Lett (1971) 2187

t-BuNCH$_3$ $\xrightarrow[\text{2 PhMgBr Et}_2\text{O}]{\text{1 Ph}_3\text{C}^+ \text{ClO}_4^- \text{CH}_2\text{Cl}_2}$ t-BuNCH$_2$Ph   75%
| |
Me   Me

Annalen (1971) 752 86

BuNH$_2$ $\xrightarrow{\text{MeI Bu}_3\text{N DMF}}$ BuNMe$_3^+$ I$^-$ 92%

JOC (1971) 36 824

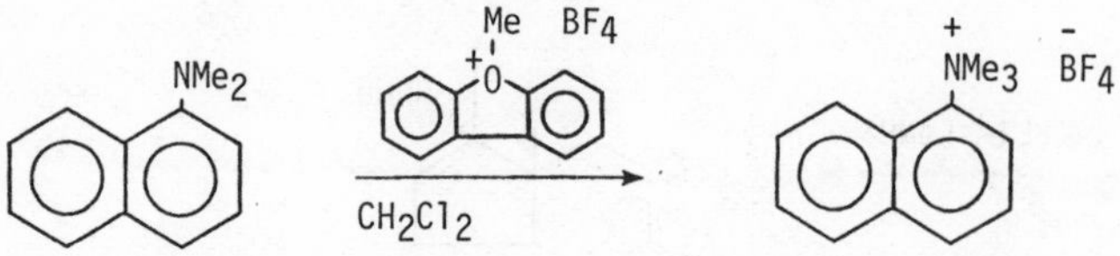

Chem Comm (1972) 315   100%

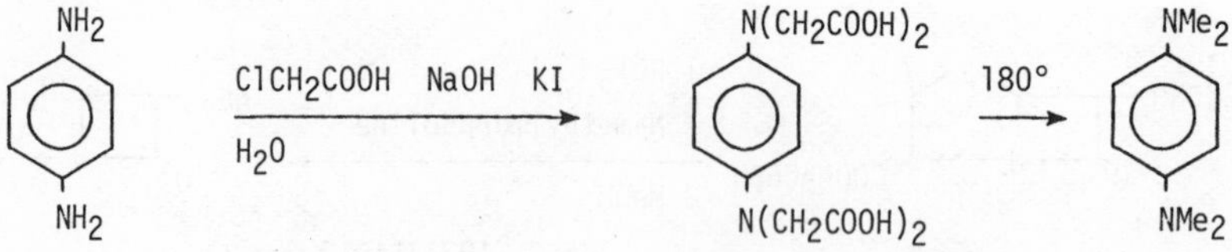

JOC (1964) 29 488

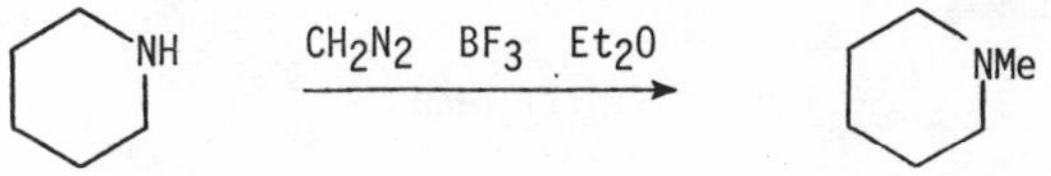

NH
CH_2N_2 BF_3 Et_2O
NMe

30%

Annalen (1959) 623 34

$PhCH_2N(Et)H$ —[HCHO $NaBH_3CN$ / MeCN H_2O]→ $PhCH_2N(Et)Me$ 98%

JOC (1972) 37 1673

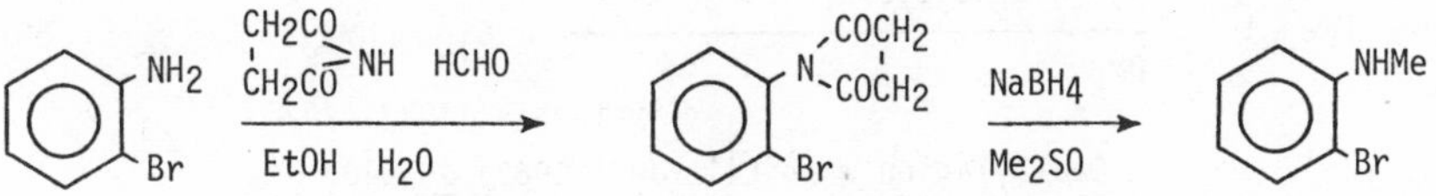

NH_2
Br
CH_2CO
CH_2CO
NH HCHO
EtOH H_2O
N
$COCH_2$
$COCH_2$
Br
$NaBH_4$
Me_2SO
NHMe
Br

JOC (1973) 38 1348

72%

$BuNH_2$ —[PhBr $NaNH_2$]→ BuNHPh

JOC (1971) 36 1841

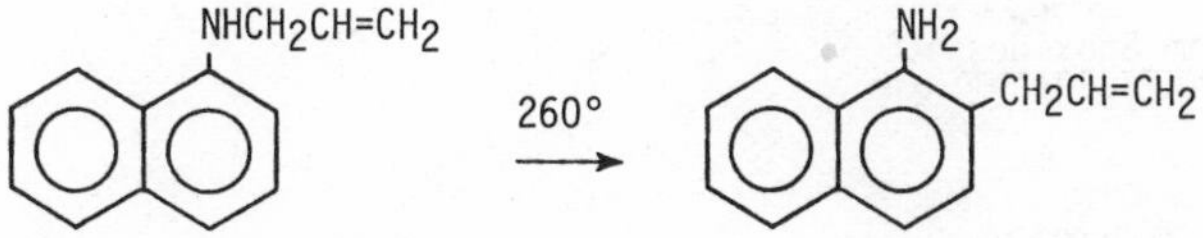

$NHCH_2CH=CH_2$
260°
NH_2
$CH_2CH=CH_2$

59%

Tetrahedron (1961) 14 208
Tetr Lett (1971) 4661 4665

PhNMe$_2$ $\xrightarrow[\text{2 HCl H}_2\text{O}]{\text{1 O}_3\text{ CH}_2\text{Cl}_2}$ PhNHMe   41%

JCS C (1971) 1369

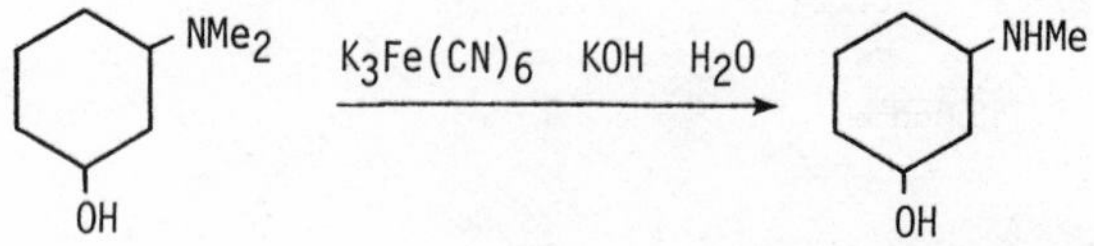

K$_3$Fe(CN)$_6$ KOH H$_2$O   93%

JOC (1951) 16 1303

$\overset{+}{C_{12}H_{25}N}\overset{-}{Me_3}$ I $\xrightarrow[\text{DMF}]{\text{1,4-Diazabicyclo[2.2.2]octane}}$ C$_{12}$H$_{25}$NMe$_2$ 68%

Synthesis (1972) 702

Dealkylation with LiI   JOC (1963) 28 2407

PrSLi   JOC (1973) 38 1961

OAc$^-$ Tetrahedron (1968) 24 5493

Ph$_3$P, (NH$_2$)$_2$CO or NaN$_3$ Synth Comm (1973) 3 99

Section 98   Amines from Esters
          ○○○○○○○○○○○○○○○○○○○

No additional examples

Section 99   Amines from Epoxides
          ○○○○○○○○○○○○○○○○○○○

PhCHCH$_2$ (epoxide O) $\xrightarrow{\text{HN}_3 \text{ SnCl}_4 \text{ C}_6\text{H}_6}$ PhNH$_2$ 17%

JACS (1952) 74 1168

Section 100   Amines from Halides

BuBr $\xrightarrow[\text{2 Et}_2\text{NCCl}_3]{\text{1 Mg   Et}_2\text{O}}$ Bu$_3$CNEt$_2$ Synthesis (1972) 611   37%

RX $\xrightarrow{\text{Me}_2\text{C=NOH   EtMgBr   toluene}}$ RNH$_2$ 62% (R=Ph)
48% (R=PhCH$_2$CH$_2$)

Tetr Lett (1972) 1007

RBr $\xrightarrow[\text{2 EtOCH=NSO}_2\text{Ph}]{\text{1Mg}}$ R$_2$CHNHSO$_2$Ph --→ R$_2$CHNH$_2$ <92% (R=Ph)
<74% (R=Pr)

Ber (1969) <u>102</u> 1641

BuI $\xrightarrow{\text{Me}_2\text{NNO   (i-Pr)}_2\text{NLi}}$ BuCH$_2$NNO --→ BuCH$_2$NHMe
                                  |
                                  Me

Angew (1972) <u>84</u> 350
(Internat Ed <u>11</u> 301)

PhCH$_2$Cl $\xrightarrow{\text{EtN} \begin{array}{c}\text{Et}\\ \text{N}\\ \end{array} \text{NEt   MeOH   H}_2\text{O}}$ PhCH$_2$NEt   65%
                                              |
                                              Me

Synthesis (1971) 89

C$_{12}$H$_{25}$I $\dashrightarrow[]{\text{EtNMe}_2}$ C$_{12}$H$_{25}\overset{+}{\underset{\text{Et}}{\text{NMe}_2}}$ I$^-$ $\xrightarrow{\text{PrSLi   HMPA}}$ C$_{12}$H$_{25}\underset{\text{Et}}{\text{NMe}}$

JOC (1973) <u>38</u> 1961

$CH_2=CHCH_2I$ $\xrightarrow[\text{2 Na-Hg}]{\text{1 }(PhCH_2)_2NMe}$ $CH_2=CHCH_2\underset{Me}{NCH_2Ph}$ $\xrightarrow[\text{2 Na-Hg}]{\text{1 PrI}}$ $CH_2=CHCH_2\underset{Me}{NPr}$

Arch Pharm (1911) <u>249</u> 111

RBr $\xrightarrow[\text{2 NaOH}]{\text{1 }(NH_2)_2C=NH\quad EtOH\quad H_2O}$ RNH_2

Bull Soc Chim Fr (1970) 1938

$BuBr$ $\xrightarrow[\text{TsNNa\quad DMF}]{\overset{COOEt}{|}}$ $\overset{COOEt}{\underset{}{TsNBu}}$ $\xrightarrow{NaOH}$ $TsNHBu$ $\dashrightarrow$ $BuNH_2$ $\quad<60\%$

JOC (1971) <u>36</u> 4102

RBr $\xrightarrow[\text{2 HCl\quad Et}_2O\quad H_2O]{\text{1 }(PhS)_2NLi\quad THF}$ RNH_2

Bull Chem Soc Jap (1971) <u>44</u> 2797

$C_7H_{15}Br$ $\xrightarrow[\text{base}]{PhCH_2NHSO_2CF_3}$ $\underset{CH_2Ph}{C_7H_{15}NSO_2CF_3}$ $\xrightarrow[\text{2 HCl\quad H}_2O\quad THF]{\text{1 NaH\quad DMF}}$ $C_7H_{15}NH_2$ $\quad80\%$

Tetr Lett (1973) 3839

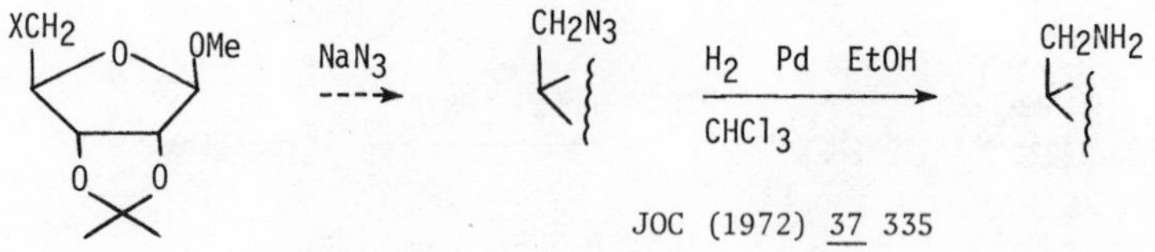

JOC (1972) <u>37</u> 335

Section 101   Amines from Hydrides
              ○○○○○○○○○○○○○○○○○○○○○○○

$$ArH \xrightarrow{\quad Me_2NCl \quad H_2SO_4 \quad Na_2SO_4 \quad} ArNMe_2$$

Synthesis (1973) 1

Section 102   Amines from Ketones
              ○○○○○○○○○○○○○○○○○○○○○○○

$$R_2CO \xrightarrow[\text{MeOH} \quad \text{pH 6-8}]{\text{Me}_2\text{NH} \quad \text{NaBH}_3\text{CN}} R_2CHNMe_2$$

JACS (1971) <u>93</u> 2897

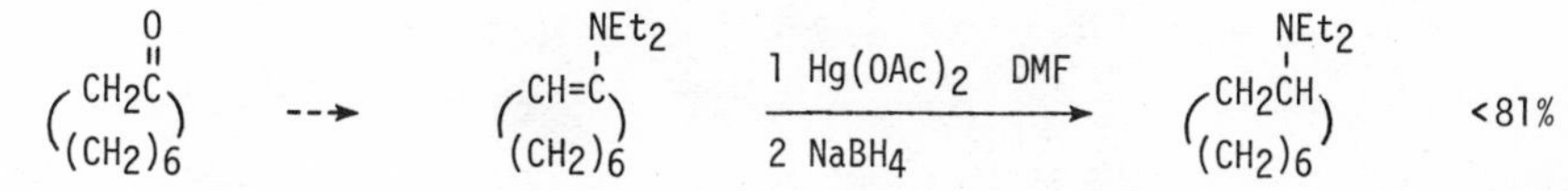

$$\begin{pmatrix} CH_2\overset{O}{\overset{\|}{C}} \\ (CH_2)_6 \end{pmatrix} \dashrightarrow \begin{pmatrix} CH=\overset{NEt_2}{C} \\ (CH_2)_6 \end{pmatrix} \xrightarrow[\text{2 NaBH}_4]{\text{1 Hg(OAc)}_2 \quad \text{DMF}} \begin{pmatrix} CH_2\overset{NEt_2}{CH} \\ (CH_2)_6 \end{pmatrix} \quad <81\%$$

Chem Comm (1971) 1433

$$R_2CO \dashrightarrow R_2C=NOH \xrightarrow{\quad Na \quad R'OH \quad} R_2CHNH_2$$

Chem Ind (1972) 683

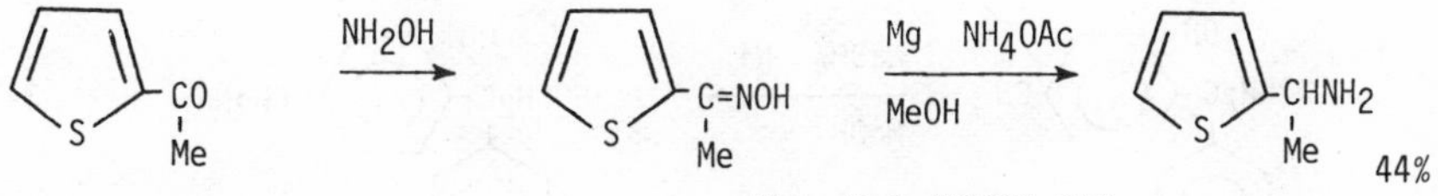

Chem Ind (1969) 260

Related methods: Amines from Aldehydes (Section 94)

Section 103   Amines from Nitriles

$$PhCN \xrightarrow{\text{EtMgBr   toluene}} \underset{\overset{|}{NH_2}}{PhCEt_2}$$ 60%

Tetr Lett (1973) 1057

$$C_{12}H_{25}CN \xrightarrow{\text{H}_2 \ \ \text{Ni} \ \ \text{Ac}_2\text{O} \ \ \text{NaOAc}} C_{12}H_{25}CH_2NHAc \dashrightarrow C_{12}H_{25}CH_2NH_2$$

JOC (1960) 25 1658   ~100%

$$PrCN \xrightarrow{\text{H}_2 \ \ \text{PtO}_2 \ \ \text{EtOH} \ \ \text{CHCl}_3} PrCH_2NH_2$$ 96%

JOC (1972) 37 335

$$PrCN \xrightarrow{\text{N}_2\text{H}_4 \cdot \text{H}_2\text{O} \ \ \text{Ni} \ \ \text{EtOH}} PrCH_2NH_2$$ 75%

JOC (1971) 36 3539

$$PhCH=CHCN \xrightarrow{\text{AlH}_3 \ \ \text{THF}} PhCH=CHCH_2NH_2$$ 78%

JACS (1968) 90 2927

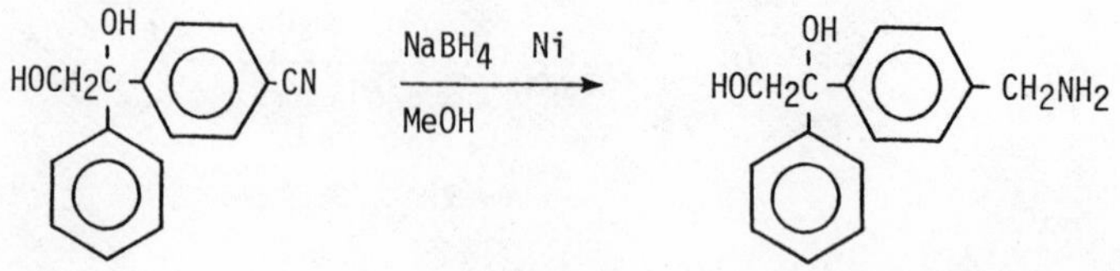

34%

Annalen (1972) 756 14

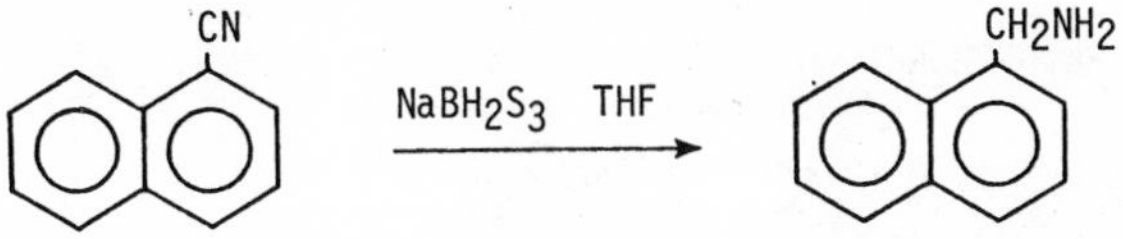

CN
$NaBH_2S_3$ THF
CH_2NH_2

40%

Can J Chem (1971) 49 2990

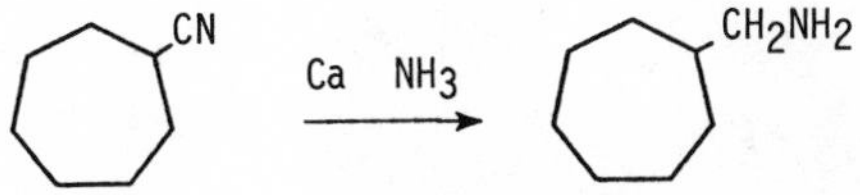

CN
Ca NH_3
CH_2NH_2

63%

JOC (1972) 37 508

Section 104 Amines from Olefins

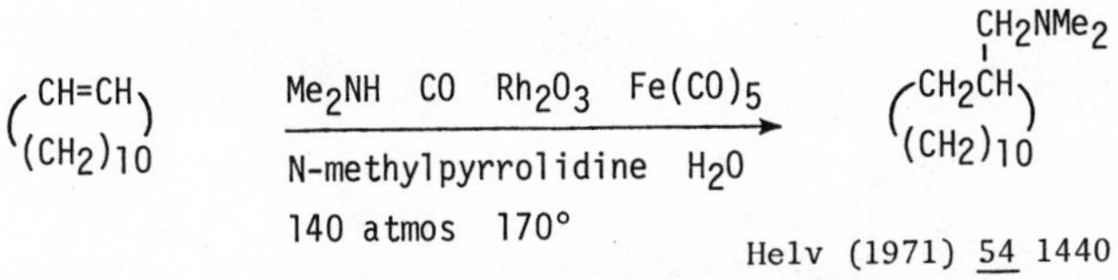

CH=CH
$(CH_2)_{10}$
Me_2NH CO Rh_2O_3 $Fe(CO)_5$
N-methylpyrrolidine H_2O
140 atmos 170°
CH_2NMe_2
CH_2CH
$(CH_2)_{10}$

90%

Helv (1971) 54 1440

1 $LiBH_4$ BCl_3 Et_2O
2 BuN_3
3 KOH H_2O
NHBu

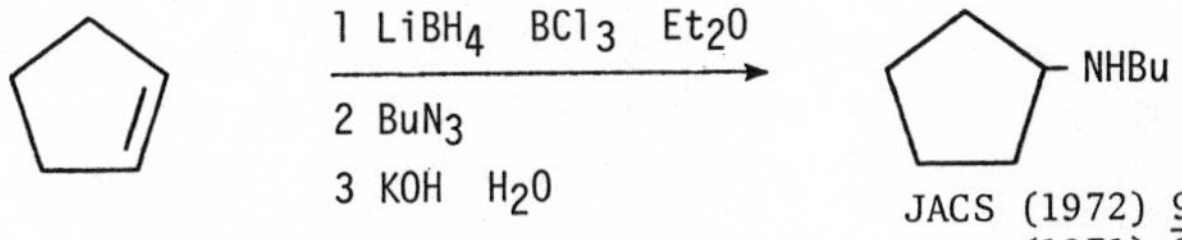

64%

JACS (1972) 94 2114
(1971) 93 4329
(1973) 95 2394 2396

$EtCH=CH_2 \xrightarrow{B_2H_6} (EtCH_2CH_2)_3B \xrightarrow{Me_2NCl} EtCH_2CH_2NMe_2$

J Organometallic Chem (1970) 23 C11

$PhCH=CH_2$ $\xrightarrow{\text{PrNH}_2 \quad \text{Na}}$ $PhCH_2CH_2NHPr$ 18%

Ber (1950) 83 1
JOC (1972) 37 4243

$C_{14}H_{29}CH=CH_2$ $\xrightarrow{\begin{array}{l}1 \ O_3 \quad MeOH \\ 2 \ H_2 \quad Ni \\ 3 \ Me_2NH\end{array}}$ $C_{14}H_{29}CH_2NMe_2$ 65%

Tetr Lett (1971) 3591

$Ph_2C=CHEt$ $\xrightarrow{\text{NaN}_3 \quad \text{H}_2\text{SO}_4}$ $PhNH_2$ 57%

JACS (1950) 72 5777

Also via: Amides (Section 89)

Section 105   Amines from Miscellaneous Compounds
○○

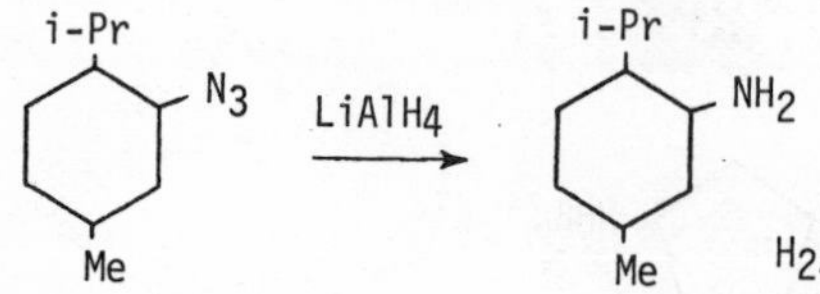

JOC (1962) 27 2925   34%
H_2/Pd   Ber (1972) 105 1524
" "   JOC (1972) 37 335

$ArNO_2 \longrightarrow ArNH_2$

Sn, HCl   JACS (1969) 91 3544
H_2/Pd   JOC (1972) 37 335
$NaBH_4$, $CoCl_2$ Tetr Lett (1969) 4555
$NaBH_2S_3$ Can J Chem (1971) 49 2990
CO, $RhCl_3$ Tetr Lett (1971) 3385
$Fe_3(CO)_{12}$ JOC (1972) 37 930
$Fe(CO)_5$ Can J Chem (1970) 48 1543

Section 105A Protection of Amines

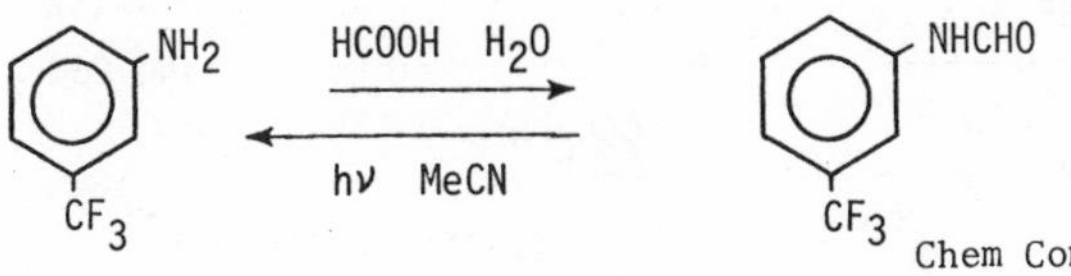
NH2
HCOOH H2O
hν MeCN
CF3
NHCHO
CF3

Chem Comm (1972) 758

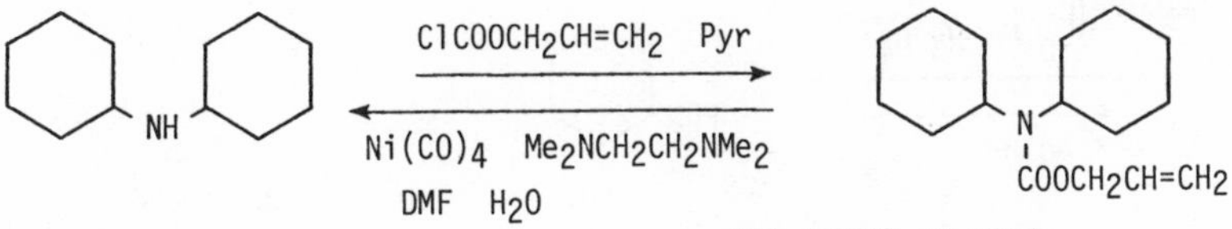
NH
ClCOOCH2CH=CH2 Pyr
Ni(CO)4 Me2NCH2CH2NMe2
DMF H2O
N
COOCH2CH=CH2

JOC (1973) 38 3223

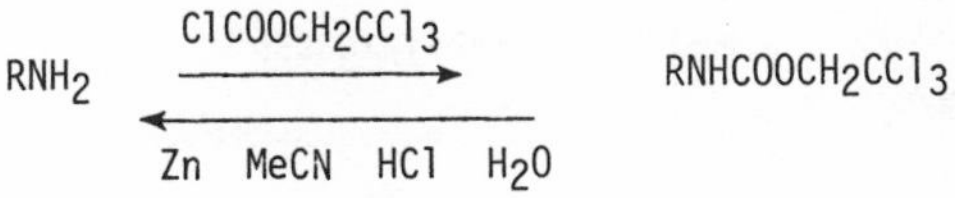
RNH2
ClCOOCH2CCl3
Zn MeCN HCl H2O
RNHCOOCH2CCl3

JOC (1971) 36 1259

Selective removal of haloethoxycarbonyl groups by electrolysis
JACS (1972) 94 5139

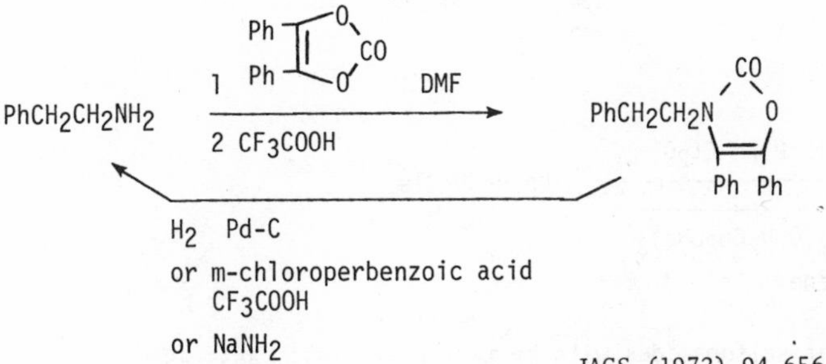
PhCH2CH2NH2
1 Ph Ph O CO O DMF
2 CF3COOH
PhCH2CH2N CO O
Ph Ph
H2 Pd-C
or m-chloroperbenzoic acid
CF3COOH
or NaNH2

JACS (1972) 94 6561

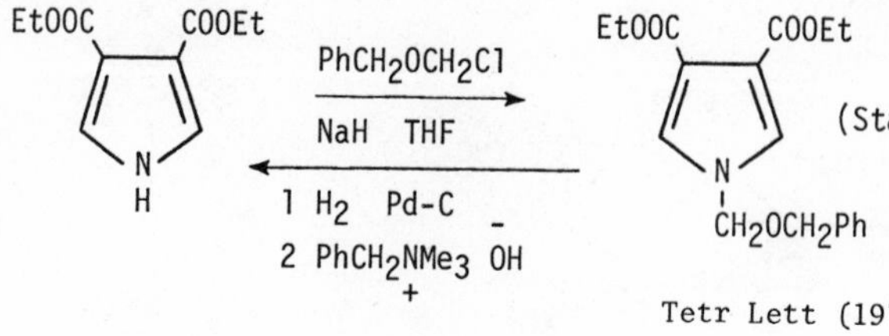

EtOOC
COOEt
N
H
$PhCH_2OCH_2Cl$
NaH THF
1 H_2 Pd-C
2 $PhCH_2\overset{+}{N}Me_3$ $\overset{-}{O}H$
EtOOC
COOEt
N
CH_2OCH_2Ph

(Stable to HCl, Ac_2O, KOH and $CdMe_2$)

Tetr Lett (1971) 3165

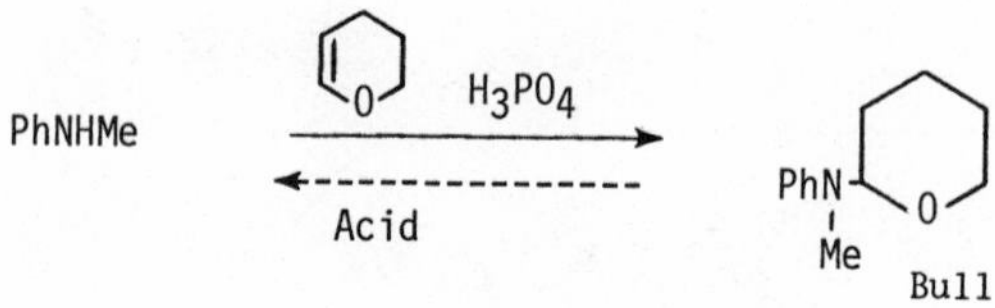

PhNHMe
O
H_3PO_4
Acid
PhN
Me
O

Bull Soc Chim Fr (1954) 575

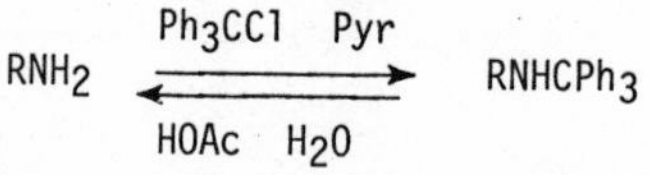

RNH_2
Ph_3CCl Pyr
HOAc H_2O
$RNHCPh_3$

J Med Chem (1973) 16 959

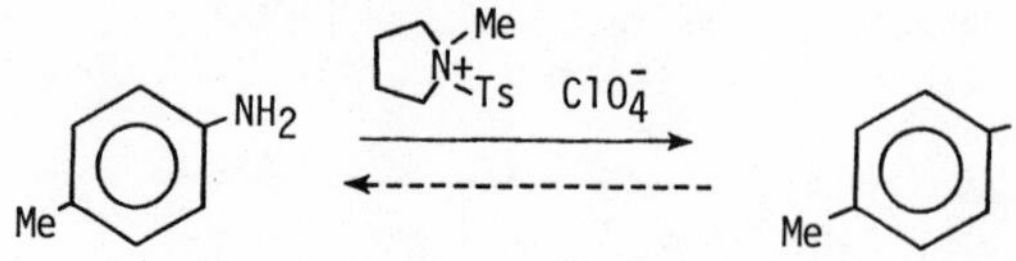

Me
NH_2
Me
N+
Ts
ClO_4^-
Me
NHTs

(Selective for NH in presence of OH)

Chem Comm (1972) 1148

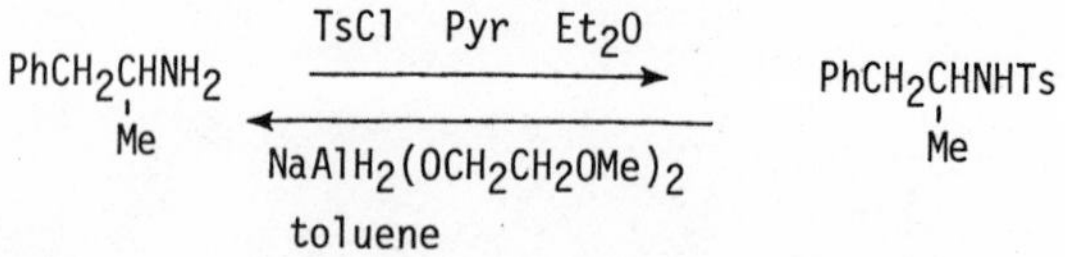

$PhCH_2CHNH_2$
Me
TsCl Pyr Et_2O
$NaAlH_2(OCH_2CH_2OMe)_2$
toluene
$PhCH_2CHNHTs$
Me

JOC (1972) 37 2208

Cleavage of sulfonamides with hν Tetr Lett (1971) 4555 4559

Chem Pharm Bull (1970) 18 182

H_2SO_4, HOAc Chem Comm (1973) 664

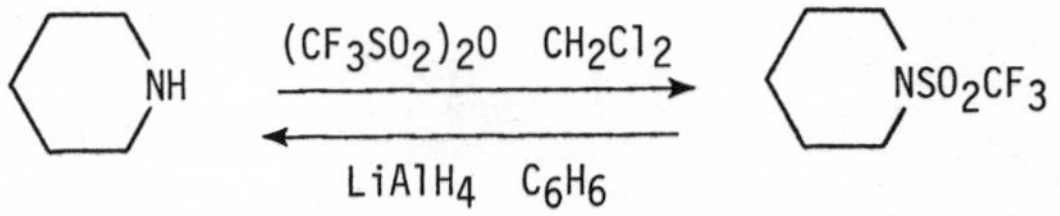

$(CF_3SO_2)_2O$ CH_2Cl_2

$LiAlH_4$ C_6H_6

NSO_2CF_3

(Stable to Zn/acid)

Tetr Lett (1973) 3839

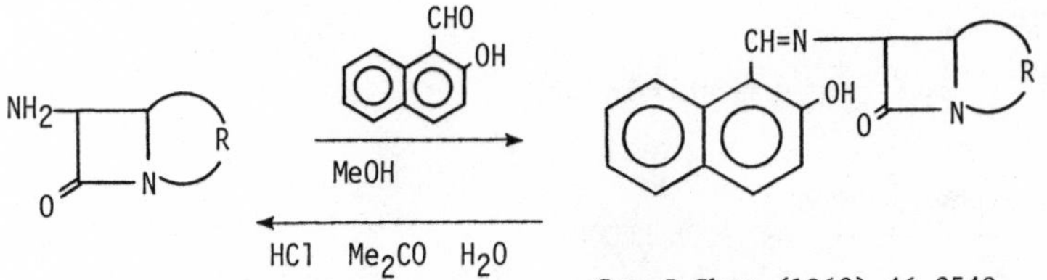

NH_2

CHO
OH

MeOH

HCl Me_2CO H_2O

CH=N

OH

R

Can J Chem (1968) <u>46</u> 2549
JCS <u>C</u> (1971) 3864

Chapter 8   PREPARATION OF ESTERS

Section 106   <u>Esters from Acetylenes</u>
ᵒᵒᵒᵒᵒᵒᵒᵒᵒᵒᵒᵒᵒᵒᵒᵒᵒᵒᵒᵒᵒᵒ

$$PhC\equiv CEt \xrightarrow{\quad Tl(NO_3)_3 \quad MeOH \quad} Ph\overset{Et}{\underset{|}{C}HCOOMe}$$

    82%

JACS (1973) <u>95</u> 1296

Also via: Carboxylic Acids (Section 16)

Section 107   <u>Esters from Carboxylic Acids and Acid Halides</u>
ᵒᵒ

Homologation of carboxylic acids. . . . . page   108
Esters by reaction of carboxylic acids, acid halides
                            etc. with alcohols 109-110
Esters by reaction of carboxylic acids with halides,
                          sulfates and sulfites 110-111
Esters by reaction of carboxylic acids and acid halides
                   with miscellaneous reagents 111-112

$$RCOCl \xrightarrow[\text{2 PhCOOAg} \quad Et_3N \quad EtOH]{\text{1 } CH_2N_2 \quad Et_3N \quad Et_2O} RCH_2COOEt$$

    85-92%

Org Synth (1970) <u>50</u> 77

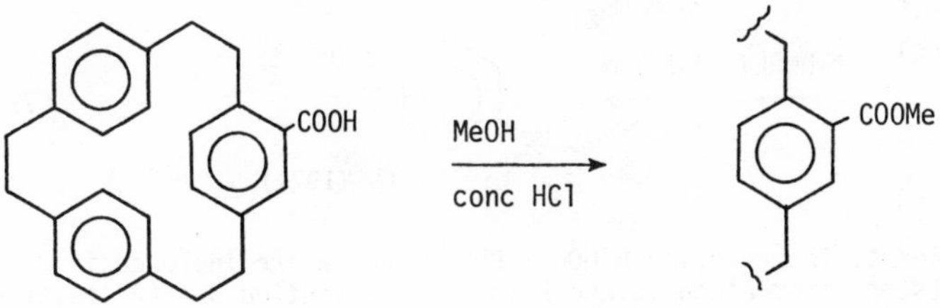

COOH
MeOH
conc HCl
COOMe

44%

Tetrahedron (1971) 27 4845

$$\mathrm{RCOOH} \xrightarrow{\mathrm{R'OH}\quad \mathrm{BF_3 \cdot Et_2O}} \mathrm{RCOOR'}$$

Synthesis (1971) 316

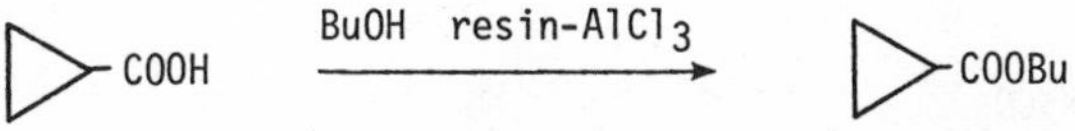

COOH
BuOH resin-AlCl3
COOBu

56%

Tetr Lett (1973) 1823

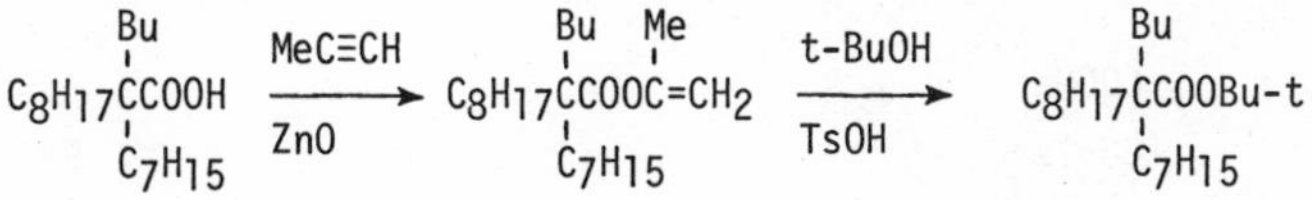

Bu
C8H17CCOOH
C7H15
MeC≡CH
ZnO
Bu Me
C8H17CCOOC=CH2
C7H15
t-BuOH
TsOH
Bu
C8H17CCOOBu-t
C7H15

67%

JOC (1972) 37 3551

CH2CONH
S
S
O
N
CH2OAc
COOH
1 (COCl)2
2 t-BuOH Et3N
COOBu-t

79%

J Med Chem (1971) 14 420

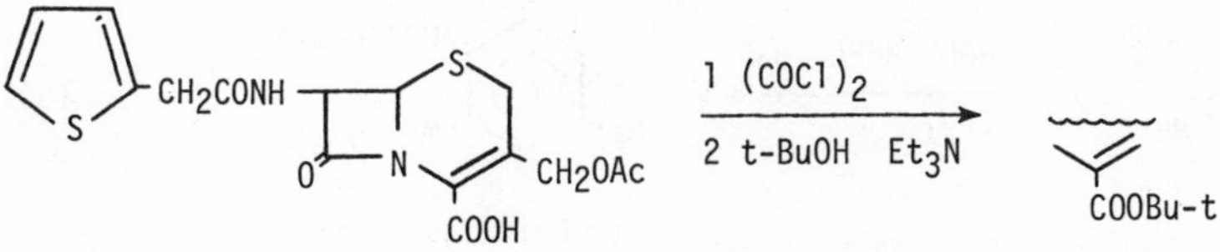

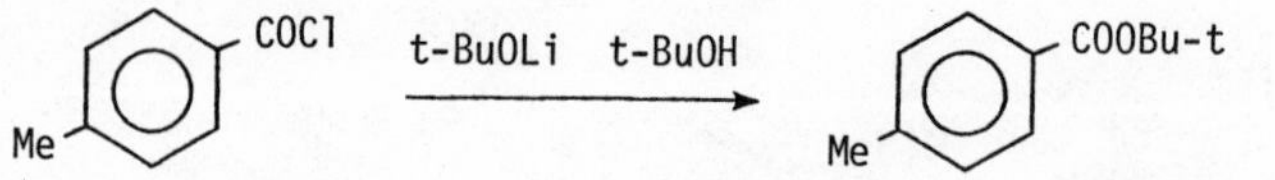

COCl → t-BuOLi   t-BuOH → COOBu-t

79-82%

Org Synth (1971) <u>51</u> 96

Further examples of the reaction RCOOH + ROH → RCOOR are included in section 108 (Esters from Alcohols and Phenols) and section 30A (Protection of Carboxylic Acids)

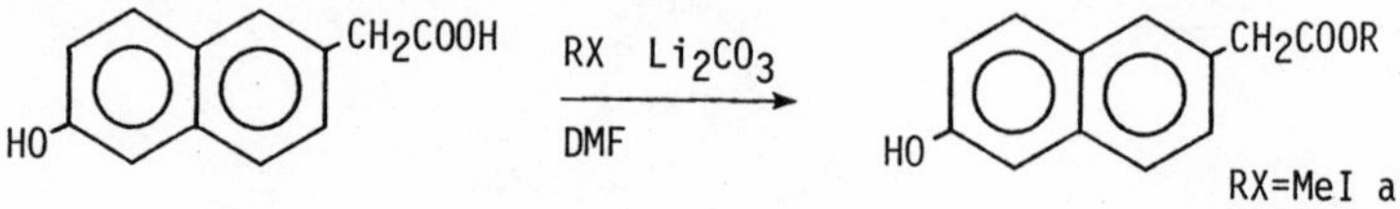

CH₂COOH → RX   Li₂CO₃ / DMF → CH₂COOR

100%

RX=MeI at 25°
RX=C₆H₁₃Br at 90°

J. Riegl, I. T. Harrison (Syntex Research) unpublished

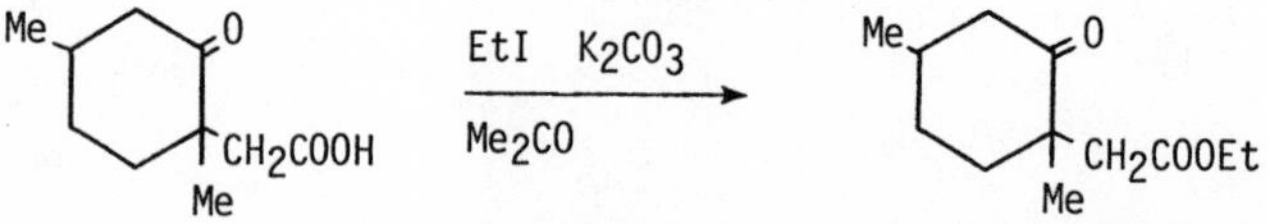

CH₂COOH → EtI   K₂CO₃ / Me₂CO → CH₂COOEt

Tetr Lett (1972) 1853
JOC (1969) <u>34</u> 3717

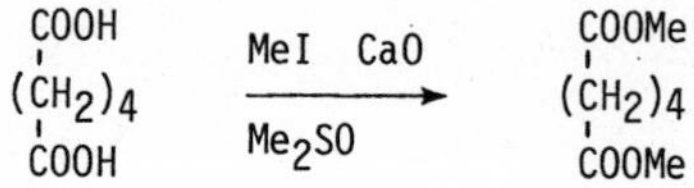

COOH
(CH₂)₄   → MeI   CaO / Me₂SO →   COOMe
COOH   (CH₂)₄
                                    COOMe

91%

Synthesis (1972) 262

COOH → 1 NaOH   HMPA   H₂O / 2 i-PrI → COOPr-i

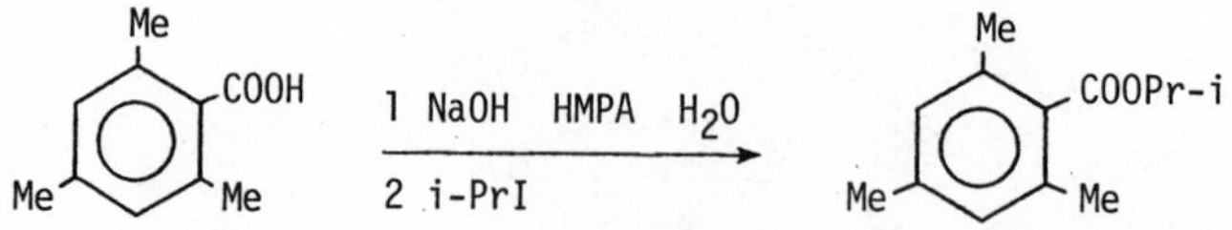

100%

Tetr Lett (1973) 689
          (1972) 4063

RCOOK $\xrightarrow{\text{PhCH}_2\text{Br} \quad \text{DMF}}$ RCOOCH$_2$Ph

JOC (1962) 27 1381

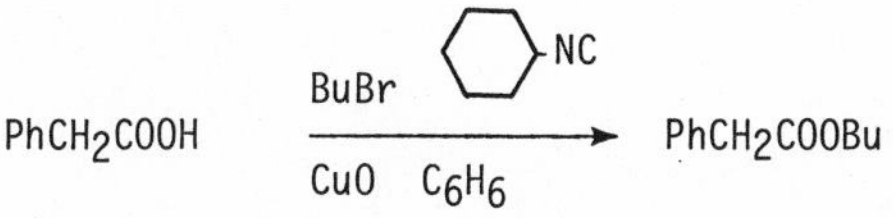

PhCH$_2$COOH $\xrightarrow[\text{CuO} \quad \text{C}_6\text{H}_6]{\text{BuBr} \quad \bigcirc\text{-NC}}$ PhCH$_2$COOBu   86%

Synth Comm (1972) 2 1
JOC (1973) 38 1753

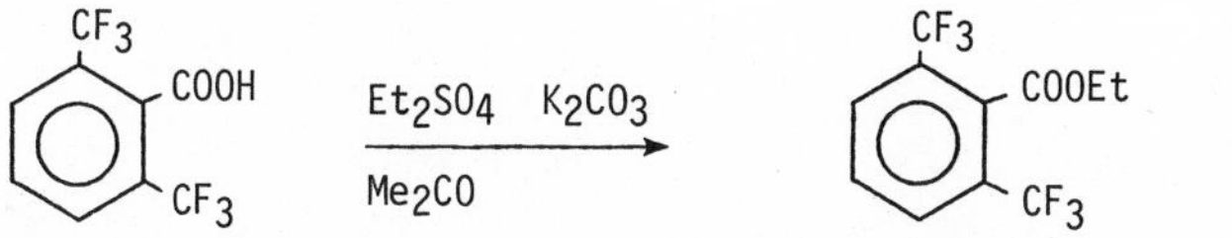

$\xrightarrow[\text{Me}_2\text{CO}]{\text{Et}_2\text{SO}_4 \quad \text{K}_2\text{CO}_3}$

~90%

Chem Ind (1972) 691
Tetr Lett (1972) 757

PhCH=CHCOOH $\xrightarrow[150°]{\text{Et}_2\text{SO}_4\text{-HCONMe}_2}$ PhCH=CHCOOEt   94%

Ber (1971) 104 3711

SH
Me$_2$CCHCOOH $\xrightarrow{\text{Me}_2\text{SO}_3 \quad \text{HCl} \quad \text{MeOH}}$
NH$_2$

SH
Me$_2$CCHCOOMe   83%
NH$_2$

J Med Chem (1971) 14 868

$\xrightarrow{\text{t-BuOAc} \quad \text{HClO}_4}$

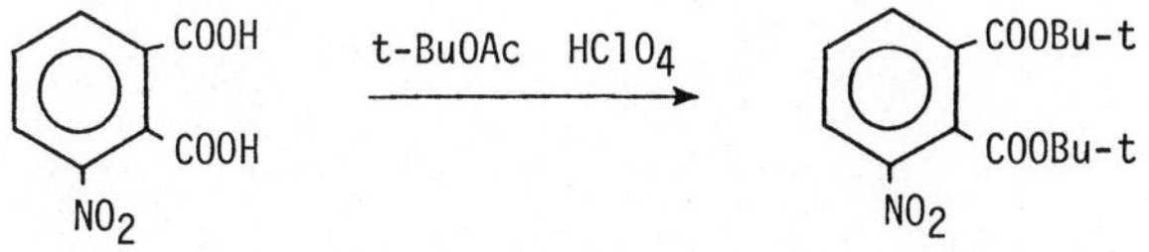

22%

Annalen (1961) 646 119 134

BuCOF $\xrightarrow{\text{Et}_3\text{SiH}}$ BuCOOCH$_2$Bu

JOC (1971) 36 2547

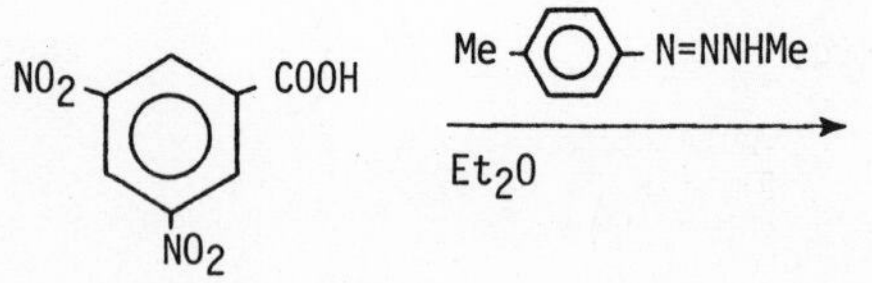

70-90%

Org Synth (1968) 48 102

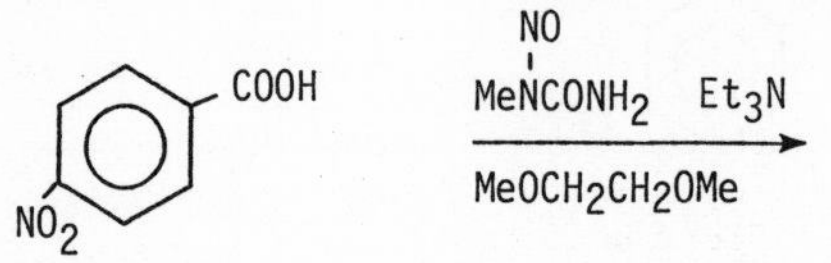

100%

Tetr Lett (1973) 1397

Et$_3$CCOOH $\xrightarrow[\text{Et}_3\text{O}^+ \text{BF}_4^- \ \ (i\text{-Pr})_2\text{NEt}]{}$ Et$_3$CCOOEt   90%

Tetr Lett (1971) 4741

C$_{13}$H$_{27}$COOH $\xrightarrow[\text{2 Toluene (reflux)}]{1 \ \text{PhNMe}_3^+ \ \text{OH}^- \ \ \text{MeOH}}$ C$_{13}$H$_{27}$COOMe   >90%

Synthesis (1973) 494

RCOOH $\xrightarrow{\text{Pb(OAc)}_4}$ ROAc

Org React (1972) 19 279

Section 108   Esters from Alcohols and Phenols

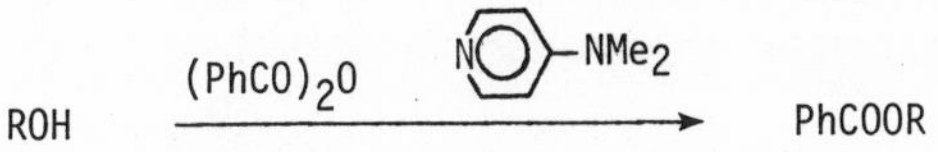

$$ROH \xrightarrow[\text{(PhCO)}_2O]{\text{N}\bigcirc\text{-NMe}_2} PhCOOR$$

Synthesis (1972) 619

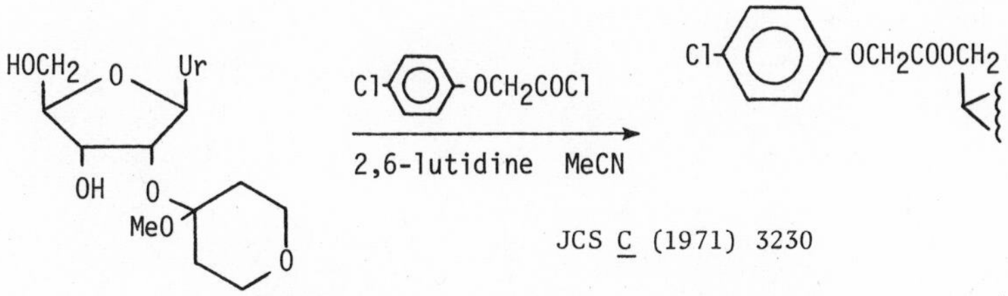

2,6-lutidine   MeCN

JCS C (1971) 3230

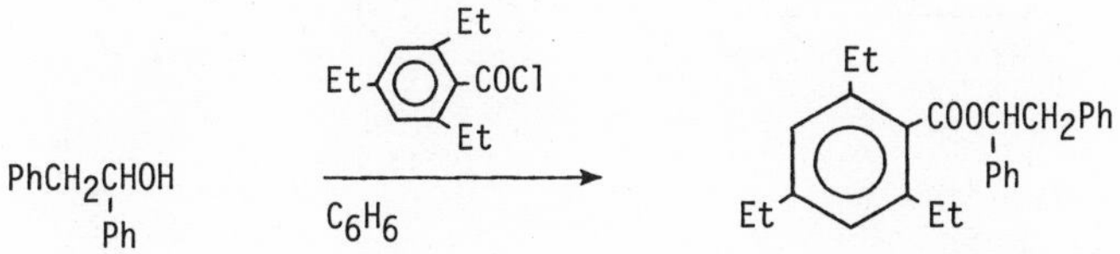

78%

$$PhCH_2CHOH \atop Ph \qquad C_6H_6$$

JACS (1953) 75 6011

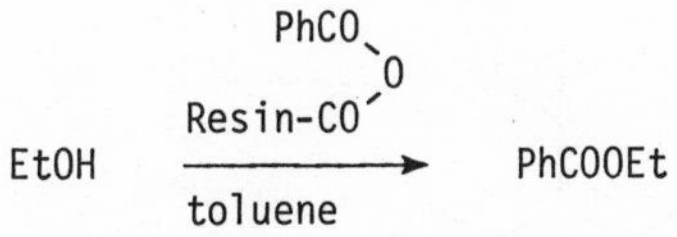

$$EtOH \xrightarrow[\text{toluene}]{\overset{\text{PhCO}}{\underset{\text{Resin-CO}}{\diagdown}}\text{O}} PhCOOEt$$

Tetr Lett (1973) 1627

EtCOCl   SO$_2$

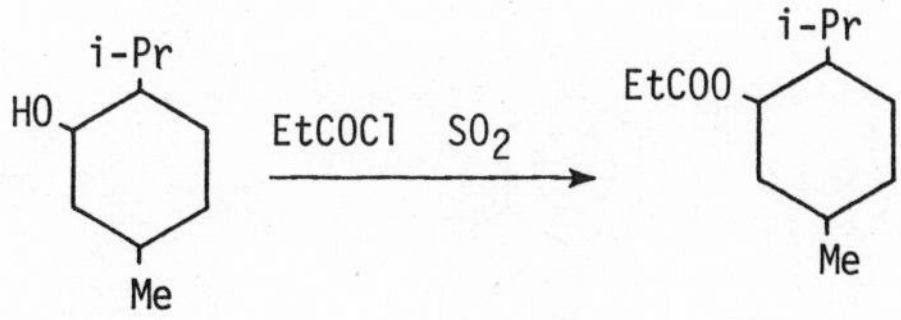

92%

Synthesis (1971) 639

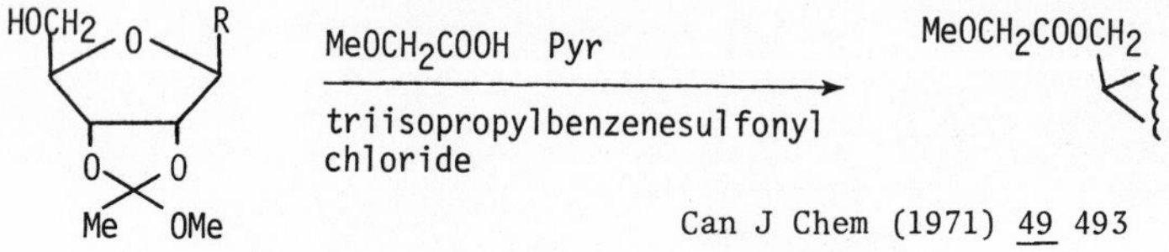

HOCH2
R
O
O
O
Me
OMe
MeOCH2COOH Pyr
triisopropylbenzenesulfonyl chloride
MeOCH2COOCH2

Can J Chem (1971) 49 493

ROH $\xrightarrow{\text{MeCOCl Mg}}$ ROAc 92%

Tetrahedron (1971) 27 753

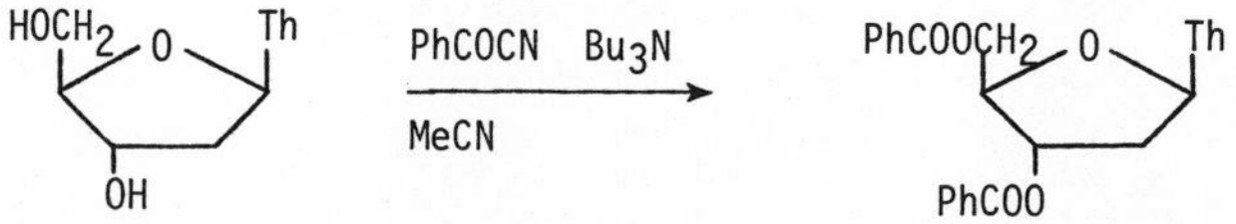

HOCH2
O
Th
OH
PhCOCN Bu3N
MeCN
PhCOOCH2
O
Th
PhCOO

Tetr Lett (1971) 185

OH
HO
PhCOOH
NCOOEt
NCOOEt
Ph3P
OCOPh
PhCOO

Tetr Lett (1973) 1619

HO
Ph
(EtCO)2O NaOH H2O
EtCOO
Ph

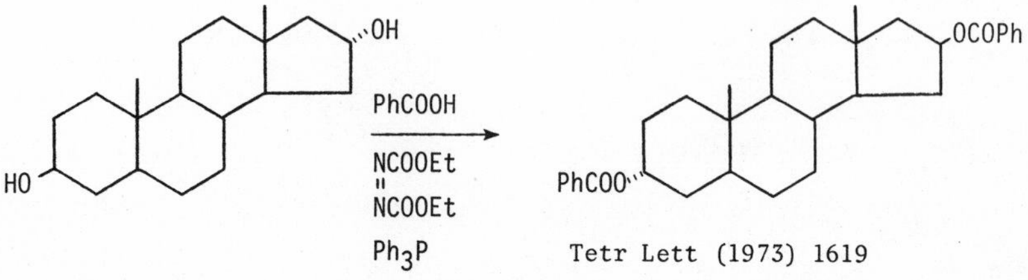

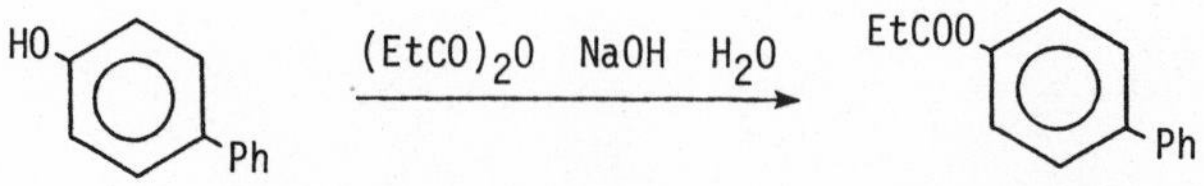

73%

JACS (1942) 64 2271 3163

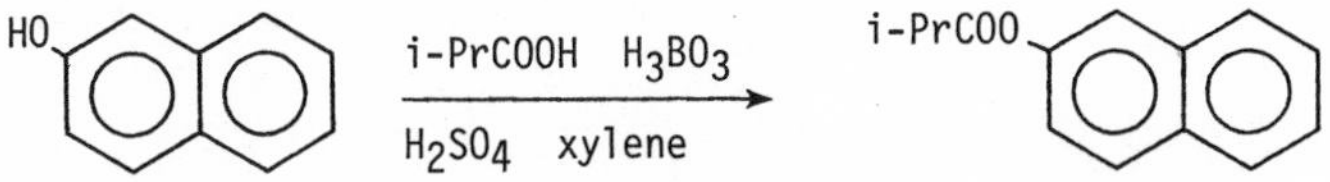

i-PrCOOH　H$_3$BO$_3$
──────────────→
H$_2$SO$_4$　xylene

58%

Tetr Lett (1971) 3453

Further examples of the reaction ROH ──→ R'COOR are included in section 107 (Esters from Carboxylic Acids and Acid Halides) and section 45A (Protection of Alcohols and Phenols)

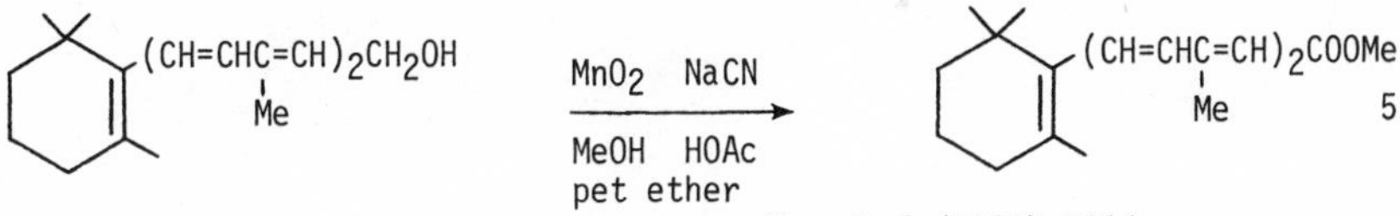

(CH=CHC=CH)$_2$CH$_2$OH
　　　|
　　　Me

MnO$_2$　NaCN
──────────────→
MeOH　HOAc
pet ether

(CH=CHC=CH)$_2$COOMe
　　　　|
　　　　Me　　52%

Chem Ind (1971) 1016

C$_5$H$_{11}$CH$_2$OH

O$_2$　Co(OAc)$_2$　HOAc
──────────────→
CoBr$_2$

C$_5$H$_{11}$COOCH$_2$C$_5$H$_{11}$

Can J Chem (1965) 43 1306

Section 109　　Esters from Aldehydes

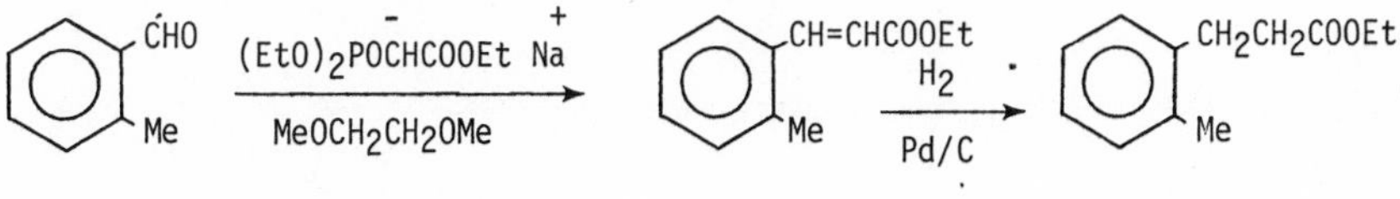

$\overset{\cdot}{C}$HO

(EtO)$_2$POCHCOOEt Na$^+$
──────────────→
MeOCH$_2$CH$_2$OMe

CH=CHCOOEt
H$_2$
──→
Pd/C

CH$_2$CH$_2$COOEt

Me

JOC (1971) 36 1040

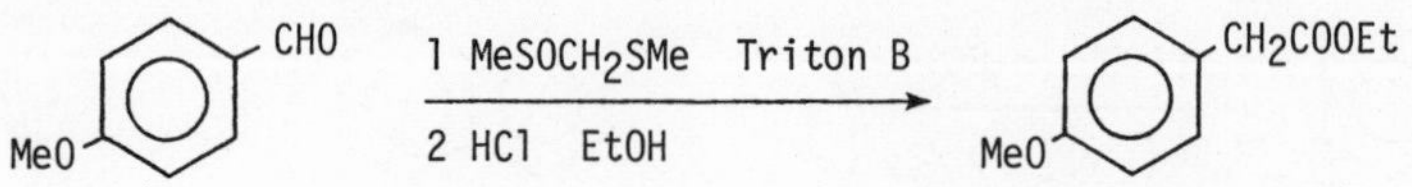

MeO—C₆H₄—CHO $\xrightarrow[\text{2 HCl EtOH}]{\text{1 MeSOCH}_2\text{SMe Triton B}}$ MeO—C₆H₄—CH₂COOEt   93%

Tetr Lett (1972) 1383

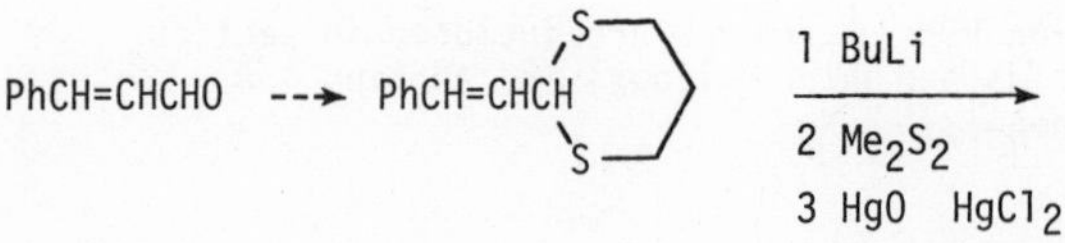

PhCH=CHCHO ⇢ PhCH=CHCH(dithiane) $\xrightarrow[\substack{\text{2 Me}_2\text{S}_2 \\ \text{3 HgO HgCl}_2}]{\text{1 BuLi}}$ PhCH=CHCOOEt   < 84%

JOC (1972) 37 2757

$C_6H_{13}CHO$ ⇢ $C_6H_{13}CH(OMe)_2$ $\xrightarrow[\text{CH}_2\text{Cl}_2]{\text{O}_3}$ $C_6H_{13}COOMe$ < 90%

Can J Chem (1971) 49 2465

$PhCHO$ ⇢ $PhCH(OEt)_2$ $\xrightarrow{\text{NBS}}$ $PhCOOEt$

JACS (1951) 73 973

BuCHCHO $\xrightarrow[250°]{\text{H}_3\text{BO}_3 \text{ heptane}}$ BuCHCOOCH₂CHBu   47%
| Et | Et   Et

JOC (1973) 38 1433

Related methods: Esters from Ketones (Section 117)

Also via: Olefinic esters (Section 362)

Section 110   <u>Esters from Alkyls, Methylenes and Aryls</u>

No examples of the reaction RR $\longrightarrow$ RCOOR' or R'COOR (R,R'=alkyl, aryl
etc.) occur in the literature. For the reaction RH $\longrightarrow$ RCOOR' or R'COOR
see section 116 (Esters from Hydrides)

Section 111   <u>Esters from Amides</u>

No additional examples

Section 112   <u>Esters from Amines</u>

$$C_6H_{13}NH_2 \quad \dashrightarrow \quad C_6H_{13}NTs_2 \quad \xrightarrow[\text{HMPA}]{\text{KI \quad NaOAc}} \quad C_6H_{13}OAc \qquad\qquad < 95\%$$

Synth Comm (1972) <u>2</u> 297

Section 113   <u>Esters from Esters</u>

$$BuCH_2COOBu\text{-}t \quad \xrightarrow[\text{2 MeI \quad Me}_2\text{SO}]{1 \quad LiN\langle\bigcirc\rangle^{\text{i-Pr}} \quad THF} \quad \underset{\overset{|}{Me}}{BuCHCOOBu\text{-}t} \qquad 83\%$$

JACS (1971) <u>93</u> 2318
Tetr Lett (1973) 2425

$$\begin{matrix} EtCOOCH_2 \\ | \\ C(SEt)_2 \\ | \\ EtCOOCH_2 \end{matrix} \quad \xrightarrow{\;C_{17}H_{35}COOMe \quad MeONa\;} \quad \begin{matrix} C_{17}H_{35}COOCH_2 \\ | \\ C(SEt)_2 \\ | \\ C_{17}H_{35}COOCH_2 \end{matrix}$$

Can J Chem (1955) <u>33</u> 717

PrCOOPr $\xrightarrow[\longleftarrow]{\text{EtOH \quad Bu}_3\text{SnOEt}}$ PrCOOEt

Bull Soc Chim Fr (1969) 262

KCN Synthesis (1973) 790

PrCOOEt $\xrightarrow{\text{BuOH \quad CO}_2}$ PrCOOBu   27%

Bull Chem Soc Jap (1971) $\underline{44}$ 852

$C_{11}H_{23}COOMe$ $\xrightarrow[\text{2 Ac}_2O]{\text{1 (i-Bu)}_2\text{AlH \quad octane}}$ $C_{11}H_{23}CH_2OAc$

Synthesis (1971) 309

$Pr_2C(COOEt)_2$ $\xrightarrow[178\text{-}183°]{\text{NaCl \quad Me}_2\text{SO \quad H}_2O}$ $Pr_2CHCOOEt$

Tetr Lett (1973) 957

Section 114   Esters from Ethers
°°°°°°°°°°°°°°°°°°°°

i-PrCH$_2$OR $\xrightarrow[\text{H}_2O]{\text{Trichloroisocyanuric acid}}$ i-PrCOOR

Tetrahedron (1971) $\underline{27}$ 2671

R_2O $\xrightarrow{\text{TsOAc \quad MeCN}}$ ROAc

JOC (1971) $\underline{36}$ 532 540

Section 115   <u>Esters from Halides</u>

BuBr $\xrightarrow[\text{2 } H_2SO_4 \quad HgO \quad EtOH]{\text{1 } NaC{\equiv}COEt \quad NH_3}$ BuCH$_2$COOEt

JCS (1954) 1860
Advances in Org Chem (1960) $\underline{2}$ 117

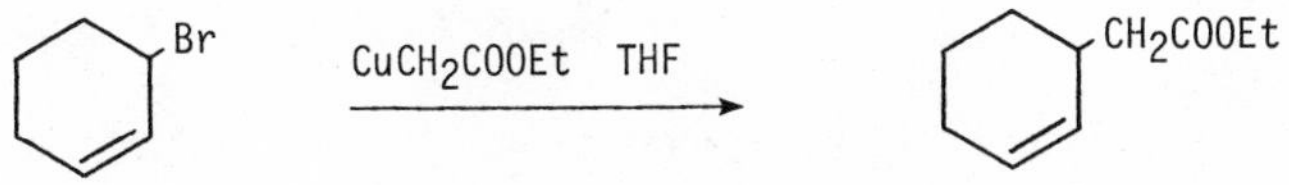

$\xrightarrow{\text{CuCH}_2\text{COOEt} \quad \text{THF}}$

69%

Tetr Lett (1972) 1163

ArBr $\xrightarrow[\text{2 } CdCl_2]{\text{1 } Mg}$ Ar$_2$Cd $\xrightarrow{\text{BrCH}_2\text{COOEt}}$ ArCH$_2$COOEt

JOC (1973) $\underline{38}$ 3189

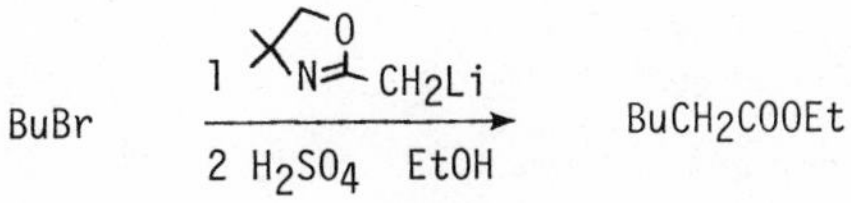

BuBr $\xrightarrow[\text{2 } H_2SO_4 \quad EtOH]{\text{1}}$ BuCH$_2$COOEt

JACS (1970) $\underline{92}$ 6644 6646

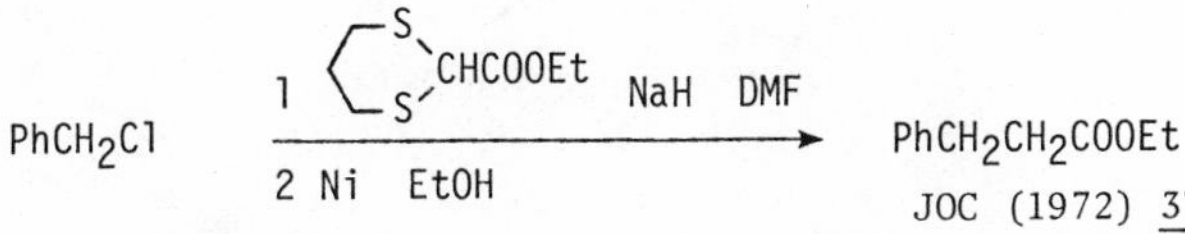

PhCH$_2$Cl $\xrightarrow[\text{2 Ni \quad EtOH}]{\text{1}}$ PhCH$_2$CH$_2$COOEt   48%

JOC (1972) $\underline{37}$ 505

RI $\xrightarrow[\text{EtONa}]{\text{CH}_2(\text{COOEt})_2}$ R$_2$C(COOEt)$_2$ $\xrightarrow[\text{H}_2\text{O} \quad 178\text{-}183°]{\text{NaCl} \quad Me_2SO}$ R$_2$CHCOOEt

Tetr Lett (1973) 957

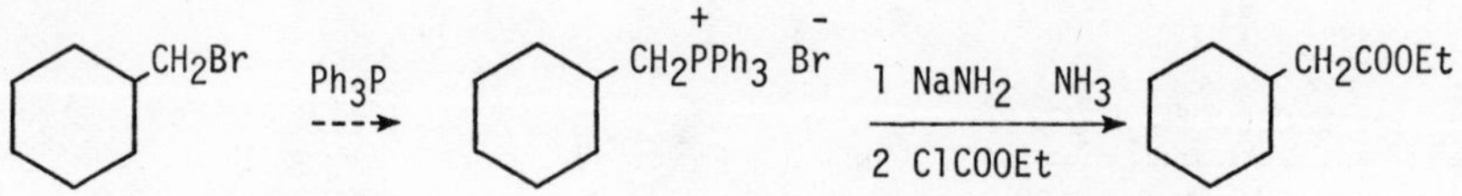

Annalen (1964) <u>674</u> 11

$C_8H_{17}Cl$ $\xrightarrow[\text{2 } I_2 \quad \text{N-methylpyrrolidone} \quad \text{EtOH}]{\text{1 } Na_2Fe(CO)_4}$ $C_8H_{17}COOEt$ 89%

JACS (1973) <u>95</u> 249

t-BuCl $\xrightarrow[\text{SO}_2 \quad \text{CH}_2\text{Cl}_2]{\text{CO} \quad \text{SbCl}_5 \quad \text{EtOH}}$ t-BuCOOEt   66%

Synthesis (1971) 639
Bull Chem Soc Jap (1971) <u>44</u> 2001

RBr $\xrightarrow{\text{CuOAc   Pyr}}$ ROAc   (R is 1ry, 2ry or 3ry, alkyl, allylic or vinylic)

Tetr Lett (1972) 491

Related methods: Carboxylic Acids from Halides (Section 25)

Also via: Section
        Acetylenic esters   306
         Olefinic esters   362

Section 116   Esters from Hydrides

This section contains examples of the reaction RH $\longrightarrow$ RCOOR' or R'COOR
(R=alkyl, allyl, aryl etc.) and ArH $\longrightarrow$ Ar-X-COOR (X=alkyl chain)

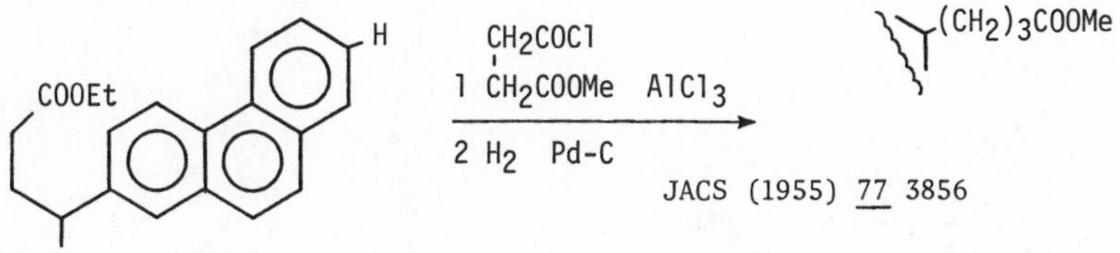

COOEt

CH_2COCl
1 CH_2COOMe $AlCl_3$

2 H_2 Pd-C

$(CH_2)_3COOMe$

JACS (1955) 77 3856

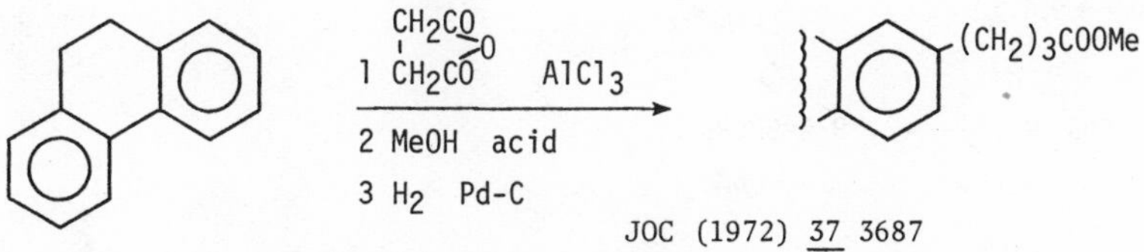

CH_2CO
1 CH_2CO >O $AlCl_3$

2 MeOH acid

3 H_2 Pd-C

$(CH_2)_3COOMe$

JOC (1972) 37 3687

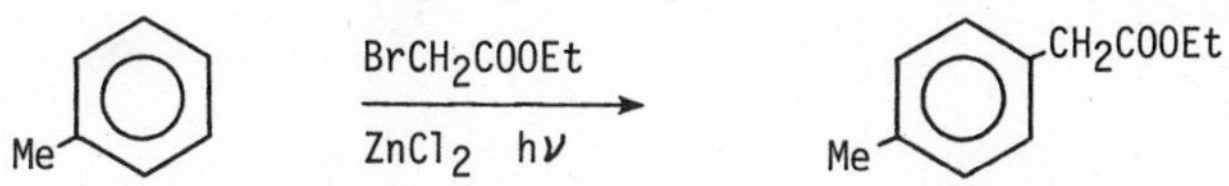

$BrCH_2COOEt$
Me ⎯⎯⎯⎯⎯→
$ZnCl_2$ hν

Me CH_2COOEt

Tetrahedron (1972) 28 211

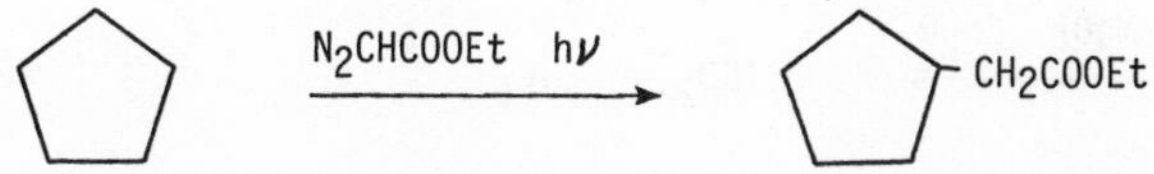

$N_2CHCOOEt$ hν
⎯⎯⎯⎯⎯→

CH_2COOEt

JACS (1956) 78 4947

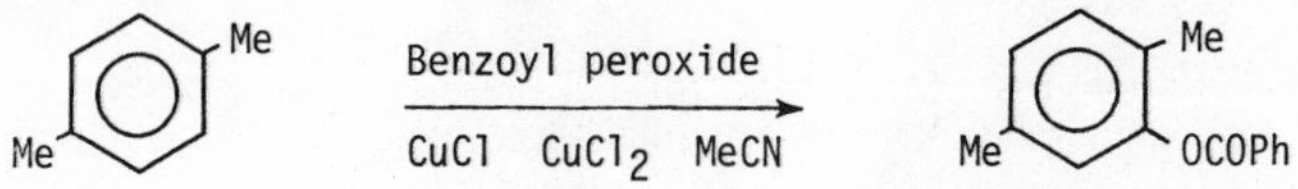

Synthesis (1972) 1

24%

$PhH \xrightarrow[\text{LiOAc \quad HOAc}]{\text{Pd(OAc)}_2 \quad K_2Cr_2O_7} PhOAc$

JOC (1971) 36 1886
Synthesis (1973) 567

$R_2C=CHCHR_2 \longrightarrow R_2C=CHCR_2$
                                $\overset{|}{OAc}$

Hg(OAc)$_2$	Synthesis (1971) 527
PhCOO$_2$Bu-t or MeCOO$_2$Bu-t, CuX	Synthesis (1972) 1
MnO$_2$, HOAc	JCS C (1971) 2355
Co(OAc)$_2$, O$_2$	Can J Chem (1965) 43 1306
HOAc, electrolysis	JACS (1972) 94 7892

Section 117   Esters from Ketones

$Cl(CH_2)_3CO \atop \overset{|}{Me} \xrightarrow[\text{2 Ac}_2\text{O}]{\text{1 C}_5\text{H}_{11}\text{MgBr \quad Et}_2\text{O}} Cl(CH_2)_3\overset{OAc}{\underset{Me}{\overset{|}{C}}}C_5H_{11}$ >75%

JOC (1948) 13 239

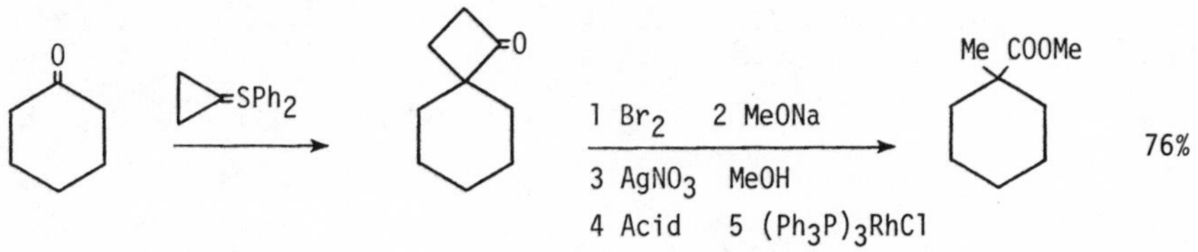
O
=SPh2
=O
1 Br2 2 MeONa
3 AgNO3 MeOH
4 Acid 5 (Ph3P)3RhCl
Me COOMe
76%

JACS (1973) 95 2038

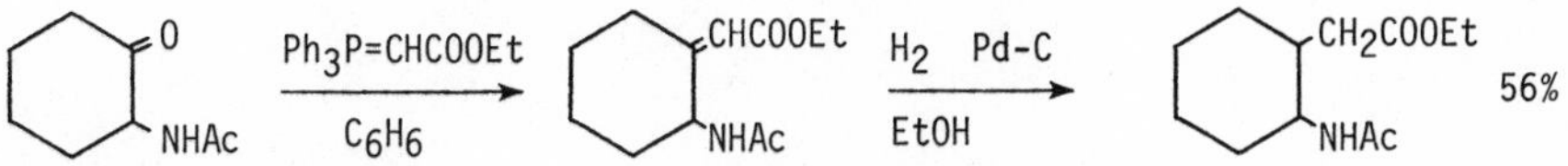
O
NHAc
Ph3P=CHCOOEt
C6H6
CHCOOEt
NHAc
H2 Pd-C
EtOH
CH2COOEt
NHAc
56%

JACS (1962) 84 4951

PhCOMe $\xrightarrow{Tl(NO_3)_3 \quad HClO_4 \quad MeOH}$ $PhCH_2COOMe$ 84%

JACS (1971) 93 4919

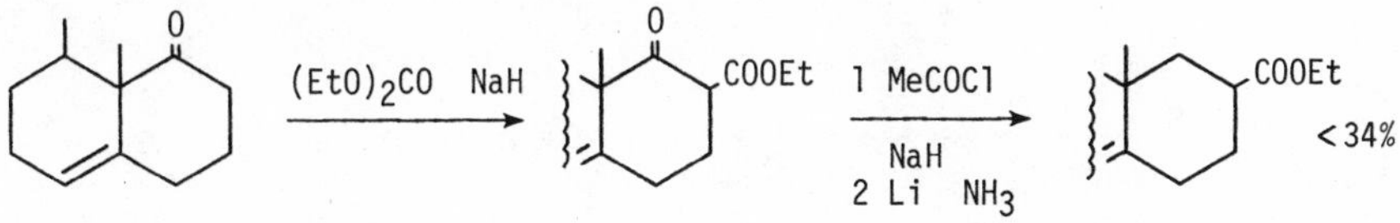
O
(EtO)2CO NaH
O
COOEt
1 MeCOCl
NaH
2 Li NH3
COOEt
<34%

Tetr Lett (1968) 5405

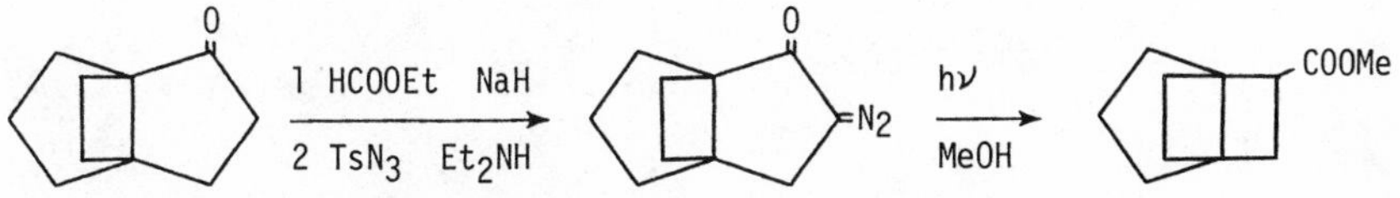
O
1 HCOOEt NaH
2 TsN3 Et2NH
O
=N2
hν
MeOH
COOMe

JACS (1971) 93 2786

$$\text{PhCOCH}_3 \xrightarrow[\text{2 Me}_2\text{SO}_4 \quad \text{NaOH} \quad \text{H}_2\text{O}]{\text{1 i-PrCH}_2\text{CH}_2\text{ONO} \quad \text{HCl} \quad \text{Et}_2\text{O}} \text{PhCOOMe} \qquad >90\%$$

Compt Rend (1942) <u>214</u> 113

Also via: Section
            Carboxylic acids   27
               Ketoesters   360
            Olefinic esters   362

Section 118   Esters from Nitriles

$$\text{RCN} \xrightarrow{\text{HCl} \quad \text{MeOH} \quad \text{Et}_2\text{O}} \text{RCOOMe}$$

Ber (1972) <u>105</u> 1778   84%

Section 119   Esters from Olefins

 $\xrightarrow[\text{2 N}_2\text{CHCOOEt}]{\text{1 BH}_2\text{Cl}\cdot\text{Et}_2\text{O}}$ 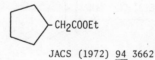 78%

JACS (1972) <u>94</u> 3662

Also via: Section
            Carboxylic acids   29
            Olefinic esters   362

Section 120   Esters from Miscellaneous Compounds
oooooooooooooooooooooooooooooooooooooooo

MeCH=CHCOOCHEt $\xrightarrow[\text{Et}_2\text{O}]{\text{BuMgBr \quad CuCl}}$ MeCHCH$_2$COOCHEt   85%
  | | |
  Me   Bu Me

              Acta Chem Scand (1961) 15 271
              Org React (1972) 19 1

PhCH=CHCOOMe $\xrightarrow{\text{NaBH}_4 \quad \text{NiCl}_2 \quad \text{MeOH}}$ PhCH$_2$CH$_2$COOMe   100%

           Chem Pharm Bull (1971) 19 817

BuCHCOOEt $\xrightarrow{\text{NaBH}_3\text{CN} \quad \text{HMPA}}$ BuCH$_2$COOEt   90%
 |
 Br   Chem Comm (1971) 1097

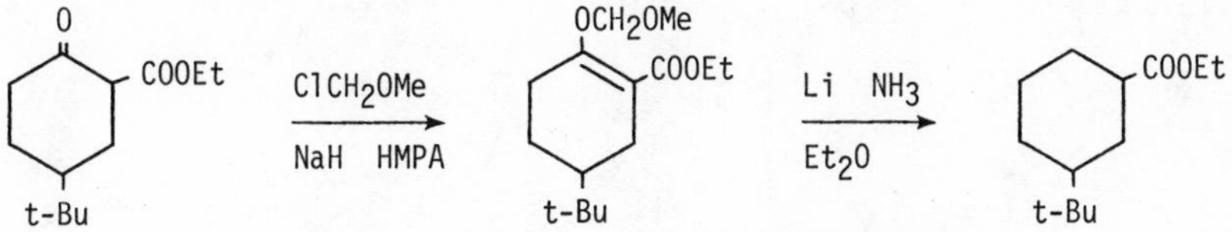

JOC (1970) 35 2601 2597
   (1971) 36 1151

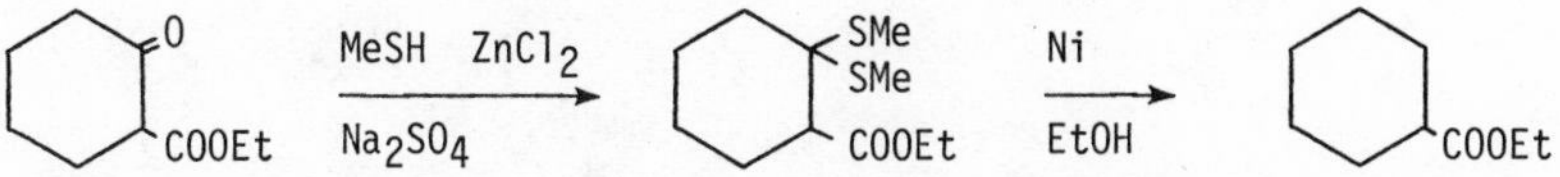

                  32%

      JACS (1950) 72 4296

$$C_6H_{13}COCH_2OH \xrightarrow{\text{Tl(NO}_3)_3 \quad \text{MeOH}} C_6H_{13}COOMe \qquad\qquad 76\%$$

JACS (1973) $\underline{95}$ 1296

Section 120A   Protection of Esters

$$C_{16}H_{33}CH_2COOMe \underset{\text{HgO} \quad \text{BF}_3 \quad \text{THF} \quad H_2O}{\overset{\overset{\displaystyle SCH_2}{\diagup\,|}(Me_2Al)_2\text{-}SCH_2 \quad CH_2Cl_2}{\rightleftharpoons}} C_{16}H_{33}CH=C{\overset{\displaystyle S}{\underset{\displaystyle S}{\diagup}}}]$$

(Stable to HOAc, KOH and LiAlH$_4$)

JACS (1973) $\underline{95}$ 5829

Chapter 9   PREPARATION OF ETHERS AND EPOXIDES

Section 121   Ethers and Epoxides from Acetylenes

No examples

Section 122   Ethers and Epoxides from Carboxylic Acids

No additional examples

Section 123   Ethers and Epoxides from Alcohols and Phenols

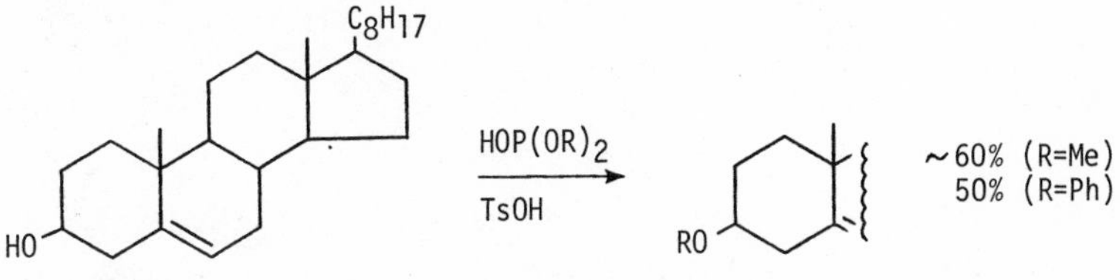

HOP(OR)$_2$ / TsOH

~60% (R=Me)
50% (R=Ph)

JOC (1972) <u>37</u> 912

t-BuOH $\xrightarrow[\text{2 (i-Pr)}_3\text{CCOOMe \quad Me}_2\text{SO}]{\text{1 K}}$ t-BuOMe   ~50%

Tetr Lett (1964) 2969

C$_7$H$_{15}$OH $\xrightarrow{\text{MeI \quad NaH \quad THF}}$ C$_7$H$_{15}$OMe   40%

Tetr Lett (1973) 21

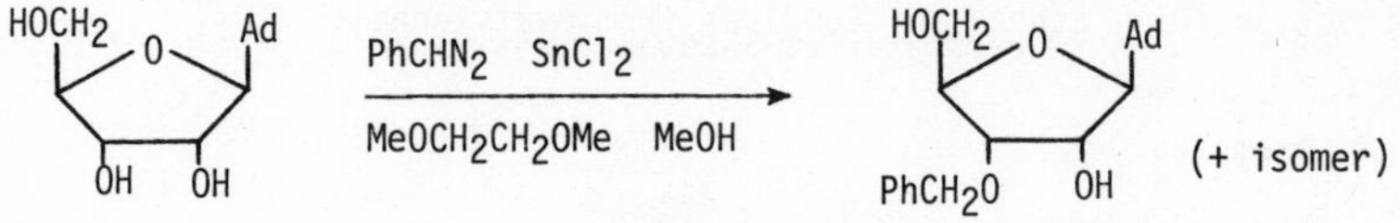

$\xrightarrow[\text{MeOCH}_2\text{CH}_2\text{OMe \quad MeOH}]{\text{PhCHN}_2 \quad \text{SnCl}_2}$ 38%

(+ isomer)

JOC (1972) 37 3398

ROH $\xrightarrow{\text{CH}_2\text{N}_2 \quad \text{HBF}_4 \quad \text{CH}_2\text{Cl}_2}$ ROMe   Org Synth (1961) 41 9

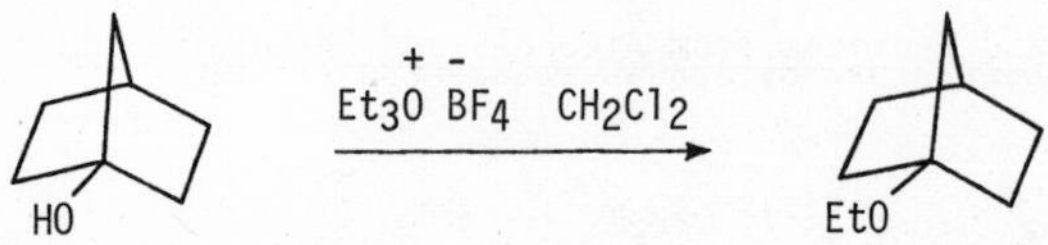

$\xrightarrow{\text{Et}_3\text{O}^+ \text{ BF}_4^- \quad \text{CH}_2\text{Cl}_2}$ 95%

Tetr Lett (1971) 1999

$\xrightarrow[\text{Me}_2\text{CO}]{\text{Me}_2\text{SO}_4 \quad \text{K}_2\text{CO}_3}$

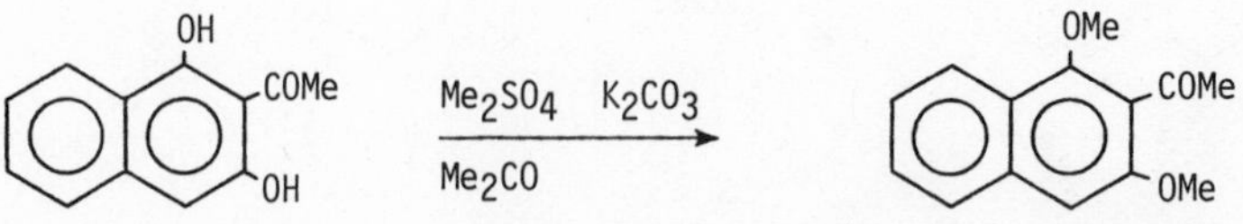

Chem Pharm Bull (1964) 12 312

MeI   NaH   THF

Tetr Lett (1973) 21

86%

1 NaOH

2 MeNNO   MeOCH$_2$CH$_2$OMe
  |
  CONH$_2$

Tetr Lett (1973) 1397

ArOH $\xrightarrow[\Delta]{\overset{+ \quad -}{\text{PhNMe}_3 \text{ OH}}}$ ArOMe   (Gas chrom scale)

J Pharm Sci (1969) 58 370

Related methods: Protection of Alcohols and Phenols (Section 45A)

C$_5$H$_{11}$OH $\xrightarrow[\text{HOAc}]{\text{NaOCl}}$ C$_5$H$_{11}$OCl $\xrightarrow[\text{h}\nu \text{ Freon-113}]{\text{CHCl=CHCl}}$ MeCH(CH$_2$)$_3$OH $\xrightarrow{\text{Lutidine}}$
                                                                          |
                                                                          Cl

JOC (1972) 37 3514

R(CH$_2$)$_4$OH $\xrightarrow{\text{Pb(OAc)}_4}$

Synthesis (1971) 501

$$\underset{\substack{| \;\; | \\ \text{MeCHCHMe}}}{\text{HO OH}} \;\; \xrightarrow{\text{PhCHO}} \;\; \underset{\substack{| \;\; | \\ \text{MeCHCHMe}}}{\overset{\displaystyle \overset{\text{PhCH}}{\diagup \diagdown}}{\overset{\text{O O}}{}}} \;\; \xrightarrow[\substack{\text{2 NaOH HOCH}_2\text{CH}_2\text{OH}}]{\text{1 NBS CCl}_4} \;\; \underset{\text{MeCHCHMe}}{\overset{\diagup\diagdown}{\overset{O}{}}} \qquad 80\%$$

JOC (1973) **38** 1691

$$\underset{\substack{| \;\; | \\ \text{MeCHCHMe}}}{\text{HO OH}} \;\; \xrightarrow[\substack{\text{ion exch resin} \\ \text{C}_6\text{H}_6}]{\text{MeCOCOOH}} \;\; \underset{\substack{| \;\; | \\ \text{MeCHCHMe}}}{\overset{\displaystyle \overset{\text{MeCCOOH}}{\diagup \diagdown}}{\overset{\text{O O}}{}}} \;\; \xrightarrow[\substack{\text{2 Base}}]{\text{1 PCl}_5 \;\; \text{CH}_2\text{Cl}_2} \;\; \underset{\text{MeCHCHMe}}{\overset{\diagup\diagdown}{\overset{O}{}}}$$

JOC (1973) **38** 1173

Chimia (1969) **23** 267

Section 124   Ethers and Epoxides from Aldehydes

$$\text{PhCHO} \;\; \xrightarrow{\text{Ph}_2\text{PONa } 180°} \;\; \underset{\text{PhCHCHPh}}{\overset{\diagup\diagdown}{\overset{O}{}}} \qquad 88\%$$

JCS Perkin I (1972) 1113

$$\text{C}_7\text{H}_{15}\text{CHO} \;\; \xrightarrow[\text{THF}]{\text{MeCHBr}_2 \;\; \text{BuLi (or Li)}} \;\; \underset{\text{C}_7\text{H}_{15}\text{CHCHMe}}{\overset{\diagup\diagdown}{\overset{O}{}}} \qquad 80\%$$

Tetrahedron (1972) **28** 3009
(1971) **27** 6109

Section 125   Ethers and Epoxides from Alkyls, Methylenes and Aryls

No examples of the preparation of ethers and epoxides by replacement of alkyl, methylene and aryl groups occur in the literature. For the conversion RH ⟶ ROR' (R,R'=alkyl) see section 131 (Ethers from Hydrides)

Section 126   Ethers and Epoxides from Amides

No additional examples

Section 127   Ethers and Epoxides from Amines

No additional examples

Section 128   Ethers and Epoxides from Esters

No additional examples

Section 129   Ethers and Epoxides from Ethers and Epoxides

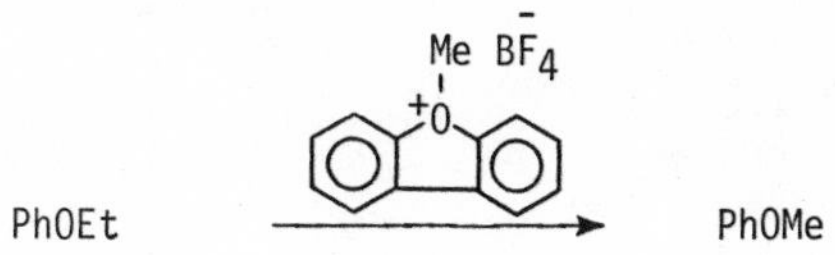

PhOEt $\xrightarrow{\hspace{2cm}}$ PhOMe

Chem Comm (1972) 315

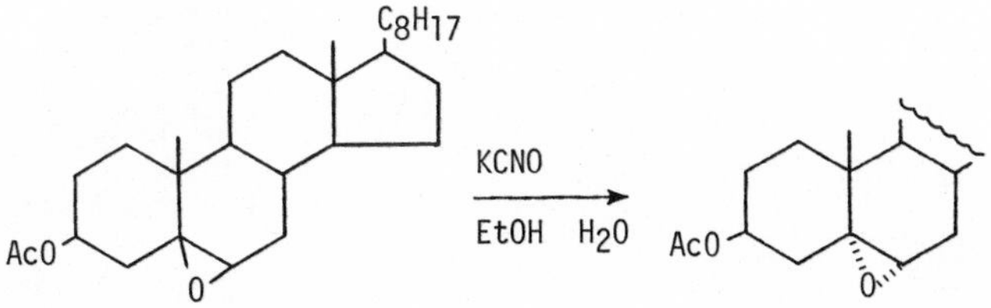

48%

Synthesis (1971) 32

Section 130   Ethers from Halides

PhBr $\xrightarrow[\begin{array}{l}\text{2 CdCl}_2 \\ \text{3 ClCH}_2\text{OMe}\end{array}]{\text{1 Mg Et}_2\text{O}}$ PhCH$_2$OMe

JOC (1973) <u>38</u> 3189

PhBr $\xrightarrow[\text{2 (BuO)}_2\text{P(O)OOBu-t}]{\text{1 Mg Et}_2\text{O}}$ PhOBu-t   95%

JOC (1972) <u>37</u> 2267

Section 131   Ethers from Hydrides

This section lists examples of the reaction RH ⟶ ROR

R$_2$C=CHCH$_2$R $\xrightarrow[\text{MeOH}]{\text{Electrolysis   Et}_4\text{N}^+\text{ OTs}^-}$ R$_2$C=CHCHR
                                                              |
                                                             OMe

JACS (1972) <u>94</u> 7892

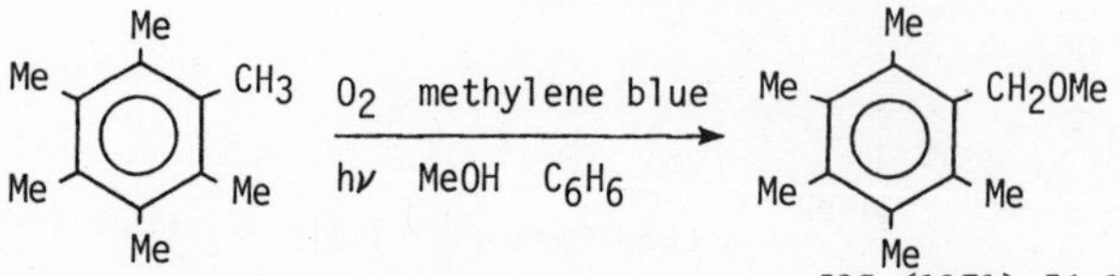

O$_2$ methylene blue
hν MeOH C$_6$H$_6$

JOC (1971) <u>36</u> 1765   31%

Section 132   Epoxides from Ketones

R$_2$CO $\xrightarrow[\text{2 BuLi}]{\text{1 LiCHCl}_2\text{ Et}_2\text{O}}$ R$_2$C$\overset{\displaystyle O}{\overset{\displaystyle \diagup\!\diagdown}{-}}$CHBu

Ber (1973) <u>106</u> 2620

R_2CO $\xrightarrow{\text{MeCHBr}_2 \quad \text{BuLi} \quad \text{THF}}$ $R_2\overset{\overset{\displaystyle O}{\diagup\diagdown}}{C}\text{-CHMe}$

Tetrahedron (1972) <u>28</u> 3009

(1971) <u>27</u> 6109

$(t\text{-Bu})_2CO$ $\xrightarrow[\substack{\text{2 Me}_3\text{O}^+ \text{ BF}_4^- \quad \text{CH}_2\text{Cl}_2 \\ \text{3 NaOH} \quad \text{H}_2\text{O}}]{\text{1 PhSCH}_2\text{Li} \quad \text{THF}}$ $(t\text{-Bu})_2\overset{\overset{\displaystyle O}{\diagup\diagdown}}{C}\text{-CH}_2$

JACS (1973) <u>95</u> 3429

Section 133   Ethers and Epoxides from Nitriles
°°°

No examples

Section 134   Ethers and Epoxides from Olefins
°°°

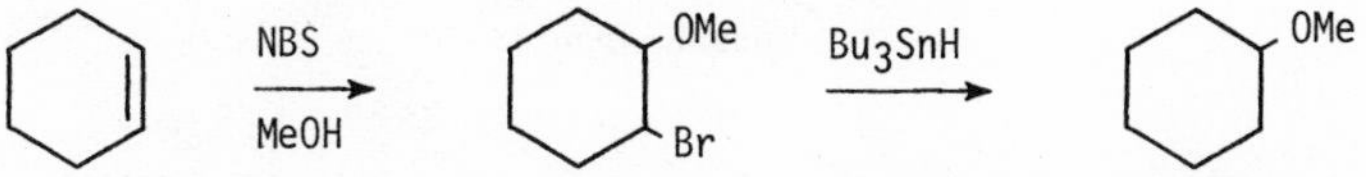

Synthesis (1972) 483

42%

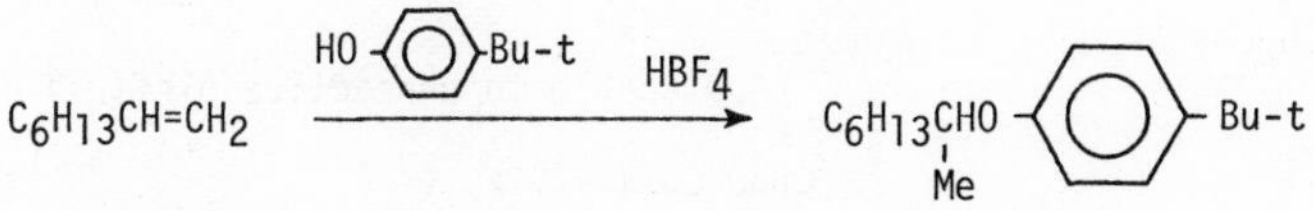

JCS <u>C</u> (1971) 1174

$Me_2C=CH_2$ $\xrightarrow[\text{H}_2\text{O} \quad \text{THF}]{\text{Tl(OAc)}_3 \quad \text{HOAc}}$ $Me_2\overset{O}{\overset{/\backslash}{C}}\text{-}CH_2$ < 82%

JOC (1971) 36 1154

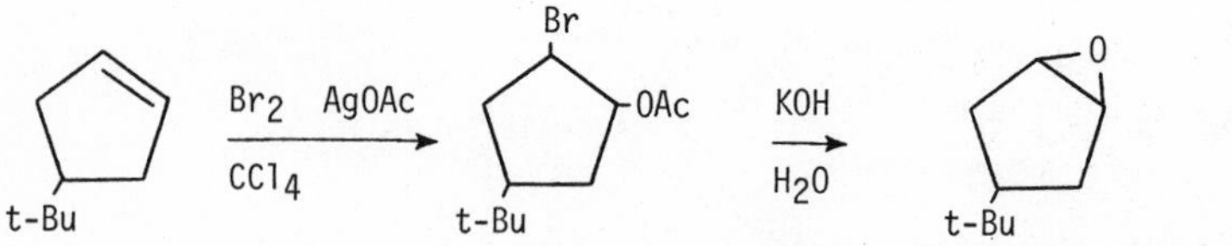

Tetrahedron (1972) 28 3475

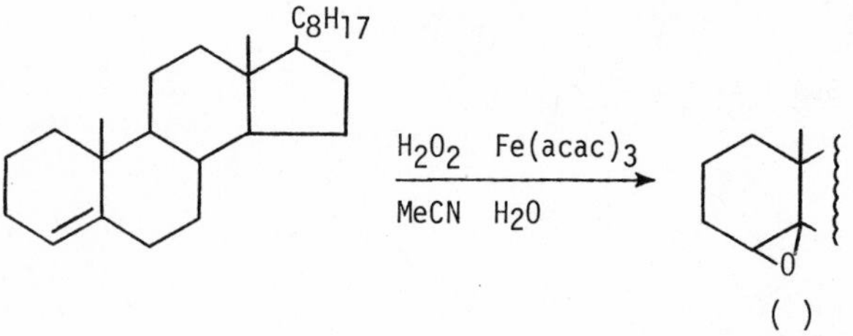

$\xrightarrow[\text{MeCN} \quad \text{H}_2\text{O}]{\text{H}_2\text{O}_2 \quad \text{Fe(acac)}_3}$

74%

( )

Tetr Lett (1973) 4359

$C_6H_{13}CH=CHMe$ $\xrightarrow{\text{[Mo(O}_2\text{)}_2\text{O]}\cdot\text{HMPA}}$ $C_6H_{13}\overset{O}{\overset{/\backslash}{C}}HCHMe$

Tetrahedron (1970) 26 37

$C_6H_{13}CH=CH_2$ $\xrightarrow[\substack{\text{4,4'-thiobis-(6-t-butyl-3-methylphenol)} \\ \text{ClCH}_2\text{CH}_2\text{Cl}}]{\text{m-Chloroperbenzoic acid}}$ $C_6H_{13}\overset{O}{\overset{/\backslash}{C}}HCH_2$

(Applicable to unreactive olefins)

Chem Comm (1972) 64

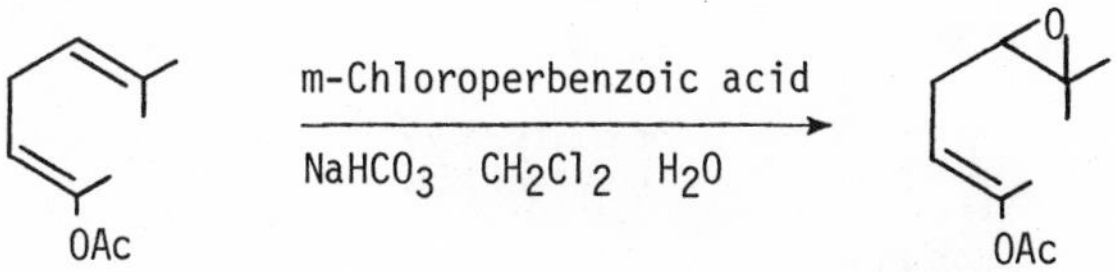

m-Chloroperbenzoic acid

———————————————————→

NaHCO$_3$ CH$_2$Cl$_2$ H$_2$O

OAc

86%

OAc

(Method for acid sensitive compounds)

JOC (1973) $\underline{38}$ 2267

Peracid   O
                ╱ ╲
R$_2$C=CR$_2$ ———————→   R$_2$C-CR$_2$

p-Methoxycarbonylperbenzoic acid JOC (1972) $\underline{37}$ 4210

disuccinoyl peroxide Synthesis (1973) 156

o-Sulfoperbenzoic acid Tetr Lett (1971) 691

Preparation of RCOO$_2$H from RCOCl + (EtO)$_2$P(O)OAg   JOC (1971) $\underline{36}$ 2162

Section 135   Epoxides from Miscellaneous Compounds

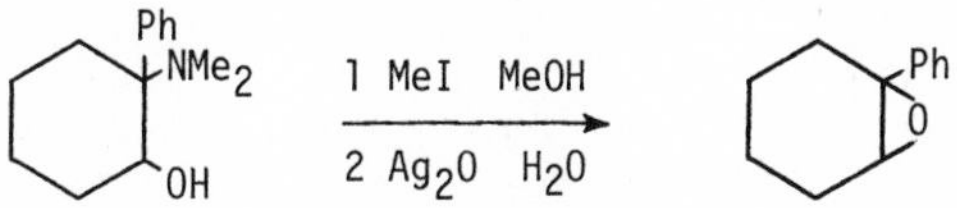

Ph
  NMe$_2$ 1 MeI MeOH   Ph
              ———————————————→   O
  OH   2 Ag$_2$O H$_2$O

JOC (1968) $\underline{33}$ 4045

Chapter 10   PREPARATION

OF

HALIDES

AND SULFONATES

Section 136   Halides and Sulfonates from Acetylenes

No additional examples

Section 137   Halides from Carboxylic Acids

RCOOH $\xrightarrow{\text{Br}_2 \quad \text{HgO} \quad \text{CCl}_4}$ RBr

Org Synth (1971) 51 106
JOC (1972) 37 669 664

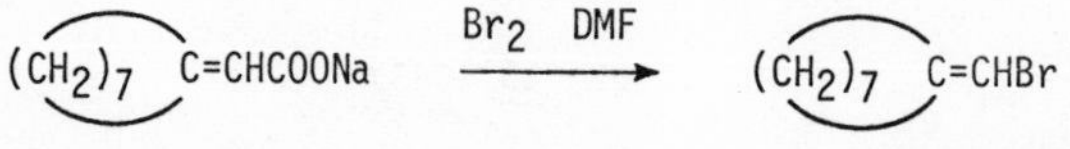

$(CH_2)_7$ C=CHCOONa $\xrightarrow{\text{Br}_2 \quad \text{DMF}}$ $(CH_2)_7$ C=CHBr   45%

JOC (1965) 30 2208
JCS C (1971) 2352

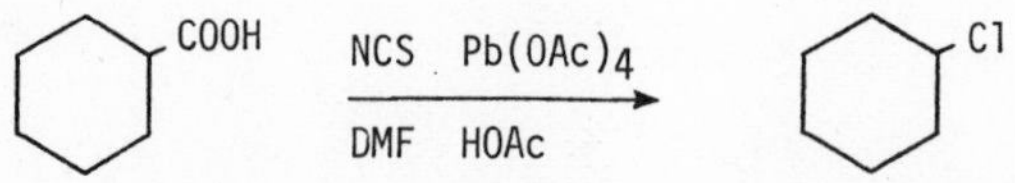

[cyclohexane-COOH] $\xrightarrow[\text{DMF} \quad \text{HOAc}]{\text{NCS} \quad \text{Pb(OAc)}_4}$ [cyclohexane-Cl]   83%

Synthesis (1973) 493

$$RCOOH \quad \xrightarrow{\text{I}_2 \quad \text{Pb(OAc)}_4} \quad RI$$

Org React (1972) <u>19</u> 279

$$C_9H_{19}COOH \quad \xrightarrow[\text{2 Br}_2 \quad \text{CCl}_4]{\text{1 TlOEt \quad pet ether}} \quad C_9H_{19}Br \qquad\qquad 89\%$$

JOC (1969) <u>34</u> 1172

Section 138   Halides and Sulfonates from Alcohols
○○○

$$ROH \quad \xrightarrow{\text{(PhO)}_3P \cdot \text{MeI}} \quad RI$$

Org Synth (1971) <u>51</u> 44

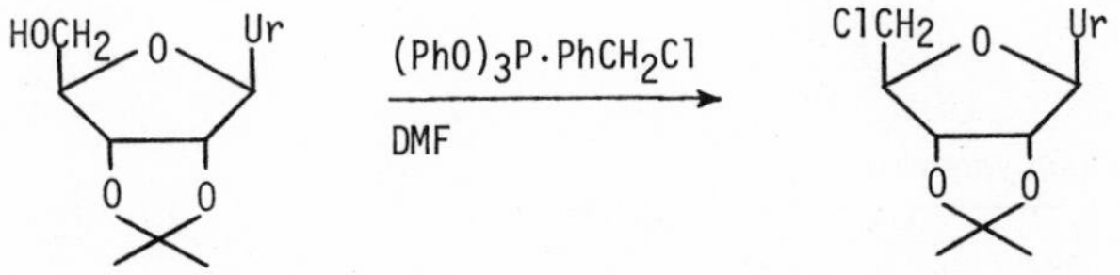

$$\xrightarrow[\text{DMF}]{\text{(PhO)}_3P \cdot \text{PhCH}_2\text{Cl}}$$

70%

JOC (1972) <u>37</u> 2289

$$ROH \quad \xrightarrow{\text{Ph}_3P \quad \text{CCl}_4} \quad RCl$$

JOC (1972) <u>37</u> 1466

BuOH $\xrightarrow{\text{Ph}_3\text{P·Cl}_2}$ BuCl   99%

JACS (1964) <u>86</u> 964

ROH $\xrightarrow{\text{NBS   Ph}_3\text{P}}$ RBr

Chem Comm (1970) 602
Carbohydrate Res (1971) <u>18</u> 342
Tetr Lett (1973) 3937

Ph$_2$CHOH $\xrightarrow{\substack{\text{CH}_2\text{CO} \\ \quad\quad\text{N-SMe}_2\ \text{Cl} \\ \text{CH}_2\text{CO} \\ \text{CH}_2\text{Cl}_2}}$ Ph$_2$CHCl   > 95%

(Specific for allylic and benzylic alcohols)
Tetr Lett (1972) 4339

$-$CH$_2$OH $\xrightarrow{\text{Me}_2\text{S·Br}_2}$ $-$CH$_2$Br   78%

Chem Comm (1973) 212

C$_5$H$_{11}$OH $\xrightarrow{\text{Trichloroisocyanuric acid}}$ C$_5$H$_{11}$Cl   44%

JOC (1970) <u>35</u> 3967
Chem Ind (1971) 1416

$\xrightarrow[\substack{+ \quad\quad - \\ \text{PhCH}_2\text{NEt}_3\ \text{Cl}}]{\text{CHCl}_3 \quad \text{NaOH} \quad \text{H}_2\text{O}}$

JACS (1971) <u>93</u> 1820

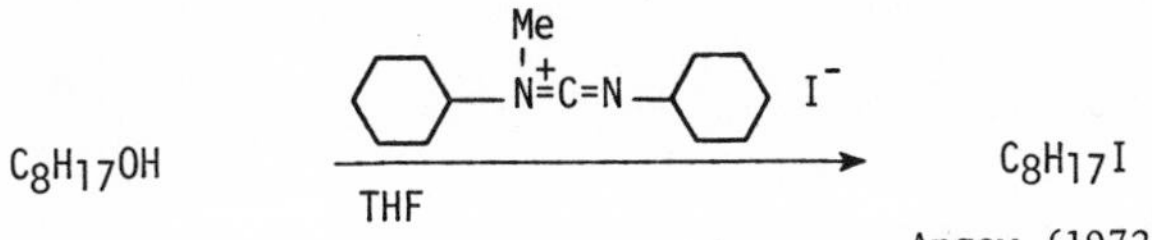

$C_8H_{17}OH$ $\xrightarrow[\text{THF}]{\underset{\text{Me}}{\overset{|+}{N}}=C=N-\bigcirc \ I^-}$ $C_8H_{17}I$ 89%

Angew (1972) <u>84</u> 158
(Internat Ed <u>11</u> 229)

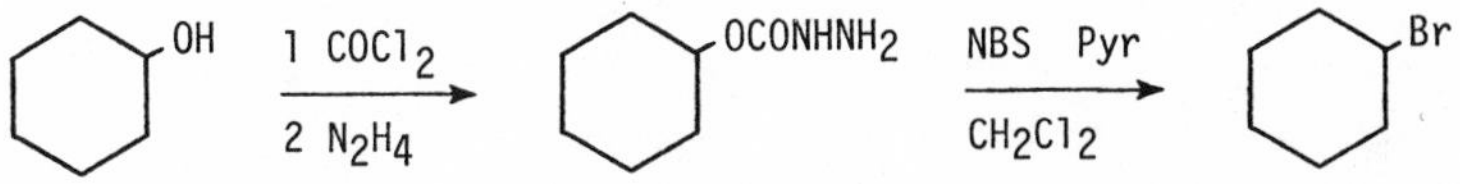

$\bigcirc-OH$ $\xrightarrow[\text{2 N}_2\text{H}_4]{\text{1 COCl}_2}$ $\bigcirc-OCONHNH_2$ $\xrightarrow[\text{CH}_2\text{Cl}_2]{\text{NBS Pyr}}$ $\bigcirc-Br$ 42%

(+38% dibromide)

Chem Comm (1971) 1112

i-PrOH $\xrightarrow{\text{Br}_2 \quad \text{SO}_2}$ i-PrBr   38%

Synthesis (1971) 639

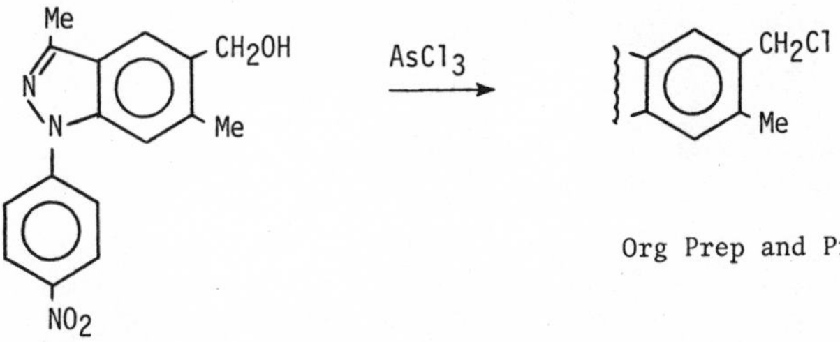

$\xrightarrow{\text{AsCl}_3}$ 68%

Org Prep and Procedures (1970) <u>2</u> 189

$Pr_2C=CHCH_2OH$ $\xrightarrow{\text{MsCl} \cdot \text{ LiCl DMF}}$ $Pr_2C=CHCH_2Cl$ 93%

Tetrahedron (1971) <u>27</u> 5979
JOC (1971) <u>36</u> 3044

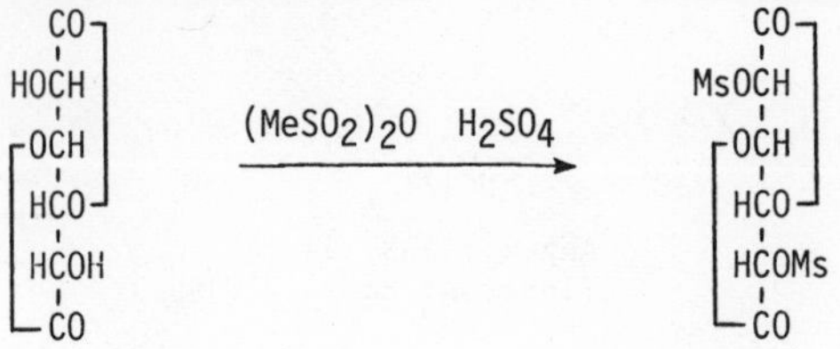

100%

JCS (1953) 1225
Chem Ind (1971) 702

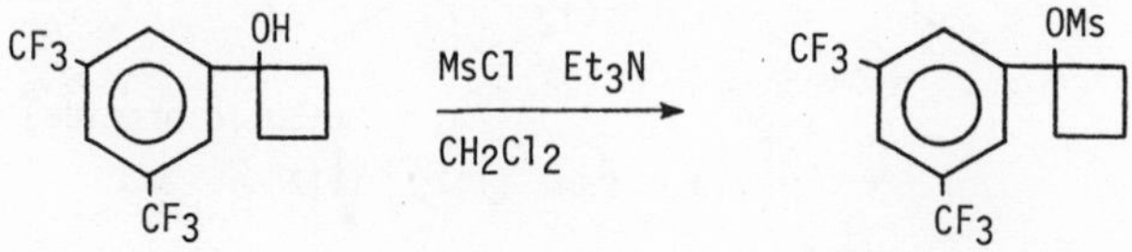

92%

JOC (1970) 35 3195

Section 139   Halides and Sulfonates from Aldehydes
○○

No additional examples

Section 140   Halides and Sulfonates from Alkyls
○○○

No additional examples

For the conversion RH ⟶ RHal see section 146 (Halides from Hydrides)

Section 141   Halides and Sulfonates from Amides
○○○

No additional examples

Section 142   Halides from Amines

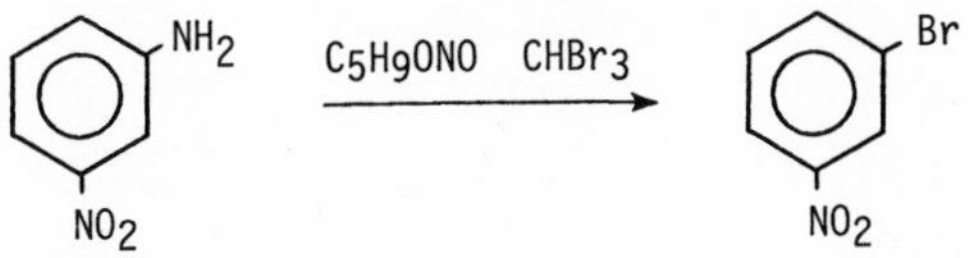

JCS C (1966) 1249

53%

RNMe$_2$ $\xrightarrow[\text{2 KOH   propylene glycol}]{\text{1 ClCOOEt   C}_6\text{H}_6}$ RCl

J Med Chem (1971) 14 982

Section 143   Halides and Sulfonates from Esters

No additional examples

Section 144   Halides from Ethers

R$_2$O $\xrightarrow{\text{Ph}_3\text{P·Br}_2 \quad \text{PhCN}}$ RBr   JOC (1972) 37 626

Section 145   Halides from Halides and Sulfonates

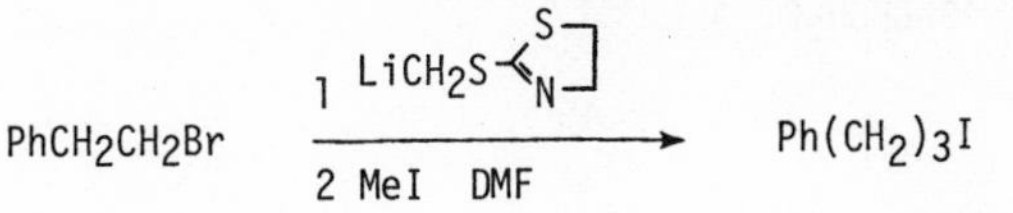

PhCH$_2$CH$_2$Br $\xrightarrow[\text{2 MeI   DMF}]{\text{1 LiCH}_2\text{S}}$ Ph(CH$_2$)$_3$I   <70%

Tetr Lett (1972) 2743

$$Cl(CH_2)_6Cl \quad \xrightarrow{\text{NaI} \quad \text{Me}_2CO} \quad Cl(CH_2)_6I \qquad\qquad 50\%$$

JCS (1950) 2100

$$ROMs \quad \xrightarrow{\text{NaI} \quad \text{EtCOMe}} \quad RI$$

Tetrahedron (1971) 27 5987

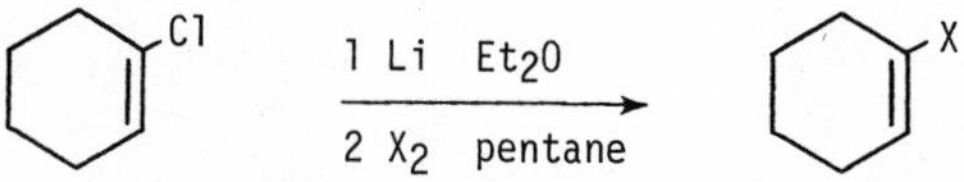

1 Li Et$_2$O

2 X$_2$ pentane

57% (X=Br)
69% (X=I)

Tetrahedron (1972) 28 4883
Ber (1971) 104 2412

$$ArBr \quad \xrightarrow{\text{CuCl} \quad \text{picoline}} \quad ArCl$$

JACS (1958) 80 1716

Section 146   Halides from Hydrides
○○○○○○○○○○○○○○○○○○○○○○

$$PhH \quad \xrightarrow[\text{BF}_3 \quad \text{P}_2\text{O}_5]{\overset{\displaystyle OH}{\underset{|}{\text{MeCHCH}_2\text{Cl}}}} \quad \overset{\displaystyle Me}{\underset{|}{\text{PhCHCH}_2\text{Cl}}} \qquad\qquad 37\%$$

JACS (1948) 70 1772

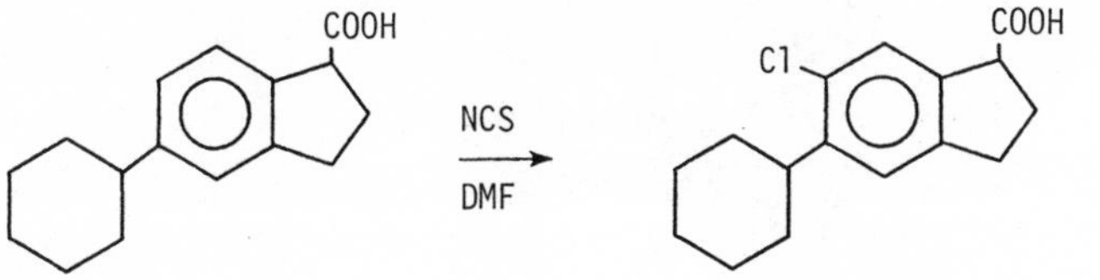

NCS

DMF

53%

J Med Chem (1972) $\underline{15}$ 1297

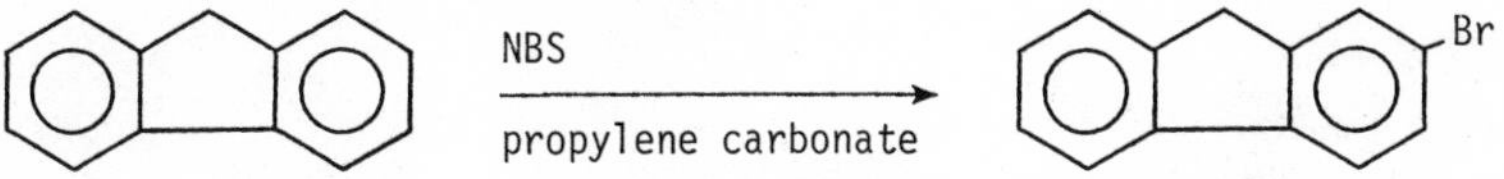

NBS

propylene carbonate

JACS (1958) $\underline{80}$ 4327

NBS H_2SO_4 H_2O

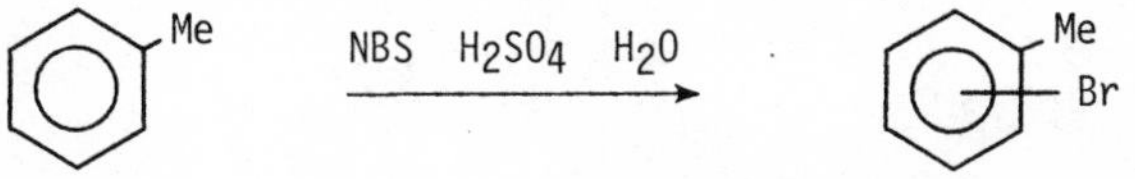

95%

JOC (1965) $\underline{30}$ 304

Other Ar brominating agents: IBr JACS (1938) $\underline{60}$ 256

Br_2-SO_2 Bull Soc Chim Fr (1971) 1785

Br_2-Ag_2SO_4 Rec Trav Chim (1960) $\underline{79}$ 1022

$Tl(OAc)_3$-Br_2 JOC (1972) $\underline{37}$ 88

Ar chlorinating agents: $TiCl_4$-CF_3COO_2H Can J Chem (1972) $\underline{50}$ 1233

Ar iodinating agents: $Tl(OCOCF_3)_3$-KI JACS (1971) $\underline{93}$ 4841

I_2-HIO_4 Org Synth (1971) $\underline{51}$ 94

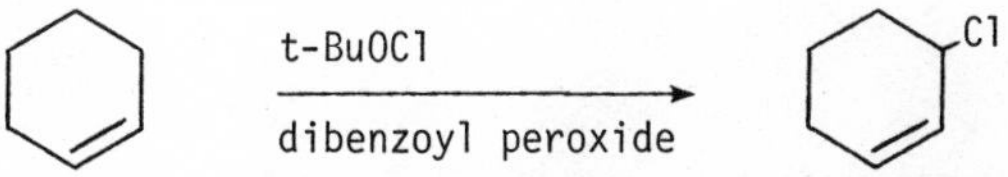

t-BuOCl

dibenzoyl peroxide

77%

Helv (1957) $\underline{40}$ 130

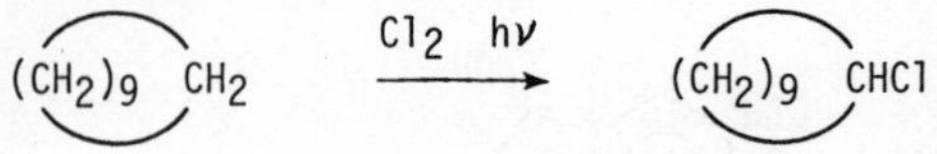

$(CH_2)_9$ CH_2 $\xrightarrow{Cl_2 \quad h\nu}$ $(CH_2)_9$ $CHCl$

JOC (1967) $\underline{32}$ 510

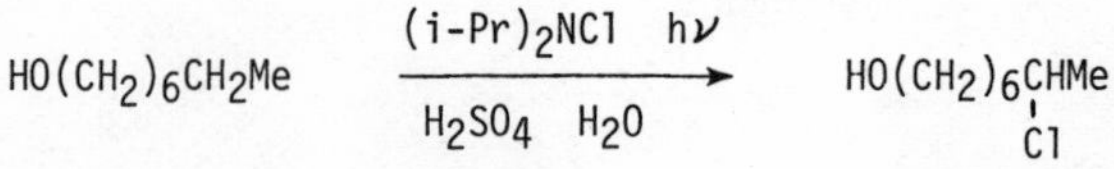

$HO(CH_2)_6CH_2Me$ $\xrightarrow[H_2SO_4 \quad H_2O]{(i\text{-}Pr)_2NCl \quad h\nu}$ $HO(CH_2)_6\underset{Cl}{CHMe}$ 60-80%

JACS (1971) $\underline{93}$ 438
Synthesis (1973) 1

$\underset{t\text{-}BuCMe_2}{\overset{CH_3}{|}}$ $\xrightarrow{Br_2 \quad HgO \quad CCl_4}$ $\underset{t\text{-}BuCMe_2}{\overset{CH_2Br}{|}}$ 32%

Can J Chem (1972) $\underline{50}$ 3109

Section 147   Halides from Ketones
○○○○○○○○○○○○○○○○○○○○

No additional examples

For the reaction C=O —▸ C(Hal)$_2$ see section 368 (Halide-Halide)

Section 148   Halides and Sulfonates from Nitriles
○○○○○○○○○○○○○○○○○○○○○○○○○○○○○○○○○

No examples

Section 149   Halides from Olefins
○○○○○○○○○○○○○○○○○○○○

For the conversion of olefins to dihalides see section 368 (Halide-Halide)
For allylic halogenation see section 146 (Halides from Hydrides)

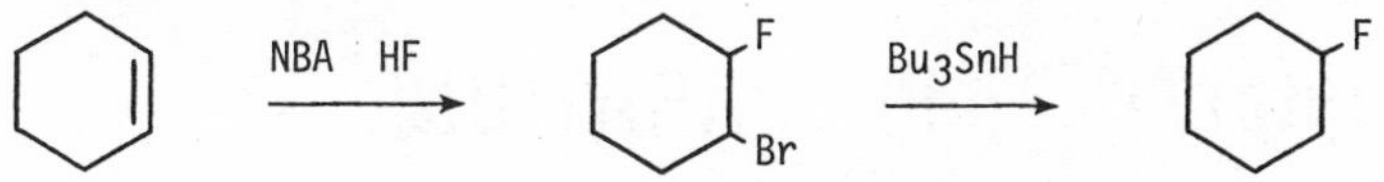

Synthesis (1971) 255

i-PrCH=CHMe $\xrightarrow{\begin{array}{c}1\ 9\text{-BBN}\quad THF \\ \hline 2\ Br_2\quad CH_2Cl_2\end{array}}$ i-PrCH$_2$CHMe   74%
                                                    |
                                                    Br

J Organometallic Chem (1971) 26 C51

BuCH=CH$_2$ $\xrightarrow{\begin{array}{c}1\ B_2H_6\quad THF \\ \hline 2\ CH_2=CHCH_2I\quad air\end{array}}$ BuCH$_2$CH$_2$I

JACS (1971) 93 1508

Section 150   Halides from Miscellaneous Compounds
oooooooooooooooooooooooooooooooooooooooo

CH$_2$=CHCH$_2$Cl ⟍ $\xrightarrow[\text{Rh-Al}_2O_3]{H_2\ (500\ psi)}$ MeCH$_2$CH$_2$Cl   42-58%
MeCH=CHCl ⟋

JOC (1964) 29 194

C$_7$H$_{15}$SH $\xrightarrow{\text{ClCOSCl}\quad Ph_3P}$ C$_7$H$_{15}$Cl

Chem Comm (1972) 773

Chapter 11 PREPARATION OF HYDRIDES

This chapter lists hydrogenolysis and related reactions by which functional groups are replaced by hydrogen, e.g. $RCH_2X \longrightarrow RCH_2\text{-}H$ or $R\text{-}H$

Section 151   Hydrides from Acetylenes

No examples of the reaction $RC\equiv CR \longrightarrow RH$ occur in the literature

Section 152   Hydrides from Carboxylic Acids

This section lists examples of the decarboxylation of acids, $RCOOH \longrightarrow RH$

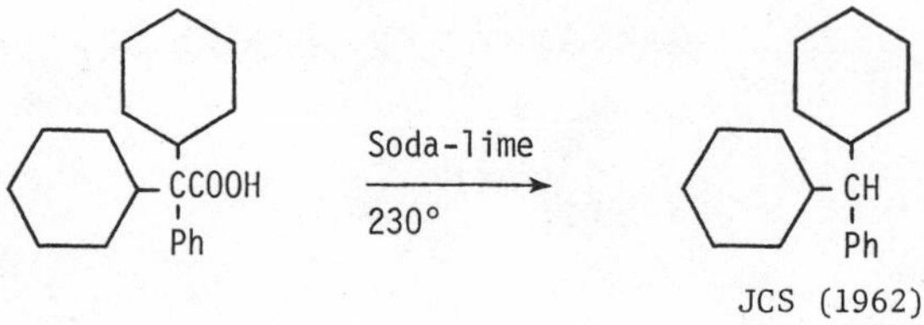

Soda-lime
230°

JCS (1962) 1445
JACS (1950) 72 1849

Section 153   Hydrides from Alcohols and Phenols

This section lists examples of the hydrogenolysis of alcohols and phenols, $ROH \longrightarrow RH$

Et_3COH $\xrightarrow[CH_2Cl_2]{Ph_3SiH \quad CF_3COOH}$ Et_3CH 78%

(3ry only)

JOC (1971) 36 758

$C_{10}H_{21}OH$ $\xrightarrow[HMPA]{(PhO)_3P\cdot MeI \quad NaBH_3CN}$ $C_{10}H_{22}$ 99%

Chem Comm (1971) 1097

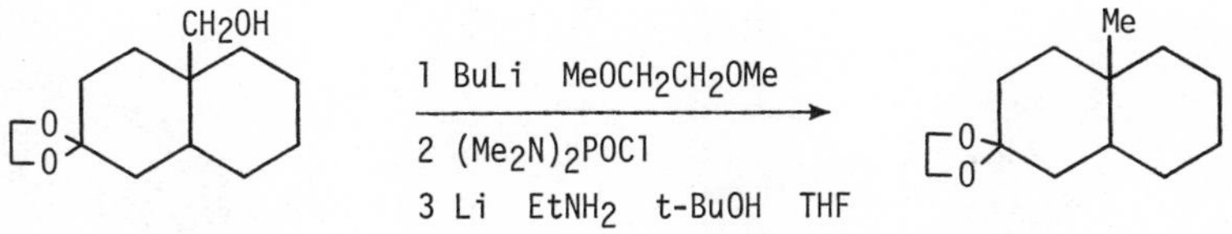

CH2OH
1 BuLi MeOCH2CH2OMe
2 (Me2N)2POCl
3 Li EtNH2 t-BuOH THF
Me

89%

JACS (1972) 94 5098

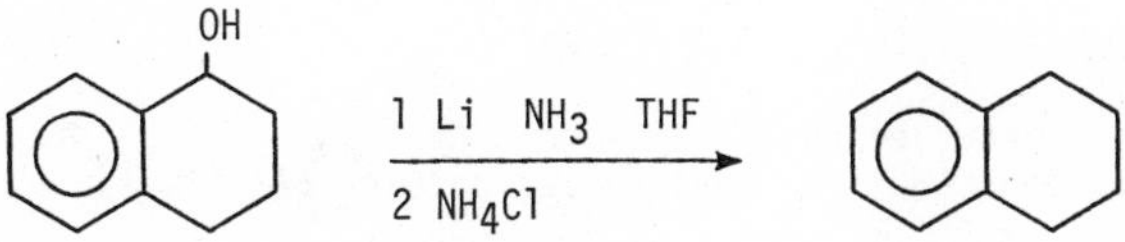

OH
1 Li NH3 THF
2 NH4Cl

Tetr Lett (1971) 1853

OH
OH
PhNCO
OCONHPh
OCONHPh
H2 Pd-C
HOAc

70%

Chem Ind (1973) 187

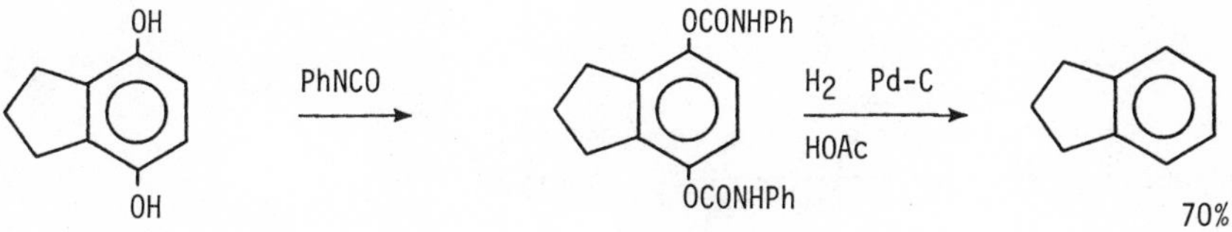

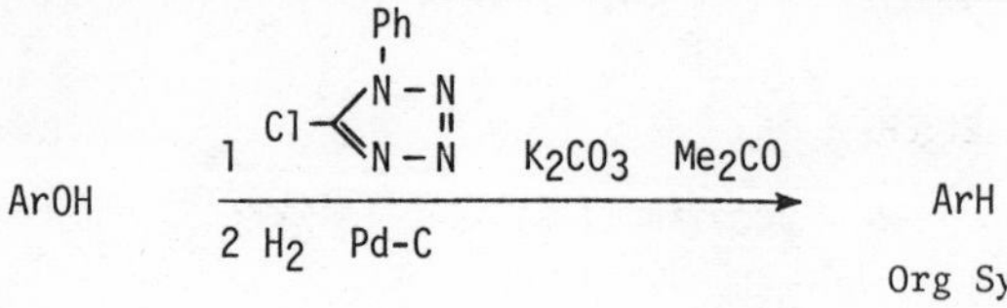

ArOH $\xrightarrow{\begin{array}{c}1 \;\; Cl-\overset{\overset{Ph}{|}}{C}\!\!\diagdown\!\!\overset{N-N}{\underset{N-N}{\parallel}}\;\; K_2CO_3 \;\; Me_2CO \\ \\ 2 \; H_2 \;\; Pd\text{-}C\end{array}}$ ArH

Org Synth (1971) $\underline{51}$ 82

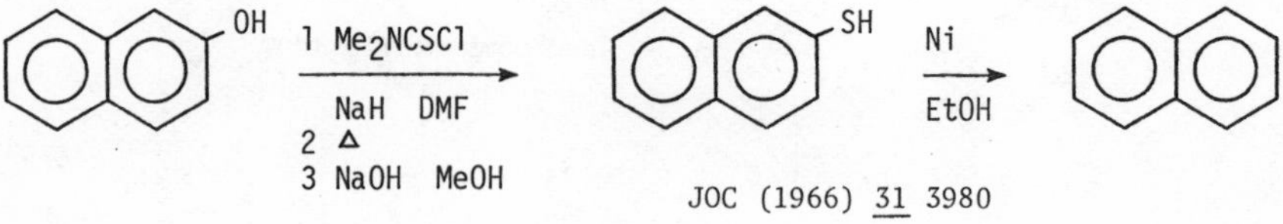

1 Me_2NCSCl

NaH   DMF
2 Δ
3 NaOH   MeOH

JOC (1966) $\underline{31}$ 3980

Also via: Section
            Halides and Sulfonates   160
                           Ethers   159

Section 154   Hydrides from Aldehydes

This section lists examples of the decarbonylation of aldehydes,
RCHO ⟶ RH. For the conversion RCHO ⟶ RMe etc. see section 64 (Alkyls
from Aldehydes)

RCHO $\xrightarrow{\text{Pd-C} \;\; 190°}$ RH

JOC (1968) $\underline{33}$ 923

$C_6H_{13}CHO \xrightarrow{h\nu \;\; PhCH_2SH \;\; PhCOMe} C_6H_{14}$

JACS (1963) $\underline{85}$ 4010

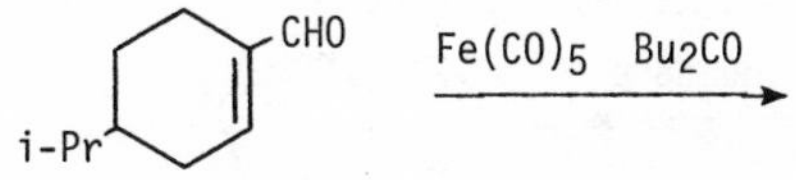

Fe(CO)$_5$ Bu$_2$CO   36%

Tetr Lett (1973) 447

Section 155   Hydrides from Alkyls, Methylenes and Aryls

No additional examples

Section 156   Hydrides from Amides

This section lists examples of the conversion RCONH$_2$ ⟶ RH

NaNH$_2$ NH$_3$

t-BuC≡CCONH$_2$ ⟶   t-BuC≡CH

JCS (1963) 4402

Also via: Carboxylic acids (Section 152)

Section 157   Hydrides from Amines

This section lists examples of the conversion RNH$_2$ ⟶ RH

C$_5$H$_{11}$ONO   THF

PhN=N—⟨◯⟩—NH$_2$ ⟶   PhN=N—⟨◯⟩   66%

JCS Perkin I (1973) 541

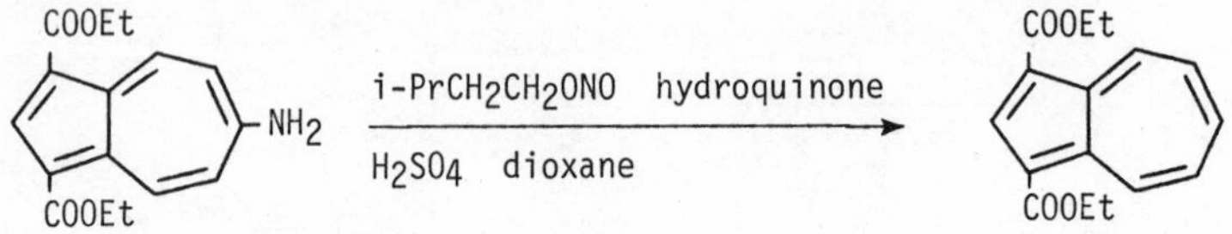

Chem Comm (1973) 605

$ArNH_2 \longrightarrow ArN_2^{+} X^{-} \xrightarrow{Y} ArH$

$Y = RhCl(PPh)_3$ JOC (1971) 36 1725

Bu_3SnH Tetrahedron (1970) 26 4609

Section 158 Hydrides from Esters

This section lists examples of the reaction RCOOR' ⟶ R'H

$PhCOOCHPh_2 \xrightarrow{Bu_2O_2 \quad Bu_3SnH} Ph_2CH_2$

Tetr Lett (1968) 4351

Section 159 Hydrides from Ethers

$PhOMe \xrightarrow{Ni \quad MeOH} PhH$ 62%

Aust J Chem (1963) 16 20

$COOH(CH_2)_{10}OMe \xrightarrow{KOH \quad 360°} COOH(CH_2)_8H$

JCS C (1971) 1840

Section 160   Hydrides from Halides and Sulfonates
000000000000000000000000000000000000000

This section lists the reduction of halides and sulfonates RX $\longrightarrow$ RH

$$C_{12}H_{25}X \xrightarrow{\text{NaBH}_3\text{CN} \quad \text{HMPA}} C_{12}H_{26}$$

96% (X=I)
97% (X=Br)
78% (X=OTs)

Chem Comm (1971) 1097
JOC (1971) 36 1568
JACS (1973) 95 6131

$$C_8H_{17}X \xrightarrow{\text{CuI-LiAlH(OMe)}_3 \quad \text{THF}} C_8H_{18}$$

98% (X=Br)
95% (X=OMs)

JACS (1973) 95 6452

LiEt$_3$BH   THF

99%

JACS (1973) 95 1669

$$PhCH=CHCH_2Cl \xrightarrow[\text{Me}_2\text{SO}]{\text{NaH} \quad \text{TsNHNH}_2} PhCH=CHCH_2NHNHTs \xrightarrow{\text{HOAc}} PhCH_2CH=CH_2 \qquad 60\%$$

Tetr Lett (1969) 871

CuI-LiAlH(OMe)$_3$

THF

99%

JACS (1973) 95 6452

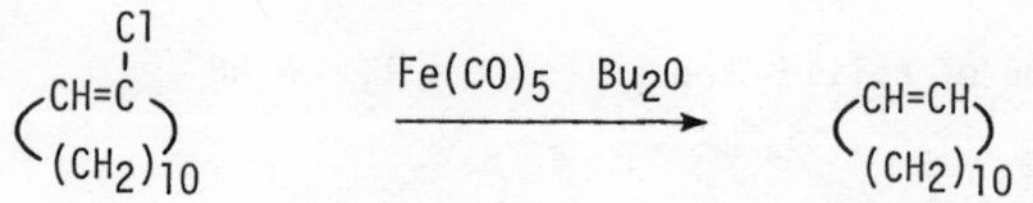

$$\underset{(CH_2)_{10}}{\overset{\overset{\displaystyle Cl}{|}}{CH=C}} \xrightarrow{\quad Fe(CO)_5 \quad Bu_2O \quad} \underset{(CH_2)_{10}}{CH=CH} \qquad 30\%$$

Tetr Lett (1973) 447

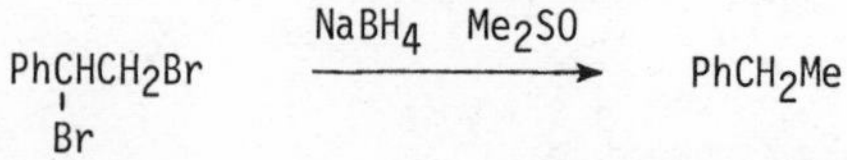

$$\underset{\underset{Br}{|}}{PhCHCH_2Br} \xrightarrow{\quad NaBH_4 \quad Me_2SO \quad} PhCH_2Me \qquad 64\%$$

Tetr Lett (1969) 3495

$$\underset{NMe_3 \ Br}{\overset{Br}{\underset{+}{CH_2Br}}} \xrightarrow{\quad H_2 \quad Pd\text{-}BaSO_4 \quad \atop H_2O} \underset{NMe_3 \ Br}{\overset{Me}{\underset{+}{}}}$$

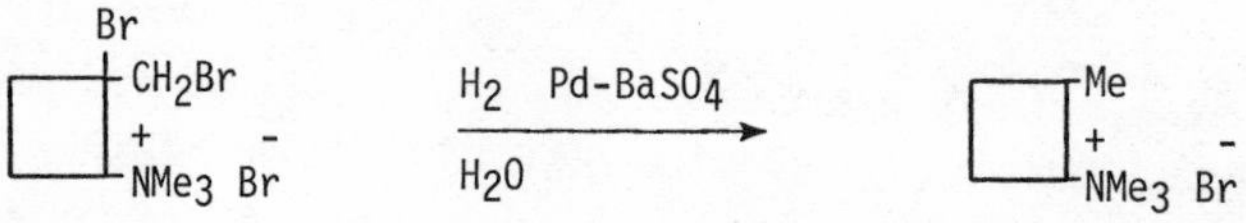

JACS (1948) 70 2517
(1947) 69 2555

Section 161   Hydrides from Hydrides
oooooooooooooooooooooooo

No additional examples

Section 162   Hydrides from Ketones
oooooooooooooooooooooooo

This section lists examples of the conversion $R_2CO \longrightarrow RH$. For the conversion $R_2CO \longrightarrow R_2CH_2$ or R_2CHR' see section 72 (Alkyls and Methylenes from Ketones)

$$\text{[naphthalene-COPh]} \xrightarrow[\text{anisole}]{\text{t-BuOK \quad H}_2\text{O}} \text{[naphthalene]} \qquad 95\%$$

Ber (1971) <u>104</u> 2637

Section 163   <u>Hydrides from Nitriles</u>

This section lists examples of the conversion RCN ⟶ RH

$$\text{RCN} \xrightarrow{\text{Fe(acac)}_3\text{-Na} \quad C_6H_6} \text{RH} \qquad\qquad \begin{array}{l}46\% \ (R=Ph) \\ 100\% \ (R=C_8H_{17})\end{array}$$

JACS (1971) <u>93</u> 7113

$$C_{12}H_{25}CN \xrightarrow{\text{Li} \quad EtNH_2} C_{12}H_{26} \qquad\qquad 35\%$$

JACS (1969) <u>91</u> 2059
Tetr Lett (1968) 1975

$$\text{RCN} \xrightarrow{\text{Na} \quad NH_3} \text{RH}$$

JOC (1972) <u>37</u> 508

Section 164   <u>Hydrides from Olefins</u>

No additional examples

Section 165 Hydrides from Miscellaneous Compounds

This section lists examples of the replacement of miscellaneous functional groups by hydrogen (RX ⟶ RH)

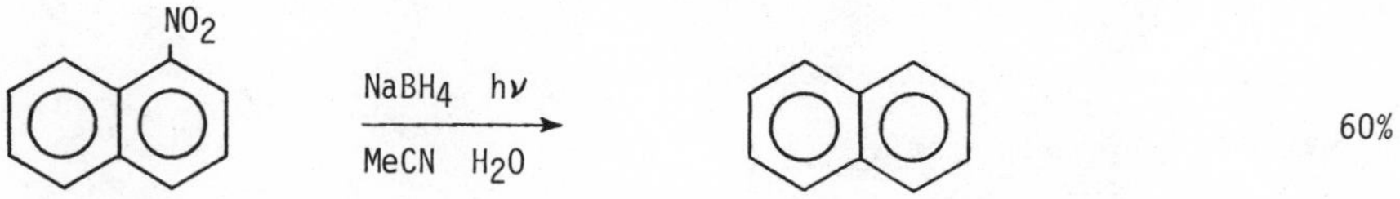

NO2
NaBH4 hν
MeCN H2O
60%

Tetr Lett (1971) 2197

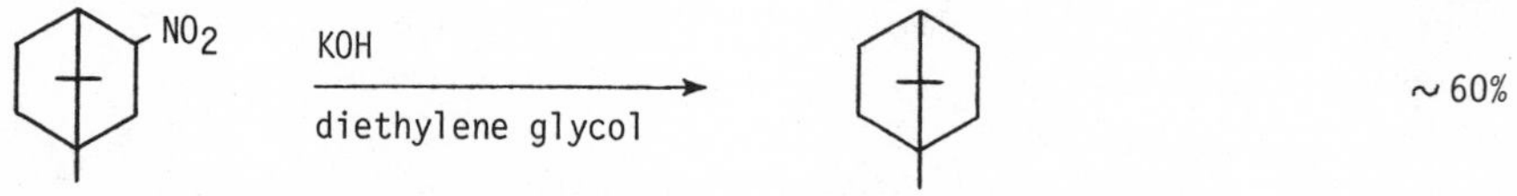

NO2
KOH
diethylene glycol
~60%

Tetr Lett (1971) 3203

Me
Me
SH
Me
Me
Ni EtOH
Me
Me
Me
Me
70%

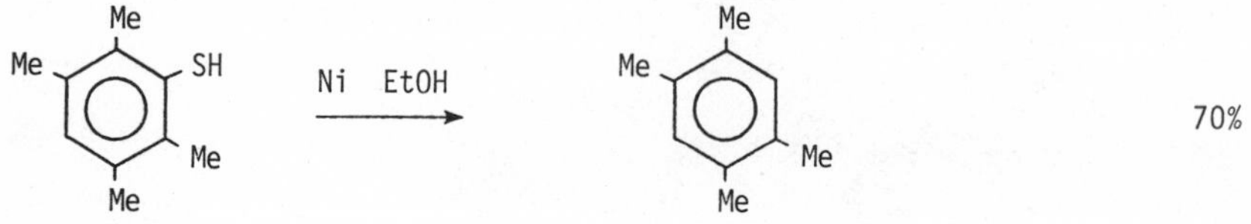

JOC (1966) 31 3980

Chapter 12   PREPARATION
OF
KETONES

BuC≡CH
$$\xrightarrow{\begin{array}{l}1 \text{ RLi} \\ 2 \text{ (C}_8\text{H}_{17}\text{)}_3\text{B} \quad \text{hexane} \\ 3 \text{ Me}_2\text{SO}_4 \quad \text{diglyme} \\ 4 \text{ NaOH} \quad \text{H}_2\text{O}_2 \quad \text{H}_2\text{O}\end{array}}$$
BuCHCOC$_8$H$_{17}$ 85%
$\quad\quad\quad$ |
$\quad\quad\quad$ Me

Chem Comm (1973) 544

RC≡CR
$$\xrightarrow{\begin{array}{l}1 \text{ BH} \\ 2 \text{ H}_2\text{O}_2 \quad \text{NaOH}\end{array}}$$
RCH$_2$COR

BH = BH$_2$Cl   JOC (1973) 38 1617
BH = catechol borane   JACS (1972) 94 4370

PhC≡CC$_6$H$_{13}$
$$\xrightarrow{\begin{array}{l}1 \text{ NaNHNH}_2 \quad \text{N}_2\text{H}_4 \quad \text{Et}_2\text{O} \\ 2 \text{ H}_2\text{SO}_4\end{array}}$$
PhCH$_2$COC$_6$H$_{13}$ 76%

Ber (1966) 99 1843

Also via: Section
$\quad$ Acetylenic ketones   309
$\quad$ Olefinic ketones   374

Section 167   Ketones from Acid Halides

RCOCl $\xrightarrow{R_2'CuLi}$ RCOR'

Tetr Lett (1971) 829
Synthesis (1972) 701

PhCOF $\xrightarrow{Bu_2CuLi \quad Et_2O}$ PhCOBu   87%

JACS (1972) <u>94</u> 5106

Br(CH$_2$)$_{10}$COCl $\xrightarrow{\overset{Me}{(t\text{-}BuOCuCHEt)Li}}$ Br(CH$_2$)$_{10}$COCHEt$\overset{Me}{}$ 83%

Tetr Lett (1973) 1815

C$_5$H$_{11}$COCl $\xrightarrow{\substack{1 \ Na_2Fe(CO)_4 \quad HMPA \quad THF \\ 2 \ EtI}}$ C$_5$H$_{11}$COEt   87%

JACS (1972) <u>94</u> 1788

PhCH=CHCOCl $\xrightarrow{Li[PhCOFe(CO)_4]}$ PhCH=CHCOPh   22%

Synthesis (1971) 55

PhCH=CHCOCl $\xrightarrow{\substack{RhCl(CO)(PPh_3)_2 \quad PhLi \\ Et_2O \quad THF}}$ PhCH=CHCOPh   85%

JACS (1973) <u>95</u> 3040

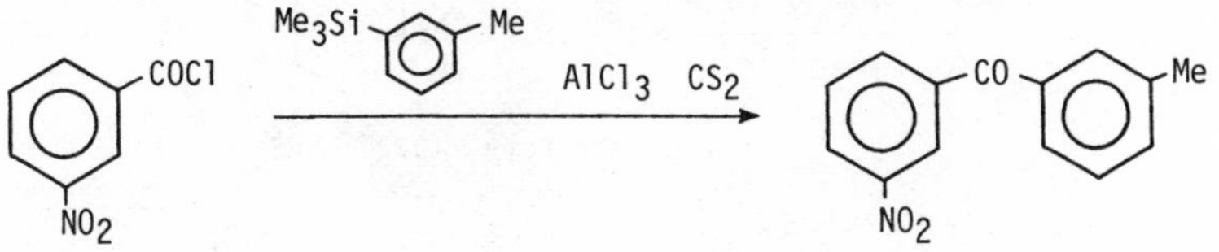

50%

Organometallics in Chem Synth (1971) 1 151

Also via: Olefinic ketones (Section 374)

Section 168 Ketones from Alcohols and Phenols

$$Ph_2CHOH \xrightarrow[CCl_4 \quad H_2SO_4]{C_6H_{13}C{\equiv}CH} Ph_2CHCH_2COC_6H_{13}$$ 23%

JCS Perkin I (1972) 2100

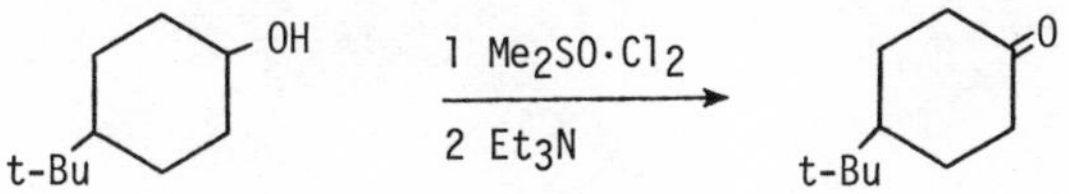

>97%

Tetr Lett (1973) 919

$$C_6H_{13}\underset{\displaystyle Me}{CHOH} \xrightarrow[Et_3N \quad toluene]{\begin{matrix}CH_2CO\\ |\quad\;\; \\ CH_2CO\end{matrix}\!\!>\overset{+}{N}SMe_2\ \overset{-}{Cl}} C_6H_{13}\underset{\displaystyle Me}{CO}$$ 91%

JACS (1972) 94 7586

67%

Tetr Lett (1972) 3285

$C_5H_{11}\overset{|}{C}HOH$ $\xrightarrow[\text{(gas phase)}]{\text{Cu}_2\text{O} \quad 250\text{-}300°}$ $C_5H_{11}\overset{|}{C}O$ 100%
$\quad\overset{|}{Pr}$ $\quad\overset{|}{Pr}$

Tetr Lett (1972) 257

$C_6H_{13}\overset{|}{C}HOH$ $\xrightarrow{\text{CrO}_3\cdot\text{Pyr} \quad \text{HOAc}}$ $C_6H_{13}\overset{|}{C}O$ 68%
$\quad\overset{|}{Me}$ $\quad\overset{|}{Me}$

$\qquad\qquad\qquad\qquad\qquad\qquad\qquad$ Acta Chem Scand (1971) <u>25</u> 1125
CrO_3-3,5-dimethylpyrazole Tetr Lett (1973) 4499

R_2CHOH $\xrightarrow{\text{RuO}_4}$ R_2CO

Tetr Lett (1970) 4003

Ph_2CHOH $\xrightarrow{\text{PhICl}_2 \quad \text{Pyr} \quad \text{CHCl}_3}$ Ph_2CO 49%

Tetr Lett (1973) 3635

$Ph\overset{|}{C}HOH$ $\xrightarrow[C_6H_6]{Ph_2C=NBr \quad h\nu}$ $Ph\overset{|}{C}O$ 94%
$\overset{|}{Me}$ $\overset{|}{Me}$

JOC (1970) <u>35</u> 4245

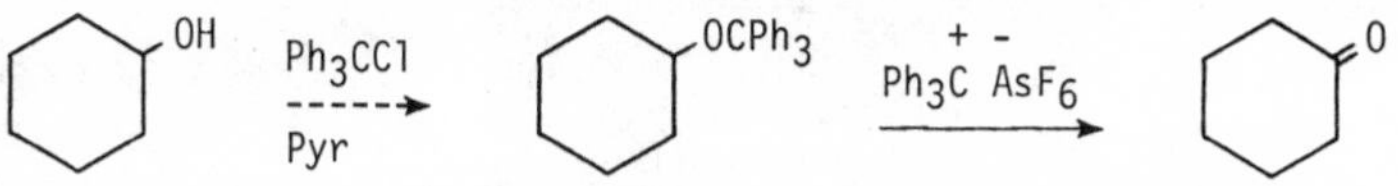

JOC (1973) 38 625

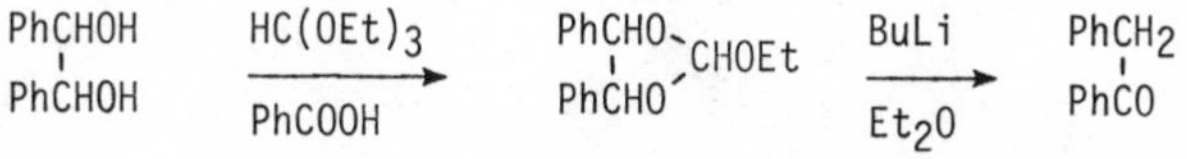

Aust J Chem (1968) 21 2013

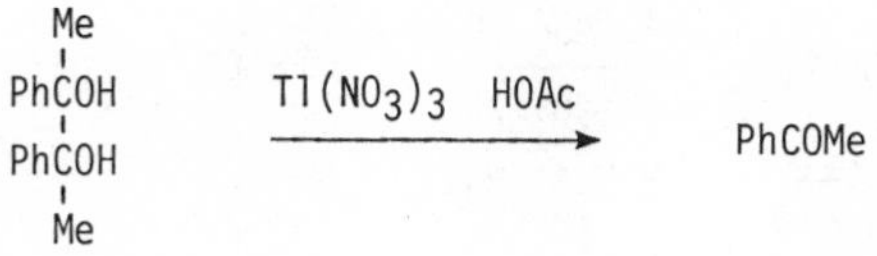

JOC (1972) 37 4204

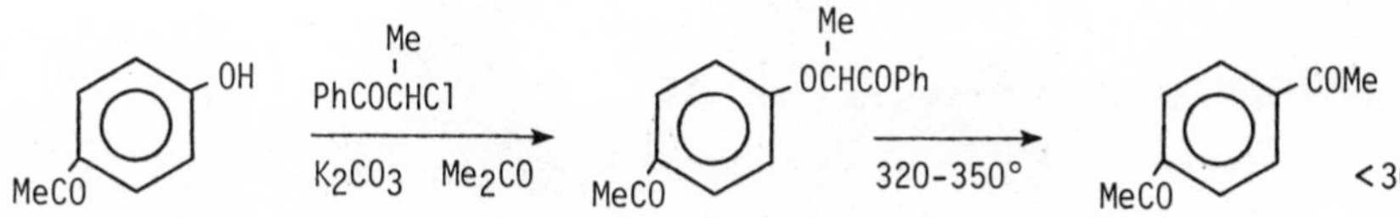

JOC (1961) 26 4308

Related methods: Aldehydes from Alcohols and Phenols (Section 48)

Section 169   Ketones from Aldehydes

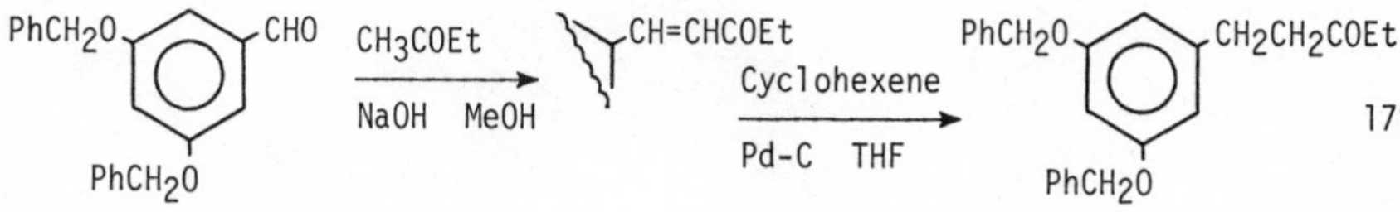

JOC (1972) 37 2204

PhCHO $\xrightarrow[\substack{\text{HOAc} \\ \text{2 NaH MeI} \\ \text{DMF}}]{\text{1 PhSH ZnCl}_2}$ PhC(SPh)$_2$ $\xrightarrow[\text{HgO}]{\text{HgCl}_2}$ PhCO

With Me on PhC(SPh)$_2$ and Me on PhCO.

JOC (1963) 28 961
    (1971) $\overline{36}$ 3553
J Med Chem (1973) 16 749

CH$_2$=CHCHO $\xrightarrow[\text{2 EtOCH=CH}_2]{\text{1 HCN}}$ CH$_2$=CHCHCN $\xrightarrow[\substack{\text{2 C}_6\text{H}_{13}\text{Br} \\ \text{3 H}_2\text{SO}_4 \text{ MeOH} \\ \text{4 NaOH H}_2\text{O}}]{\text{1 (i-Pr)}_2\text{NLi THF}}$ CH$_2$=CHCOC$_6$H$_{13}$

On CH$_2$=CHCHCN: OCHMe, OEt

~75%

JACS (1971) 93 5286
Synthesis (1973) 777

PhCHO $\dashrightarrow$ PhCHCN $\xrightarrow[\text{HCl Et}_2\text{O}]{\text{Dihydropyran}}$ PhCHCN $\xrightarrow[\substack{\text{2 C}_7\text{H}_{15}\text{Br} \\ \text{3 HCl H}_2\text{O}}]{\text{1 NaH Me}_2\text{SO}}$ PhCOC$_7$H$_{15}$ 56%

On PhCHCN: OH

Synthesis (1973) 358

Related methods: Aldehydes from Aldehydes (Section 49)

Also via: Olefinic Ketones (Section 374)

Section 170   Ketones from Alkyls and Methylenes
            ∘∘

$\xrightarrow[\text{2 Na}_2\text{CO}_3 \text{ H}_2\text{O}]{\text{1 NBS h}\nu}$

Chem Comm (1972) 350

PhBu-t $\xrightarrow[\text{HOAc \quad PhCl}]{\text{O}_2 \quad \text{Co(OAc)}_2 \quad \text{HCl}}$ PhCOMe   23%

Chem Comm (1971) 1166

Section 171   Ketones from Amides
○○○○○○○○○○○○○○○○○○○○

PhCONEt$_2$ $\xrightarrow{\text{EtBr \quad Li \quad THF}}$ PhCOEt   (79%)

(One-step procedure)

Synthesis (1973) 160

Section 172   Ketones from Amines
○○○○○○○○○○○○○○○○○○○○

PhNMe$_3$ I (+ −) $\xrightarrow[\text{2 CrO}_3 \quad \text{H}_2\text{SO}_4]{\text{1 MeCOCH}_2\text{K} \quad \text{K} \quad \text{NH}_3}$ PhCH$_2$COMe   56%

JACS (1972) 94 683

EtCHNH$_2$ $\xrightarrow[\text{H}_2\text{O}]{[(\pi\text{-C}_5\text{H}_5)_2\text{Mo(SMe}_2)\text{Br]}^+ \text{PF}_6^-}$ EtCO
| |
Me   Me

Chem Comm (1971) 1274

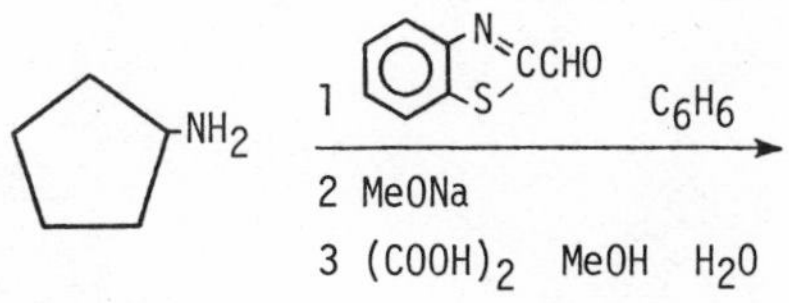

86%

JCS Perkin I (1972) 1652

PhCHNH$_2$ $\xrightarrow[\text{2 NaNO}_2 \quad \text{HOAc}]{\text{1 TsCl}}$ PhCHNTs $\xrightarrow{\text{h}\nu \quad \text{MeOH}}$ PhCO   < 25%
| | |
Me   Me   Me

Rec Trav Chim (1971) <u>90</u> 901

JOC (1972) <u>37</u> 1254

Section 173   Ketones from Esters
○○○○○○○○○○○○○○○○○○○○

t-BuCOOMe $\xrightarrow{\text{PrMgCl} \quad \text{HMPA}}$ t-BuCOPr   98%

$\xrightarrow[\text{2 MeI}]{\text{1 EtCH}_2\text{MgCl} \quad \text{HMPA}}$ t-BuCOCHEt   81%
|
Me

Tetrahedron (1973) <u>29</u> 479

Section 174   Ketones from Ethers and Epoxides
○○○○○○○○○○○○○○○○○○○○○○○○○○○○○○○○○○

R$_2$CHOR $\xrightarrow{\text{t-Butyl perbenzoate}}$ R$_2$COR $\dashrightarrow$ R$_2$CO
                                                       PhCOO

Synthesis (1972) 1

(Me$_2$CH)$_2$O $\xrightarrow{\text{Ph}_3\text{C}^+ \text{BF}_4^-}$ Me$_2$CO

JACS (1967) <u>89</u> 3550

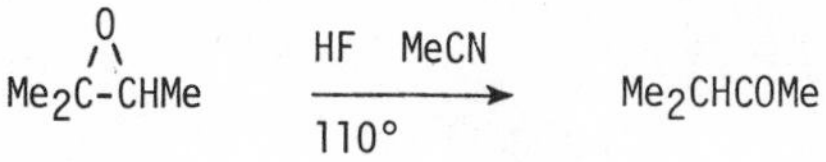

$$\underset{Me_2C-CHMe}{\overset{O}{\triangle}} \quad \xrightarrow[110°]{HF \quad MeCN} \quad Me_2CHCOMe \qquad\qquad 67\%$$

Z Chem (1967) **7** 229

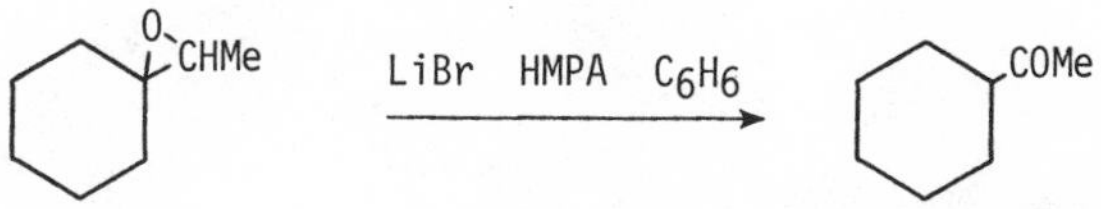

CHMe (on cyclohexane epoxide) $\xrightarrow{LiBr \quad HMPA \quad C_6H_6}$ COMe (on cyclohexane)

JACS (1971) **93** 1693
Ber (1973) **106** 1365

Also via: Olefinic Ketones (Section 374)

Section 175　　**Ketones from Halides**

$$BuBr \quad \xrightarrow[base]{MeCOCH_2COOEt} \quad \underset{BuCHCOMe}{\overset{COOEt}{|}} \quad \xrightarrow[Me_2SO]{NaCl \quad H_2O} \quad BuCH_2COMe$$

Tetr Lett (1973) 957
Cleavage of ß-ketoesters with LiH-BuLi　Tetr Lett (1971) 4585
B_2O_3　Tetr Lett (1970) 3903

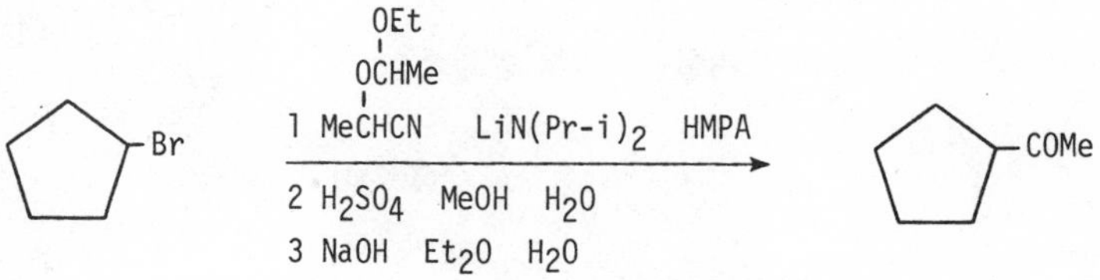

$$\underset{}{\overset{}{\bigcirc}}\text{-Br} \quad \xrightarrow[\substack{2\ H_2SO_4\ \ MeOH\ \ H_2O \\ 3\ NaOH\ \ Et_2O\ \ H_2O}]{\substack{OEt \\ | \\ OCHMe \\ | \\ 1\ MeCHCN\ \ \ LiN(Pr-i)_2\ \ \ HMPA}} \quad \bigcirc\text{-COMe} \qquad 80\%$$

JACS (1971) **93** 5286

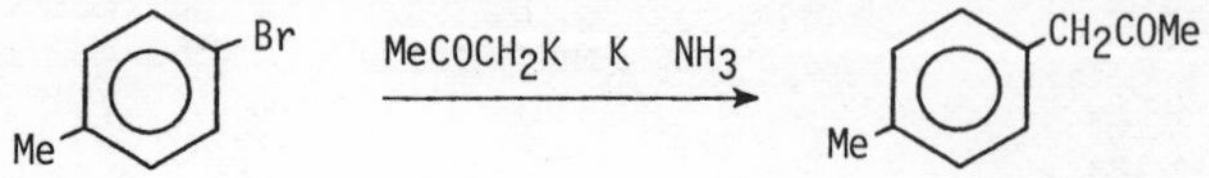
Br
MeCOCH2K K NH3
CH2COMe
Me
Me

57%

JACS (1972) 94 683

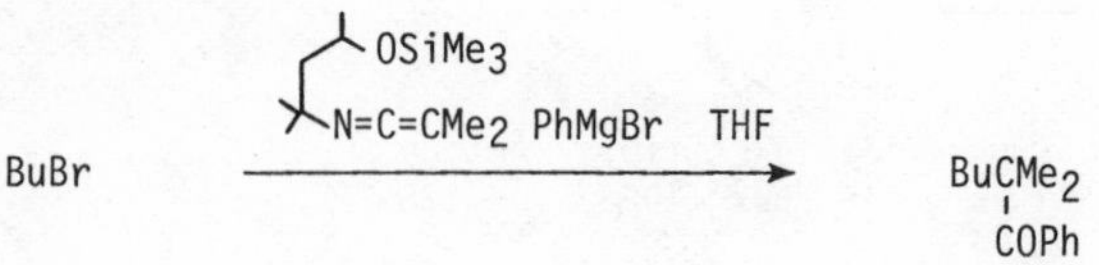
OSiMe3
N=C=CMe2 PhMgBr THF
BuBr
BuCMe2
COPh

81%

JOC (1973) 38 2129

BuBr —(1 EtS(SOEt)CHEt BuLi THF; 2 $HClO_4$)→ BuCOEt

Tetr Lett (1973) 3267

$C_8H_{17}Br$ —(1 $MeSCH{=}CHC_8H_{17}$ EtCH(Me)Li HMPA THF; 2 $HgCl_2$ MeCN H_2O)→ $C_8H_{17}COC_9H_{19}$ 82%

JACS (1973) 95 2694

PrBr —(1 $MeSOCH_2COPh$ NaH DMF; 2 Hg-Al THF H_2O)→ $PrCH_2COPh$ 44%

JOC (1966) 31 2355
Tetr Lett (1974) 107

$C_5H_{11}Br$ →[1 $CH_3COCH=PPh_3$ BuLi THF][2 H_2O EtOH] $C_5H_{11}CH_2COMe$ 82%

JOC (1973) <u>38</u> 4082

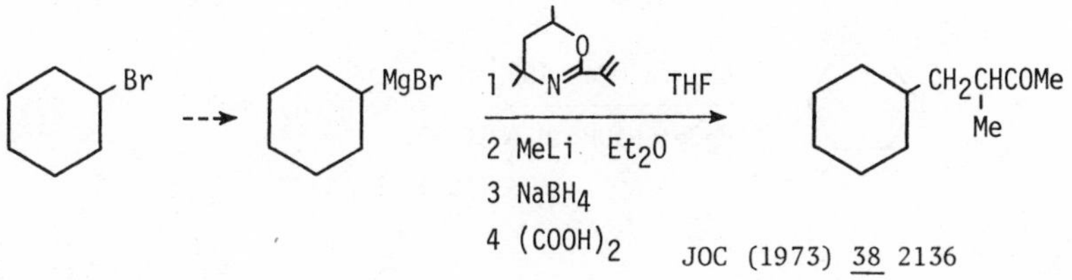

1 (oxazine) THF
2 MeLi Et_2O
3 $NaBH_4$
4 $(COOH)_2$ JOC (1973) <u>38</u> 2136

BuBr --→ Bu_2CuLi →[CO Et_2O] Bu_2CO 70%

Tetr Lett (1972) 2803

i-PrCl →[1 Mg Et_2O][2 EtCOOMe HMPA] i-PrCOEt   43%

Tetrahedron (1973) <u>29</u> 479

PhBr →[PhCONEt$_2$ Li THF] Ph_2CO 76%

(One-step procedure)

Synthesis (1973) 160

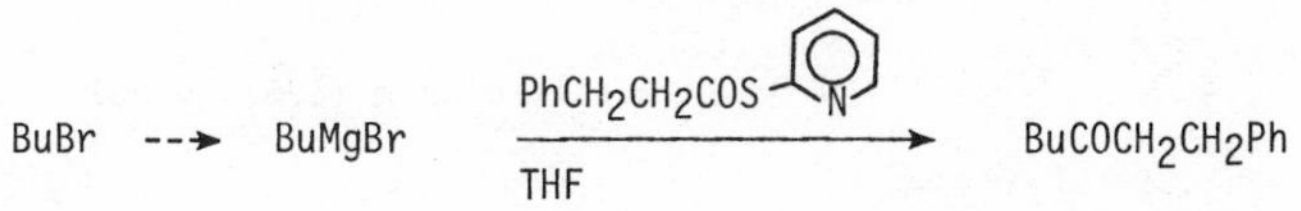

BuBr --→ BuMgBr →[PhCH$_2$CH$_2$COS-pyridyl][THF] BuCOCH$_2$CH$_2$Ph   <97%

JACS (1973) <u>95</u> 4763

BuBr $\xrightarrow{\begin{array}{c}1 \ Mg \\ 2\end{array}}$ BuCOPh   55%

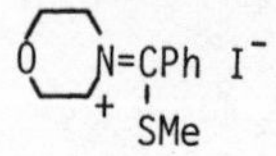

O⟨ ⟩N=CPh I⁻
     + |
        SMe

Chem Ind (1972) 380

t-BuCl --→ t-BuB⟨ ⟩ $\xrightarrow{\begin{array}{c}1 \ CH_2=CHCOMe \ THF \\ 2 \ H_2O_2 \ NaOH \ H_2O\end{array}}$ t-BuCH$_2$CH$_2$COMe   <90%

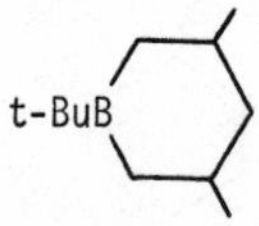

JACS (1971) 93 3777

RHal --→ R$_2$Hg $\xrightarrow{\begin{array}{c}Co_2(CO)_8 \ CO \\ h\nu \ THF\end{array}}$ R$_2$CO

Synthesis (1971) 55

C$_8$H$_{17}$Br $\xrightarrow{\begin{array}{c}1 \ Na_2Fe(CO)_4 \ N-methylpyrrolidone \\ 2 \ EtI\end{array}}$ C$_8$H$_{17}$COEt   80%

JACS (1972) 94 1788 2516

RBr $\xrightarrow{Li[R'COFe(CO)_4] \ Et_2O \ C_6H_6}$ RCOR'

Synthesis (1971) 55

PhCH$_2$Cl $\xrightarrow{Fe_2(CO)_9 \ C_6H_6}$ (PhCH$_2$)$_2$CO   52%

J Organometallic Chem (1967) 9 361

PhCH$_2$I $\xrightarrow{\text{Fe(CO)}_5 \quad \text{BuLi} \quad \text{Et}_2\text{O}}$ PhCH$_2$COBu   50%

Tetr Lett (1969) 5189

MeCH$_2$Br $--\rightarrow$ MeCH=PPh$_3$ $\xrightarrow[\text{2 HCl} \quad \text{H}_2\text{O}]{\text{1 Ph}\overset{\text{OMe}}{\underset{|}{\text{C}}}=\text{W(CO)}_5 \quad \text{Et}_2\text{O}}$ MeCH$_2$COPh

JACS (1972) <u>94</u> 6543

$\underset{\overset{|}{\text{Me}}}{\text{PhCHBr}}$ $\xrightarrow[\text{Me}_2\text{SO}]{\text{NaNO}_2}$ $\underset{\overset{|}{\text{Me}}}{\text{PhCHNO}_2}$ $\xrightarrow[\text{Me}_2\text{SO}]{\text{NaNO}_2 \quad \text{PrONO}}$ $\underset{\overset{|}{\text{Me}}}{\text{PhCO}}$ 39%

JOC (1973) <u>38</u> 1418

Related methods: Ketones from Ketones (Section 177)

Also via: Olefinic ketones (Section 374)

Section 176   Ketones from Hydrides
This section lists examples of the replacement of hydrogen by ketonic groups, RH $\longrightarrow$ RCOR'. For the oxidation of methylenes R$_2$CH$_2$ $\longrightarrow$ R$_2$CO see section 170 (Ketones from Alkyls and Methylenes)

88%

(Method for reactive compounds)

Synthesis (1972) 533

Section 177   Ketones from Ketones

$$PhCH_2COMe \quad \xrightarrow[\text{NaOH} \quad H_2O]{\overset{+}{BuBr} \quad \overset{-}{Et_3NCH_2Ph} \ Cl}} \quad \underset{\underset{Bu}{|}}{PhCHCOMe} \qquad 90\%$$

Tetr Lett (1971) 1351

$$RCOCHR_2 \quad \xrightarrow[\text{2 BuI}]{\text{1 BuMgBr \quad HMPA}} \quad \underset{\underset{Bu}{|}}{RCOCR_2}$$

Bull Soc Chim Fr (1969) 160

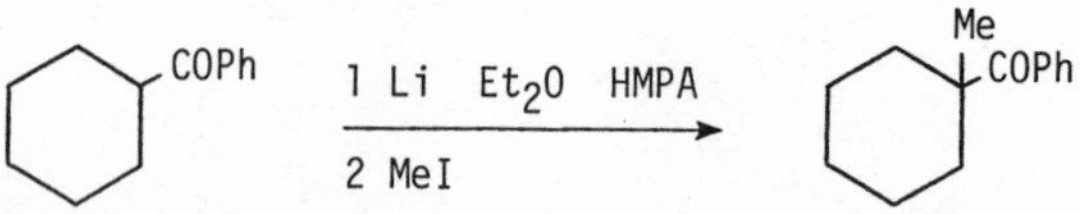

1 Li   Et$_2$O   HMPA
2 MeI

72%

Bull Soc Chim Fr (1968) 595

$$PhCOCH_2Me \quad \xrightarrow[\text{2 BuBr}]{\text{1 }(i\text{-Pr})_2NLi \quad MeOCH_2CH_2OMe} \quad \underset{\underset{Bu}{|}}{PhCOCHMe} \qquad 75\%$$

JOC (1973) 38 2756

$$PhCOCH_2Me \quad \xrightarrow[\text{KOH \quad EtOH}]{\text{HCHO \quad Fe(CO)}_5} \quad PhCOCHMe_2 \qquad 85\%$$

Tetr Lett (1973) 2491

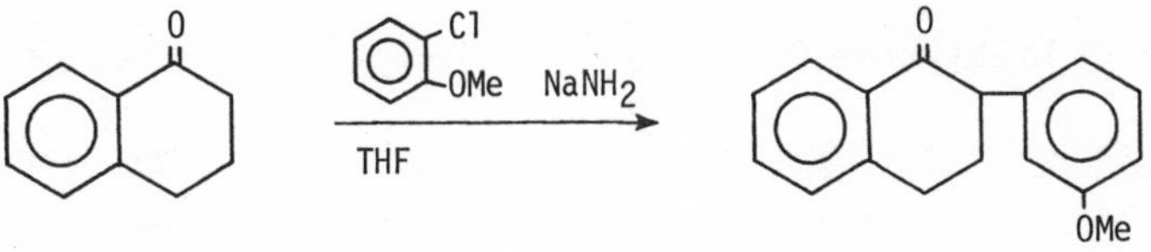

Chem Pharm Bull (1971) 19 1150

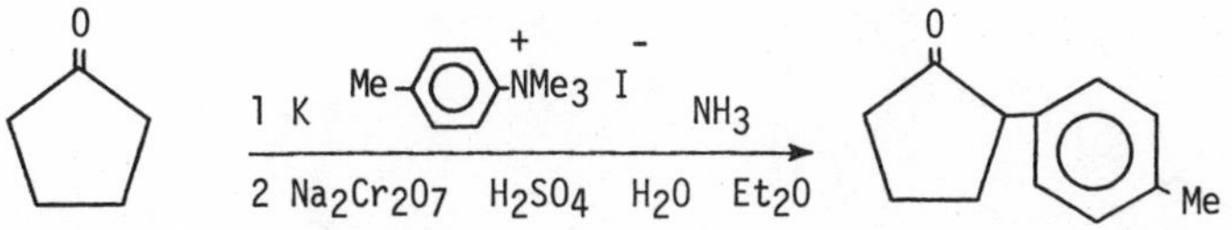

60%

JACS (1972) 94 683

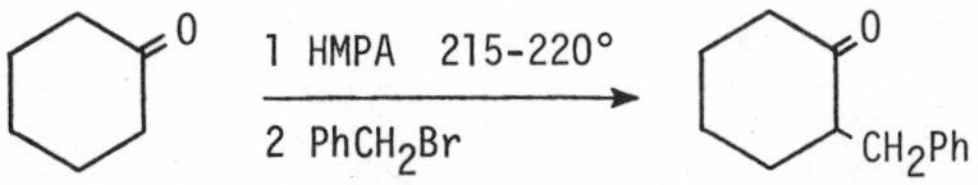

34%

Tetr Lett (1972) 929

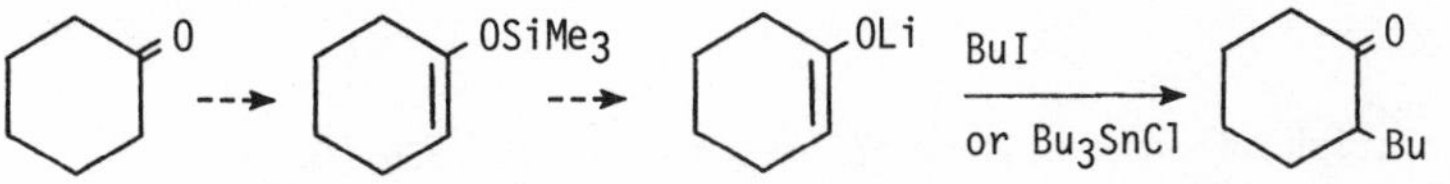

52%

Tetr Lett (1969) 1117 505

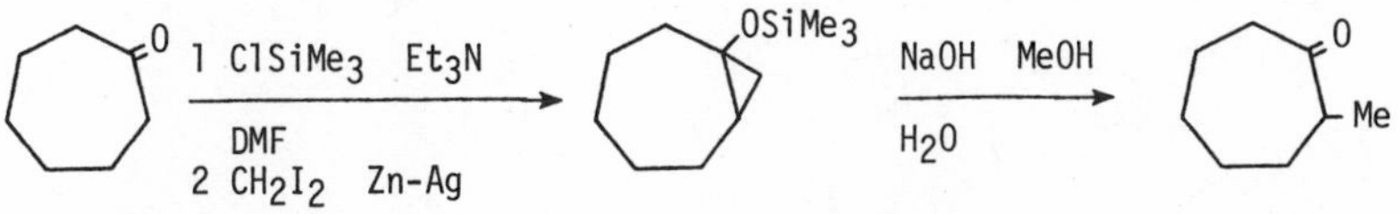

Tetr Lett (1973) 2767

64%

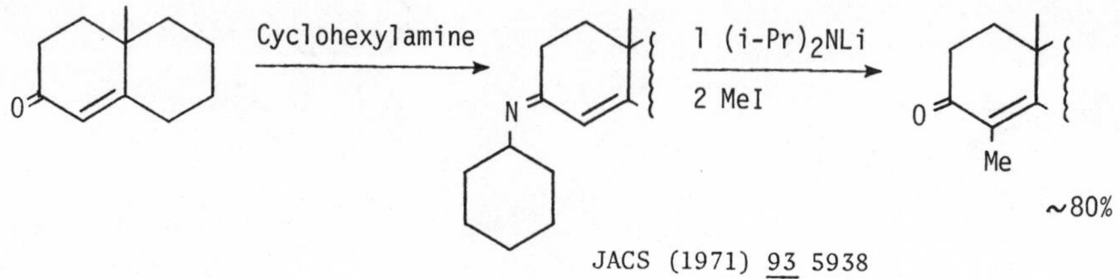

JACS (1971) <u>93</u> 5938

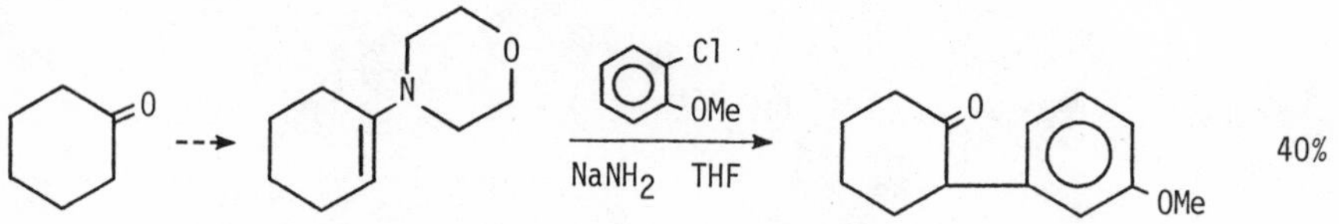

40%

JCS <u>C</u> (1971) 1047

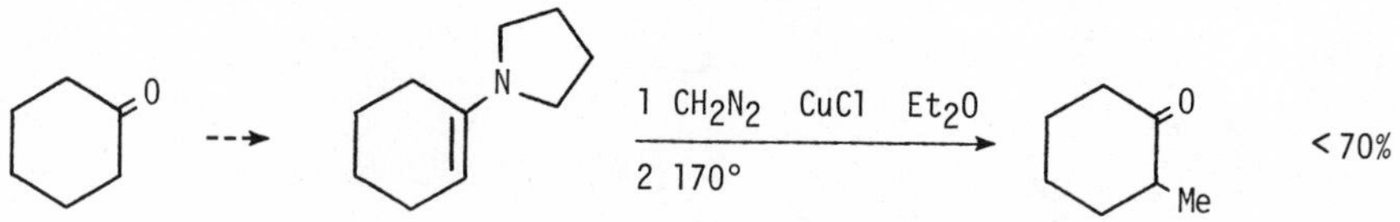

JOC (1973) <u>38</u> 304

For the preparation of enamines from ketones see section 356 (Amine - Olefin)

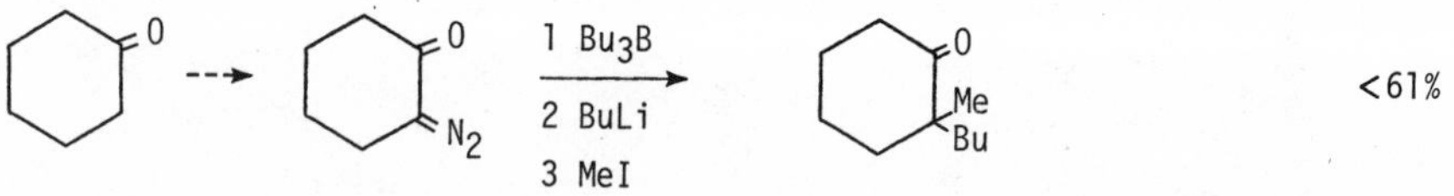

<61%

JOC (1971) <u>36</u> 1790

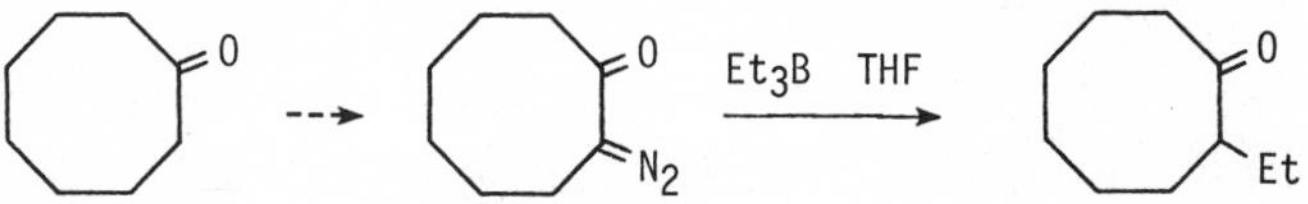

Can J Chem (1971) 49 2371

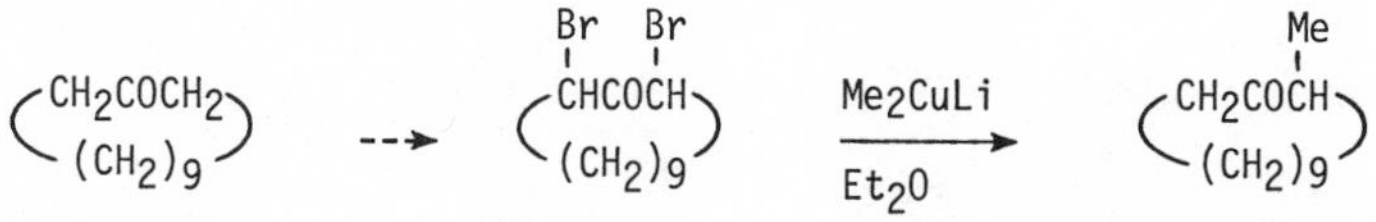

JACS (1973) 95 3076
Tetr Lett (1973) 4349

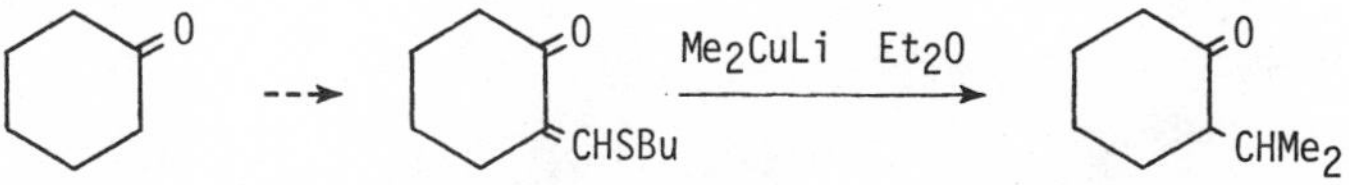

<97%

JACS (1971) 93 1027
JOC (1974) 39 275

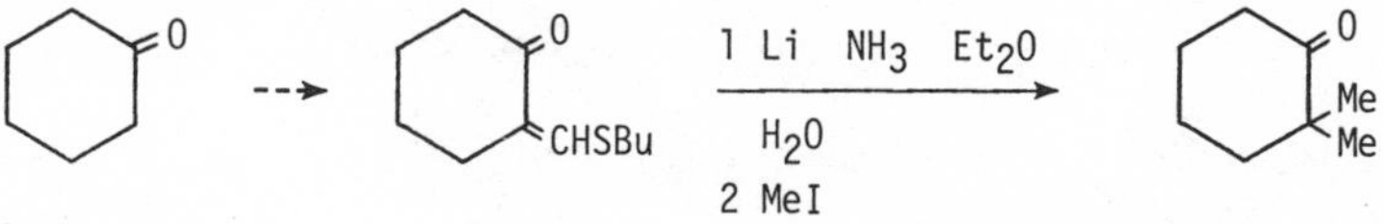

JACS (1971) 93 1027

<83%

Ketones may also be alkylated and homologated via olefinic ketones
(Section 374)

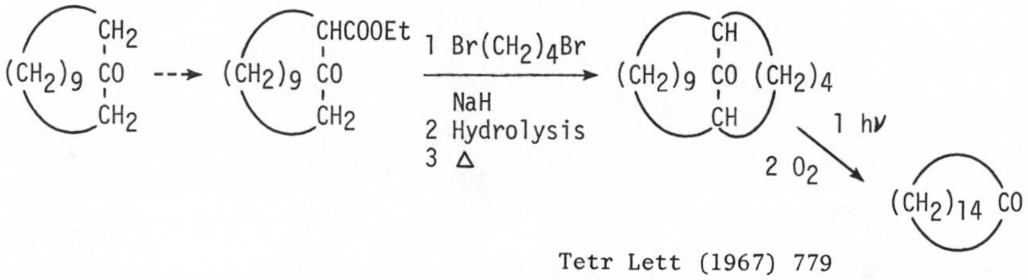

Tetr Lett (1967) 779

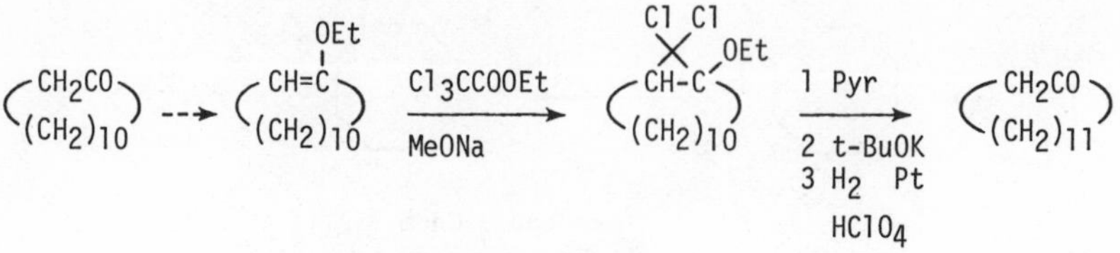

JOC (1967) <u>32</u> 926

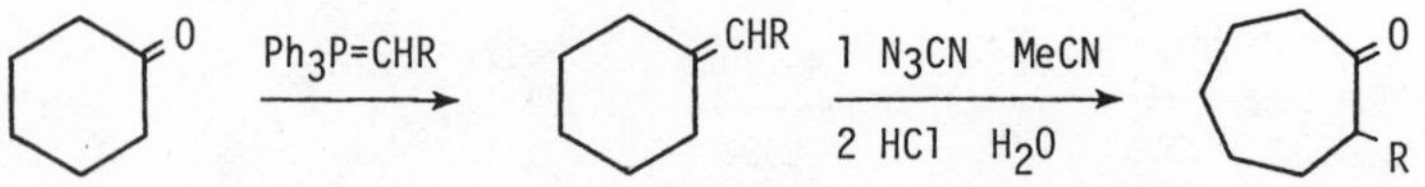

JOC (1973) <u>38</u> 2821 3862

R = H or Me

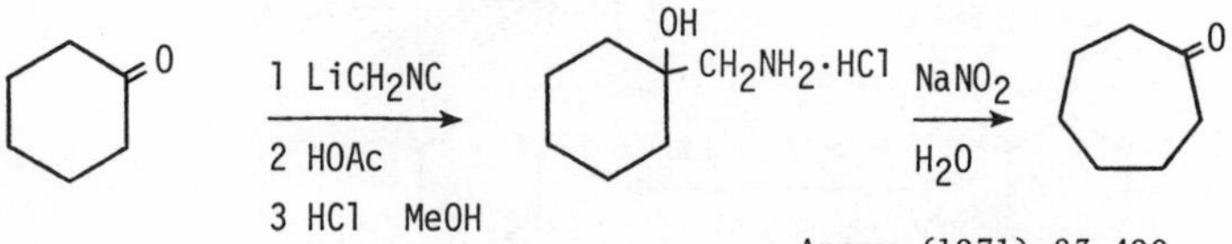

35%

Angew (1971) <u>83</u> 490
(Internat Ed <u>10</u> 491)

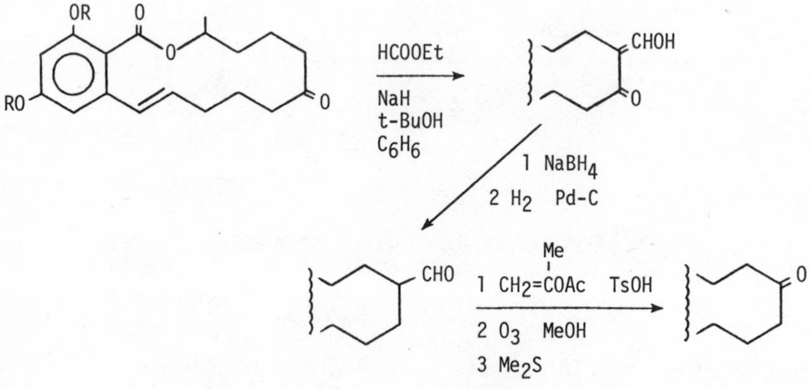

JOC (1972) <u>37</u> 1639

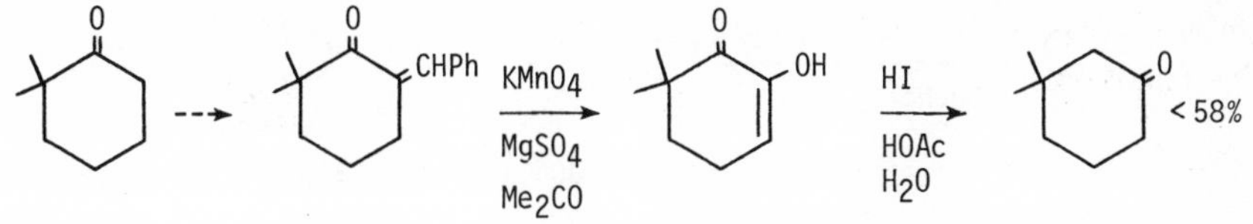

JACS (1964) <u>86</u> 3068

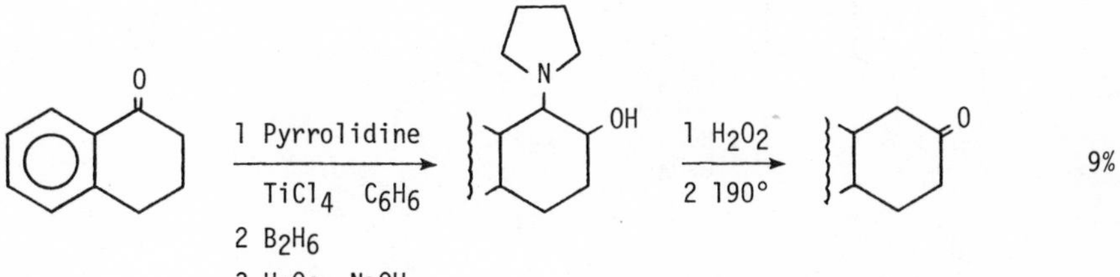

Bull Soc Chim Fr (1971) 1649

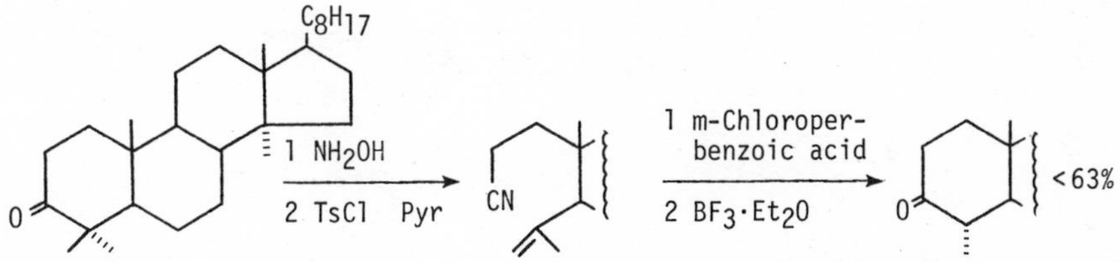

9%

JCS Perkin I (1973) 2076

< 63%

$$RCOCH_2R \xrightarrow{h\nu} RCOMe$$

Acc Chem Res (1971) <u>4</u> 168

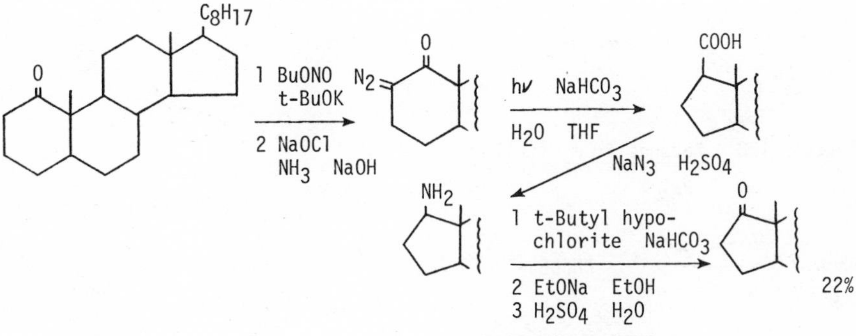

JOC (1965) <u>30</u> 3775
JACS (1971) <u>93</u> 2786

22%

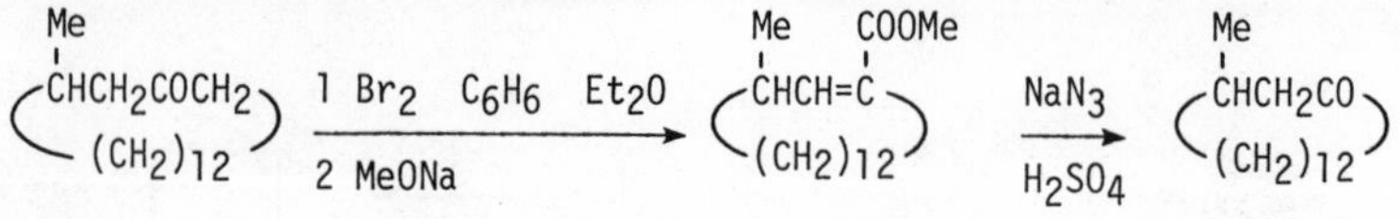

Me
CHCH₂COCH₂⟩   1 Br₂ C₆H₆ Et₂O   Me COOMe   Me
(CH₂)₁₂   2 MeONa   CHCH=C⟩   NaN₃   CHCH₂CO⟩
                                (CH₂)₁₂   H₂SO₄   (CH₂)₁₂

$$Me\text{-}CHCH_2COCH_2 / (CH_2)_{12} \xrightarrow[\text{2 MeONa}]{\text{1 } Br_2 \ C_6H_6 \ Et_2O} Me\text{-}CHCH=C(COOMe)/(CH_2)_{12} \xrightarrow[H_2SO_4]{NaN_3} Me\text{-}CHCH_2CO/(CH_2)_{12}$$

JOC (1971) 36 3266

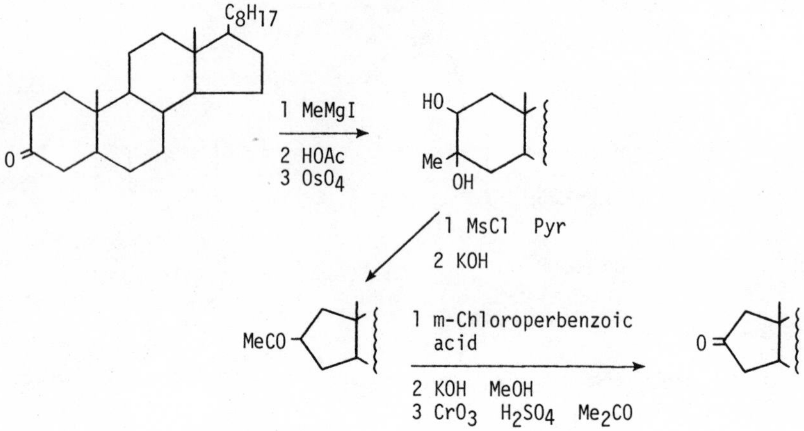

1 MeMgI
2 HOAc
3 OsO₄

1 MsCl Pyr
2 KOH

MeCO—   1 m-Chloroperbenzoic acid
        2 KOH MeOH
        3 CrO₃ H₂SO₄ Me₂CO   → O=

JOC (1971) 36 2400

CrO₃
HOAc

HOOC
HOOC

1 CH₂N₂
2 t-BuOK
3 Na₂CO₃ MeOH
H₂O   44%

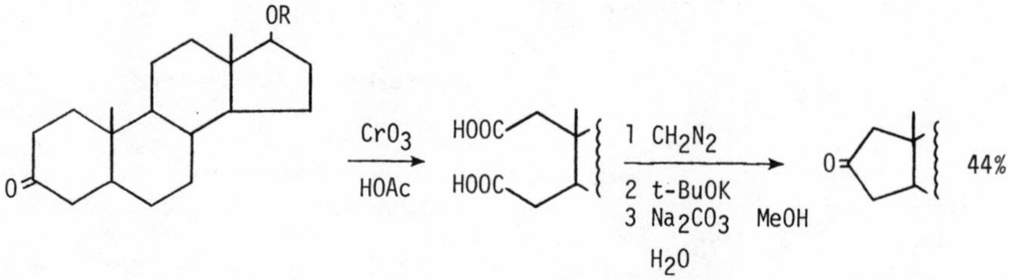

JOC (1971) 36 81

Section 178   Ketones from Nitriles

PhCN $\xrightarrow{(i\text{-Bu})_2Be}$ PhCOBu-i

Chem Ind (1972) 689

t-BuCN $\xrightarrow[\text{pet ether}]{\text{Na}}$ $(t\text{-Bu})_2C\dot=NH$ $\xrightarrow[\text{H}_2\text{O}]{\text{H}_2\text{SO}_4}$ $(t\text{-Bu})_2CO$

JACS (1971) <u>93</u> 4527
Tetr Lett (1972) 2071

Section 179   <u>Ketones from Olefins</u>

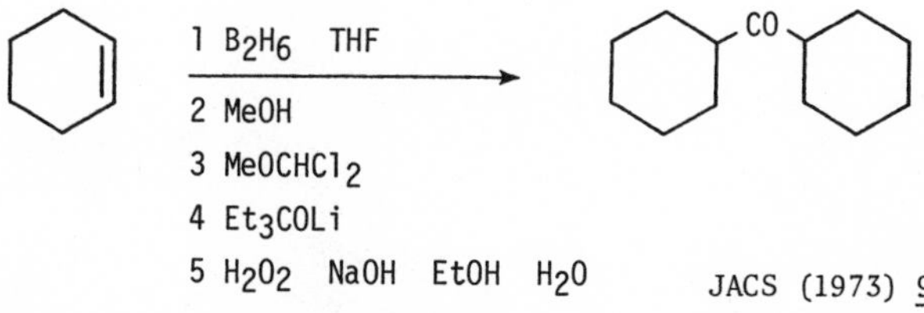

1 B_2H_6 THF
2 MeOH
3 MeOCHCl$_2$
4 Et$_3$COLi
5 H_2O_2 NaOH EtOH H_2O

JACS (1973) <u>95</u> 6876

85%

1 BH ... THF
2 RCH=CHCOMe
3 H_2O_2 NaOH H_2O

R
CHCH$_2$COMe

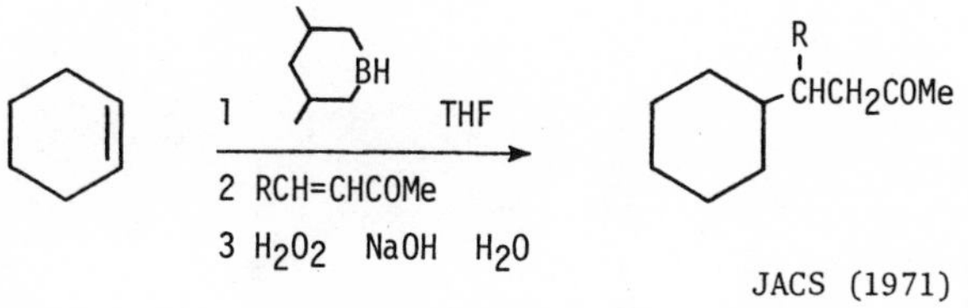

<81% (R=H)
<71% (R=Me)

JACS (1971) <u>93</u> 3777

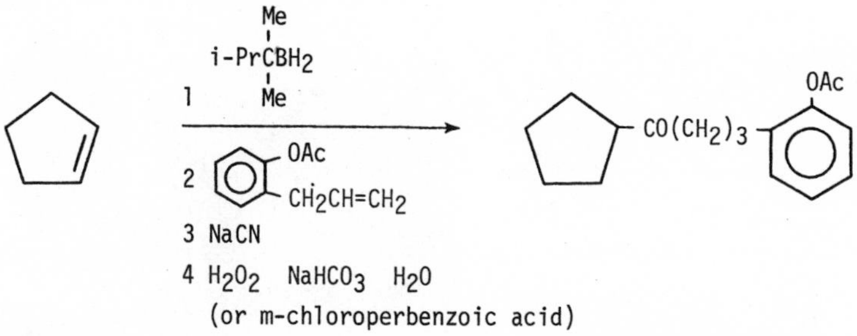

Me
i-PrCBH$_2$
1   Me
2   OAc CH$_2$CH=CH$_2$
3 NaCN
4 H_2O_2 NaHCO$_3$ H_2O
(or m-chloroperbenzoic acid)

CO(CH$_2$)$_3$ OAc

80%

Chem Comm (1971) 1048

$PhCH_2CH=CH_2$ $\xrightarrow{\text{PhCSOMe} \quad h\nu}$ $Ph(CH_2)_2COPh$

Tetr Lett (1972) 4993

$C_{10}H_{21}CH=CH_2$ $\xrightarrow{\begin{array}{l}1 \text{ :CBr}_2 \\ 2 \text{ Me}_2S_2 \\ 3 \text{ AgClO}_4 \quad \text{MeOH} \\ 4 \text{ MeSH} \quad \text{CF}_3\text{COOH}\end{array}}$ $C_{10}H_{21}\overset{\overset{\displaystyle C(SMe)_2}{|}}{CHCH_2}$ $\xrightarrow[H_2O]{CF_3COOH}$ $C_{10}H_{21}CH_2COMe$ 51%

Tetr Lett (1973) 3509

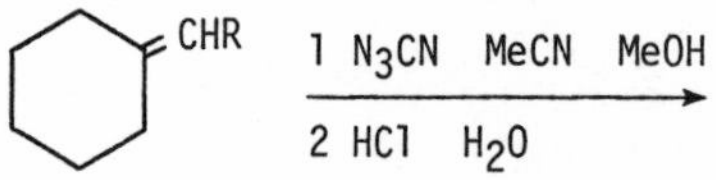

$\xrightarrow{\begin{array}{l}1 \text{ N}_3\text{CN} \quad \text{MeCN} \quad \text{MeOH} \\ 2 \text{ HCl} \quad \text{H}_2\text{O}\end{array}}$

80% (R=H or Me)

JOC (1973) **38** 2821

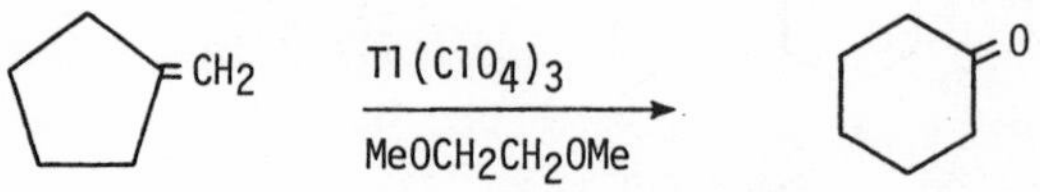

$\xrightarrow[MeOCH_2CH_2OMe]{Tl(ClO_4)_3}$

JOC (1973) **38** 3455

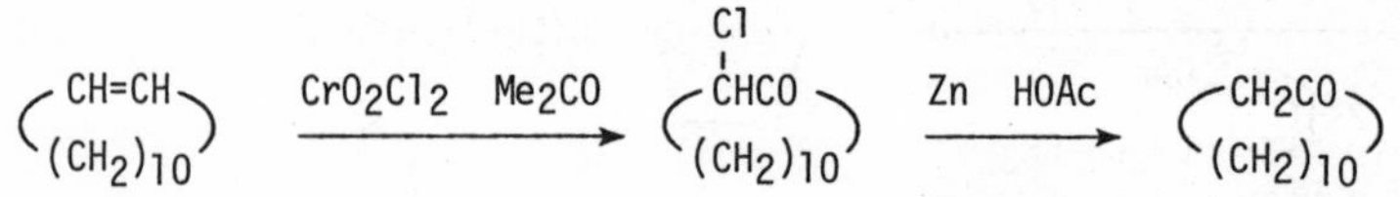

JOC (1973) **38** 185

Cl
NO
NOCl
i-PrOH
$NaBH_4$
i-PrOH H_2O
NOH
O

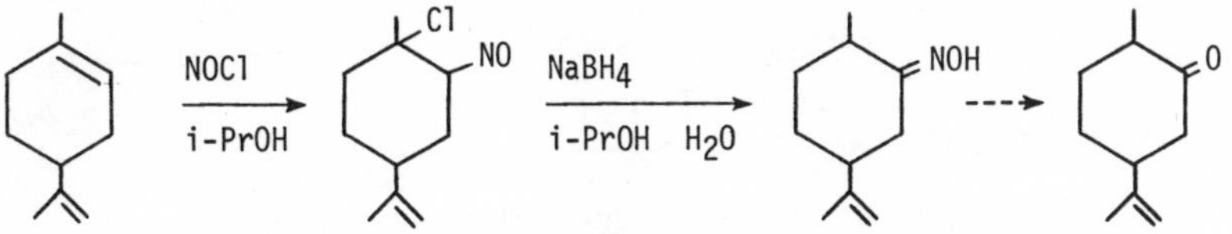

Aust J Chem (1971) 24 1089

$C_5H_{11}CH{=}CH_2$ $\xrightarrow{O_2\ \ Rh(Ph_3P)_3Cl\ \ C_6H_6}$ $C_5H_{11}COMe$ 20-25%

Tetr Lett (1972) 5273

$BuCH{=}CH_2$ $\xrightarrow[2\ Li_2PdCl_4\ \ CuCl_2]{1\ Hg(OAc)_2\ \ MeOH}$ BuCOMe 100%

Chem Comm (1971) 818
Tetr Lett (1972) 3595
Synthesis (1971) 527

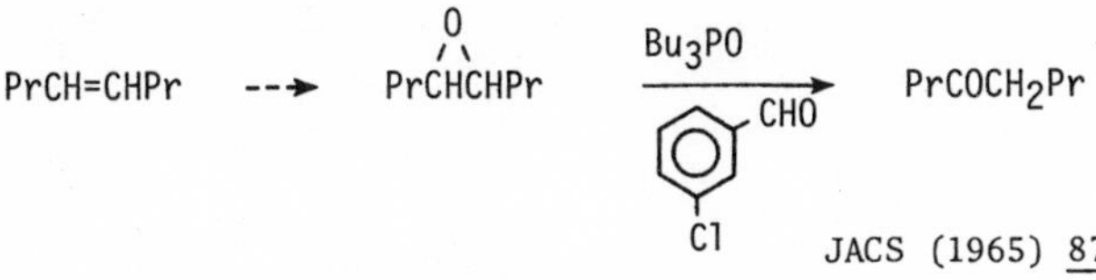

PrCH=CHPr
O
PrCHCHPr
Bu_3PO
CHO
Cl
$PrCOCH_2Pr$

<85%

JACS (1965) 87 1405

See also section 134 (Ethers and Epoxides from Olefins) and section 174 (Ketones from Ethers and Epoxides)

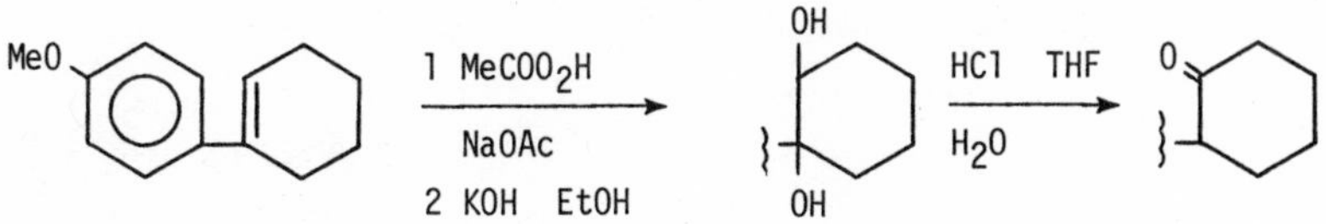

MeO
1 $MeCOO_2H$
NaOAc
2 KOH EtOH
OH
OH
HCl THF
H_2O
O

35%

Tetrahedron (1971) 27 3013

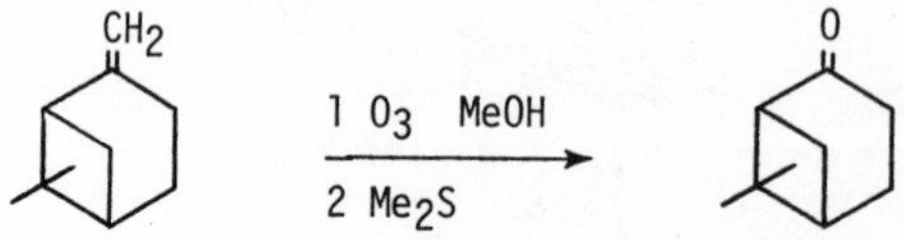

70%

JCS Perkin I (1972) 50

Reduction of ozonides with EtCHO   JACS (1971) 93 3042

Section 180   Ketones from Miscellaneous Compounds
ooooooooooooooooooooooooooooooooooooooo

$$RCH=CHCOR \xrightarrow{R'_2CuLi} \underset{R'}{RCHCH_2COR} \qquad (R'=alkyl\ or\ vinyl)$$

Org React (1972) 19 1

R'CuC≡CPr   JACS (1972) 94 7210

R'Li - t-BuOCu   JACS (1972) 94 658

$\underset{CN}{R'CuLi}$ Chem Comm (1973) 88

$$PhCH=CHCOMe \xrightarrow{HCOOH\quad RuCl_2(PPh_3)_3} PhCH_2CH_2COMe \qquad 90\%$$

Tetr Lett (1972) 1015

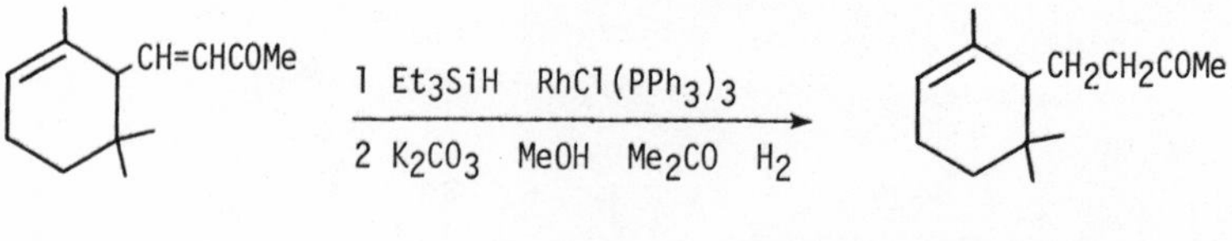

96%

Tetr Lett (1972) 5035

$Me_2C{=}CHCOMe$ —(H_2 nickel boride EtOH)→ Me_2CHCH_2COMe

JOC (1971) <u>36</u> 2018

$RCH{=}CHCOR$ —(Bu_3SnH)→ RCH_2CH_2COR

Organometallics in Chem Synth (1970) <u>1</u> 57
JCS <u>C</u> (1971) 1241

$PhCH{=}CHCOMe$ —($Fe(CO)_5$ NaOH / MeOH H_2O)→ $PhCH_2CH_2COMe$ 90%

JOC (1972) <u>37</u> 1542

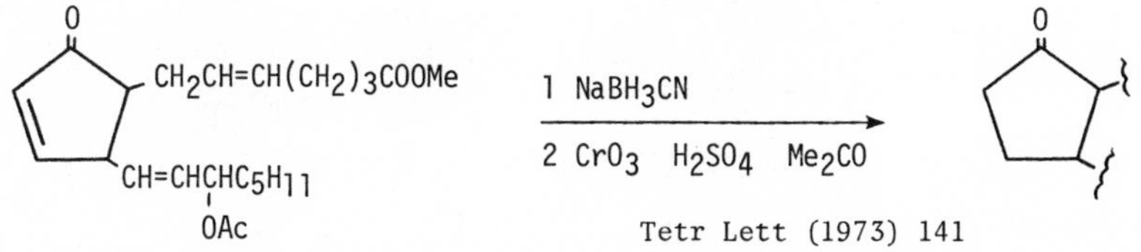

O
CH2CH=CH(CH2)3COOMe
CH=CHCHC5H11
OAc
1 NaBH3CN
2 CrO3 H2SO4 Me2CO
O

Tetr Lett (1973) 141

$\text{cyclo-}[COCH(OAc)(CH_2)_{10}]$ —($Fe(CO)_5$ Bu_2O)→ $\text{cyclo-}[COCH_2(CH_2)_{10}]$ 42%

Tetr Lett (1973) 447

$LiCuMe_2$ Tetr Lett (1973) 4349

$PhCOCH(Cl)Ph$ —(EtSNa EtOH)→ $PhCOCH_2Ph$ 100%

Bull Chem Soc Jap (1971) <u>44</u> 828

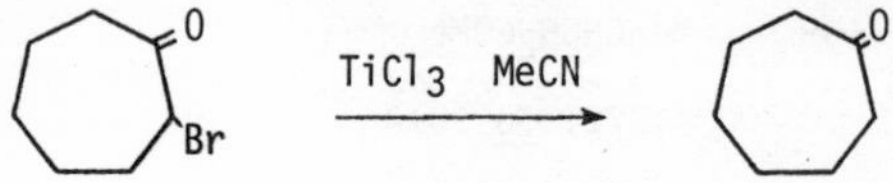

$TiCl_3$ MeCN

86%

Synth Comm (1973) 3 237

$LiCuMe_2$ Tetr Lett (1973) 4349

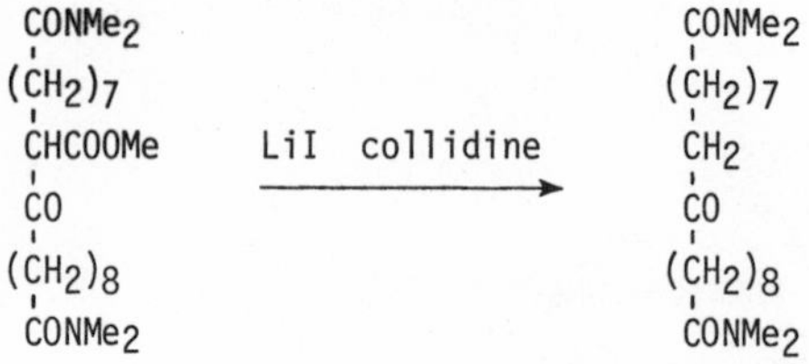

```
CONMe2
 |
(CH2)7
 |
CHCOOMe        LiI   collidine
 |
CO
 |
(CH2)8
 |
CONMe2
```

→

```
CONMe2
 |
(CH2)7
 |
CH2
 |
CO
 |
(CH2)8
 |
CONMe2
```

63%

JOC (1973) 38 1424

Org Synth (1965) 45 7

Decarboxylation with NaCl in Me_2SO Tetr Lett (1973) 957

NaCN in HMPA   Tetr Lett (1973) 3565

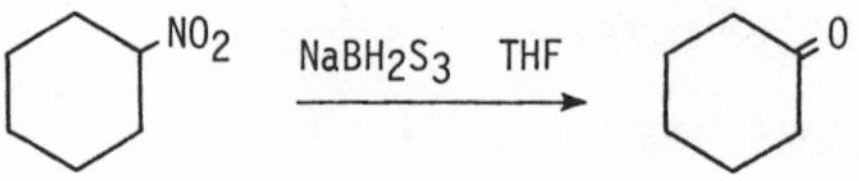

$NaBH_2S_3$ THF

(+ oxime)

75%

Can J Chem (1971) 49 2990

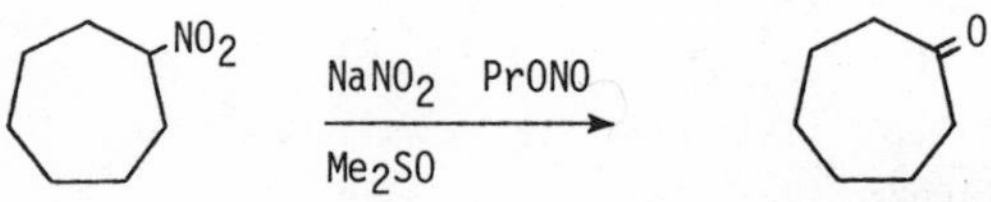

$NaNO_2$ PrONO

Me_2SO

88%

JOC (1973) 38 1418

O_3 JOC (1974) 39 259

$EtC≡C(CH_2)_2CH(CH_2)_2COMe$

NO_2

$\xrightarrow[MeOCH_2CH_2OMe]{TiCl_3 \quad H_2O}$

$EtC≡C(CH_2)_2CO(CH_2)_2COMe$

JACS (1971) 93 5309

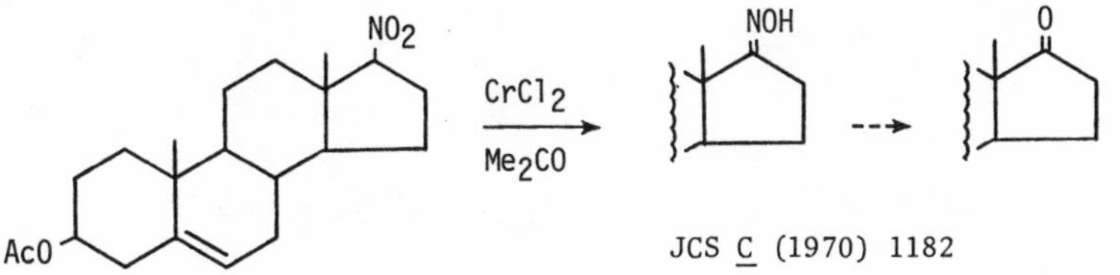

NO2
CrCl2
Me2CO
NOH
O
AcO

<73%

JCS C (1970) 1182

Me
N
CONH2
NaOCl
MeOH
Me
N
O

40%

JACS (1960) 82 4642

OH
COOH
OH
Pb(OAc)4 HOAc
O
OH

92%

JCS C (1971) 1292

OH
C≡CH
MeO
Ag2CO3-Celite
toluene
O

100%

Chem Comm (1972) 468

Section 180A Protection of Ketones

See also section 363 (Ether-Ether) for ketals and section 367 (Ether-Olefin) for enol ethers. Some of the methods in section 60A (Protection of Aldehydes) are also applicable to ketones

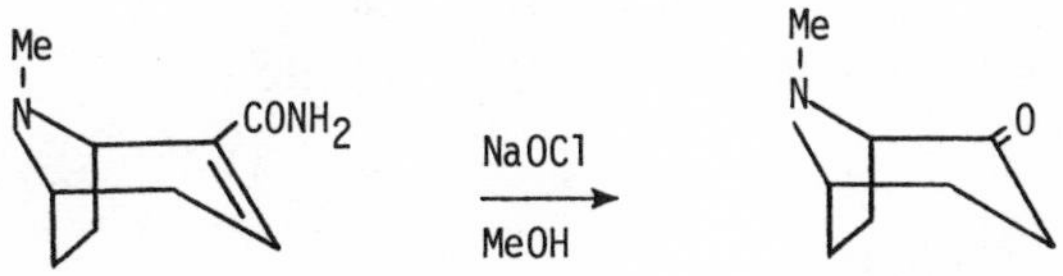

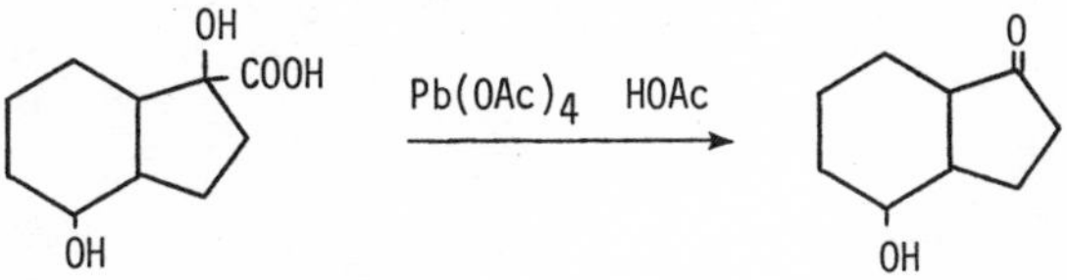

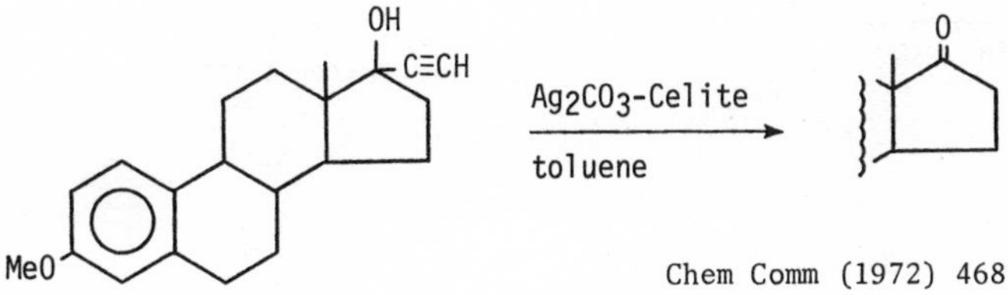

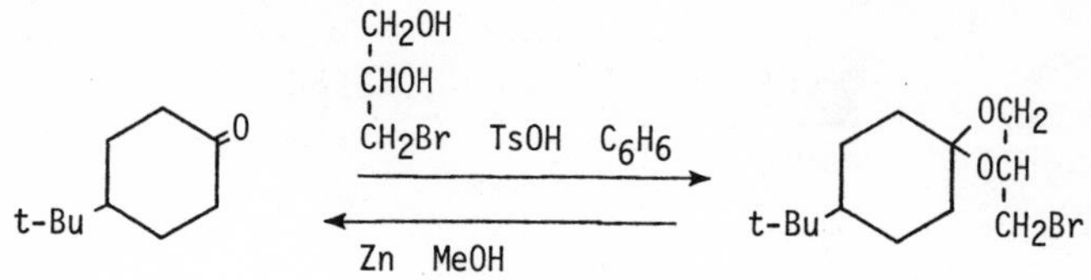

(Stable to MeLi, NaBH$_4$, NH$_3$, CrO$_3$
and peracids)

JOC (1973) <u>38</u> 834

Trichloroethyl ketals JOC (1973) <u>38</u> 554

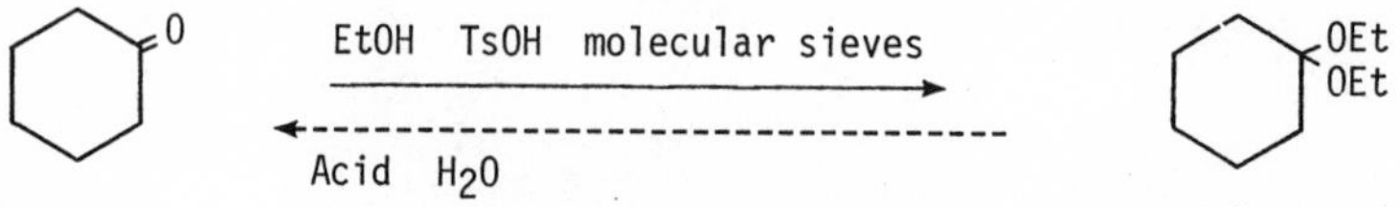

Rec Trav Chim (1971) <u>90</u> 1141

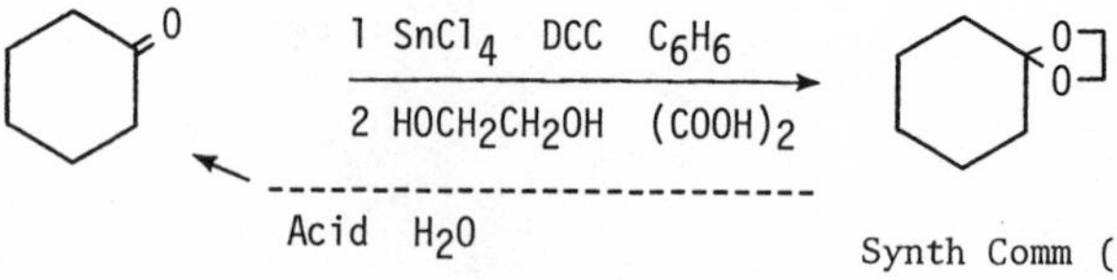

Synth Comm (1973) <u>3</u> 125

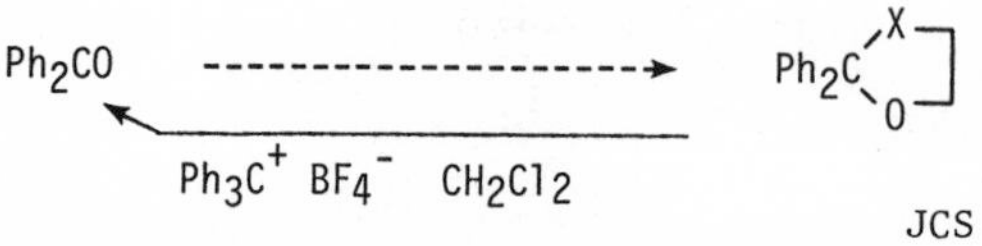

X=0 or S

JCS Perkin I (1972) 542

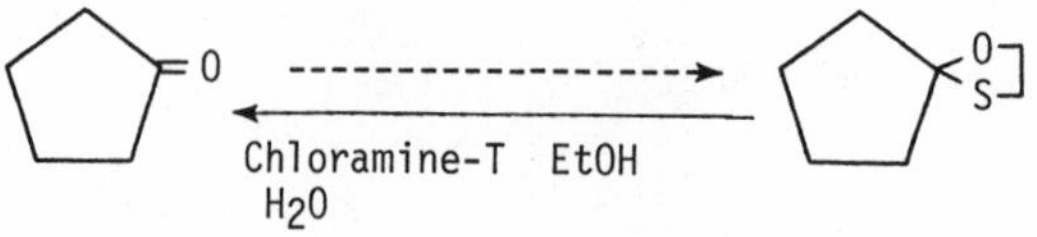

Tetr Lett (1971) 3445 3449

Nickel boride JCS Perkin I (1973) 654

PhCO   PhC
| --------------------> |
Me   <-----------------   Me
      1 Et$_3$O$^+$BF$_4^-$ CH$_2$Cl$_2$
      2 NaOH H$_2$O   Tetr Lett (1972) 1085

The following reagents may also be used for the cleavage of thioketals:

CuCl$_2$-CuO	Bull Chem Soc Jap (1972) 45 3724
	JACS (1972) 94 8641
Ce(NH$_4$)$_2$(NO$_3$)$_6$	Chem Comm (1972) 791
MeOSO$_2$F	Synthesis (1972) 561
MeI	Chem Comm (1972) 382
Ag$_2$O	Chem Comm (1972) 1323
AgNO$_3$	Tetrahedron (1968) 24 4249
NCS, AgNO$_3$	JOC (1968) 33 298
NBS	JOC (1971) 36 3553
Chloramine-T	Tetr Lett (1971) 3445 3449
I$_2$, Me$_2$SO	Tetr Lett (1973) 3735
Tl(OCOCF$_3$)$_3$	Can J Chem (1972) 50 3740
Et$_3$O$^+$ BF$_4^-$	Tetr Lett (1974) 11

1-Chlorobenzotriazole
or m-chloroperbenzoic acid

NaOH

Chem Comm (1971) 750

PhCO   NH$_2$OH
| -------------------->   PhC=NOH   (Not applicable to hindered ketones)
Me   <----- |
       1 Al(OPr-i)$_3$ Me
       2 HCl H$_2$O

Chem Ind (1972) 680

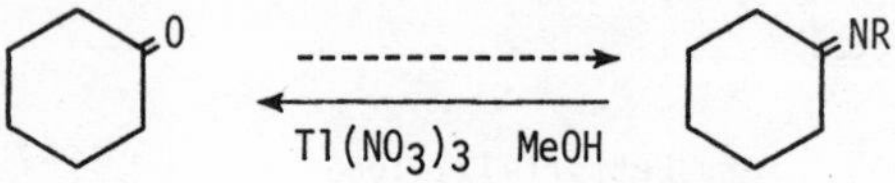

R=OH, $NHCONH_2$ or NHPh

JACS (1971) 93 4918

$$R_2CO \underset{Me_2CO}{\overset{TsNHNH_2}{\rightleftharpoons}} R_2C{=}NNHTs$$

Chem Ind (1964) 153 footnote 3

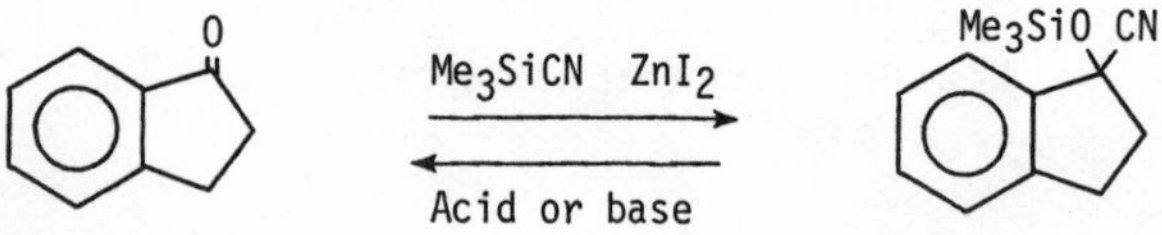

Chem Comm (1973) 55
Tetr Lett (1973) 4929

Chapter 13   PREPARATION
OF
NITRILES

Section 181   Nitriles from Acetylenes

No additional examples

Section 182   Nitriles from Carboxylic Acids and Acid Halides

$$C_{11}H_{23}COOH \xrightarrow{\begin{array}{c} COOH \\ | \\ C_{12}H_{25}CCN \\ | \\ Me \end{array} \quad electrolysis} \begin{array}{c} C_{12}H_{25} \\ | \\ C_{11}H_{23}CCN \\ | \\ Me \end{array}$$

J Pharm Soc Jap (1944) <u>64</u> 25

$$RCOOH \xrightarrow[285°]{\begin{array}{cc} CN & Me \\ | & | \\ CH_2CH_2CHCN & H_3PO_4 \end{array}} RCN$$

JOC (1971) <u>36</u> 3050

$$RCOOH \xrightarrow[\text{2 DMF}]{1 \; ClSO_2NCO \quad CH_2Cl_2} RCN$$

Org Synth (1970) <u>50</u> 18

$C_5H_{11}COCl \xrightarrow[160°]{(PNCl_2)_3} C_5H_{11}CN$

Tetr Lett (1973) 3825

Section 183   Nitriles from Alcohols

No additional examples

Section 184   Nitriles from Aldehydes

$$PhCHO \xrightarrow{\overset{\overset{NC}{|}}{TsCHK}} \overset{\overset{NHCHO}{|}}{PhCH=CTs} \xrightarrow[MeOH]{MeONa} PhCH_2CN \qquad <81\%$$

Angew (1973) 85 402
(Internat Ed 12 407)

$$RCHO \xrightarrow[NH_3 \quad i\text{-PrOH}]{MnO_2 \quad NaCN} RCN$$

Chem Comm (1971) 733

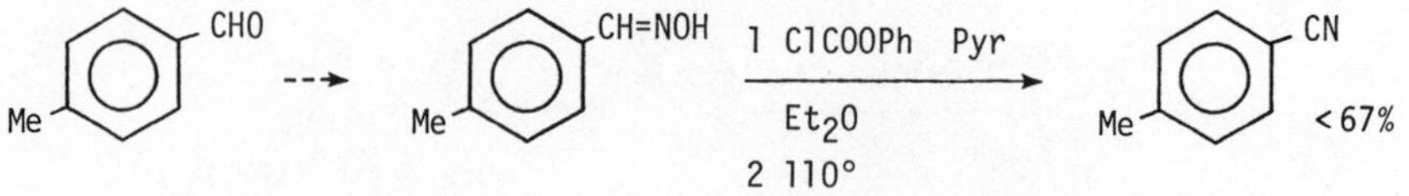

Can J Chem (1971) 49 1321

The following reagents may also be used for dehydration of oximes to nitriles:

$TiCl_4$-Pyr   Tetr Lett (1971) 559

HCOONa, HCOOH   JCS (1965) 1564

Ph_3P, CCl_4, Et_3N Ber (1971) <u>104</u> 2025

$CHCl_3$, NaOH, $PhCH_2\overset{+}{N}Et_3$ Cl^- Tetr Lett (1973) 2121

1,1'-dicarbonyldiimidazole   Chem Comm (1973) 628

DCC (190°)   Synth Comm (1973) <u>3</u> 101

CCl_3CN JOC (1973) <u>38</u> 2241

HgO   Tetr Lett (1971) 361

PhCHO $--\rightarrow$ PhCH=NN=CHPh $\xrightarrow{h\nu \quad \text{cyclohexane}}$ PhCN   85%

JACS (1963) <u>85</u> 2679

Related methods: Nitriles from Ketones (Section 192)

Also via: Olefinic nitriles (Section 376)

Section 185   <u>Nitriles from Alkyls, Methylenes and Aryls</u>

No additional examples

Section 186   <u>Nitriles from Amides</u>

$PhCH=CHCONH_2$ $\xrightarrow[\text{NaOH} \quad H_2O]{CHCl_3 \quad PhCH_2\overset{+}{N}Et_3 \; Cl^-}$ PhCH=CHCN   52%

Tetr Lett (1973) 2121

The following reagents may also be used for the conversion of amides into nitriles:

CCl_4, Ph_3P Ber (1972) <u>105</u> 244

$(PNCl_2)_3$ Can J Chem (1972) <u>50</u> 3857

HMPA (220-240°)   Can J Chem (1971) <u>49</u> 2897

$(HNSiMe_2)_3$ JOC (1970) <u>35</u> 3253

$TiCl_4$, amine   Tetr Lett (1971) 1501

Section 187   Nitriles from Amines

$BuCHCH_2NH_2$ Argentic picolinate
$|$ ————————————→ $BuCHCN$ 35%
Et H_2O $|$
                                         Et

Tetrahedron (1973) <u>29</u> 751

Section 188   Nitriles from Esters

No additional examples

Section 189   Nitriles from Ethers and Epoxides

No additional examples

Section 190   Nitriles from Halides

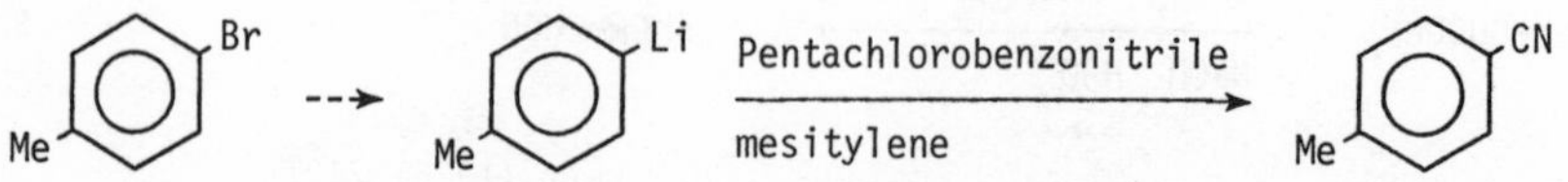

Pentachlorobenzonitrile
mesitylene

Tetr Lett (1972) 4169

Section 191   Nitriles from Hydrides

PhH $\xrightarrow{\text{Cl(CH}_2)_3\text{CN} \quad \text{AlCl}_3}$ Ph(CH$_2$)$_3$CN

Tetr Lett (1972) 1929

ArH $\dashrightarrow$ ArTlCl$_2$ $\xrightarrow{\text{CuCN} \quad \text{MeCN}}$ ArCN

Tetrahedron (1972) 28 3025

ArH $\xrightarrow{\text{KCN} \quad \text{h}\nu \quad \text{t-BuOH} \quad \text{H}_2\text{O}}$ ArCN

Tetrahedron (1972) 28 5081

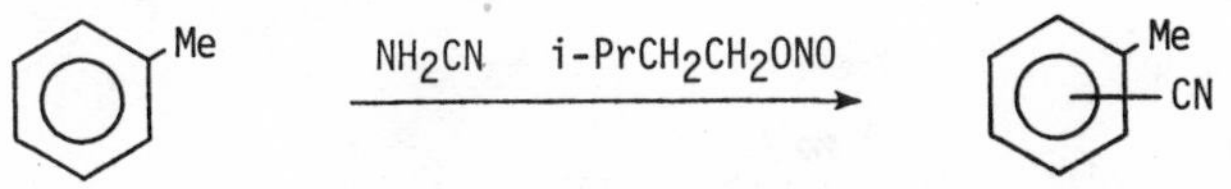

Acta Chem Scand (1972) 26 3870

Section 192   Nitriles from Ketones

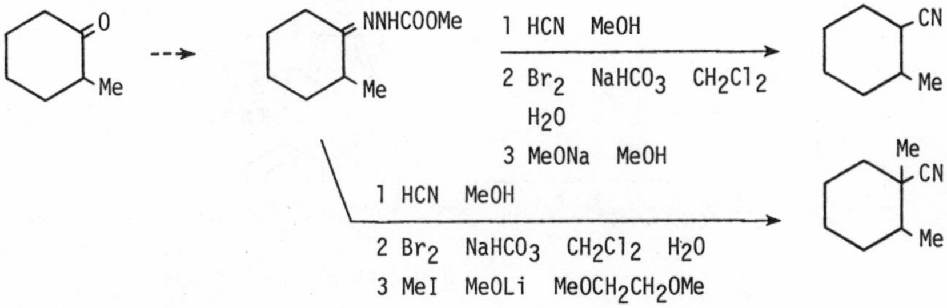

JACS (1971) 93 4318

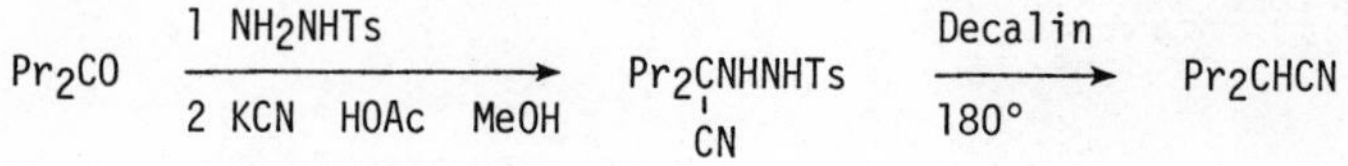

$$Pr_2CO \xrightarrow[\text{2 KCN HOAc MeOH}]{\text{1 NH}_2\text{NHTs}} Pr_2\overset{|}{C}NHNHTs \xrightarrow[\text{180°}]{\text{Decalin}} Pr_2CHCN$$

CN

Chem Ind (1972) 213

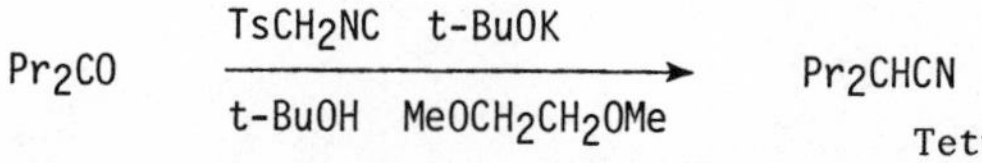

$$Pr_2CO \xrightarrow[\text{t-BuOH MeOCH}_2\text{CH}_2\text{OMe}]{\text{TsCH}_2\text{NC t-BuOK}} Pr_2CHCN \qquad 75\%$$

Tetr Lett (1973) 1357
Angew (1973) 85 402
(Internat Ed 12 407)

Related methods: Nitriles from Aldehydes (Section 184)

Also via: Olefinic nitriles (Section 376)

Section 193   Nitriles from Nitriles

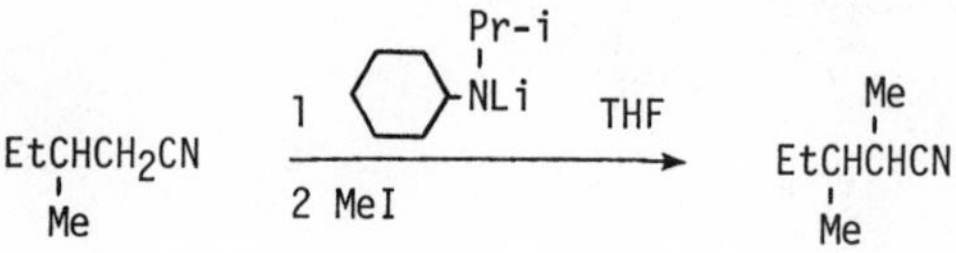

$$EtCHCH_2CN \xrightarrow[\text{2 MeI}]{\text{1}} EtCHCHCN \qquad <66\%$$

Me   Me

Ber (1973) 106 1376

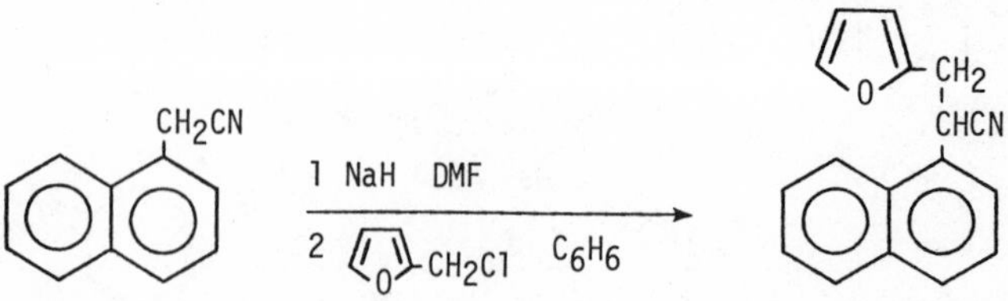

86%

J Med Chem (1965) 8 598
         (1973) 16 490
JOC (1958) 23 1346

PhCH$_2$CN $\xrightarrow[\text{PhCOOMe} \quad 210\text{-}220°]{\text{C}_7\text{H}_{15}\text{ONa} \quad \text{C}_7\text{H}_{15}\text{OH}}$ PhCHCN   73%

$\quad\quad\quad\quad\quad\quad\quad\quad\quad\quad\quad\quad\quad\quad$ $\overset{|}{\text{C}_7\text{H}_{15}}$

$\quad\quad\quad\quad\quad\quad\quad\quad\quad\quad\quad\quad\quad\quad$ JOC (1971) 36 2948

$\quad\quad\quad\quad\quad\quad\quad\quad\quad\quad\quad\quad\quad\quad\quad\quad$ (1972) 37 526

Section 194   Nitriles from Olefins
oooooooooooooooooooooo

No additional examples

Section 195   Nitriles from Miscellaneous Compounds
oooooooooooooooooooooooooooooooooooooooooo

PhCH=CHCN $\xrightarrow{\text{H}_2 \quad \text{nickel boride}}$ PhCH$_2$CH$_2$CN

$\quad\quad\quad\quad\quad\quad\quad\quad\quad\quad$ JOC (1972) 37 3552

$\xrightarrow{\text{POCl}_3 \quad \text{DMF}}$

$\quad\quad\quad\quad\quad\quad\quad\quad\quad\quad\quad\quad\quad\quad\quad\quad\quad$ 85%

$\quad\quad\quad\quad\quad\quad\quad\quad\quad$ Acta Chem Scand (1970) 24 3424

PhCH$_2$NO$_2$ $\xrightarrow{\text{NaBH}_2\text{S}_3 \quad \text{THF}}$ PhCN   80%

$\quad\quad\quad\quad\quad\quad\quad\quad\quad$ Can J Chem (1971) 49 2990

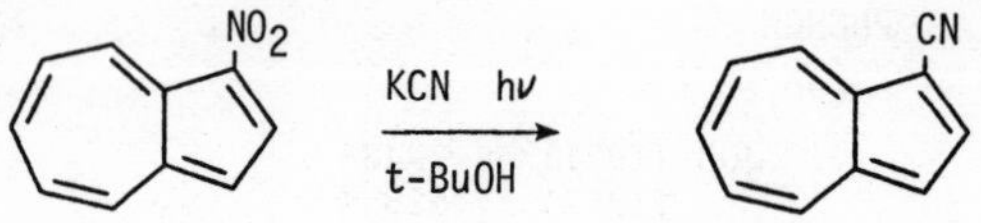

28%

Tetr Lett (1970) 4701

Chapter 14   PREPARATION
OF
OLEFINS

Section 196   Olefins from Acetylenes

Reviews: Stereoselective and Stereospecific Olefin   Synthesis
Quart Rev (1971) 25 135
     Catalytic Semihydrogenation of the Triple Bond
Synthesis (1973) 457

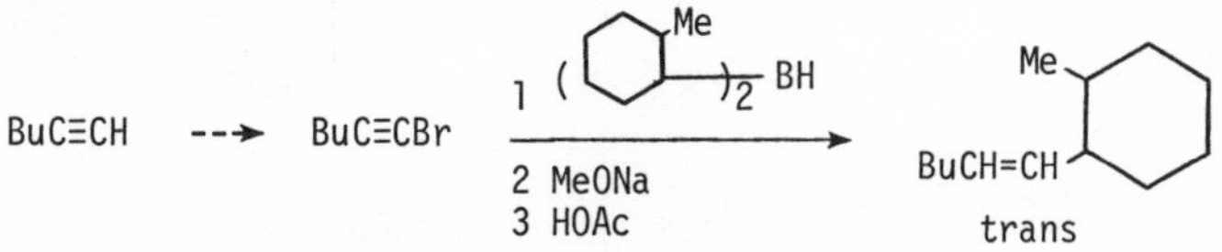

BuC≡CH   --→   BuC≡CBr

1 (⬡Me)₂ BH

2 MeONa
3 HOAc

BuCH=CH⟍⬡Me

trans

JACS (1971) 93 6309
(1972) 94 6560

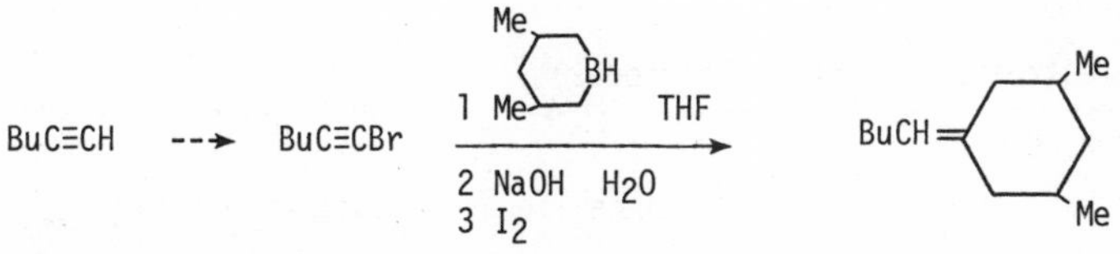

BuC≡CH   --→   BuC≡CBr

1 Me⬡Me BH   THF

2 NaOH   H₂O
3 I₂

BuCH=⬡ Me / Me

Synthesis (1972) 557

EtC≡CEt

1 BH₂Cl   Et₂O
───────────
2 HOAc   THF

EtCH=CHEt

cis

81%

JOC (1973) 38 1617
Catecholborane   JACS (1972) 94 4370

RC≡CR $\xrightarrow[\text{or Li~~~amine}]{\text{Na~~~NH}_3}$ RCH=CHR

Synthesis (1972) 391

Section 197   Olefins from Carboxylic Acids and Acid Halides

R(CH$_2$)$_5$COCl $\xrightarrow[\substack{\text{SnCl}_4 \\ 2~\text{N}_2\text{H}_4~~\text{NaOH}}]{1~~\overset{\displaystyle \diagdown\!\!\diagup}{\text{S}}\!\!-\text{C}_7\text{H}_{15}}$ R(CH$_2$)$_6$ [thiophene] C$_7$H$_{15}$ $\xrightarrow[\text{boride}]{\text{Nickel}}$ R(CH$_2$)$_7$CH=CHC$_8$H$_{17}$

Synth Comm (1972) 2 415

Me$_2$CHCOOH $\xrightarrow[\substack{\text{hexane} \\ 2~[\text{cyclohexanone}] \\ 3~\text{PhSO}_2\text{Cl}~~\text{Pyr}}]{1~(\text{i-Pr})_2\text{NLi}~~\text{THF}}$ [cyclohexane with Me$_2$C-CO / O] $\xrightarrow{140-160°}$ [cyclohexane with =CMe$_2$] 50%

JACS (1972) 94 2000

EtCH$_2$CH$_2$COOH $\xrightarrow[\text{KOH~~~H}_2\text{O}]{\text{Electrolysis (carbon electrodes)}}$ EtCH=CH$_2$ (+ isomers)

JACS (1964) 86 4686

RCH$_2$CH$_2$COOH $\xrightarrow{\text{Pb(OAc)}_4}$ RCH=CH$_2$

Org React (1972) 19 279

$C_6H_{13}(CH_2)_3COCl$ $\xrightarrow{\text{PhH AlCl}_3}$ $C_6H_{13}(CH_2)_3COPh$ $\xrightarrow{h\nu}$ $C_6H_{13}CH=CH_2$ <75%

JOC (1971) <u>36</u> 1838

RCHCOOH
| $\xrightarrow{\text{Pb(OAc)}_4}$ RCH
RCHCOOH   ‖
                                             RCH

Org React (1972) <u>19</u> 279

Section 198   Olefins from Alcohols
oooooooooooooooooooooo

OH
| $\xrightarrow{\text{I}_2 \text{ xylene}}$ $R_2C=CR_2$
R_2CHCR_2

Monatsh (1968) <u>99</u> 642

$\xrightarrow[\text{CHCl}_3]{\text{Ph}_2S[OC(CF_3)_2]}$

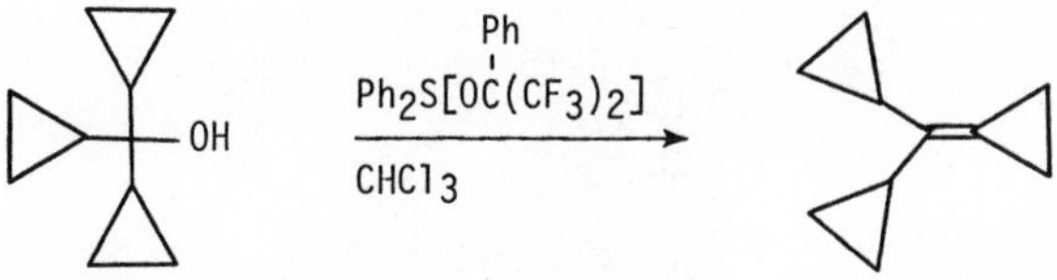

32%

(2ry and 3ry only)   JACS (1971) <u>93</u> 4327
                                       (1972) <u>94</u> 5003

$C_6H_{13}CH_2CH_2OH$ $\xrightarrow{\text{HMPA 220-240°}}$ $C_6H_{13}CH=CH_2$ 59%

Tetr Lett (1971) 567
(1972) 165
JOC (1971) <u>36</u> 3826

OH
|
BuCHCH₂Pr

(2ry only)

$\xrightarrow{\text{(PhO)}_3\text{P·MeI \quad HMPA}}$

BuCH=CHPr

JOC (1972) 37 4190

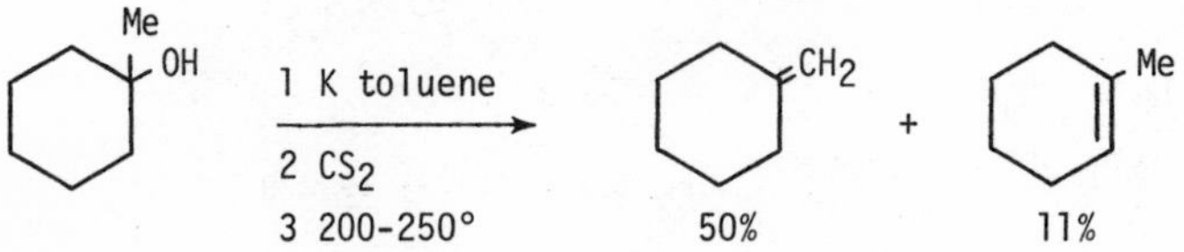

88%

---

1 K toluene

2 CS₂

3 200-250°

50%   11%

JCS C (1971) 582

---

1 NaH   Me₂SO

2 Cl-[N pyridine]

3 MeI

165°

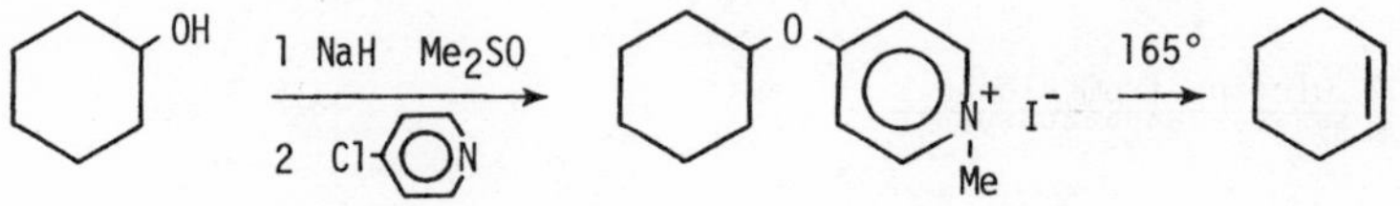

55%

Can J Chem (1972) 50 1181

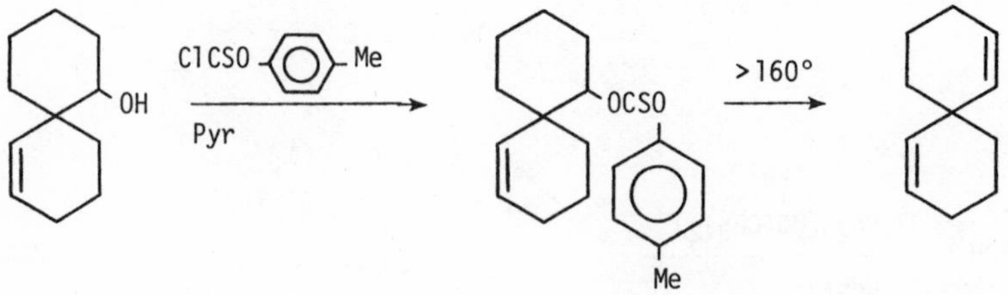

ClCSO-[C₆H₄]-Me

Pyr

>160°

<60%

Chem Comm (1972) 1215

PhCSCl

Pyr

hν

cyclohexane

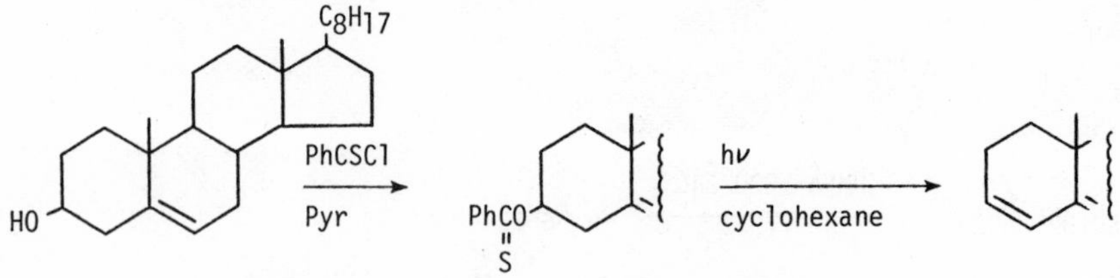

Chem Comm (1971) 1014

$$(CH_2)_{10} \begin{matrix} CMe-OH \\ | \\ CMe-OH \end{matrix} \quad \xrightarrow[\substack{2\ K_2WCl_6 \\ 3\ NaOH\ H_2O}]{1\ MeLi\quad THF\quad Et_2O} \quad (CH_2)_{10} \begin{matrix} CMe \\ \| \\ CMe \end{matrix} \qquad 58\%$$

Chem Comm (1972) 370

Bis(1,5-cycloocta-diene)nickel   DMF   82%

Tetr Lett (1973) 2667

$$\begin{matrix} RCHOH \\ | \\ RCHOH \end{matrix} \quad \xrightarrow[\text{diimidazole\quad toluene}]{\text{N,N'-Thiocarbonyl}} \quad \begin{matrix} RCHO \\ RCHO \end{matrix} CS \quad \xrightarrow{(RO)_3P} \quad \begin{matrix} RCH \\ \| \\ RCH \end{matrix}$$

Ber (1973) 106 1076
JCS C (1971) 2960
Can J Chem (1970) 48 383

Section 199   Olefins from Aldehydes
ooooooooooooooooooooooooo

Reviews: Stereoselective and Stereospecific Olefin Synthesis
                              Quart Rev (1971) 25 135
          The Wittig Reaction   Org React (1965) 14 270

$$PhCHO \quad \xrightarrow[\text{2 MeONa}]{\text{1 TsNHNH}_2} \quad PhCHN_2 \quad \xrightarrow[\text{MeCN\quad pentane}]{Ce(NH_4)_2(NO_3)_6} \quad PhCH=CHPh \qquad <74\%$$

JACS (1971) 93 2086

MeO—C$_6$H$_4$—CHO $\xrightarrow[\text{}]{\text{WCl}_6\text{-BuLi \quad THF}}$ MeO—C$_6$H$_4$—CH=CH—C$_6$H$_4$—OMe   47%

JACS (1972) 94 6538

PhCHO $\xrightarrow[\text{BuLi \quad THF}]{\text{BuCH}_2\text{CH[B(}\bigcirc\text{)}_2\text{]}_2}$ PhCH=CHCH$_2$Bu   45-50%

Tetr Lett (1966) 4315

PhCHO $\xrightarrow[\substack{\text{THF \quad hexane} \\ \text{2 PhSO}_2\text{Cl \quad Pyr}}]{\text{1 Me}_2\text{CHCOOH \quad (i-Pr)}_2\text{NLi}}$ PhCH-CMe$_2$ (O-CO) $\xrightarrow{140-160°}$ PhCH=CMe$_2$ 29%

JACS (1972) 94 2000

PhCHO $\xrightarrow[\text{C}_6\text{H}_6]{\substack{\text{SH} \\ \text{Ph}_2\text{CCOOH \quad TsOH}}}$ PhCHSCPh$_2$ (O-CO) $\xrightarrow[\text{(Et}_2\text{N)}_3\text{P}]{150-160°}$ PhCH=CPh$_2$ 89%

JCS Perkin I (1972) 305

PhCHO $\xrightarrow[\text{2 Al-Hg \quad H}_2\text{O \quad THF}]{\substack{\text{Ph} \\ \text{1 PhSO}_2\text{CLi}_2}}$ PhCH=CHPh   < 90%

Chem Comm (1973) 351

C$_9$H$_{19}$CHO $\xrightarrow[\text{2 Al-Hg \quad HOAc}]{\substack{\text{NMe \quad Me} \\ \text{1 PhSO—CHLi \quad THF}}}$ C$_9$H$_{19}$CH=CHMe   60%

JACS (1973) 95 6462

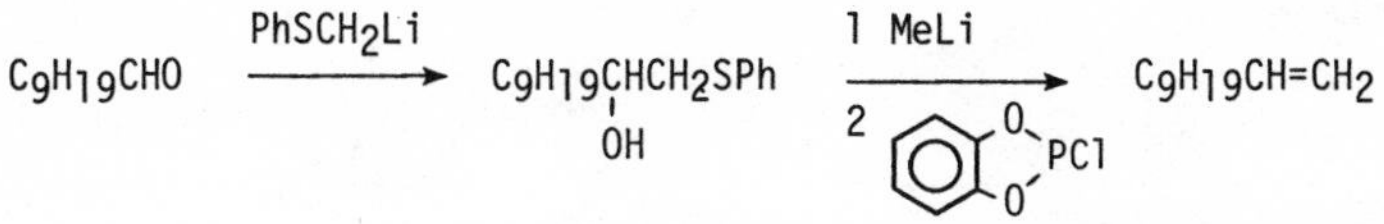

$C_9H_{19}CHO$ $\xrightarrow{\text{PhSCH}_2\text{Li}}$ $C_9H_{19}\underset{\underset{\text{OH}}{|}}{\text{C}}\text{HCH}_2\text{SPh}$ $\xrightarrow[\underset{2}{\overset{1\ \text{MeLi}}{}}]{}$ $C_9H_{19}CH=CH_2$

Tetr Lett (1972) 737

Related methods: Olefins from Ketones (Section 207)

Section 200   Olefins from Aryls

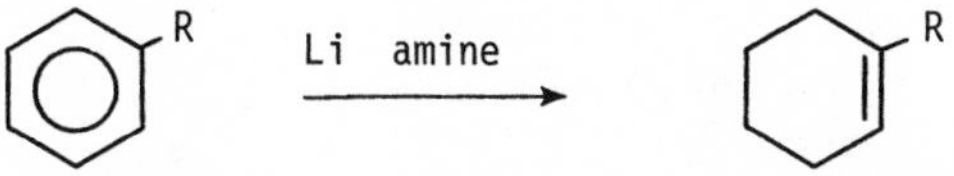

Li amine

Synthesis (1972) 391

Section 201   Olefins from Amides

No additional examples

Section 202   Olefins from Amines

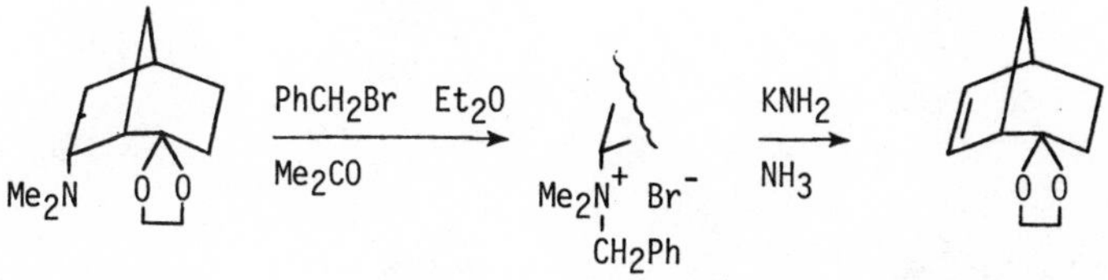

$\xrightarrow[\text{Me}_2\text{CO}]{\text{PhCH}_2\text{Br}\ \ \text{Et}_2\text{O}}$ $\xrightarrow[\text{NH}_3]{\text{KNH}_2}$ ~67%

JOC (1972) 37 2896

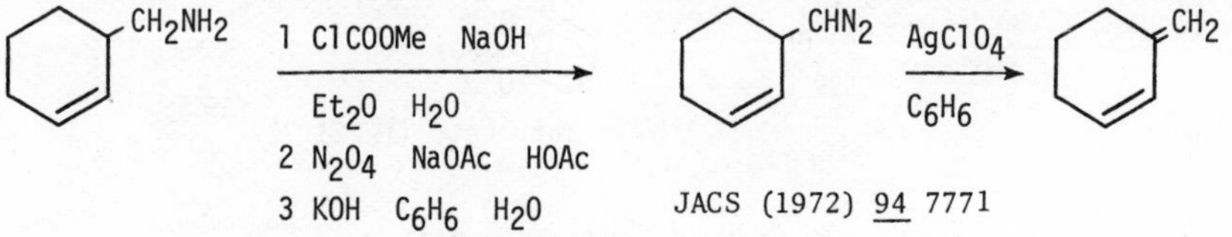

1 ClCOOMe   NaOH

Et$_2$O   H$_2$O

2 N$_2$O$_4$ NaOAc   HOAc

3 KOH   C$_6$H$_6$ H$_2$O

AgClO$_4$

C$_6$H$_6$

JACS (1972) <u>94</u> 7771

Section 203   Olefins from Esters

C$_9$H$_{19}$COOMe

1 PhSCH$_2$Li   THF

2 (PhCO)$_2$O

3 Li   NH$_3$

$$C_9H_{19}\underset{Me}{C}=CH_2$$

46%

JACS (1972) <u>94</u> 4758

Section 204   Olefins from Epoxides

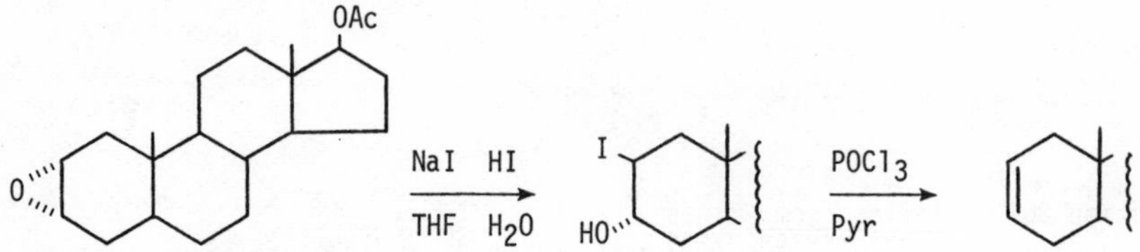

NaI   HI

THF   H$_2$O

POCl$_3$

Pyr

JCS Perkin I (1973) 91

$$\underset{\text{trans}}{\overset{O}{PrCHCHPr}}$$

Zn-Cu   EtOH

$$\underset{\text{trans}}{PrCH=CHPr}$$

JOC (1971) <u>36</u> 1187

$$C_5H_{11}\overset{O}{\overset{\diagup\backslash}{C}H}CHMe \quad \xrightarrow[\text{2 MeI}]{\text{1 Ph}_2\text{PLi THF}} \quad C_5H_{11}CH=CHMe \qquad 75\%$$

cis (left, under formula)   trans (right, under product)

JACS (1973) 95 822
Tetr Lett (1973) 1983

$$\xrightarrow{\text{WCl}_6\text{-BuLi THF}}$$

75%

JACS (1972) 94 6538

$$Et\overset{O}{\overset{\diagup\backslash}{C}H}CH_2 \quad \xrightarrow[\begin{array}{l}\text{THF}\\ \text{2 HBF}_4\\ \text{3 NaI Me}_2\text{CO}\end{array}]{\text{1 Sodium (cyclopentadienyl)dicarbonylferrate}} \quad EtCH=CH_2 \qquad \sim 91\%$$

JACS (1972) 94 7170

$$C_6H_{13}\overset{O}{\overset{\diagup\backslash}{C}H}CH_2 \quad \xrightarrow[\text{CH}_2\text{Cl}_2]{\text{Ph}_3\text{PSe CF}_3\text{COOH}} \quad C_6H_{13}CH=CH_2 \qquad 71\%$$

Chem Comm (1973) 253

Section 205   Olefins from Halides and Sulfonates
ooooooooooooooooooooooooooooooooooooooooo

$$PhCH_2Cl \xdashrightarrow{\text{Ph}_3\text{P}} PhCH_2PPh_3 \overset{+}{} \overset{-}{Cl} \xrightarrow[\text{toluene}]{\text{BuLi}} PhCH=PPh_3 \xrightarrow{\text{S}} PhCH=CHPh$$

Ber (1970) 103 2995

65%

Organometallics in Chem Synth (1972) <u>1</u> 375

JACS (1971) <u>93</u> 4316

PhCH₂Br $\xrightarrow{\text{EtCHO} \quad \text{Ph}_3\text{P} \quad \overset{\displaystyle O}{\overset{\diagup\ \diagdown}{\text{CH}_2\text{-CH}_2}}}$ PhCH=CHEt

Angew (1968) <u>80</u> 535
(Internat Ed <u>7</u> 536)

BuBr $\xrightarrow[\begin{array}{l}2\ \text{NaBH}_4\\3\ \text{MsCl}\quad\text{Na}_2\text{CO}_3\end{array}]{1\ \overset{\overset{\displaystyle \text{Me}}{|}}{\text{BuCOCHCOOEt}}\quad\text{NaH}}$ $\underset{\text{O-CO}}{\overset{\overset{\displaystyle\text{Me}}{|}}{\text{BuCHCBu}}}$ $\xrightarrow[\text{(reflux)}]{\text{Collidine}}$ $\overset{\overset{\displaystyle\text{Me}}{|}}{\text{BuCH=CBu}}$

Tetr Lett (1968) 4569

EtBr $\xrightarrow[\begin{array}{l}2\ \text{PhC-CHPh}\\ \quad\ \ \overset{}{\underset{N}{\diagdown\!\!\diagup}}\end{array}]{1\ \text{Mg}}$ $\underset{\overset{}{\underset{NH}{\diagdown\!\!\diagup}}}{\overset{\overset{\displaystyle\text{Ph}}{|}}{\text{EtC-CHPh}}}$ $\xrightarrow{\text{BuONO}}$ $\overset{\overset{\displaystyle\text{Ph}}{|}}{\text{EtC=CHPh}}$

Tetr Lett (1969) 4001

C₅H₁₁Br --→ C₅H₁₁MgBr $\xrightarrow[2\ \text{NaOAc}\quad\text{HOAc}]{1\ \text{Me}_3\text{SiCH}_2\text{COMe}\quad\text{Et}_2\text{O}}$ $\underset{\text{Me}}{\overset{}{\text{C}_5\text{H}_{11}\text{C=CH}_2}}$ ∼76%

Tetr Lett (1972) 1785

$$C_6H_{13}Br \quad \dashrightarrow \quad C_6H_{13}MgBr \quad \xrightarrow[\text{THF}]{BrCH=CHMe \quad FeCl_3} \quad C_6H_{13}CH=CHMe \qquad \sim 67\%$$

JACS (1971) <u>93</u> 1487
Synthesis (1971) 303

$$PhCH_2I \quad \xrightarrow[\text{air \quad THF}]{ICH_2CH=CH_2 \quad Et_3B} \quad PhCH_2CH_2CH=CH_2 \qquad 72\%$$

(Benzylic and allylic halides only)

JACS (1971) <u>93</u> 1508

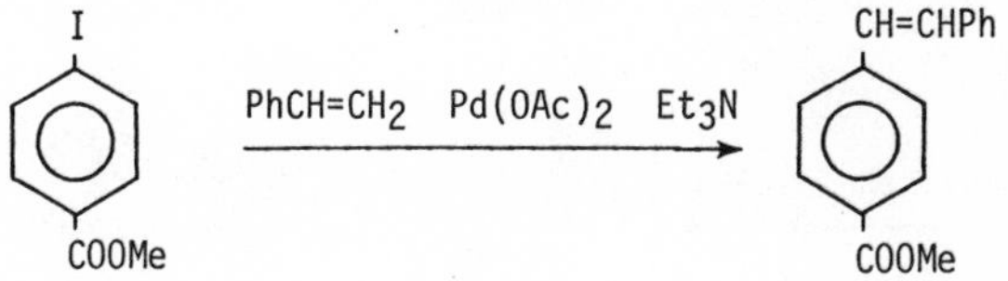

$$\xrightarrow{PhCH=CH_2 \quad Pd(OAc)_2 \quad Et_3N}$$

75%

JOC (1972) <u>37</u> 2320

$$C_8H_{17}Cl \quad \dashrightarrow \quad C_8H_{17}MgCl \quad \xrightarrow[\text{Et}_2O \quad C_6H_6 \quad THF]{CH_2=CHCl \quad [NiCl_2(dpe)]} \quad C_8H_{17}CH=CH_2 \qquad 95\%$$

JACS (1972) <u>94</u> 4374

$$C_7H_{15}Br \quad \xrightarrow{MeCH=CHMgBr \quad THF} \quad C_7H_{15}CH=CHMe \qquad 71\%$$

Bull Soc Chim Fr (1963) 1888

$$RI \quad \xrightarrow{(R'CH=CH)_2CuLi} \quad RCH=CHR'$$

Helv (1971) <u>54</u> 1939

$$\begin{pmatrix} \overset{\displaystyle X}{\underset{\displaystyle |}{CHCH_2}} \\ (CH_2)_{10} \end{pmatrix} \quad \xrightarrow[\text{2 } H_2O_2 \quad H_2O \quad THF]{\text{1 PhSeNa \quad EtOH}} \quad \begin{pmatrix} CH=CH \\ (CH_2)_{10} \end{pmatrix} \qquad <94\%$$

(X=halide or OTs)

Tetr Lett (1973) 1979

$$R_2\overset{\displaystyle X}{\underset{\displaystyle |}{CHCR_2}} \quad \xrightarrow[\text{or diazabicyclononene}]{\text{Diazabicycloundecene}} \quad R_2C=CR_2$$

(X=halide or OTs)

Synthesis (1972) 591

$$\begin{pmatrix} \overset{\displaystyle OMs}{\underset{\displaystyle |}{CH}} - \overset{\displaystyle OMs}{\underset{\displaystyle |}{CH}} \\ (CH_2)_{10} \end{pmatrix} \quad \xrightarrow[\text{MeOCH}_2CH_2OMe]{\alpha\text{-Dimethylaminonaphthalene-Na}} \quad \begin{pmatrix} CH=CH \\ (CH_2)_{10} \end{pmatrix} \qquad 67\%$$

Tetr Lett (1973) 2097
(1972) 3447

Section 206   Olefins from Hydrides

This section lists examples of the replacement of hydrogen by olefinic groups

—CH$_3$ CH$_2$=CHCH=CH$_2$ Na $\longrightarrow$ —CH$_2$CH$_2$CH=CHMe   19%

JOC (1965) 30 82

Section 207   Olefins from Ketones

Conversion of ketones into olefins with longer carbon chains   page 205-207
Conversion of ketones into olefins of the same chain length . . .   207
Conversion of ketones into olefins with shorter carbon chains . .   208

Reviews:　Stereoselective and Stereospecific Olefin Synthesis
　　　　　　　　　　　　　　　　　　　　　Quart Rev (1971) 25 135
　　　　　　　The Wittig Reaction　　　　　Org React (1965) 14 270

EtCO　　　$\xrightarrow{\text{WCl}_6\text{-BuLi　THF}}$　　EtC=CEt　　　　　　　　10%
|　　　　　　　　　　　　　　　　　|　　\
Me　　　　　　　　　　　　　　　Me　　Me

　　　　　　　　　　　　　　　　　JACS (1972) 94 6538

R_2CO　$\xrightarrow[\text{H}_2\text{S}]{\text{N}_2\text{H}_4}$　$R_2CNHNHCR_2$　$\xrightarrow[\text{Ph}_3\text{P}]{\text{Pb(OAc)}_4}$　$R_2C=CR_2$
　　　　　　　　　　　　\ S /

　　　　　　　　　　　　　　　　　JCS Perkin I (1972) 305

　　　　　　　　　　　SH
　　　　　　　　　　　|
　　1 Ph$_2$CCOOH　TsOH　C$_6$H$_6$
$\xrightarrow{\hspace{3cm}}$
　　2 (Et$_2$N)$_3$P　160-200°

68%

　　　　　　　　　　JCS Perkin I (1972) 305

Ph$_2$CO　$\xrightarrow[\text{2　PhSO}_2\text{Cl}]{\text{1　Me}_2\text{CHCOOH　LiN(Pr-i)}_2}$　Ph$_2$C-CMe$_2$　$\xrightarrow{140\text{-}160°}$　Ph$_2$C=CMe$_2$　47%
　　　　　　　　　　　　　　　　　　　　　　　|　|
　　　　　　　　　　　　　　　　　　　　　　O-CO

　　　　　　　　　　　　　　　　　JACS (1972) 94 2000
　　　　　　　　　　　　　　　　　　JOC (1971) 36 1149

　　1 CCl$_3$COCl　Zn
$\xrightarrow{\hspace{2cm}}$
　　　Et$_2$O

　　2 Δ

　　Na　NH$_3$
$\xrightarrow{\hspace{1.5cm}}$

　　　　　　　Synthesis (1972) 565

Resin-Ph$_3$P=CHPr
                              → cis   100%

PhCO
| 1 Resin-Ph$_3$P=CHPr   PhC=CHPr
Me   → trans   Me   59%
           2 BuLi
           3 LiClO$_4$
           4 HCl Et$_2$O
           5 t-BuOK
                                            Annalen (1973) 227
                                            Chem Comm (1972) 134

PhCH$_2$Cl   Ph$_3$P $\overset{O}{\overset{\wedge}{CH_2CH_2}}$
─────────────────────────→   69%
        90°

                            Angew (1972) 84 1173
                            (Internat Ed 11 1041)

1 PhSCH$_2$Li THF
─────────────────────   PhCOO CH$_2$SPh   Li
2 BuLi (PhCO)$_2$O   ─────   31%
                                               NH$_3$

                            JACS (1972) 94 4758

Ph$_2$CO $\xrightarrow{\text{PhSCH}_2\text{Li}}$ Ph$_2$CCH$_2$SPh $\xrightarrow[\text{2 o-Phenylene phosphochloridite}]{\text{1 MeLi}}$ Ph$_2$C=CH$_2$
                                   |
                                   OH
                                        Tetr Lett (1972) 737

PhCO $\xrightarrow[\text{MeLi}]{\text{t-BuSOCH}_2\text{Et}}$ PhC-CHEt $\xrightarrow{\text{NCS CH}_2\text{Cl}_2}$ PhC=CHEt   73%
| | |
Me   Me   Me

                            JACS (1973) 95 3420
                            Tetr Lett (1972) 649

$$C_{15}H_{31}CO \quad \xrightarrow[\text{2 Al-Hg HOAc}]{\overset{\overset{\displaystyle NMe}{\|}}{\text{1 PhSOCH}_2\text{Li} \quad \text{THF}}} \quad C_{15}H_{31}\underset{Me}{\overset{|}{C}}=CH_2 \qquad 90\%$$
$$\underset{Me}{|}$$

JACS (1973) 95 6462

$$Ph_2CO \quad \xrightarrow{\overset{\overset{\displaystyle Ph}{|}}{PhSO_2CLi_2}} \quad Ph_2C=\overset{\overset{\displaystyle Ph}{|}}{C}SO_2Ph \quad \xrightarrow[\text{THF}]{\text{Al-Hg} \quad H_2O} \quad Ph_2C=CHPh \qquad <85\%$$

Chem Comm (1973) 351

$$\left(\underset{(CH_2)_{10}}{\overset{COCH_2}{\Big<}}\right) \quad \text{--->} \quad \left(\underset{(CH_2)_{10}}{\overset{\overset{\overset{\displaystyle OAc}{|}}{C=CH}}{\Big<}}\right) \quad \xrightarrow[\text{Bu}_2\text{O}]{\text{Fe(CO)}_5} \quad \left(\underset{(CH_2)_{10}}{\overset{CH=CH}{\Big<}}\right) \qquad <40\%$$

Tetr Lett (1973) 447

$$\xrightarrow[\text{Et}_2\text{O}]{\text{Me}_3\text{SiCl} \quad \text{Zn}} \qquad 48\%$$

Chem Comm (1973) 935

$$\xrightarrow[\text{HOAc}]{\overset{\text{PhCH}_2\text{SH}}{\text{BF}_3\cdot\text{Et}_2\text{O}}} \quad PhCH_2S \cdots \xrightarrow[\text{EtOH}]{\text{Nickel boride}} \cdots \qquad 67\%$$

JCS Perkin I (1973) 654

$$BuCH_2CH_2CH_2COPh \xrightarrow{h\nu} BuCH=CH_2 \qquad 74\%$$

JOC (1971) 36 1838

Related methods: Olefins from Aldehydes (Section 199)

Section 208   Olefins from Nitriles
ooooooooooooooooooooo

No additional examples

Section 209   Olefins from Olefins
oooooooooooooooooo

Homologation, alkylation and arylation of olefins . . . . . page 208-209
Migration of double bonds 209-210
Cis-trans interconversion and equilibration 210-211

1 B$_2$H$_6$ THF
2 HC≡CCH$_2$Cl
3 MeLi   Et$_2$O
  HOAc

69%

Synthesis (1973) 672

```
            Me
            |
       i-PrCBH2
1          |
           Me
```

EtCH=CHEt   ──────────→   EtCHCH$_2$Et   78%

trans   2 BuC≡CBr   CH=CHBu   Synthesis (1972) 555
             3 MeONa MeOH   JACS (1971) 93 6309
             4 i-PrCOOH   trans   (1972) 94 4013

1 BH$_3$ THF
2 BuC≡CH
3 BrCN CH$_2$Cl$_2$

CH=CHBu
trans   69%

JACS (1972) 94 6560

$$EtCH=CH_2 \quad \xrightarrow[\substack{2 \text{ MeC=CH} \quad \text{MeOH} \\ \diagdown / \\ CH_2}]{1 \text{ BH}_3} \quad Et(CH_2)_3\underset{\underset{Me}{|}}{C}=CH_2 \qquad \text{91\%}$$

Tetr Lett (1972) 4627

$$BuCH=CH_2 \quad \xrightarrow[\text{Li}_2\text{PdCl}_4 \quad \text{MeOH}]{MeCo(dmg)_2\cdot Pyr} \quad BuCH=CHMe \qquad \text{32\%}$$

Chem Comm (1971) 849

$$\underset{EtCH=CPr}{\overset{CH_3}{\underset{|}{}}} \quad \xrightarrow[\substack{2 \text{ MeSO}_2\text{CHCOOMe} \\ | \\ Na \\ 3 \text{ NaOH} \\ 4 \text{ } 180°}]{1 \text{ PdCl}_2 \quad Na_2CO_3 \quad CH_2Cl_2} \quad \underset{EtCH=CPr}{\overset{CH_2CH_2SO_2Me}{\underset{|}{}}} \xrightarrow[EtNH_2]{Li} \quad \underset{EtCH=CPr}{\overset{CH_2Me}{\underset{|}{}}}$$

JACS (1973) <u>95</u> 292

PhI　Pd(OAc)$_2$

Bu$_3$N

85%

JOC (1972) <u>37</u> 2320

$$EtCH_2CH=CH_2 \quad \underset{\longleftarrow}{\xrightarrow{\text{Ni-HCN} \quad C_6H_6}} \quad EtCH=CHMe$$

Chem Ind (1971) 1465

t-BuOK　18-crown-6

Me$_2$SO

Tetr Lett (1972) 1797

$$PhCH_2CH=CH_2 \quad \xrightarrow[\text{t-BuOOH}]{RuCl_2(Ph_3P)_3} \quad PhCH=CHMe$$

Chem Comm (1971) 562

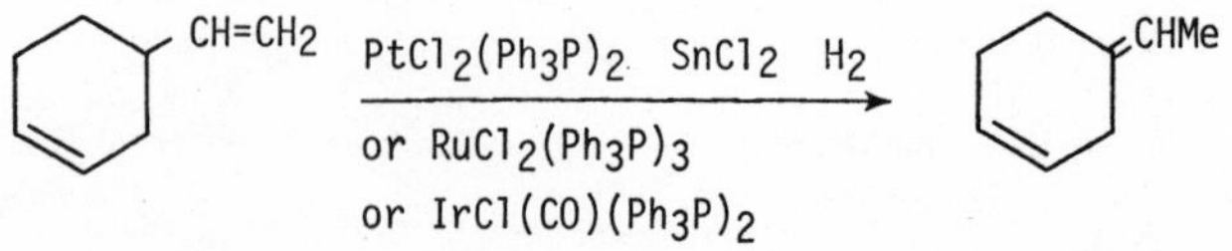

PtCl$_2$(Ph$_3$P)$_2$ SnCl$_2$ H$_2$
or RuCl$_2$(Ph$_3$P)$_3$
or IrCl(CO)(Ph$_3$P)$_2$

JOC (1971) 36 2497

PtClH(PMe$_2$Ph)$_2$-MeSO$_3$F   Chem Comm (1973) 766

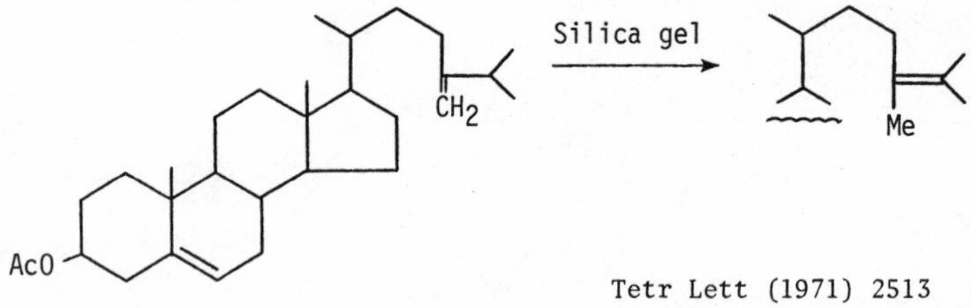

Silica gel

Tetr Lett (1971) 2513

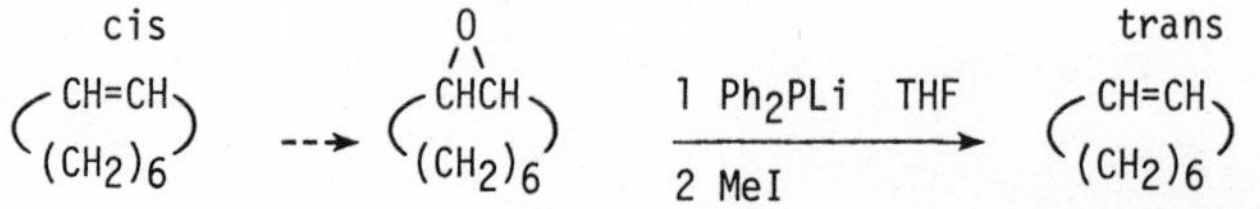

JACS (1971) 93 4070
(1973) 95 822
Tetr Lett (1973) 1983

< 75%

2-Acetonaphthone hν

JOC (1973) 38 1247

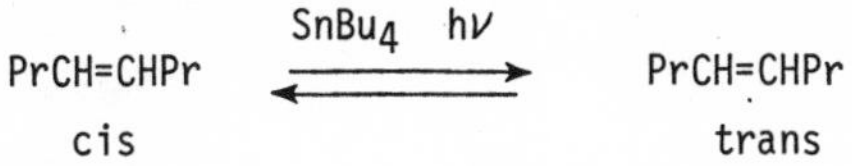

$$\text{PrCH=CHPr} \underset{\xrightarrow{\hspace{2cm}}}{\overset{\text{SnBu}_4 \quad h\nu}{\xleftarrow{\hspace{2cm}}}} \text{PrCH=CHPr}$$

cis   trans

Ber (1973) <u>106</u> 505

The following catalysts may also be used for cis-trans equilibration:

CuCl_2-hν JOC (1972) <u>37</u> 3561

Ph_2S-hν JACS (1972) <u>94</u> 2124

HSCH_2COOH Chem Ind (1965) 507

$(\text{i-Bu})_3\text{Al}$ γ-rays   Ber (1972) <u>105</u> 88

Section 210   Olefins from Miscellaneous Compounds

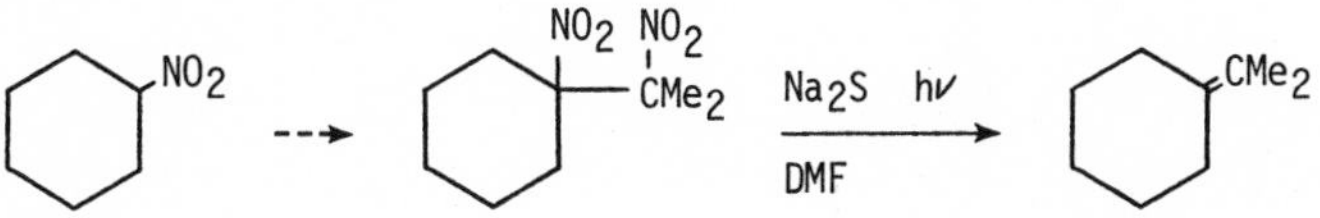

∘∘

JACS (1971) <u>93</u> 4316

Section 210A   Protection of Olefins

∘∘∘∘∘∘∘∘∘∘∘∘∘∘∘∘∘∘∘∘∘∘∘∘∘∘

The protection of isolated double bonds is considered in this section

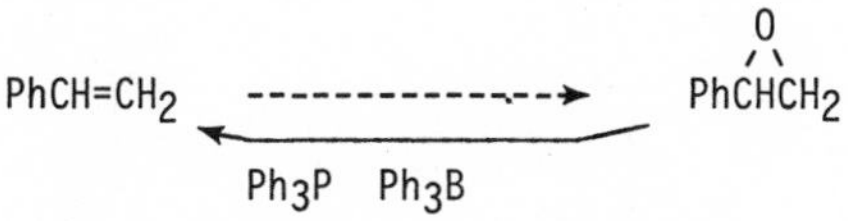

$$\text{PhCH=CH}_2 \underset{\text{Ph}_3\text{P} \quad \text{Ph}_3\text{B}}{\overset{}{\rightleftharpoons}} \text{PhCHCH}_2 \quad (\text{epoxide})$$

Ber (1955) <u>88</u> 1654

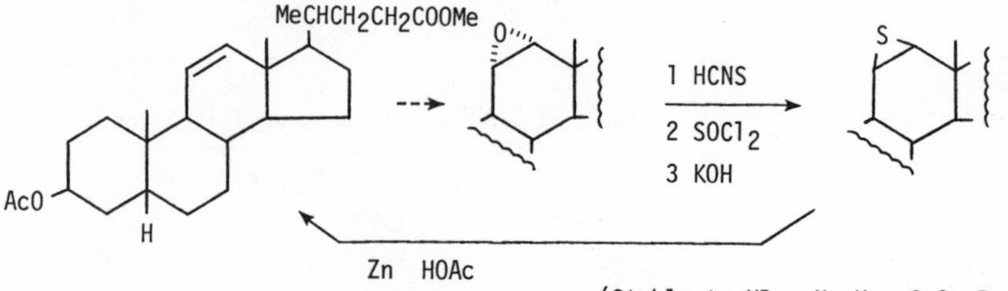
MeCHCH2CH2COOMe
AcO
H
O
1 HCNS
2 SOCl2
3 KOH
S
Zn HOAc

(Stable to HBr, Na-Hg, CrO_3-Pyr, $CrCl_2$ and H_2/Pt)

Chem Pharm Bull (1960) 8 621

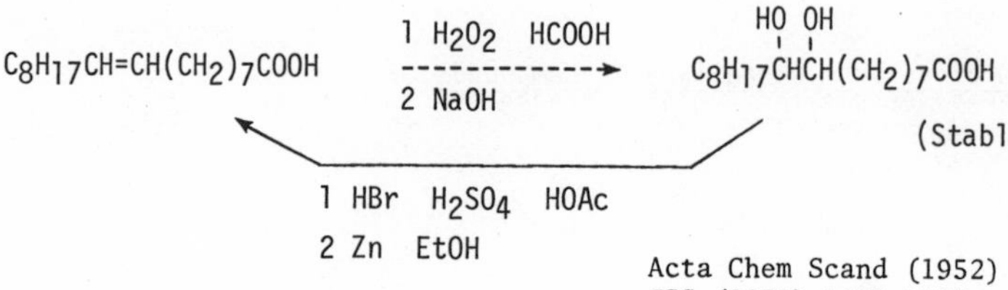
C8H17CH=CH(CH2)7COOH
1 H2O2 HCOOH
2 NaOH
HO OH
C8H17CHCH(CH2)7COOH
1 HBr H2SO4 HOAc
2 Zn EtOH

(Stable to Br_2)

Acta Chem Scand (1952) 6 1127
JCS (1951) 1079 1087

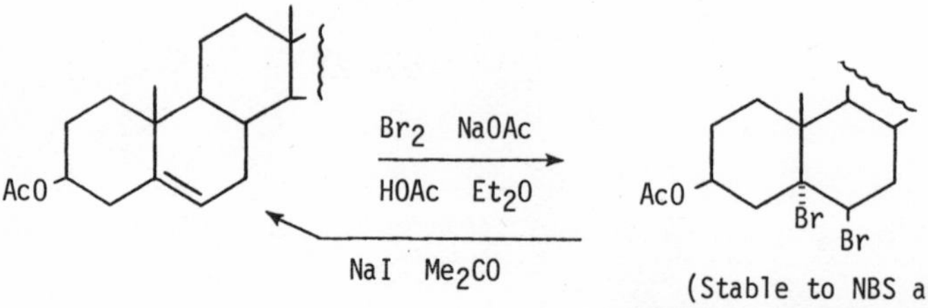
AcO
Br2 NaOAc
HOAc Et2O
NaI Me2CO
AcO
Br
Br

(Stable to NBS and H_2/Pd)

JOC (1965) 30 1658
J Biol Chem (1958) 230 447
Org Synth (1963) Coll Vol 4 195

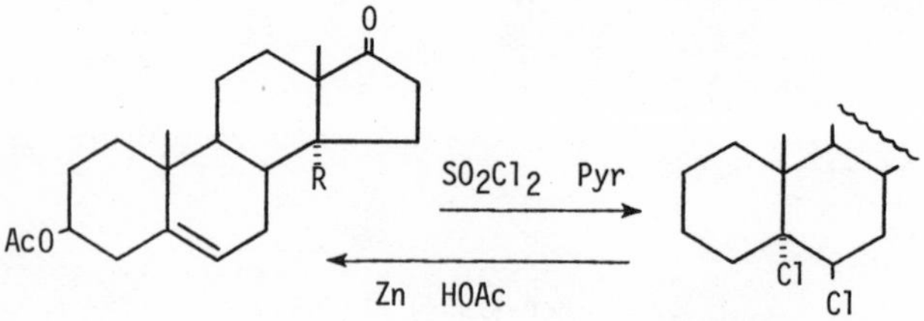
O
AcO
R
SO2Cl2 Pyr
Zn HOAc
Cl
Cl

(Stable to HBr, CrO_3, $SOCl_2$-Pyr and H_2/Pt)

Rec Trav Chim (1971) 90 549
JCS (1961) 4547

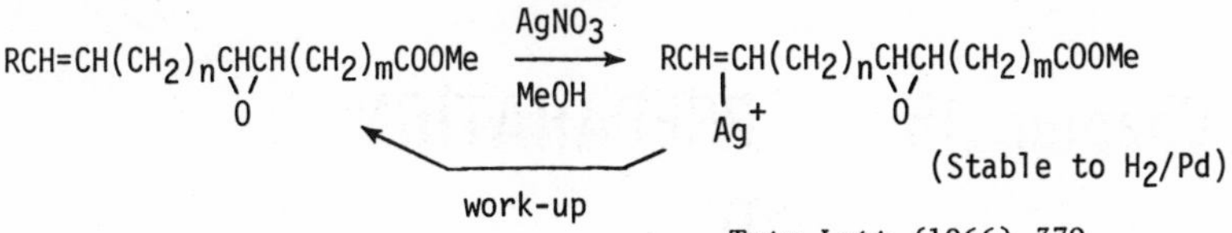

$$\text{RCH=CH(CH}_2)_n\text{CHCH(CH}_2)_m\text{COOMe} \underset{\text{MeOH}}{\overset{\text{AgNO}_3}{\longrightarrow}} \text{RCH=CH(CH}_2)_n\text{CHCH(CH}_2)_m\text{COOMe}$$

(Stable to H_2/Pd)

work-up

Tetr Lett (1966) 379

ooooooooooooooooooooo

Sections 211 to 299 are reserved for future additions (e.g. the preparation of nitro compounds)

Chapter 15   PREPARATION
OF
DIFUNCTIONAL
COMPOUNDS

Section 300   <u>Acetylene — Acetylene</u>

Diacetylenes

Review: The Coupling of Acetylenic Compounds

Advances in Org Chem (1963) <u>4</u> 225

$$PhC{\equiv}CH \xrightarrow{\text{Cu(OAc)}_2 \quad \text{Pyr} \quad \text{MeOH}} PhC{\equiv}C\text{-}C{\equiv}CPh \qquad 70\text{-}80\%$$

Org Synth (1965) <u>45</u> 39
JCS (1959) 889
   (1960) 3614
JACS (1972) <u>94</u> 658

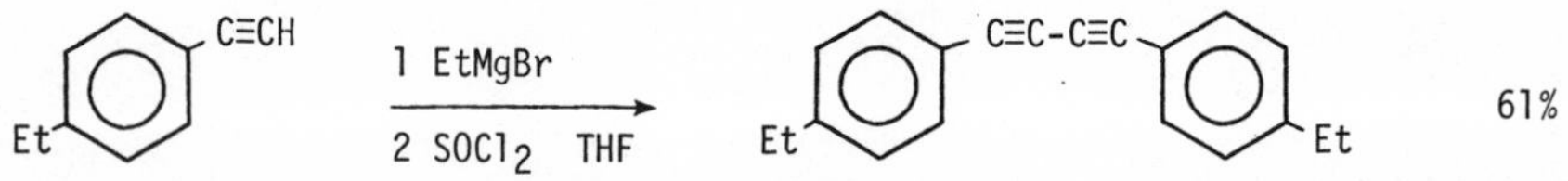

61%

JOC (1972) <u>37</u> 3749

$BuC\equiv CH$ $\xrightarrow[\substack{2\ CuCl_2 \\ 3\ I_2\ Et_2O}]{1\ EtMgBr\ Et_2O}$ $BuC\equiv C-C\equiv CBu$ 71%

JCS (1954) 1704

$PhC\equiv CH$ $\xrightarrow[THF\ H_2O]{Br_2\ NaOH}$ $PhC\equiv CBr$ $\xrightarrow[C_6H_6]{CuCl_2\ BuNH_2}$ $PhC\equiv C-C\equiv CPh$ 84%

Advances in Org Chem (1963) 4 318
JCS (1962) 3055

$PhC\equiv CH$ $\xrightarrow[\substack{NH_2OH\cdot HCl\ EtNH_2 \\ 2\ Copper\ tetrammine\ sulfate \\ Me_2CO}]{1\ BrC\equiv CCOOH\ CuCl_2}$ $PhC\equiv C-C\equiv CH$ 41%

Ber (1964) 97 2586

$PhC\equiv CH$ $\xrightarrow[\substack{EtNH_2\ DMF \\ 2\ NaOH\ MeOH\ H_2O}]{1\ BrC\equiv CSiEt_3\ CuCl_2\ NH_2OH\cdot HCl}$ $PhC\equiv C-C\equiv CH$ ~50%

Tetrahedron (1972) 28 4591 5221

$C_5H_{11}C\equiv CH$ $\xrightarrow[\substack{2\ CuCl_2 \\ 3\ BrCH_2C\equiv CC_5H_{11}}]{1\ EtMgBr\ Et_2O}$ $C_5H_{11}C\equiv CCH_2C\equiv CC_5H_{11}$ 70%

JACS (1951) 73 4601
Org Synth (1970) 50 97

$BuBr$ $\xrightarrow{ClCH_2C\equiv CCH_2Cl\ NaNH_2}$ $BuC\equiv C-C\equiv CBu$ ~ 50%

JCS (1951) 44

$C_5H_{11}Br$ $\xrightarrow{\text{ClCH}_2\text{C}\equiv\text{CCH}_2\text{Cl} \quad \text{NaNH}_2}$ $C_5H_{11}C\equiv C-C\equiv CH$

JCS (1955) 1007

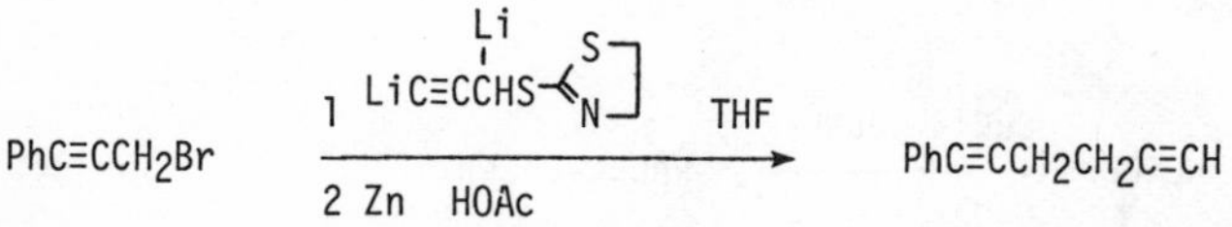

$PhC\equiv CCH_2Br$ $\xrightarrow[\text{2 Zn HOAc}]{\text{1 LiC}\equiv\text{CCHS-}\overset{\text{Li}}{\underset{\text{N}}{\diagdown}}\text{S}\rfloor \quad \text{THF}}$ $PhC\equiv CCH_2CH_2C\equiv CH$

Tetr Lett (1972) 2117

$\xrightarrow[\text{2 NaNH}_2 \quad \text{NH}_3]{\text{1 PCl}_5 \quad C_6H_6}$

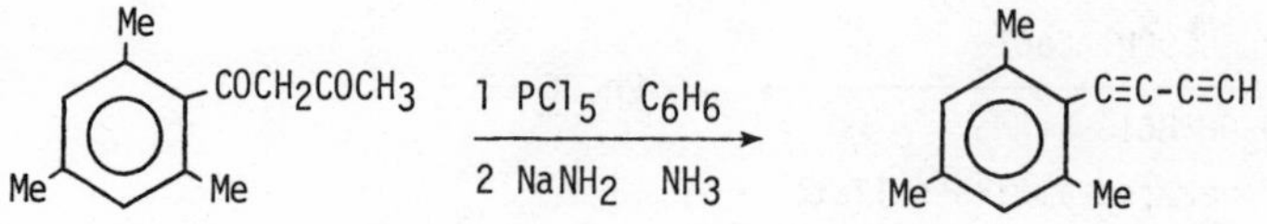

Tetrahedron (1972) 28 5221

$PhCOCH_2CH_2COPh$ $\xrightarrow[\substack{\text{2 TsOH}\\ \text{3 Me}_2\text{SO}_4 \\ \text{4 NaOH}}]{\text{1 EtSH HCl dioxane}}$ $PhC\equiv C-C\equiv CPh$

Z Chem (1969) 9 229

Section 301   Acetylene — Carboxylic Acid

Acetylenic acids

Review: Recent Developments in the Synthesis of Fatty Acids

Chem Rev (1957) 57 191

$C_{11}H_{23}C\equiv CH$ $\xrightarrow[\text{2 Br(CH}_2)_4\text{COOH THF}]{\text{1 LiNH}_2 \quad \text{NH}_3}$ $C_{11}H_{23}C\equiv C(CH_2)_4COOH$ 49%

JCS (1965) 894

$C_{12}H_{25}C\equiv CH$ $\xrightarrow[\text{2 } CO_2]{\text{1 EtMgBr Et}_2O}$ $C_{12}H_{25}C\equiv CCOOH$ 76%

Annalen (1965) <u>685</u> 154

LiAlH$_4$, CO$_2$ Izv (1964) 2066
(Chem Abs <u>62</u> 7629)

PhCOCl $\dashrightarrow$ PhCOCH(COOBu-t)$_2$ $\xrightarrow[\begin{array}{l}\text{2 } \beta\text{-Naphthalene}\\ \text{sulfonic anhydride}\\ \text{3 Acid}\\ \text{4 NaHCO}_3\end{array}]{\text{1 t-BuOK}}$ PhC$\equiv$CCOOH

JCS <u>C</u> (1971) 2013

PhCHO $\xrightarrow[C_6H_6]{\overset{\displaystyle Br}{Ph_3P=CCOOMe}}$ $\overset{\displaystyle Br}{PhCH=CCOOMe}$ $\xrightarrow{\text{KOH MeOH}}$ PhC$\equiv$CCOOH

Org React (1965) <u>14</u> 270

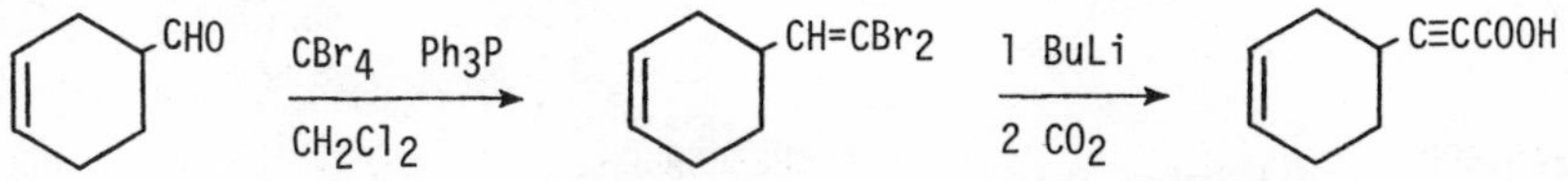

CBr$_4$ Ph$_3$P / CH$_2$Cl$_2$ 1 BuLi / 2 CO$_2$

Tetr Lett (1972) 3769

1 CuC$\equiv$CCOOEt Pyr / 2 NaOH

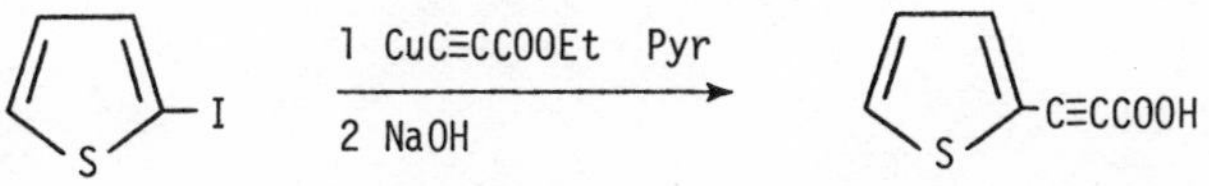

JCS <u>C</u> (1969) 2173

PhCH=CHBr $\xrightarrow[\text{2 } CO_2]{\text{1 BuLi Et}_2O}$ PhC$\equiv$CCOOH   42%

JACS (1940) <u>62</u> 2327

$$HC \equiv CCH_2CH_2OH \xrightarrow[\text{Me}_2CO \quad H_2O]{CrO_3 \quad H_2SO_4} HC \equiv CCH_2COOH \qquad 28\%$$

JCS (1949) 604

$$MeCOCH_2COOEt \xrightarrow[\text{2 Br}_2]{\text{1 N}_2H_4 \quad EtOH} \underset{\underset{NH}{\overset{N \quad CO}{\diagdown \diagup}}}{MeC\text{-}CBr_2} \xrightarrow[]{NaOH \quad H_2O} MeC \equiv CCOOH \qquad 29\%$$

JACS (1958) <u>80</u> 599
JOC (1966) <u>31</u> 2867

Also via: Acetylenic esters   Section
           Acetylenic amides   306
                                304

Section 302   Acetylene — Alcohol
              ○○○○○○○○○○○○○○○○○○○○○○

Acetylenic alcohols, hydroxyacetylenes

$$PrCHC \equiv CH \xrightarrow[\text{2 HCHO}]{\text{1 EtMgBr} \quad THF} PrCHC \equiv CCH_2OH \qquad 80\%$$

JCS Perkin I (1973) 720
JCS (1950) 2100

$$BuCH_2C \equiv CH \xrightarrow[]{SeO_2 \quad EtOH} \underset{OH}{BuCHC \equiv CH} \qquad 27\%$$

Compt Rend (1933) <u>196</u> 706

$$C_5H_{11}C \equiv CH \xrightarrow[\underset{O}{\overset{2 \quad CH_2CH_2}{\diagdown \diagup}}]{\text{1 Li} \quad NH_3} C_5H_{11}C \equiv CCH_2CH_2OH$$

Chem Comm (1967) 1055

$C_6H_{13}CHO$ $\xrightarrow[\text{Mg Et}_2\text{O or Zn THF}]{\text{BrCH}_2C{\equiv}CH}$ $C_6H_{13}\underset{\underset{OH}{|}}{C}HCH_2C{\equiv}CH$

JCS (1955) 1740 1007

i-PrCHO $\xrightarrow{\text{HC}{\equiv}\text{CMgBr THF}}$ i-Pr$\underset{\underset{OH}{|}}{C}HC{\equiv}$CH   86%

Bull Soc Chim Fr (1964) 3218

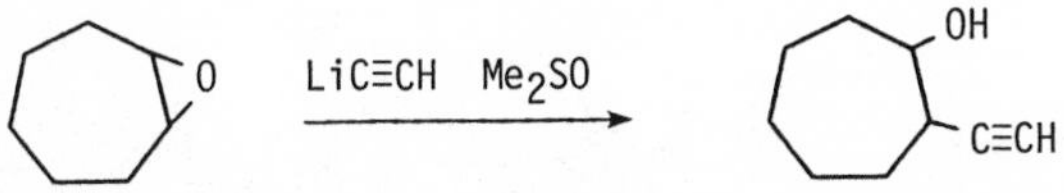

$\xrightarrow{\text{LiC}{\equiv}\text{CH Me}_2\text{SO}}$

Chem Comm (1969) 674

$\xrightarrow[\text{Et}_3\text{N toluene}]{\text{Et}_2\text{AlC}{\equiv}\text{CC}_6\text{H}_{13}}$

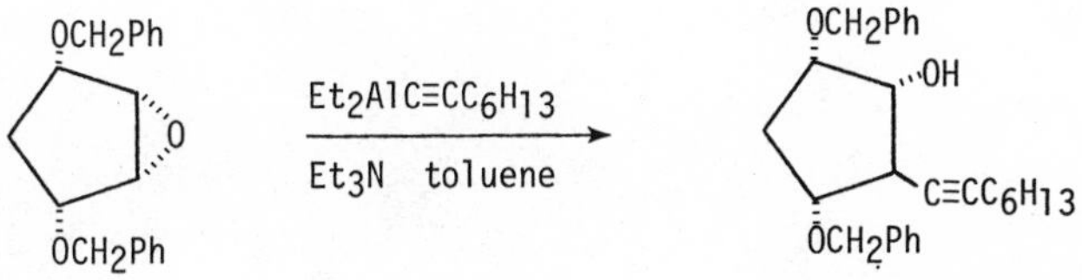

Chem Comm (1968) 634
Tetr Lett (1970) 2695

$C_{11}H_{23}Br$ $\xrightarrow[\text{NH}_3 \text{ THF}]{\text{HC}{\equiv}\text{CCH}_2\text{OH LiNH}_2}$ $C_{11}H_{23}C{\equiv}CCH_2OH$ 87%

JCS (1963) 5889
    (1965) 4373
JCS C (1969) 2173

COOH
|
(CH₂)₇Br   HC≡C(CH₂)₈OH   COOH
                                    |
              ⟶   (CH₂)₇C≡C(CH₂)₈OH
              LiNH₂   THF

                                    JCS C (1967) 1556
                                        (1965) 894

[cyclohexanone] ═O   HC≡CH   Na   NH₃   [cyclohexane with] OH
                        ⟶ |—C≡CH

                                                                            65-75%

                                              Org Synth (1955) Coll Vol 3 416
                        LiC≡CH·NH₂CH₂CH₂NH₂   JOC (1964) 29 1872

[cyclohexanone] ═O   HC≡CC₆H₁₃   MeMgBr   [cyclohexane] OH
                        ⟶ |—C≡CC₆H₁₃
                        CH₂Cl₂
                                                                            ~74%

                                              Ber (1962) 95 2557
                                              JCS (1957) 573

[cyclohexane]—C≡CCOOMe   i-Bu₂AlH   hexane   [cyclohexane]—C≡CCH₂OH
                            ⟶
                                                                            >80%

                            Tetr Lett (1973) 1495

Section 303   Acetylene ⎯ Aldehyde
               °°°°°°°°°°°°°°°°°°°°°°°°°°

Acetylenic aldehydes

C≡CH   C≡CCH(OEt)₂   C≡CCHO
| 1 EtMgBr   Et₂O | (COOH)₂   H₂O |
(CH₂)₇ ⟶   (CH₂)₇   ⟶   (CH₂)₇   <82%
| 2 HC(OEt)₃ | |
C≡CH   C≡CCH(OEt)₂   C≡CCHO

                                  JCS (1951) 2693
                                  Org Synth (1963) Coll Vol 4 801

BuC≡CH
$\xrightarrow{\begin{array}{l}\text{1 NaNH}_2\quad\text{NH}_3\\ \text{2 HCOOEt}\quad\text{Et}_2\text{O}\\ \text{3 HOAc}\end{array}}$
BuC≡CCHO   24%

JCS (1950) 3361

BuC≡CH
$\xrightarrow{\begin{array}{l}\text{1 EtMgBr}\quad\text{Et}_2\text{O}\\ \text{2 HCONMe}_2\\ \text{3 H}_2\text{SO}_4\quad\text{H}_2\text{O}\end{array}}$
BuC≡CCHO   51%

JCS (1958) 1054

$\xrightarrow{\begin{array}{l}\text{1 CuC≡CCH(OEt)}_2\quad\text{Pyr}\\ \text{2 H}_2\text{SO}_4\quad\text{H}_2\text{O}\end{array}}$
49%

JCS C (1969) 2173

—C≡CCH₂OH
$\xrightarrow[\text{C}_6\text{H}_6]{\text{Nickel peroxide}}$
C≡CCHO   72%

JCS C (1969) 2173
MnO₂   JCS (1958) 1054
          Ber (1973) 106 2755
CrO₃-3,5-dimethylpyrazole   Tetr Lett (1973) 4499

PhCH=CHCHO
$\xrightarrow{\begin{array}{l}\text{1 Br}_2\quad\text{HOAc}\\ \text{2 K}_2\text{CO}_3\\ \text{3 HC(OEt)}_3\quad\text{NH}_4\text{Cl}\\ \quad\text{EtOH}\\ \text{4 KOH}\quad\text{EtOH}\end{array}}$
PhC≡CCH(OEt)₂
$\xrightarrow[\text{H}_2\text{O}]{\text{H}_2\text{SO}_4}$
PhC≡CCHO

Org Synth (1955) Coll Vol 3 731

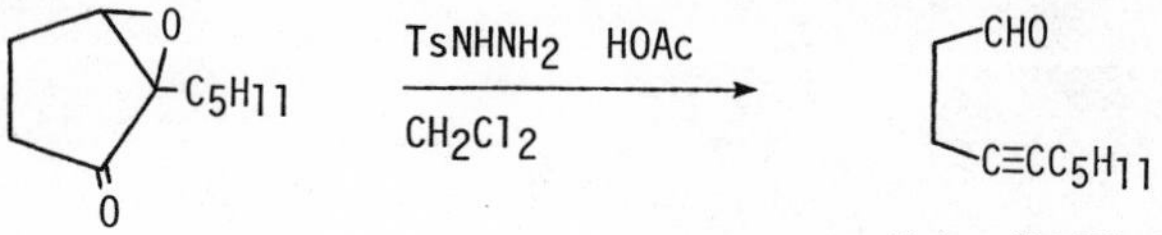

27%

Helv (1967) 50 2101
     (1972) 55 1276

Section 304   Acetylene — Amide
                ○○○○○○○○○○○○○○○○○○○○

Acetylenic amides

$$BuC\equiv CH \xrightarrow[\text{2 Me}_2\text{NCOCl}]{\text{1 Na   NH}_3} BuC\equiv CCONMe_2$$

60%

Compt Rend (1955) 240 989

Section 305   Acetylene — Amine
                ○○○○○○○○○○○○○○○○○○○○

Acetylenic amines

Review: Synthesis and Reactions of the Alkynylamines
                              Angew (1967) 79 744
                              (Internat Ed 6 767)

$$MeC\equiv CH \xrightarrow[\text{Zn(OAc)}_2 \quad 120°]{\text{Et}_2\text{NH}\quad\text{Cd(OAc)}_2} MeC\equiv CCMe_2$$
$$\underset{NEt_2}{}$$

JACS (1961) 83 216

$$PhC\equiv CH \xrightarrow[\text{dioxane}]{\text{Et}_2\text{NH   HCHO}} PhC\equiv CCH_2NEt_2$$

80%

Ber (1933) 66 418

PhC≡CH $\xrightarrow[\text{2 ClNEt}_2]{\text{1 EtMgBr \quad Et}_2\text{O}}$ PhC≡CNEt$_2$　　　　　　2%

Annalen (1960) <u>638</u> 33

PhC≡CH $\xrightarrow[\text{C}_6\text{H}_6]{\text{Me}_2\text{NH \quad O}_2 \quad \text{Cu(OAc)}_2}$ PhC≡CNMe$_2$　　　　　> 40%

Tetr Lett (1968) 5357

BuBr $\xrightarrow[\quad\quad\quad]{\overset{\text{Me}}{\overset{|}{\text{LiC≡CNPh}}} \text{HMPA}}$ $\overset{\text{Me}}{\overset{|}{\text{BuC≡CNPh}}}$　　　　　75%

Bull Soc Chim Fr (1965) 2787

BuC≡COEt $\xrightarrow[\text{2 110-120°}]{\text{1 LiNEt}_2 \quad \text{Et}_2\text{O}}$ BuC≡CNEt$_2$　　　　　65%

Rec Trav Chim (1964) <u>83</u> 1211

PhC≡CCl $\xrightarrow{\text{LiNMe}_2 \quad \text{Et}_2\text{O}}$ PhC≡CNMe$_2$　　　　　87%

Angew (1964) <u>76</u> 537
(Internat Ed <u>3</u> 506)

t-BuC≡CCl $\xrightarrow{\text{Me}_3\text{N} \quad 135°}$ t-BuC≡CNMe$_2$　　　　　44%

Angew (1964) <u>76</u> 537
(Internat Ed <u>3</u> 506 582)
Ber (1970) <u>103</u> 564

PhCH=CClF $\xrightarrow{\text{LiNEt}_2 \quad \text{Et}_2\text{O}}$ PhC≡CNEt$_2$　　　　　86%

Angew (1963) <u>75</u> 638
(Internat Ed <u>2</u> 477)

Section 306   Acetylene — Ester

Acetylenic esters

$C_5H_{11}C{\equiv}CH$ $\xrightarrow[\begin{array}{l}2\ BF_3\\3\ N_2CHCOOEt\end{array}]{1BuLi\quad THF}$ $C_5H_{11}C{\equiv}CCH_2COOEt$ 87%

Can J Chem (1972) <u>50</u> 1105

$MeC{\equiv}CH$ $\dashrightarrow$ $MeC{\equiv}CLi$ $\xrightarrow{ClCOOEt}$ $MeC{\equiv}CCOOEt$ 71%

JOC (1973) <u>38</u> 3588
Annalen (1958) <u>614</u> 37

$PhC{\equiv}CH$ $\xrightarrow{C(OEt)_4\quad ZnCl_2}$ $PhC{\equiv}CC(OEt)_3$ $\dashrightarrow$ $PhC{\equiv}CCOOEt$ < 14%

JACS (1958) <u>80</u> 4607

$\xrightarrow[2\ 270°]{1\ Ph_3P{=}CHCOOEt\quad C_6H_6}$ 70%

J Med Chem (1973) <u>16</u> 72
Ber (1961) <u>94</u> 3005

$C_6H_{13}CHO$ $\xdashrightarrow{N_2CHCOOEt}$ $C_6H_{13}\overset{OH}{\underset{N_2}{CHCCOOEt}}$ $\xrightarrow[MeCN]{BF_3\cdot Et_2O}$ $C_6H_{13}C{\equiv}CCOOEt$

Synth Comm (1972) 331

$\xrightarrow{LiCH_2CONMe_2}$ $\xrightarrow[2\ Tl(NO_3)_3]{1\ N_2H_4}$ 57%

Tetr Lett (1973) 1495

$\xrightarrow{\text{CuC} \equiv \text{CCOOEt}\quad \text{Pyr}}$ —C≡CCOOEt   ~61%

JCS C (1969) 2173

$C_8H_{17}C \equiv C(CH_2)_7COOH$ $\xrightarrow[\text{electrolysis}]{\overset{\displaystyle \text{COOH}}{\underset{\displaystyle (CH_2)_4COOMe}{|}}}$ $C_8H_{17}C \equiv C(CH_2)_{11}COOMe$ 28%

JCS (1955) 2218

$C_5H_{11}COCH_2COOMe$ $\xrightarrow[\text{2 Tl}(NO_3)_3]{\text{1 N}_2H_4 \quad \text{MeOH}}$ $C_5H_{11}C \equiv CCOOMe$ ~90%

Angew (1971) 84 60
(Internat Ed 11 48)

Also via: Acetylenic acids (Section 301)

Section 307   Acetylene — Ether, Epoxide
°°°°°°°°°°°°°°°°°°°°°°°°°°°°°°°°°°°°°°°

Acetylenic ethers and acetylenic epoxides

Review: Ethynyl Ethers and Thioethers as Synthetic Intermediates
                              Advances in Org Chem (1960) 2 117

—C≡CH $\xrightarrow[\text{2 ClCH}_2\text{OEt}]{\text{1 BuLi}\quad \text{Et}_2\text{O}}$ —C≡CCH₂OEt

                                                                        90%

Rec Trav Chim (1965) 84 31
Bull Soc Chim Fr (1969) 4514

$$EtCH_2CHO \xrightarrow[\begin{array}{c}C_6H_6\\ 2\ Br_2\ \ C_6H_6\end{array}]{1\ PhCH_2OH\ \ HCl\ \ MgSO_4} EtCH\underset{Br}{-}CHOCH_2Ph \xrightarrow[\begin{array}{c}C_6H_6\\ 2\ NaNH_2\end{array}]{1\ PhNEt_2} EtC{\equiv}COCH_2Ph\quad 60\%$$

Note: the intermediate is written $EtCH\text{-}CHOCH_2Ph$ with Br below each of the two central carbons.

Rec Trav Chim (1964) **83** 301

$$EtBr \xrightarrow[\hphantom{xx}]{HC{\equiv}COBu\text{-}t\ \ LiNH_2\ \ NH_3} EtC{\equiv}COBu\text{-}t$$

Rec Trav Chim (1961) **80** 810
JCS (1954) 1860
Org Prep and Procedures (1972) **4** 89

$$BuC{\equiv}CH \xrightarrow[\begin{array}{c}2\ BrCH_2COMe\\ 3\ KOH\ \ Et_2O\end{array}]{1\ EtMgBr} BuC{\equiv}C\underset{Me}{\overset{O}{C}}\text{-}CH_2 \qquad 40\%$$

Note: the product is drawn as $BuC{\equiv}CC(=O)\text{-}CH_2$ with an epoxide (O bridging) and Me below the carbon.

Zh Obshch Khim (1960) **30** 1194
(Chem Abs **55** 499)

Section 308   <u>Acetylene — Halide</u>

Acetylenic halides

$$BuC{\equiv}CH \xrightarrow[2\ I(CH_2)_4Cl]{1\ NaNH_2\ \ NH_3} BuC{\equiv}C(CH_2)_4Cl \qquad 53\%$$

JACS (1948) **70** 1699
JCS (1950) 2100
JOC (1951) **16** 1405

$$C_5H_{11}C{\equiv}CH \xrightarrow[2\ Cl_2\ \ Et_2O]{1\ KNH_2\ \ NH_3} C_5H_{11}C{\equiv}CCl$$

JACS (1937) **59** 1307
TsCl   Annales de Chimie (1931) **16** 309

$PhC\equiv CH$ $\xrightarrow[\text{THF \quad hexane}]{\text{BuMgCl \quad SO}_2\text{Cl}_2}$ $PhC\equiv CCl$ 　　　　48%

JCS $\underline{C}$ (1968) 1265

ClNEt$_2$ or NBS　Annalen (1960) $\underline{638}$ 33
JACS (1961) $\underline{83}$ $\overline{4663}$

$PhC\equiv CH$ $\xrightarrow[\text{THF \quad H}_2\text{O}]{\text{Br}_2 \quad \text{NaOH}}$ $PhC\equiv CBr$ 　　　　89%

JCS (1963) 2295
Org Synth (1965) $\underline{45}$ 86

$\xrightarrow[\substack{\text{2 Br}_2 \quad \text{C}_6\text{H}_6}]{\text{1 HgCl}_2 \quad \text{KI} \quad \text{NaOH} \\ \text{EtOH} \quad \text{H}_2\text{O}}$ 　　79%

JCS (1963) 2295

$C_5H_{11}C\equiv CH$ $\xrightarrow[\text{2 Br}_2]{\text{1 EtMgBr \quad Et}_2\text{O}}$ $C_5H_{11}C\equiv CBr$ 　　　70%

JACS (1937) $\underline{59}$ 1307

BrCN　Annales de Chimie (1926) $\underline{5}$ 5

$PrC\equiv CH$ $\xrightarrow[\text{2 I}_2]{\text{1 MeMgBr \quad Et}_2\text{O}}$ $PrC\equiv CI$

Annales de Chimie (1926) $\underline{5}$ 5

$C_5H_{11}C\equiv CH$ $\xrightarrow[\substack{\text{or 1 Na \quad NH}_3 \\ \text{2 I}_2}]{\text{I}_2 \quad \text{NH}_3}$ $C_5H_{11}C\equiv CI$

JACS (1933) $\underline{55}$ 2150

$C_6H_{13}C\equiv CH$ $\xrightarrow[\text{CHCl}_3 \quad \text{Pyr}]{(I\cdot Pyr_2)^+ \ NO_3^-}$ $C_6H_{13}C\equiv CI$ 40%

Can J Chem (1971) $\underline{49}$ 403

Section 309   <u>Acetylene — Ketone</u>

Acetylenic ketones

$BuC\equiv CH$ $\xrightarrow[\begin{array}{l}\text{2 Et}_2\text{AlCl} \quad \text{Et}_2\text{O} \quad \text{pet ether}\\ \text{3}\end{array}]{\text{1 BuLi}}$

79%

JACS (1971) $\underline{93}$ 7320

$t\text{-}BuC\equiv CH$ $\xrightarrow[\begin{array}{l}\text{2 EtCH-C=NOH}\\ \quad \ | \quad \ \ |\\ \quad \text{Me} \ \text{Cl}\end{array}]{\text{1 EtMgBr}}$ $t\text{-}BuC\equiv CCCHEt$ (with $\overset{Me}{|}$ and $\overset{\|}{NOH}$) $\dashrightarrow$ $t\text{-}BuC\equiv CCOCHEt$ (with $\overset{Me}{|}$)   <69%

Tetr Lett (1970) 2101

$PhC\equiv CH$ $\xrightarrow[\text{2}]{\text{1 EtMgBr} \quad \text{Et}_2\text{O} \quad C_6H_6}$ AcN⟨⟩ $PhC\equiv CCOMe$ 8%

Tetrahedron (1962) $\underline{18}$ 1381

$PhC\equiv CH$ $\xrightarrow{\text{MeC(OEt)}_3 \quad \text{ZnCl}_2}$ $PhC\equiv CC(OEt)_2$ (with $\overset{|}{Me}$) $\dashrightarrow$ $PhC\equiv CCO$ (with $\overset{|}{Me}$)   <34%

JACS (1958) $\underline{80}$ 4607

$C_6H_{13}C\equiv CH$ $\xrightarrow[\substack{\text{MeOH} \quad H_2O \\ \text{2 MeCOCl}}]{\text{1 AgNO}_3\text{-NH}_3}$ $C_6H_{13}C\equiv CCOMe$

JACS (1956) <u>78</u> 1675

$BuC\equiv CH$ $\dashrightarrow$ $BuC\equiv CSiMe_3$ $\xrightarrow[CS_2]{\text{MeCOCl} \quad AlCl_3}$ $BuC\equiv CCOMe$ 75%

Ber (1963) <u>96</u> 3280

$EtCOCl$ $\xrightarrow{BuC\equiv CAg \quad CCl_4}$ $BuC\equiv CCOEt$ $\sim$ 53%

JACS (1956) <u>78</u> 1675
Tetr Lett (1970) 2659

$PrCOOH$ $\xrightarrow[\text{2 PhC}\equiv\text{CLi}]{\text{1 ClCOOEt} \quad Et_3N \text{ pet ether}}$ $PhC\equiv CCOPr$ 57%

Ber (1964) <u>97</u> 1649

$MeCOCl$ $\xrightarrow[CH_2Cl_2]{Na(Et_3BC\equiv CMe)}$ $MeCOC\equiv CMe$

Angew (1967) <u>79</u> 57
(Internat Ed <u>6</u> 84)

$(PhCO)_2O$ $\xrightarrow[\text{2 250-280°}]{\text{1 Ph}_3P=CHCOPh \quad CHCl_3}$ $PhC\equiv CCOPh$ 60%

JOC (1965) <u>30</u> 1015

Synth Comm (1972) 2 331

Quart Rev (1959) 13 61
Chem Pharm Bull (1973) 21 2051

HC≡CCHOH $\xrightarrow[\text{Me}_2\text{CO}\quad\text{H}_2\text{O}]{\text{CrO}_3\quad\text{H}_2\text{SO}_4}$ HC≡CCO   80%
      | |
      Ph   Ph

JCS (1946) 39

PhC≡CCHOH $\xrightarrow[\text{Et}_2\text{O}\quad\text{H}_2\text{O}]{\text{Na}_2\text{Cr}_2\text{O}_7\cdot2\text{H}_2\text{O}\quad\text{H}_2\text{SO}_4}$ PhC≡CCO   70%
      | |
      Me   Me

Tetrahedron (1962) 18 1381

PrC≡CCH$_2$Et $\xrightarrow{\text{CrO}_3\cdot\text{Pyr}\quad\text{CH}_2\text{Cl}_2}$ PrC≡CCOEt   42%

Tetr Lett (1971) 4379

MeC≡CCH$_2$Cl $\xrightarrow[\text{2 NaOH}\quad\text{H}_2\text{O}]{\overset{\text{COMe}}{\overset{|}{\text{1 NaCHCOOEt}}}}$ MeC≡CCH$_2$CH$_2$COMe   37%

JCS (1951) 2445

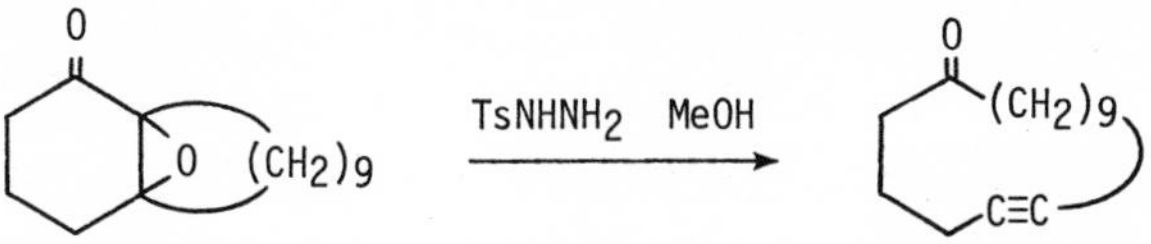

O
O (CH2)9
TsNHNH2 MeOH
O
(CH2)9
C≡C

58%

Helv (1967) 50 2101 708
(1972) 55 1276

Section 310 Acetylene — Nitrile

Acetylenic nitriles

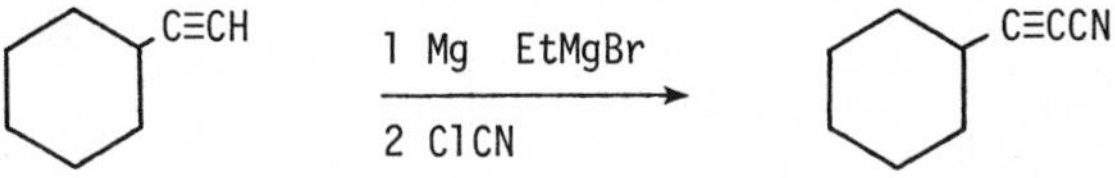

C≡CH
1 Mg EtMgBr
2 ClCN
C≡CCN

Annales de Chimie (1926) 5 5
Bull Soc Chim Fr (1915) 17 228

$PhCOCl \xrightarrow[2\ 280°]{1\ Ph_3P{=}CHCN} PhC{\equiv}CCN$ 85%

Proc Chem Soc (1961) 302
JCS Perkin I (1973) 2241

Section 311 Acetylene — Olefin

Acetylenic olefins, enynes

$C_5H_{11}C{\equiv}CH \xrightarrow{Cu_2O\ \ HOAc} C_5H_{11}C{\equiv}CCH{=}CHC_5H_{11}$ 60%

Chem Ind (1962) 1684
Proc Chem Soc (1958) 303

$PrC\equiv CH$ $\xrightarrow{\quad Et_2Zn \quad Cr(OBu-t)_4 \quad}$ $PrC\equiv CC=CH_2$
                                                    $\overset{|}{Pr}$

Bull Chem Soc Jap (1961) <u>34</u> 892

$C_5H_{11}C\equiv CH$ $\dashrightarrow$ $C_5H_{11}C\equiv CCu$ $\xrightarrow[NaCN \quad DMF]{BrCH_2CH=CH_2}$ $C_5H_{11}C\equiv CCH_2CH=CH_2$ 96%

Compt Rend (1970) <u>270</u> 354

$PhC\equiv CH$ $\dashrightarrow$ $PhC\equiv CCu$ $\xrightarrow[DMF]{PhCH=CHBr}$ $PhC\equiv CCH=CHPh$ ~75%

Chem Comm (1967) 1259

$EtCH_2C\equiv CNa$ $\xdashrightarrow[HCl]{HCHO \quad EtOH}$ $EtCH_2C\equiv CCH_2OEt$ $\xrightarrow[NH_3]{NaNH_2}$ $EtCH=CHC\equiv CH$ <42%

Rec Trav Chim (1963) <u>82</u> 1115

$BuC\equiv C-C\equiv CBu$ $\xrightarrow[2 \ HOAc]{1 \ (i\text{-}Pr\overset{\overset{\displaystyle Me}{|}}{C}H)_2BH \quad THF}$ $BuCH=CHC\equiv CBu$ 76%

JACS (1970) <u>92</u> 4068

$PhCHO$ $\xrightarrow[NH_3 \quad MeCN]{Ph_3\overset{+}{P}CH_2C\equiv CH \ \overset{-}{Br}}$ $PhCH=CHC\equiv CH$ 59%

Annalen (1965) <u>682</u> 62
Tetr Lett (1973) 1495

PhBr $\xrightarrow{Mg}$ PhMgBr $\xrightarrow[\text{THF}]{}$ PhCH=CHC≡CH   35%

Chem Comm (1969) 710

$C_6H_{13}Br$ $\xrightarrow[\text{NaNH}_2 \quad \text{NH}_3]{\text{MeCH}_2\text{C≡CSEt}}$ MeCH=C=CSEt $\underset{C_6H_{13}}{|}$ $\xrightarrow[\text{NH}_3]{\text{KNH}_2}$ $C_6H_{13}CH=CHC≡CH$ ~68%

Rec Trav Chim (1964) **83** 456

MeC=CHCH$_2$CH$_2$CO $\underset{Et}{|}$ $\underset{Et}{|}$ $\xrightarrow[\text{EtONa} \quad \text{DMF}]{\overset{+}{\text{Ph}_3\text{PCH}_2\text{CH}_2\text{CH}_2\text{C≡CH}}}$ MeC=CHCH$_2$CH$_2$C=CHCH$_2$CH$_2$C≡CH $\underset{Et}{|}$ $\underset{Et}{|}$ 36%

JCS Perkin I (1972) 361

PrC=CHCH$_2$Br $\underset{Me}{|}$ $\xrightarrow{\text{HC≡CCH}_2\text{MgBr}}$ PrC=CHCH$_2$CH$_2$C≡CH $\underset{Me}{|}$

JOC (1973) **38** 2733

Section 312   Carboxylic Acid — Carboxylic Acid
ooooooooooooooooooooooooooooooooooooooooo

Review: Synthesis of Alkylated Alkanedioic Acids

Chem Rev (1959) **59** 89

BuC≡CH $\xrightarrow[\substack{\text{2 MeLi} \\ \text{3 CO}_2}]{\text{1 (i-Bu)}_2\text{AlH} \quad \text{THF}}$ BuCH$_2$CH(COOH)$_2$ 62%

Tetr Lett (1966) 6021

$$\text{(adamantane)}-\text{COOH} \xrightarrow[\text{2 } CO_2]{\text{1 } LiN(Pr\text{-}i)_2 \quad THF \quad \text{hexane}} \text{(structure)} \begin{array}{c} COOH \\ COOH \end{array} \quad 30\%$$

Tetr Lett (1971) 2339

$$C_{12}H_{25}CH_2COOH \xrightarrow{PBr_3} C_{12}H_{25}CH_2COBr \xrightarrow[\text{2 } H_2O]{\overset{COBr}{\underset{}{1 \ \overset{|}{C}OBr \quad CCl_4}}} C_{12}H_{25}CH(COOH)_2 \quad <42\%$$

Ber (1958) __91__ 297

$$C_7H_{15}Br \xrightarrow[\text{BuONa \quad BuOH}]{CH_2(COOEt)_2} C_7H_{15}CH(COOEt)_2 \xrightarrow{KOH \quad H_2O} C_7H_{15}CH(COOH)_2$$

Org Synth (1943) Coll Vol 2 474 416
(1932) Coll Vol 1 250
Org React (1957) __9__ 107

$$\text{(cycloheptyl-phenyl)}-CHO \xrightarrow[\substack{\text{piperidine} \\ \text{HOAc} \quad C_6H_6 \\ \text{2 KCN \quad EtOH} \\ H_2O}]{\text{1 } CH_2(COOEt)_2} \overset{CN}{\underset{}{\text{(ring)}\overset{|}{C}HCH_2COOEt}} \xrightarrow[\text{HOAc}]{HCl \quad H_2O} \overset{COOH}{\underset{}{\text{(ring)}\overset{|}{C}HCH_2COOH}} \quad 15\%$$

J Med Chem (1972) __15__ 1297
Org Synth (1963) Coll Vol 4 804

$$\underset{OMe}{\overset{Cl}{\text{(ring)}}}-CH_2Br \xrightarrow[\substack{\text{MeOH} \\ \text{2 NaOH} \\ \text{3 } 160°}]{\overset{COOMe}{\underset{}{\text{1 } \overset{|}{C}H_2CH(COOMe)_2 \quad MeONa}}} \underset{OMe}{\overset{Cl}{\text{(ring)}}}-CH_2\overset{\overset{COOH}{|}}{C}HCH_2COOH \quad 85\%$$

JACS (1965) __87__ 933

$$Ph_2CO \xrightarrow[\substack{NH_4OAc \quad HOAc \\ toluene \\ 2 \ KCN \quad EtOH}]{1 \ \overset{CN}{\underset{}{CH_2COOEt}}} Ph_2C\overset{CN}{\underset{CN}{C}}HCOOEt \xrightarrow[2 \ KOH \quad H_2O]{1 \ H_2SO_4 \quad HOAc} Ph_2C\overset{COOH}{\underset{}{CH_2COOH}}$$

                                                                                        30%

JOC (1950) <u>15</u> 381

$$\substack{PhCH \\ \parallel \\ PhCH} \xrightarrow[DMF]{Electrolysis \quad CO_2 \quad Bu_4\overset{+}{N}\overset{-}{I}} \substack{PhCHCOOH \\ \mid \\ PhCHCOOH}$$

                                                                                        91%

Synthesis (1971) 285

$$PhCH_2Cl \dashrightarrow PhCH_2CH(COOEt)_2 \xrightarrow[\substack{t-BuOH \\ 2 \ HBr \quad H_2O}]{1 \ CH_2=CHCN \quad KOH} PhCH_2CHCH_2CH_2COOH$$
$$COOH$$

JCS (1950) 1683

$$\substack{COOH \\ \mid \\ (CH_2)_2 \\ \mid \\ C \\ \parallel\parallel \\ CH} \xrightarrow[NaOH \quad H_2O]{Air \quad CuO} \substack{COOH \\ \mid \\ (CH_2)_2 \\ \mid \\ C \\ \parallel\parallel \\ C \\ \mid \\ C \\ \parallel\parallel \\ C \\ \mid \\ (CH_2)_2 \\ \mid \\ COOH} \xrightarrow[MeOH]{H_2 \quad Pd-BaSO_4} \substack{COOH \\ \mid \\ (CH_2)_8 \\ \mid \\ COOH}$$

                                                                                        ~80%

Annalen (1954) <u>589</u> 222

$$\substack{COOH \\ \mid \\ Me_2C(CH_2)_2COOH} \xrightarrow[Na_2CO_3 \quad MeOH]{Electrolysis} \substack{COOH \quad COOH \\ \mid \qquad\quad \mid \\ Me_2C(CH_2)_4CMe_2}$$

                                                                                        30%

Bull Soc Chim Fr (1970) 183

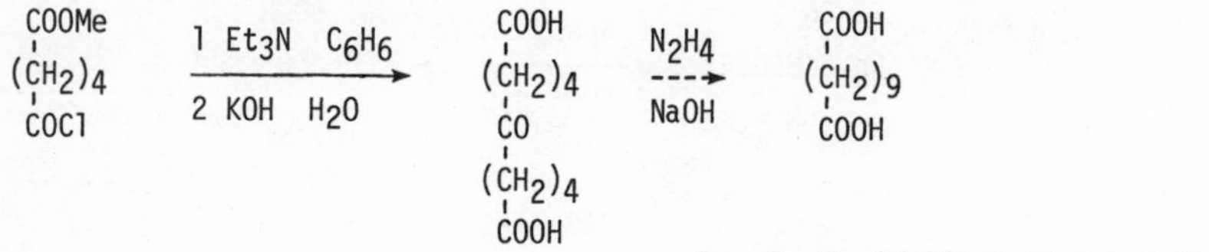

COOMe
|
(CH$_2$)$_4$ $\xrightarrow[\text{2 KOH H}_2\text{O}]{\text{1 Et}_3\text{N C}_6\text{H}_6}$ COOH
| |
COCl   (CH$_2$)$_4$ COOH
                                                          | $\xrightarrow[\text{NaOH}]{\text{N}_2\text{H}_4}$ |
                                                          CO   (CH$_2$)$_9$ <60%
                                                          | |
                                                         (CH$_2$)$_4$ COOH
                                                          |
                                                         COOH

Org Synth (1963) Coll Vol 4 555

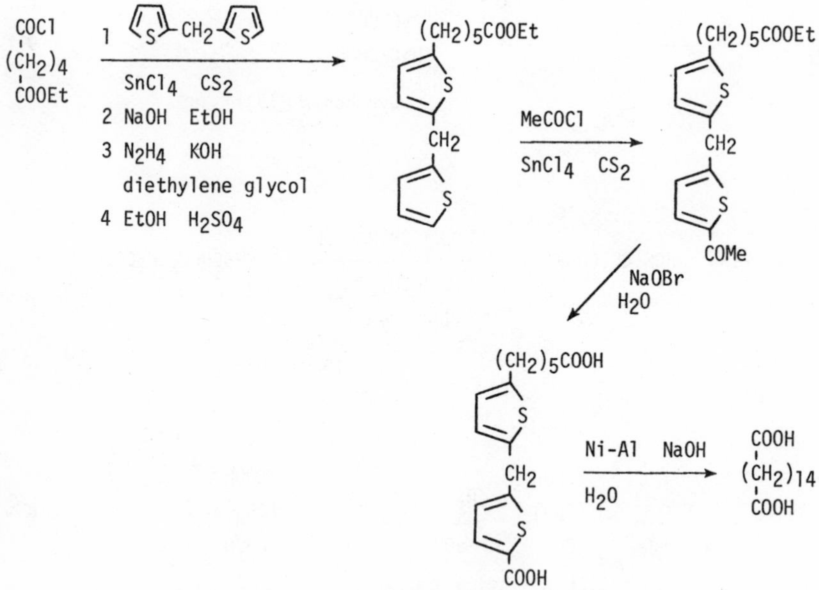

Bull Soc Chim Fr (1955) 1583

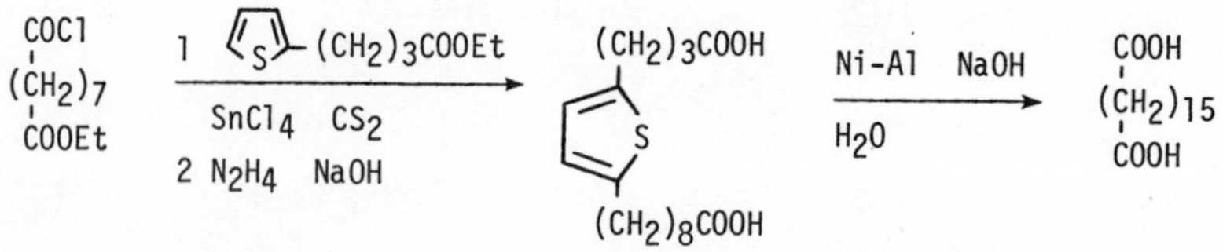

Bull Soc Chim Fr (1955) 1583

$\begin{array}{c} \text{CONMe}_2 \\ | \\ (\text{CH}_2)_8 \\ | \\ \text{COOMe} \end{array}$ $\xrightarrow{\text{MeONa}}$ $\begin{array}{c} \text{CONMe}_2 \\ | \\ (\text{CH}_2)_7 \\ | \\ \text{CHCOOMe} \\ | \\ \text{CO} \\ | \\ (\text{CH}_2)_8 \\ | \\ \text{CONMe}_2 \end{array}$ $\xrightarrow[\begin{array}{c} \text{2 Zn   HgCl}_2 \\ \text{HOAc   H}_2\text{O} \end{array}]{\text{1 HBr   H}_2\text{O}}$ $\begin{array}{c} \text{COOH} \\ | \\ (\text{CH}_2)_{17} \\ | \\ \text{COOH} \end{array}$ 32%

JOC (1973) **38** 1424

$\begin{array}{c} \text{COOEt} \\ | \\ (\text{CH}_2)_4 \\ | \\ \text{I} \end{array}$ $\xrightarrow[\text{EtONa   EtOH}]{\text{EtCH(COOEt)}_2}$ $\begin{array}{c} \text{COOEt} \\ | \\ (\text{CH}_2)_4 \\ | \\ \text{EtC(COOEt)}_2 \end{array}$ $\xrightarrow[\text{2 >140°}]{\text{1 NaOH   H}_2\text{O}}$ $\begin{array}{c} \text{COOH} \\ | \\ (\text{CH}_2)_4 \\ | \\ \text{EtCH} \\ | \\ \text{COOH} \end{array}$

JACS (1928) **50** 1967

$\text{Br(CH}_2)_9\text{Br}$ $\xrightarrow[\begin{array}{l} \text{2 CdCl}_2 \text{   C}_6\text{H}_6 \\ \text{3   COOMe} \\ \quad | \\ \quad (\text{CH}_2)_3\text{COCl} \end{array}]{\text{1 Mg   Et}_2\text{O}}$ $\begin{array}{c} \text{COOMe} \\ | \\ (\text{CH}_2)_3 \\ | \\ \text{CO} \\ | \\ (\text{CH}_2)_9 \\ | \\ \text{CO} \\ | \\ (\text{CH}_2)_3 \\ | \\ \text{COOMe} \end{array}$ $\xrightarrow[\begin{array}{c} \text{2 N}_2\text{H}_4 \text{   KOH} \\ \text{diethylene glycol} \end{array}]{\text{1 NaOH   H}_2\text{O}}$ $\begin{array}{c} \text{COOH} \\ | \\ (\text{CH}_2)_{17} \\ | \\ \text{COOH} \end{array}$ 34%

JACS (1953) **75** 3339

$\text{Br(CH}_2)_{12}\text{Br}$ $\xrightarrow{\text{NaCH(COOEt)}_2}$ $\begin{array}{c} \text{CH(COOEt)}_2 \\ | \\ (\text{CH}_2)_{12} \\ | \\ \text{CH(COOEt)}_2 \end{array}$ $\xrightarrow[\text{2 } \Delta]{\text{1 Hydrolysis}}$ $\begin{array}{c} \text{COOH} \\ | \\ (\text{CH}_2)_{14} \\ | \\ \text{COOH} \end{array}$ 83%

JACS (1947) **69** 2350

$\text{Br(CH}_2)_{12}\text{Br}$ $\xrightarrow[\text{decalin}]{\text{NaCH}_2\text{COONa}}$ $\begin{array}{c} \text{COOH} \\ | \\ (\text{CH}_2)_{14} \\ | \\ \text{COOH} \end{array}$ 34%

Annalen (1966) **691** 61

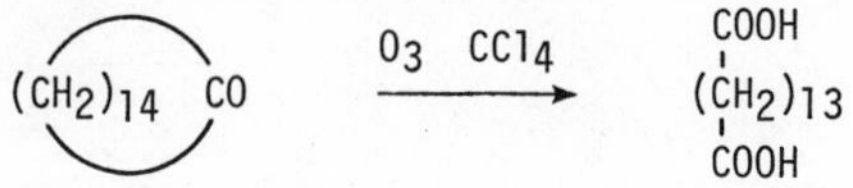

(CH₂)₁₄ CO → [O₃ CCl₄] COOH (CH₂)₁₃ COOH   25%

Helv (1930) 13 142

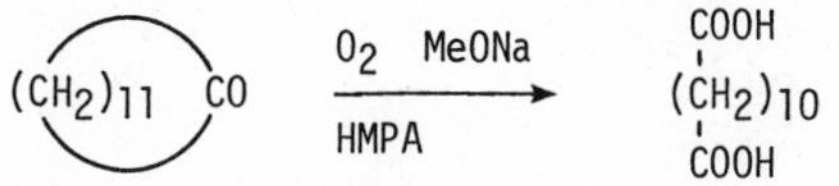

(CH₂)₁₁ CO → [O₂ MeONa / HMPA] COOH (CH₂)₁₀ COOH   40%

JOC (1965) 30 3768

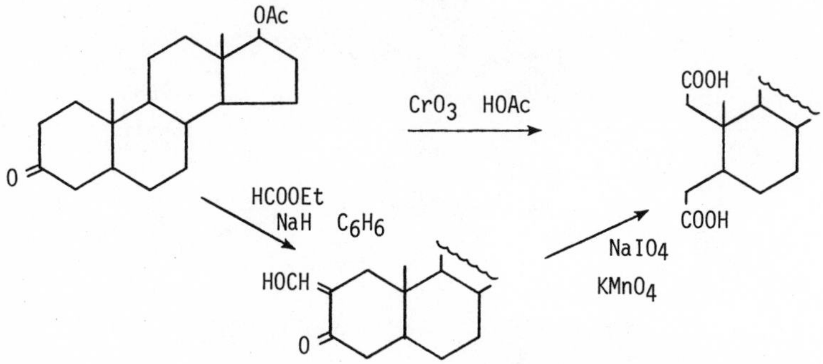

Tetrahedron (1964) 20 1271

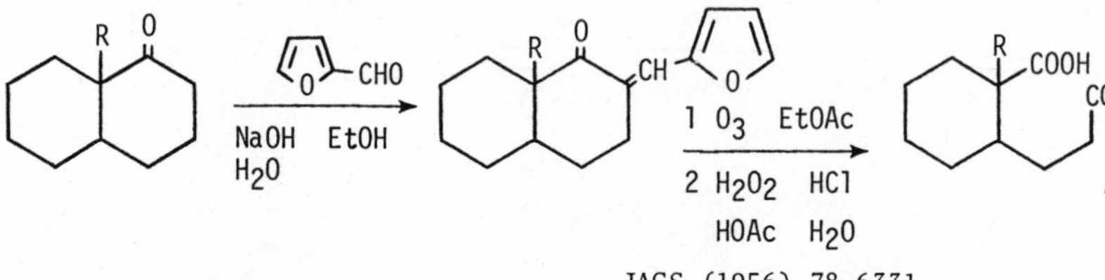

~67%

JACS (1956) 78 6331

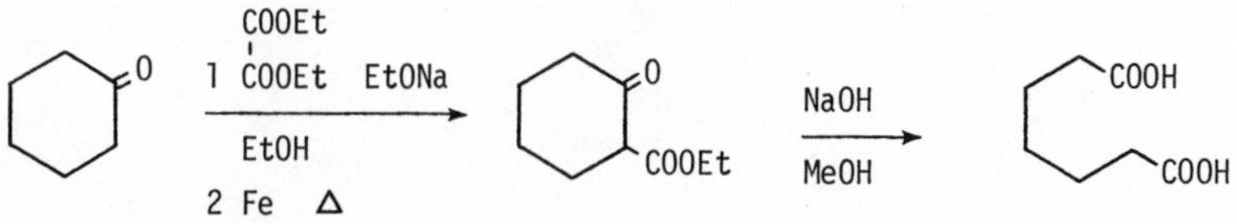

47-54%

Org Synth (1943) Coll Vol 2 531

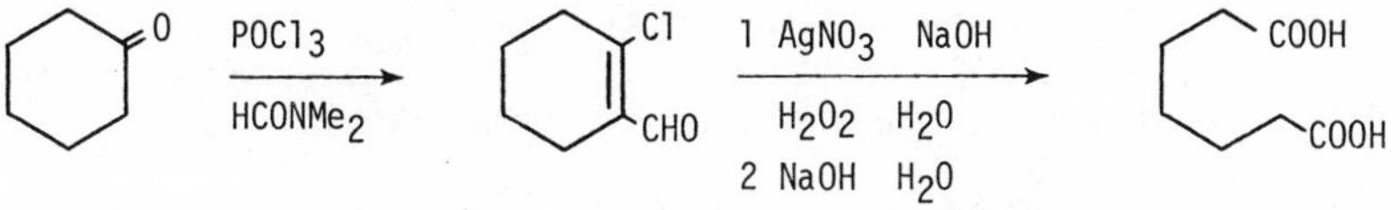

Ber (1960) 93 2743

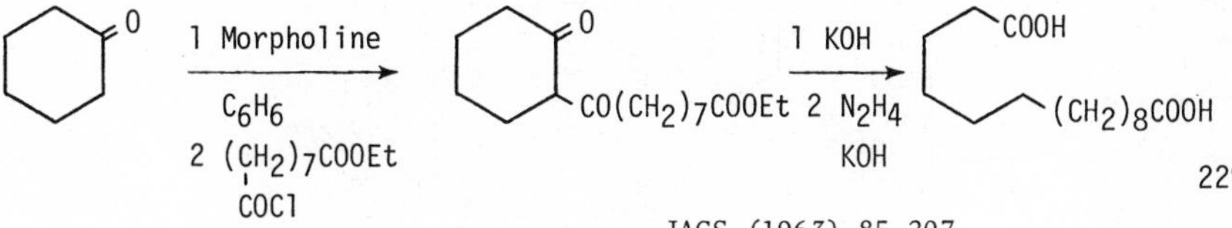

JACS (1963) 85 207

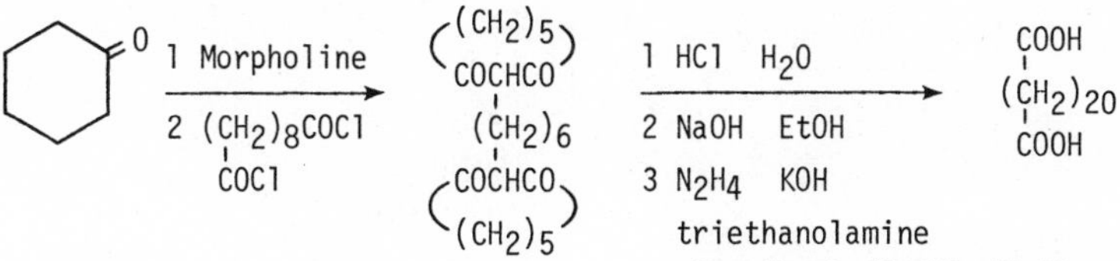

Org Synth (1963) 43 34
Ber (1967) 100 4017

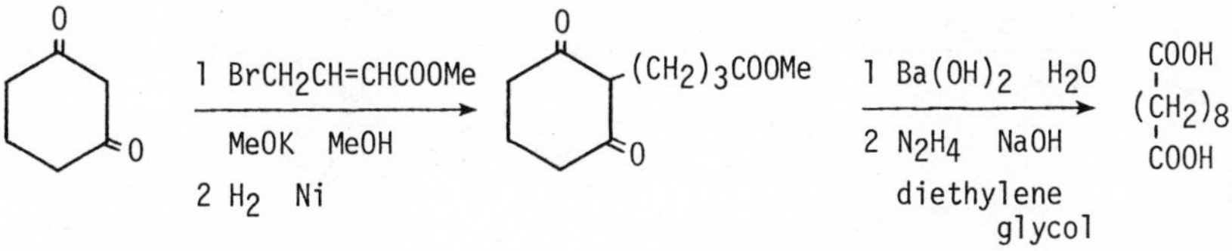

Ber (1953) 86 693

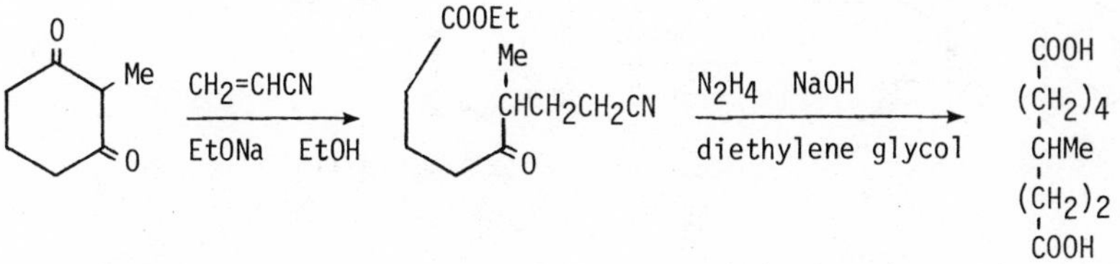

Ber (1954) 87 990
Newer Meth Prep Org Chem (1963) 2 51

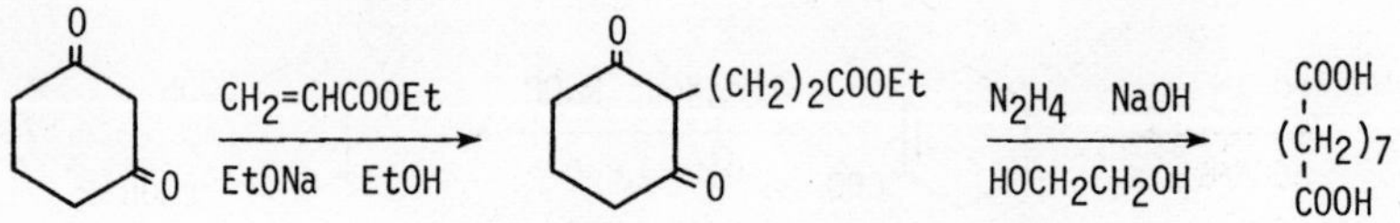

Newer Meth Prep Org Chem (1963) 2 51 96

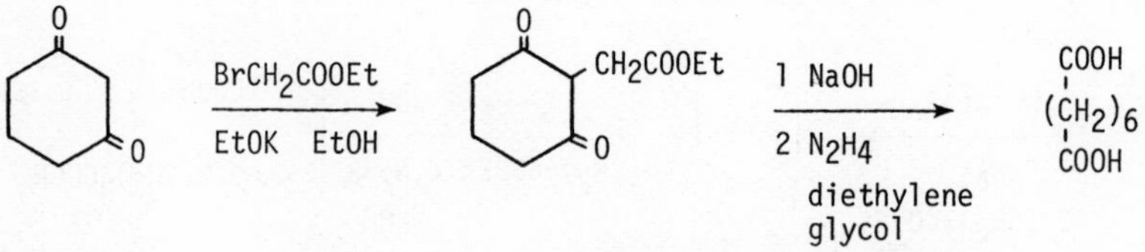

Ber (1952) 85 61 290
Newer Meth Prep Org Chem (1963) 2 51 95

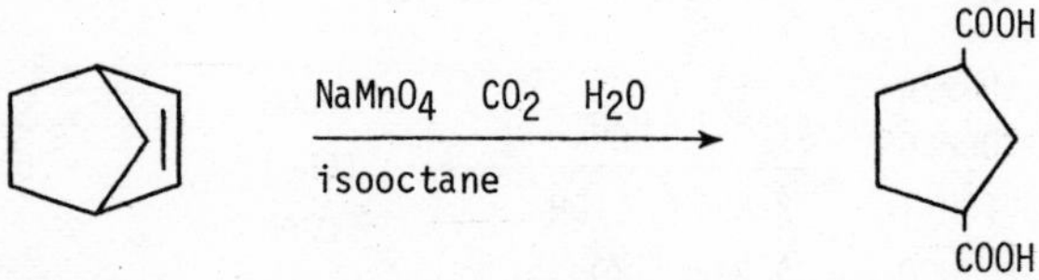

~65%

JCS (1947) 818

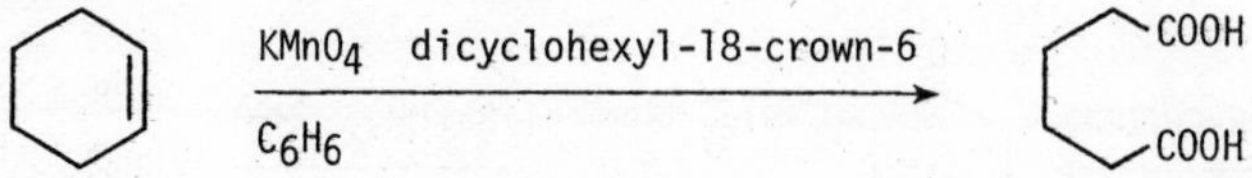

100%

JACS (1972) 94 4024

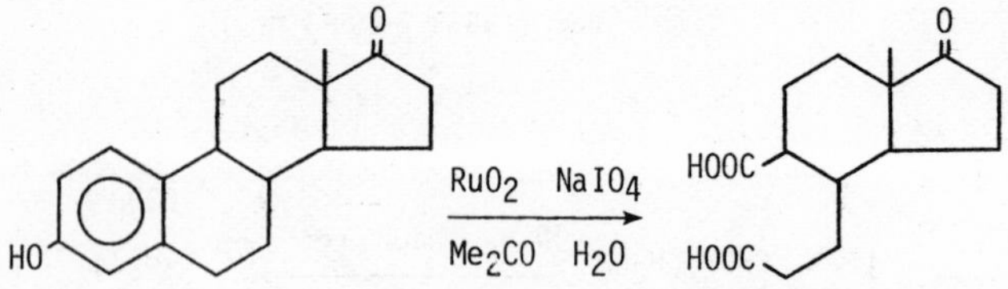

JOC (1969) 34 116

Also via: Section
        Diesters   357
        Dinitriles   375

Section 313   <u>Carboxylic Acid — Alcohol</u>

$$PhCOCl \xrightarrow[Et_2O]{CH_2N_2} PhCOCHN_2 \xrightarrow{EtSCl} \underset{Cl}{PhCOCHSEt} \xrightarrow[H_2O]{KOH} \underset{OH}{PhCHCOOH}$$

Ber (1955) <u>88</u> 1988

$$\text{cyclopentyl-}CH_2CH_2COOH \xrightarrow[\text{2 Air}]{\text{1 LiN(Pr-i)}_2 \quad THF} \text{cyclopentyl-}\underset{OH}{CH_2CHCOOH}$$ 51%

Synthesis (1971) 647

$$\text{cyclopentyl-}(CH_2)_{11}CH_2COOH \xrightarrow[\text{2 } H_2O]{\text{1 } Br_2 \quad P_4} \text{cyclopentyl-}(CH_2)_{11}\underset{}{\overset{Br}{C}HCOOH} \xrightarrow[\text{2 KOH } H_2O]{\text{1 KI EtOH}} \text{cyclopentyl-}(CH_2)_{11}\overset{OH}{C}HCOOH$$

Ber (1942) <u>75</u> 1181

$$PhCHO \xrightarrow[\text{2 } HgCl_2 \quad Me_2CO \quad H_2O]{\text{1 } (PhS)_3CLi} \underset{OH}{PhCHCOOH}$$ 47%

Angew (1967) <u>79</u> 468
(Internat Ed <u>6</u> 442)

o-chlorobenzaldehyde (CHO, Cl) $\xrightarrow[\text{dioxane } H_2O]{CHBr_3 \quad KOH \quad LiCl}$ (CHCOOH, OH, Cl substituted benzene)

JOC (1968) <u>33</u> 2565

PhCOOEt $\xrightarrow[\text{t-BuOK}]{\text{CH}_3\text{SOMe}}$ PhCOCH$_2$SOMe $\xrightarrow[\text{2 Cu(OAc)}_2 \quad \text{EtOH}]{\text{1 HCl} \quad \text{Me}_2\text{SO} \quad \text{H}_2\text{O}}$ PhCHCOOH
$\qquad\qquad\qquad\qquad\qquad\qquad\qquad\qquad\qquad\qquad\qquad\qquad$ |
$\qquad\qquad\qquad\qquad\qquad\qquad\qquad\qquad\qquad\qquad\qquad\qquad$ OH

JACS (1966) **88** 5498

PhCO $\xrightarrow[\text{Et}_2\text{O} \quad \text{H}_2\text{O}]{\text{NaCN} \quad \text{HCl}}$ PhCCN $\xrightarrow[\text{2 NaOH} \quad \text{H}_2\text{O}]{\text{1 HCl} \quad \text{H}_2\text{O}}$ PhCCOOH $\qquad$ 29-30%
|$\qquad\qquad\qquad\qquad\qquad\qquad$ |$\qquad\qquad\qquad\qquad\qquad\qquad\qquad$ |
Me$\qquad\qquad\qquad\qquad\qquad\qquad$ Me$\qquad\qquad\qquad\qquad\qquad\qquad\qquad$ Me

$\qquad\qquad\qquad\qquad\qquad\qquad\qquad$ OH$\qquad\qquad\qquad\qquad\qquad\qquad\qquad\qquad\qquad$ OH

Org Synth (1963) Coll Vol 4 58
JOC (1949) **14** 1013

$\xrightarrow[\text{Pyr}]{\substack{\text{1 HCN} \\ \text{2 Ac}_2\text{O}}}$ $\overset{\text{OAc}}{\underset{\text{CN}}{>\!\!<}}$ $\xrightarrow[\text{HOAc}]{\text{HBr}}$ $\overset{\text{OAc}}{\underset{\text{CONH}_2}{>\!\!<}}$ $\xrightarrow[\text{2 NaOH}]{\text{1 N}_2\text{O}_3}$ $\overset{\text{OH}}{\underset{\text{COOH}}{>\!\!<}}$

Angew (1971) **83** 329
(Internat Ed **10** 336)

PhCO $\xrightarrow[\substack{\text{2 K}_2\text{CO}_3 \quad \text{i-PrOH} \\ \text{H}_2\text{O}}]{\text{1 LiCHCl}_2 \quad \text{THF}}$ PhCCHO $\xrightarrow[\text{H}_2\text{O}]{\text{KMnO}_4 \quad \text{dioxane}}$ PhCCOOH $\qquad$ <43%

JOC (1972) **37** 1248

PhCO $\xrightarrow{\text{NaC}\equiv\text{CH}}$ PhCC$\equiv$CH $\xrightarrow{\text{O}_3 \quad \text{cyclohexane}}$ PhCCOOH $\qquad$ ~40%

JOC (1964) **29** 3419
Carbohydrate Res (1966) **2** 315

PhCOCH$_3$ $\xrightarrow{\text{Cl}_2 \quad \text{HOAc}}$ PhCOCHCl$_2$ $\xrightarrow{\text{NaOH} \quad \text{H}_2\text{O}}$ PhCHCOOH $\qquad$ 76-87%
$\qquad\qquad\qquad\qquad\qquad\qquad\qquad\qquad\qquad\qquad\qquad\qquad\qquad\qquad\qquad\qquad\quad$ |
$\qquad\qquad\qquad\qquad\qquad\qquad\qquad\qquad\qquad\qquad\qquad\qquad\qquad\qquad\qquad\qquad\quad$ OH

Org Synth (1955) Coll Vol 3 538

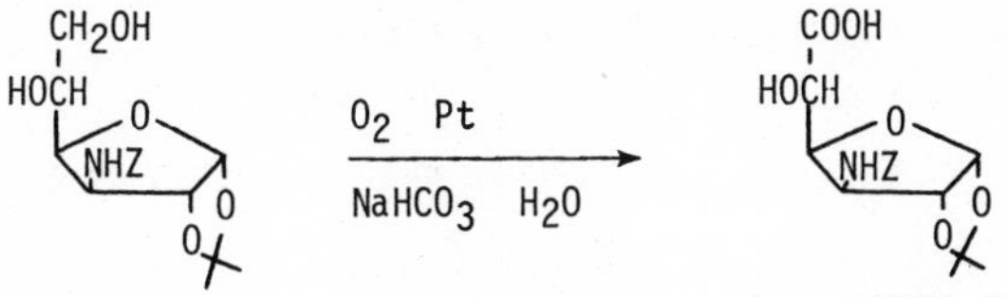

Ber (1972) <u>105</u> 1524
    (1956) <u>89</u> 1648
Angew (1957) <u>69</u> 600

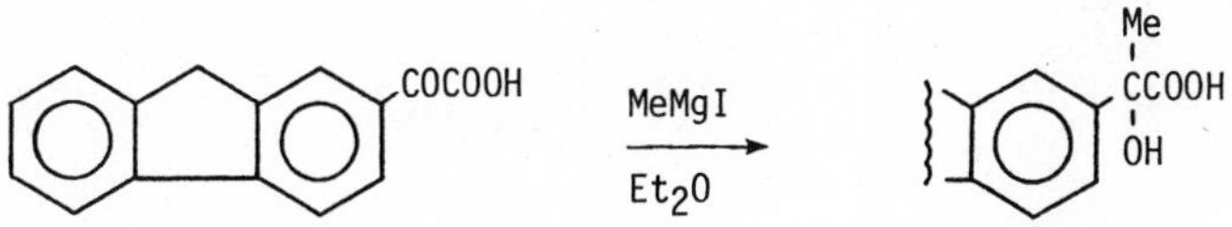

88%

J Med Chem (1972) <u>15</u> 1029

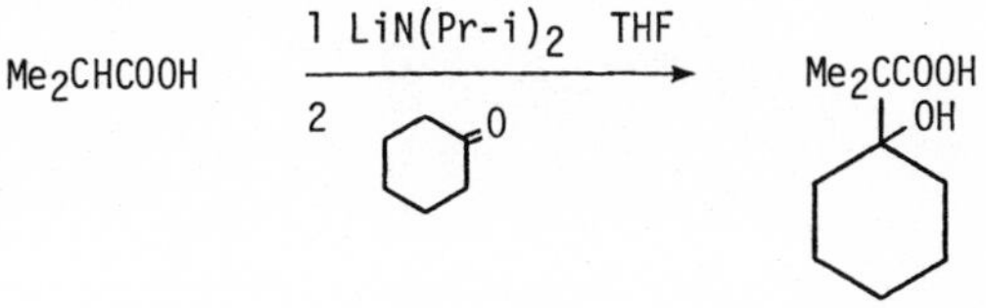

$Me_2CHCOOH$

$\xrightarrow{\begin{array}{c} 1\ LiN(Pr\text{-}i)_2\ \ THF \\ 2\ \text{(cyclohexanone)} \end{array}}$

JACS (1972) <u>94</u> 2000
    (1951) <u>73</u> 4221
JOC (1971) <u>36</u> 1149 2403
Synthesis (1970) 615

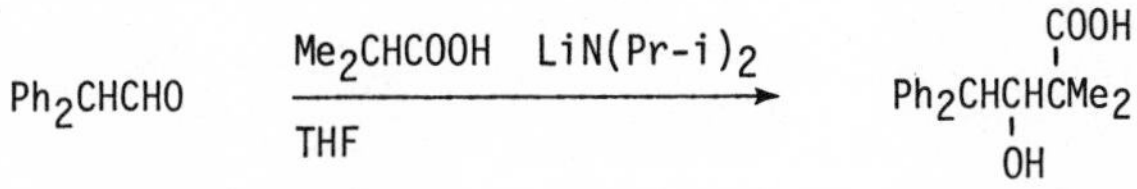

Ph_2CHCHO

$\xrightarrow[\text{THF}]{Me_2CHCOOH\ \ LiN(Pr\text{-}i)_2}$

$Ph_2CHCHCMe_2$

50%

JOC (1971) <u>36</u> 1149

$PhCHO$

$\xrightarrow[\text{2 Acid}]{1\ Me_2C=COSiMe_3\ (OH)}$

$PhCHCMe_2$

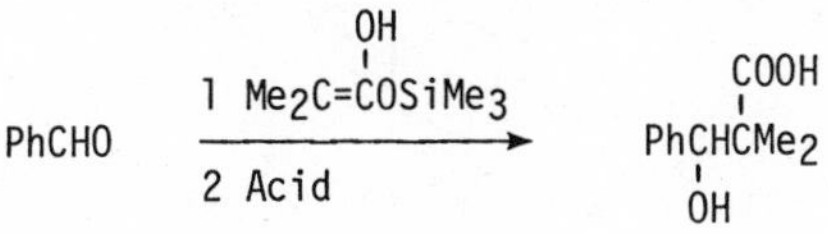

68%

Tetr Lett (1972) 79

PhCHO $\xrightarrow{\begin{array}{c}1 \ Bu_2BSBu \quad CH_2=CO \\ \hline 2 \ H_2O_2 \quad MeOH \quad H_2O\end{array}}$ PhCHCH$_2$COSBu $\dashrightarrow$ PhCHCH$_2$COOH
$\qquad\qquad\qquad\qquad\qquad\qquad\qquad$ |$\qquad\qquad\qquad\qquad$ |
$\qquad\qquad\qquad\qquad\qquad\qquad\qquad$ OH$\qquad\qquad\qquad\qquad\quad$ OH

Bull Chem Soc Jap (1971) __44__ 3215

$\xrightarrow{\begin{array}{c}1 \ BrCH_2COOBu\text{-}t \quad Zn \quad THF \\ \hline 2 \ C_6H_6 \ (reflux)\end{array}}$

76%

Chem Comm (1969) 679
JCS __C__ (1969) 2799

PhCO $\xrightarrow{\begin{array}{c}1 \ BrCH_2COOSiMe_3 \quad Zn \quad C_6H_6 \quad Et_2O \\ \hline 2 \ H_2O\end{array}}$
|
Et

$\qquad\qquad\qquad\qquad\qquad\qquad\qquad\qquad\qquad$ OH
$\qquad\qquad\qquad\qquad\qquad\qquad\qquad\qquad\qquad$ |
$\qquad\qquad\qquad\qquad\qquad\qquad\qquad\qquad$ PhCCH$_2$COOH$\qquad$ 80%
$\qquad\qquad\qquad\qquad\qquad\qquad\qquad\qquad\qquad$ |
$\qquad\qquad\qquad\qquad\qquad\qquad\qquad\qquad\qquad$ Et

Tetr Lett (1971) 3227

$\xrightarrow{\begin{array}{c}Me_2CHCOOH \quad (i\text{-}Pr)_2NH \quad BuLi \\ \hline THF\end{array}}$

71%

JOC (1971) __36__ 1149 2403

$\xrightarrow{\begin{array}{c}MeCH_2COOH \quad Na\text{-}naphthalene \\ \hline THF\end{array}}$

38%

Bull Soc Chim Fr (1970) 1848
Chem Ind (1969) 1811
$\qquad\qquad$ (1972) 80
Israel J Chem (1971) __8__ 731

$$\text{(cyclopentene with Me)} \xrightarrow[\substack{\text{2 HC}\equiv\text{CCOOEt} \quad \text{NaOH} \quad \text{H}_2\text{O} \\ \text{3 H}_2\text{O}_2 \quad \text{H}_2\text{O}}]{\substack{\text{Me} \\ \text{i-PrCBH}_2 \\ 1 \quad \text{Me} \quad \text{THF}}} \text{(product)} \quad \sim 71\%$$

trans

JACS (1973) 95 6837

$$\text{MeCH=CCOOH} \xrightarrow[\text{2 NaBH}_4 \quad \text{NaOH} \quad \text{H}_2\text{O}]{\text{1 Hg(OAc)}_2 \quad \text{MeOH} \quad \text{H}_2\text{O}} \underset{\text{Me}}{\text{MeCHCHCOOH}} \quad 56\%$$
Me

OH

JCS Perkin I (1973) 109

$$\underset{\text{Me}}{\text{BuCHCH}_2\text{CH}_2\text{COOH}} \xrightarrow[\text{H}_2\text{O}]{\text{KMnO}_4 \quad \text{NaOH}} \underset{\text{OH}}{\text{BuCCH}_2\text{CH}_2\text{COOH}} \quad 70\%$$
Me

JCS (1953) 2129 3580

$$\xrightarrow[\substack{\text{LiN(Pr-i)}_2 \\ \text{THF}}]{\text{Me}_2\text{CHCOOH}}$$

(→ lactone, 81%)

JACS (1967) 89 2500

$$\xrightarrow{\text{CH}_2\text{=CHCH}_2\text{MgBr}} \xrightarrow[\text{2 CrO}_3]{\text{1 B}_2\text{H}_6}$$

Chem Comm (1968) 122

COOEt
|
(CH$_2$)$_8$ $\xrightarrow[\text{C}_6\text{H}_6]{\text{C}_7\text{H}_{15}\text{-}\!\!\langle\text{S}\rangle\quad \text{SnCl}_4}$
|
COCl

COOEt
|
(CH$_2$)$_8$ $\xrightarrow[\substack{2\ \text{KOH} \\ 3\ \text{Ni}}]{1\ \text{NaBH}_4}$
|
CO
[thiophene ring]
C$_7$H$_{15}$

COOH
|
(CH$_2$)$_8$
|
CHOH   29%
|
(CH$_2$)$_4$
|
C$_7$H$_{15}$

JOC (1961) <u>26</u> 5217
JCS (1960) 1502

$\left(\begin{array}{c}\text{COO} \\ (\text{CH}_2)_{11}\end{array}\right)$ $\xrightarrow[\text{2 Ac}_2\text{O}\quad\text{Pyr}]{\text{1 Me}_2\text{SO}\quad\text{NaH}}$ $\left(\begin{array}{c}\text{COCH}_2\text{SOMe} \\ (\text{CH}_2)_{11}\text{OAc}\end{array}\right)$

$\xrightarrow[\substack{\text{NaH}\quad\text{Me}_2\text{SO} \\ \text{2 Al-Hg}\quad\text{H}_2\text{O}\quad\text{THF} \\ \text{3 NaOH}\quad\text{EtOH}\quad\text{H}_2\text{O} \\ \text{4 Zn-Hg}\quad\text{HCl}\quad\text{H}_2\text{O}}]{\text{1 BrCH}_2\text{COOMe}}$

$\left(\begin{array}{c}(\text{CH}_2)_3\text{COOH} \\ (\text{CH}_2)_{11}\text{OH}\end{array}\right)$

46%

Can J Chem (1968) <u>46</u> 3767
Tetr Lett (1966) 5669

Also via: Section
         Hydroxyesters   327
         Hydroxyamides   325
         Carboxyesters   317

Section 314   Carboxylic Acid — Aldehyde
                ○○○○○○○○○○○○○○○○○○○○○○○○○○○○○○

Carboxyaldehydes, aldehydic acids

CH$_2$CO
| \
CH$_2$ O $\xrightarrow[130°\quad 200\ \text{kg/cm}^2]{\text{H}_2\quad\text{CO}\quad\text{Co}_2(\text{CO})_8}$
| /
CH$_2$CO

CH$_2$COOH
|
CH$_2$ 32%
|
CH$_2$CHO

Bull Chem Soc Jap (1971) <u>44</u> 288

$\begin{array}{l} CH_2CO \\ \qquad >0 \\ CH_2C\bar{O} \end{array}$ $\xrightarrow{\ Na_2Fe(CO)_4 \quad THF\ }$ $\begin{array}{l} CH_2COOH \\ | \\ CH_2CHO \end{array}$ 81%

Tetr Lett (1973) 3535

$\begin{array}{l} BuCHCHO \\ \quad | \\ \quad Et \end{array}$ $\xrightarrow[KOH \quad H_2O]{CH_2=CHCN}$ $\begin{array}{l} CHO \\ | \\ Bu\overset{|}{C}CH_2CH_2CN \\ | \\ Et \end{array}$ $\xrightarrow[H_2O]{NaOH}$ $\begin{array}{l} CHO \\ | \\ Bu\overset{|}{C}CH_2CH_2COOH \\ | \\ Et \end{array}$

JACS (1944) <u>66</u> 56

$CH_2(COOMe)_2$ $\xrightarrow[\begin{array}{c} NaH \quad THF \\ 2\ HClO_4 \quad H_2O \quad MeCN \end{array}]{\begin{array}{c} SOMe \\ | \\ 1\ CH_2=CSMe \end{array}}$ $\begin{array}{l} CHO \\ | \\ CH_2CH(COOMe)_2 \end{array}$ $\dashrightarrow$ $\begin{array}{l} CHO \\ | \\ CH_2CH_2COOH \end{array}$

Tetr Lett (1973) 4711 4715

$C_6H_{13}CH(COOEt)_2$ $\xrightarrow[EtONa \quad EtOH]{CH_2=CHCHO}$ $\begin{array}{l} C_6H_{13}\overset{|}{C}(COOEt)_2 \\ \quad | \\ \quad CH_2CH_2CHO \end{array}$ $\dashrightarrow$ $\begin{array}{l} C_6H_{13}CHCOOH \\ \qquad | \\ \qquad CH_2CH_2CHO \end{array}$

JACS (1948) <u>70</u> 3470
Org React (1959) <u>10</u> 179

$\xrightarrow{\ HIO_4 \quad H_2O\ }$ 72%

Ber (1950) <u>83</u> 390

Also via: Aldehydic esters (Section 336)

Section 315   Carboxylic Acid — Amide

Carboxyamides

Synthesis (1973) 40   19%

JOC (1965) 30 1321
    (1973) 38 457
Org React (1957) 9 107   55%

Also via: Aminoacids (Section 316)

Section 316   Carboxylic Acid — Amine

α-Aminoacids, β-aminoacids and higher aminoacids

PhCH$_2$COOH $\xrightarrow{\begin{array}{c}1 \text{ LiN(Pr-i)}_2 \text{ HMPA THF}\\ \hline 2 \text{ NH}_2\text{OMe}\end{array}}$ PhCHCOOH   55%
                                                        |
                                                       NH$_2$

Chem Comm (1972) 623

i-PrCH$_2$CH$_2$COOH $\xrightarrow{\begin{array}{c}\text{Br}_2 \text{ PCl}_3\\ \hline \end{array}}$ i-PrCH$_2$CHCOOH $\xrightarrow{\begin{array}{c}\text{NH}_3 \text{ EtOH}\\ \hline \text{H}_2\text{O}\end{array}}$ i-PrCH$_2$CHCOOH   ~28%
                                                    | |
                                                    Br   NH$_2$

Org Synth (1955) Coll Vol 3 523 848

Me
|
EtCHCH(COOH)$_2$ $\xrightarrow[\text{2 130°}]{\text{1 Br}_2 \quad \text{Et}_2\text{O}}$ EtCHCHCOOH $\xrightarrow{\text{NH}_3 \quad \text{H}_2\text{O}}$ EtCHCHCOOH   49%
                                         | |
                                         Br   NH$_2$

Org Synth (1955) Coll Vol 3 495 705

PhCHO $\xrightarrow[\text{NaOAc} \quad \text{Ac}_2\text{O}]{\text{PhCONHCH}_2\text{COOH}}$ PhCH=C $\xrightarrow[\text{Ac}_2\text{O}]{\text{HI} \quad \text{P}_4}$ PhCH$_2$CHCOOH   ~41%
                                              | |
                                           N=CPh   NH$_2$

Org Synth (1943) Coll Vol 2 489
Org React (1942) $\underline{1}$ 210

Catalytic reduction of azlactones   JOC (1972) $\underline{37}$ 2916

$\xrightarrow[\substack{\text{2 Na-Hg} \quad \text{H}_2\text{O} \\ \text{3 Ba(OH)}_2 \quad \text{H}_2\text{O}}]{\substack{\text{CH}_2\text{NMe} \\ \text{C=NH} \\ \text{1 CONH}}}$ ~51%

Org Synth (1955) Coll Vol 3 586
Org React (1942) $\underline{1}$ 210

MeSCH$_2$CH$_2$CHO $\xrightarrow[\text{Et}_3\text{N} \quad \text{EtOH} \quad \text{H}_2\text{O}]{\text{NaCN} \quad \text{(NH}_4\text{)}_2\text{CO}_3}$ MeSCH$_2$CH$_2$CHCO $\xrightarrow[\substack{\text{EtOH} \\ \text{H}_2\text{O}}]{\text{Ba(OH)}_2}$ MeSCH$_2$CH$_2$CHCOOH
                                                        | NH |
                                                       NHCO   NH$_2$
                                                                                                        ~83%

JCS (1952) 3403

$\xrightarrow[\text{2 HI} \quad \text{P}_4 \quad \text{H}_2\text{O}]{\substack{\text{CH}_2\text{NH} \\ \text{CS} \\ \text{1 COS} \quad \text{HOAc}}}$ 83%

JOC (1952) $\underline{17}$ 1459
Org React (1942) $\underline{1}$ 210

PhCHO $\xrightarrow[\text{2 NaCN \quad H}_2\text{O}]{\text{1 \quad NaHSO}_3 \quad \text{H}_2\text{O}}$ PhCHCN $\xrightarrow[\text{2 H}_2\text{SO}_4 \quad \text{H}_2\text{O}]{\text{1 H}_2\text{SO}_4}$ PhCHCOOH   76%

JOC (1961) <u>26</u> 4741

i-PrCHO $\xrightarrow[\text{t-BuOH}]{\text{CHCl}_3 \quad \text{t-BuOK}}$ i-PrCHCCl$_3$ $\underset{\text{OH}}{}$ $\xrightarrow[\text{2 HCl \quad H}_2\text{O}]{\text{1 \quad KNH}_2 \quad \text{NH}_3}$ i-PrCHCOOH $\underset{\text{NH}_2}{}$

JOC (1964) <u>29</u> 1148

PhCH$_2$CONHBu-t $\xrightarrow[\text{t-BuOK}]{\text{t-BuOCl}}$ PhCH$_2$CONBu-t $\underset{\text{Cl}}{}$ $\xrightarrow[\substack{\text{t-BuOH} \\ \text{2 HCl \quad H}_2\text{O}}]{\text{1 \quad t-BuOK}}$ PhCHCOOH $\underset{\text{NHBu-t}}{}$ 72%

JACS (1961) <u>83</u> 4469

PhNH$_2$ $\dashrightarrow$ PhN$_2^+$ Cl$^-$ $\xrightarrow[\substack{\text{Cu}^{2+} \\ \text{2 HCl \quad HCOOH \quad H}_2\text{O}}]{\text{1 \quad CH}_2\text{=CHCOOMe}}$ PhCH$_2$CHCOOH $\underset{\text{Cl}}{}$ $\xrightarrow{\text{NH}_3}$ PhCH$_2$CHCOOH $\underset{\text{NH}_2}{}$

Proc Chem Soc (1962) 117

PhCH$_2$NH$_2$ $\dashrightarrow$ PhCH$_2$NC $\xrightarrow[\text{DMF}]{\text{(EtO)}_2\text{CO} \quad \text{NaH}}$ PhCHCOOEt $\underset{\text{NC}}{}$ $\xrightarrow[\text{2 NaOH}]{\text{1 HCl}}$ PhCHCOOH $\underset{\text{NH}_2}{}$ 57%

JOC (1973) <u>38</u> 2094

PhCH₂Cl --→ PhCH₂CH(COOEt)₂ $\xrightarrow[\text{2 NaN}_3]{\text{1 Br}_2}$ PhCH₂C(COOEt)₂ $\xrightarrow[\text{2 HBr}]{\text{1 Ph}_3\text{P}}$ PhCH₂CHCOOH
                                                                            |N₃ |NH₂
                                                                                                                38%

Annalen (1955) <u>591</u> 117
Z Naturforsch (1960) <u>15b</u> 811

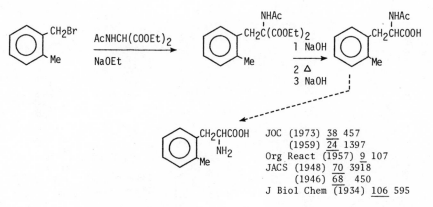

JOC (1973) <u>38</u> 457
    (1959) <u>24</u> 1397
Org React (1957) <u>9</u> 107
JACS (1948) <u>70</u> 3918
    (1946) <u>68</u> 450
J Biol Chem (1934) <u>106</u> 595

COOEt
|
CH₂Cl $\xrightarrow{1 \quad 2 \text{ HCl} \quad \text{HOAc} \quad \text{H}_2\text{O}}$ COOH
                                                                |
                                                              CH₂CHCOOH   ~41%
                                                                |
                                                              NH₂

Org Synth (1963) Coll Vol 4 55

           COOEt
           |
PrBr --→ PrCHCOOK $\xrightarrow{\begin{array}{l}1\ \text{N}_2\text{H}_4\cdot\text{H}_2\text{O}\\2\ \text{HNO}_2\\3\ \text{CCl}_4\ (\text{reflux})\\4\ \text{HCl}\quad\text{H}_2\text{O}\end{array}}$ PrCHCOOH
                                                                                            |
                                                                                          NH₂

J Prakt Chem (1930) <u>125</u> 211
JOC (1956) <u>21</u> 1182

[o-Cl-C₆H₄-O(CH₂)₃Cl] $\xrightarrow[\text{CH}_2\text{CN}]{\text{COOEt}}$ [o-Cl-C₆H₄-O(CH₂)₃CHCN] $\xrightarrow{\begin{array}{l}1\ \text{N}_2\text{H}_4\\2\ \text{NaNO}_2\\ \ \ \text{HCl}\\3\ \Delta\\4\ \text{HCl}\end{array}}$ [o-Cl-C₆H₄-O(CH₂)₃CHCOOH]
                                                                                                                                        with NH₂

Can J Chem (1952) <u>30</u> 592

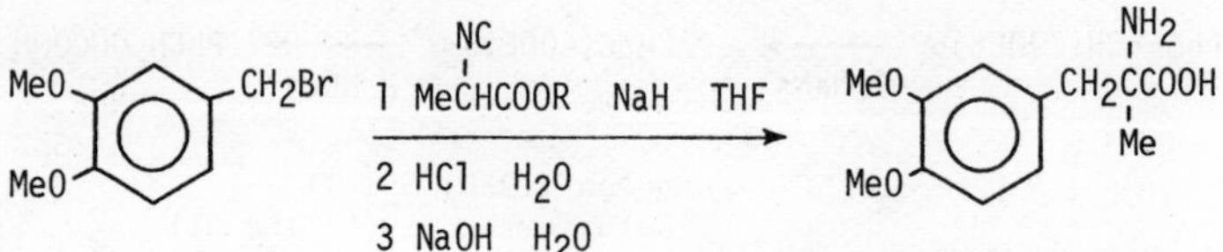

MeO—C₆H₃—CH₂Br 1 MeCHCOOR (NC) NaH THF 2 HCl H₂O 3 NaOH H₂O → MeO—C₆H₃—CH₂CCOOH (NH₂, Me)

Chem Ind (1972) 687

Et₂CO 1 NaCN NH₄Cl H₂O 2 NH₃ H₂O → Et₂CCN (NH₂) HCl H₂O → Et₂CCOOH (NH₂)   39-43%

Org Synth (1955) Coll Vol 3 66 84 88

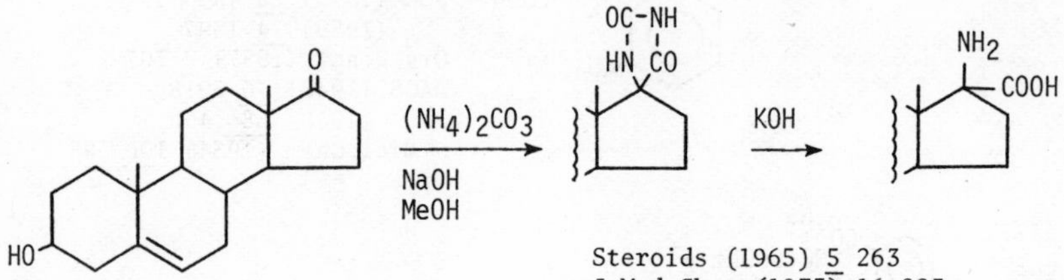

steroid + (NH₄)₂CO₃ NaOH MeOH → hydantoin → KOH → amino acid

Steroids (1965) 5 263
J Med Chem (1973) 16 823

BuCH₂COMe 1 MeONO Et₂O 2 Et₂SO₄ NaOH H₂O → BuCCOMe (NOEt) 1 NaOCl dioxane H₂O 2 Zn HOAc → BuCHCOOH (NH₂)   45%

JOC (1959) 24 1726

i-PrCN 1 Me₂CCN (OH) diisopinocampheylborane 2 MeOH 3 HCl H₂O → i-PrCHCOOH (NH₂)

Tetr Lett (1972) 3145

i-PrCOCOOH --→ i-PrCCOOH $\xrightarrow[\text{H}_2\text{O}]{\text{H}_2 \quad \text{Pd-C}}$ i-PrCHCOOH   87%
            ‖ |
           NNHPh   NH$_2$

JOC (1973) <u>38</u> 822

PhCH$_2$COCOOH $\xrightarrow{\text{NH}_2\text{COOCH}_2\text{Ph}}$ PhCH$_2$C(NHCOOCH$_2$Ph)$_2$ $\xrightarrow[\text{EtOH}]{\text{H}_2 \quad \text{Pd}}$ PhCH$_2$CHCOOH

(COOH on center carbon of PhCH$_2$C(NHCOOCH$_2$Ph)$_2$; NH$_2$ on product)   30%

JOC (1941) <u>6</u> 878

PhCOCOONa $\xrightarrow[\text{MeOH}]{\text{NH}_4\text{Br} \quad \text{LiBH}_3\text{CN}}$ PhCHCOOH   49%
                                                NH$_2$

JACS (1971) <u>93</u> 2897

PhCH$_2$COCOOH $\xrightarrow[\text{NH}_4\text{Br} \quad \text{MeOH}]{\text{NH}_3 \quad \text{PhCHNH}_2\text{-BH}_3}$ PhCH$_2$CHCOOH   66%

(Me on PhCHNH$_2$-BH$_3$; NH$_2$ on product)

JOC (1972) <u>37</u> 2347

PhCHO $\xrightarrow[\text{EtOH}]{\text{CH}_2(\text{COOH})_2 \quad \text{NH}_4\text{OAc}}$ PhCHCH$_2$COOH   48%
                                                NH$_2$

JACS (1936) <u>58</u> 299
J Prakt Chem (1965) <u>30</u> 18

Me—[thiophene]—CHO $\xrightarrow[\text{NH}_3]{\text{CH}_2(\text{COOH})_2}$ Me—[thiophene]—CHCH$_2$COOH $\xrightarrow{\text{H}_2 \quad \text{Ni}}$ Me(CH$_2$)$_4$CHCH$_2$COOH

(NH$_2$ substituent on CHCH$_2$COOH of both products)

Tetrahedron (1962) <u>18</u> 21
Zh Obshch Khim (1963) <u>33</u> 2697
(Chem Abs <u>60</u> 512)

PhCH=CHCOOH $\xrightarrow{\text{NH}_2\text{OH} \quad \text{EtOH}}$ PhCHCH$_2$COOH   34%
                                        |
                                       NH$_2$

Org Synth (1955) Coll Vol 3 91

Ts(CH$_2$)$_3$CH=CCl$_2$ $\xrightarrow[\text{H}_2\text{SO}_4]{}$ (phthalimide-N-CH$_2$OH)

Ts(CH$_2$)$_3$CHCH$_2$N(phthalimide with COOH) --►Ts(CH$_2$)$_3$CHCOOH
                                                                |
                                                              CH$_2$NH$_2$

Ber (1973) 106 2513

CN
|
(CH$_2$)$_3$Cl $\xrightarrow{\text{phthalimide-NK}}$ (phthalimide)N(CH$_2$)$_3$CN $\xrightarrow[\text{H}_2\text{O}]{\text{H}_2\text{SO}_4}$ NH$_2$(CH$_2$)$_3$COOH   47-62%

Org Synth (1943) Coll Vol 2 25

(cyclohexanone) $\xrightarrow{\text{NH}_2\text{OH}}$ (cyclohexanone oxime, =NOH) $\xrightarrow[\text{2 H}_2\text{SO}_4 \quad \text{H}_2\text{O}]{\text{1 H}_2\text{SO}_4}$ (chain with COOH and NH$_2$)

Org Synth (1943) Coll Vol 2 76

(cyclohexanone) $\xrightarrow{\text{NH}_2\text{OH}}$ (cyclohexanone oxime, =NOH) $\xrightarrow[\begin{array}{c}\text{2 } \Delta\\ \text{3 (cyclohexenyl morpholine)}\\ \text{4 HOAc } \text{H}_2\text{O}\end{array}]{\text{1 PhSO}_2\text{Cl} \quad \text{Pyr}}$ (bicyclic NH / O product)

$\downarrow$ 1 NaOH
  2 N$_2$H$_4$ KOH

NH$_2$(CH$_2$)$_{11}$COOH   48%

Ber (1967) 100 3039

                              Section
Also via: Aminoesters   351
            Amidoesters   344

Section 317   Carboxylic Acid — Ester

Carboxyesters

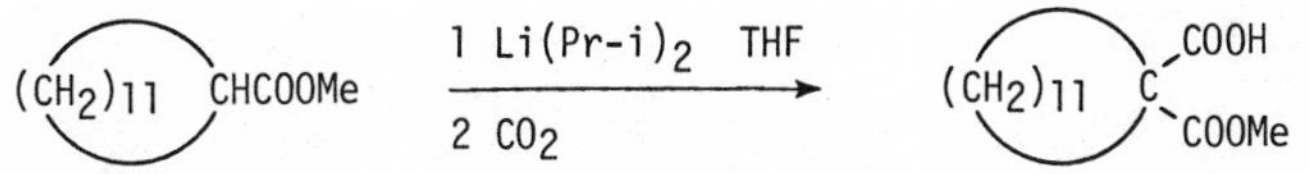

(CH₂)₁₁ CHCOOMe $\xrightarrow[\text{2 } CO_2]{\text{1 Li(Pr-i)}_2 \text{ THF}}$ (CH₂)₁₁ C⟨COOH / COOMe   94%

Tetr Lett (1971) 3001

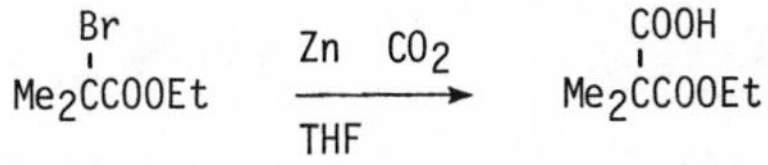

Br
|
Me₂CCOOEt $\xrightarrow[\text{THF}]{\text{Zn } CO_2}$ COOH
|
Me₂CCOOEt   50%

Chem Ind (1966) 1457

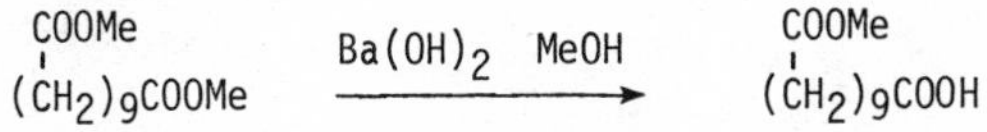

COOMe
|
(CH₂)₉COOMe $\xrightarrow{\text{Ba(OH)}_2 \text{ MeOH}}$ COOMe
|
(CH₂)₉COOH   60-64%

Org Synth (1963) Coll Vol 4 635

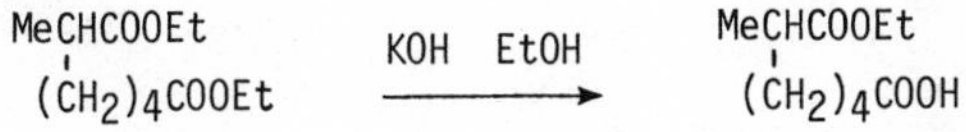

MeCHCOOEt
|
(CH₂)₄COOEt $\xrightarrow{\text{KOH EtOH}}$ MeCHCOOEt
|
(CH₂)₄COOH   59%

JACS (1948) 70 3206 364

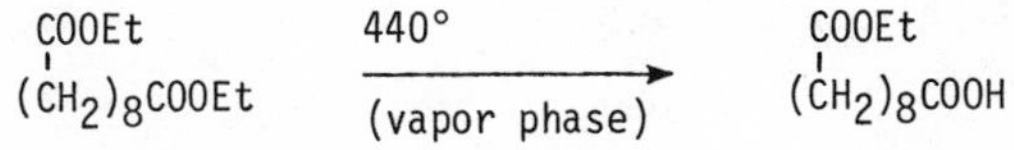

COOEt
|
(CH₂)₈COOEt $\xrightarrow[\text{(vapor phase)}]{440°}$ COOEt
|
(CH₂)₈COOH   69%

JOC (1964) 29 1252

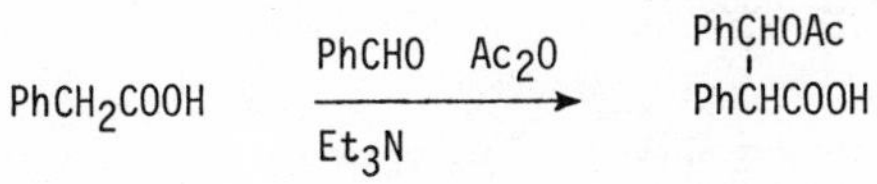

PhCH₂COOH $\xrightarrow[\text{Et}_3\text{N}]{\text{PhCHO Ac}_2\text{O}}$ PhCHOAc
|
PhCHCOOH   32%

Synthesis (1972) 263

$BuCH_2COOH$ $\xrightarrow{Tl(OAc)_3}$

$$\begin{array}{c} BuCHCOOH \\ | \\ BuCH_2COO \end{array}$$

Tetr Lett (1970) 5285

Also via: Section
            Diesters   357
        Hydroxyacids   313

Section 318   Carboxylic Acid —— Ether, Epoxide
○○○

Alkoxyacids and epoxyacids

PhCHO $\xrightarrow[MeOH]{CHCl_3 \quad KOH}$

$$\begin{array}{c} PhCHCOOH \\ | \\ OMe \end{array}$$

JACS (1960) 82 4062
Chim Ther (1967) 2 9
(Chem Abs 67 99319)
Synthesis (1971) 131

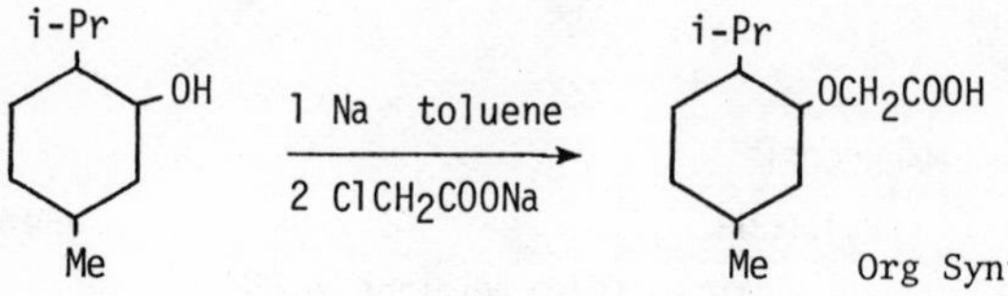

49%

1 Na   toluene
———————————————
2 ClCH$_2$COONa

78-84%

Org Synth (1955) Coll Vol 3 544
           (1943) Coll Vol 2 260

$CH_2=CHCN$ $\xrightarrow[2 \ H_2SO_4 \quad H_2O]{1 \ PrOH \quad MeONa}$ $PrOCH_2CH_2COOH$ 50%

JACS (1949) 71 3480

$$\begin{array}{c} Me \\ | \\ MeCH=CCOOH \end{array}$$ $\xrightarrow[2 \ NaBH_4]{1 \ Hg(OAc)_2 \quad MeOH}$ $$\begin{array}{c} Me \\ | \\ MeCHCHCOOH \\ | \\ OMe \end{array}$$ 72%

JCS Perkin I (1973) 109

PhCHCOOH $\xrightarrow{\text{Me}_2\text{CO}\ \text{H}_2\text{SO}_4}$ PhCHCO $\xrightarrow[\text{Et}_2\text{O}]{\text{t-BuMgCl}}$ PhCHCOOH   44%
| | | |
OH   O O   OCHMe$_2$
                                                \ /
                                               CMe$_2$ JACS (1942) $\underline{64}$ 1567

cyclohexanone $\xrightarrow[\text{Me}_2\text{SO}]{\overset{+\ -}{\text{Me}_2\text{SCHCOONa}}}$ cyclohexane-O-CHCOOH   ~60%

                                                                      JOC (1970) $\underline{35}$ 1600

MeCH=CHCOOH $\xrightarrow[\text{NaOH}\ \text{H}_2\text{O}]{\text{H}_2\text{O}_2\ \text{Na}_2\text{WO}_4}$ MeCHCHCOOH   50%
                                                            \ /
                                                             O
                                                                    JOC (1959) $\underline{24}$ 54
                                                                    JCS (1962) $\overline{1116}$

Also via: Section
               Alkoxyesters   358
               Epoxyesters   358

Section 319   Carboxylic Acid — Halide
               oooooooooooooooooooooooooo

Haloacids, Halogenoacids

BuCH$_2$COOH $\xrightarrow[\text{sulpholane}]{\text{C}_5\text{H}_{11}\text{COCl}\ \text{CuCl}_2\ \text{LiCl}}$ BuCHCOOH   <84%
                                                             |
                                                             Cl
                                                    Chem Comm (1966) 544

i-PrCH$_2$CH$_2$COOH $\xrightarrow{\text{Br}_2\ \text{PCl}_3}$ i-PrCH$_2$CHCOOH   63-66%
                                                         |
                                                         Br

                              Org Synth (1955) Coll Vol 3 523 848
                                          (1943) Coll Vol 2 74

$$\text{(cyclohexane-COOH)} \xrightarrow{\text{Cl}_2 \quad \text{PCl}_3} \text{(cyclohexane-Cl,COOH)} \qquad 93\%$$

JACS (1969) <u>91</u> 7090

$$ClCH_2CH_2COCl \xrightarrow[\text{HOAc}]{\text{NBS} \quad \text{HBr}} ClCH_2\underset{Br}{CHCOCl} \dashrightarrow ClCH_2\underset{Br}{CHCOOH} \qquad <70\%$$

Tetr Lett (1970) 3431

$$C_{10}H_{21}CH_2COCl \xrightarrow{\text{SOCl}_2 \quad h\nu} C_{10}H_{21}\underset{Cl}{CHCOCl} \dashrightarrow C_{10}H_{21}\underset{Cl}{CHCOOH} \qquad \sim90\%$$

JOC (1973) <u>38</u> 3919

$$MeCH_2(CH_2)_3COOH \xrightarrow{(i\text{-}Pr)_2NCl \quad h\nu} MeCH(CH_2)_3COOH$$
$$\underset{Cl}{}$$

JACS (1971) <u>93</u> 438
Rec Trav Chim (1964) <u>83</u> 891

$$i\text{-}PrCH(COOH)_2 \xrightarrow[\text{2 125-130°}]{\text{1 Br}_2 \quad \text{Et}_2O} i\text{-}Pr\underset{Br}{CHCOOH}$$

Org Synth (1943) Coll Vol 2 93
chloroacids   JCS (1950) 2900

$$PhCHO \dashrightarrow Ph\underset{OH}{CHCCl_3} \xrightarrow[\text{H}_2O]{\text{KOH} \quad \text{oleic acid}} Ph\underset{Cl}{CHCOOH} \qquad 50\%$$

Synthesis (1971) 131

$$\text{(3-nitrobenzyl alcohol)} \xrightarrow[\text{H}_2\text{SO}_4]{\text{ClCH=CCl}_2} \text{(product)} \qquad 82\%$$

Ber (1970) 103 3850

$$\text{(naphthylamine)} \dashrightarrow \text{(diazonium)} \xrightarrow[\text{Cu}^{2+}]{1\ \text{CH}_2=\text{CHCOOMe}} \text{(product)} \sim 51\%$$

2 HCl   HCOOH

H$_2$O

Proc Chem Soc (1962) 117

Bromoacids   Org Synth (1971) 51 1

$$\underset{\underset{F}{|}}{\text{C}_5\text{H}_{11}\text{CHCH}_2\text{Br}} \xrightarrow[\text{DMF}]{\text{NaOAc   NaI}} \underset{\underset{F}{|}}{\text{C}_5\text{H}_{11}\text{CHCH}_2\text{OAc}} \xrightarrow[\text{HOAc}]{\text{HNO}_3} \underset{\underset{F}{|}}{\text{C}_5\text{H}_{11}\text{CHCOOH}} \qquad \sim 55\%$$

Org Synth (1966) 46 37

$$\text{Me}_2\text{CO} \xrightarrow[\underset{\text{Me}_4\text{N Br}}{+\ -}]{\text{CHCl}_3 \quad \overset{O}{\overset{/\ \backslash}{\text{CH}_2-\text{CH}_2}}} \underset{\underset{\text{Cl}}{|}}{\text{Me}_2\text{CCOOCH}_2\text{CH}_2\text{Cl}} \dashrightarrow \underset{\underset{\text{Cl}}{|}}{\text{Me}_2\text{CCOOH}} \qquad <24\%$$

Ber (1968) 101 1299

$$\text{C}_5\text{H}_{11}\text{CH=CH}_2 \xrightarrow[2\ \text{HCl}\quad \text{H}_2\text{O}]{1\ \text{NOCl}\quad \text{HCl}\quad \text{Et}_2\text{O}} \underset{\underset{\text{Cl}}{|}}{\text{C}_5\text{H}_{11}\text{CHCOOH}}$$

Zh Obshch Khim (1964) 34 1227
(Chem Abs 61 2967)

$$\underset{\underset{\text{Me}}{|}}{\text{MeCH=CCOOH}} \xrightarrow[\text{CHCl}_3]{\text{HI   Ag}} \underset{\underset{\text{I\ \ Me}}{|\ \ |}}{\text{MeCHCHCOOH}}$$

JACS (1929) 51 2528

Also via: Haloesters (Section 359)

Section 320   Carboxylic Acid — Ketone

$$PhCOCl \xrightarrow{CuCN} PhCOCN \xrightarrow[H_2O]{HCl} PhCOCOOH \qquad \sim 45\%$$

Org Synth (1955) Coll Vol 3 114
Can J Chem (1971) 49 919

$$PhCHO \xrightarrow[NaOAc]{\overset{\overset{NHAc}{|}}{CH_2COOH} \quad Ac_2O} PhCH=C-CO \xrightarrow[2\ HCl\quad H_2O]{1\ H_2O \quad Me_2CO} PhCH_2COCOOH \qquad \sim 65\%$$

with the intermediate ring: PhCH=C-CO / N   O / CMe

Org Synth (1943) Coll Vol 2 519 1
Org React (1942) 1 210

$$(PhCH_2)_2CHCHO \dashrightarrow (PhCH_2)_2CHCHCN \xrightarrow[2\ CrO_3\quad HOAc]{1\ H_2SO_4 \quad t\text{-}BuOH} (PhCH_2)_2CHCOCOOH \quad 77\%$$

with OH on the CHCHCN carbon

Synthesis (1971) 538

$$PrCHO \xrightarrow[Et_2NH\quad EtOH]{CH_3COCOOH} PrCH=CHCOCOOH \xrightarrow[H_2O]{KBH_4 \quad KHCO_3} PrCH_2CH_2COCOOH$$

Bull Soc Chim Fr (1966) 1435

$$C_5H_{11}CHO \dashrightarrow C_5H_{11}CH=C(COOEt)_2 \xrightarrow[\substack{2\ KOH\quad EtOH \\ 3\ 180\text{-}200°}]{1\ H_2O_2 \quad Na_2WO_4 \quad H_2O} C_5H_{11}CH_2COCOOH$$

JOC (1964) 29 2080

PhCOOEt $\xrightarrow[\text{2 HCl H}_2\text{O Me}_2\text{SO}]{\text{1 Me}_2\text{SO t-BuOK}}$ PhCOCHSMe (OH) $\xrightarrow[\text{CHCl}_3]{\text{Cu(OAc)}_2 \cdot \text{H}_2\text{O}}$ PhCOCOOH   64-73%

JACS (1966) **88** 5498

EtCH$_2$COOEt $\xrightarrow[\text{Et}_2\text{O}]{\text{(COOEt)}_2 \quad \text{EtONa}}$ EtCHCOOEt (COCOOEt) $\xrightarrow[\text{H}_2\text{O}]{\text{H}_2\text{SO}_4}$ EtCH$_2$COCOOH

Annalen (1943) **555** 41

EtCHBr (Me) $\dashrightarrow$ EtCHLi (Me) $\xrightarrow[\text{2 CO}_2 \quad]{\text{1 t-BuCH}_2\text{CMe}_2 \text{(NC) Et}_2\text{O}}$ EtCHCOCOOH (Me)   80%

3 (COOH)$_2$ H$_2$O

JACS (1969) **91** 7778
(1970) **92** 6675

EtBr $\dashrightarrow$ EtMgBr $\xrightarrow[]{\text{COOEt} \atop \text{CONEt}_2}$ EtCOCONEt$_2$ $\xrightarrow[]{\text{NaOH}}$ EtCOCOOH   ~55%

Compt Rend (1927) **184** 825

RO-C$_6$H$_3$(MeO)-COCH$_3$ $\xrightarrow[\text{H}_2\text{O}]{\text{KMnO}_4 \quad \text{NaOH}}$ RO-C$_6$H$_3$(MeO)-COCOOH

Monatsh (1952) **83** 883

Me$_2$CO $\dashrightarrow$ Me$_2$C=CCOOEt (CN) $\xrightarrow[\text{H}_2\text{O EtOH} \atop \text{2 KOH EtOH} \atop \text{3 HCl H}_2\text{O}]{\text{1 H}_2\text{O}_2 \quad \text{Na}_2\text{WO}_4}$ Me$_2$CHCOCOOH

JOC (1963) **28** 3088

PhCHCOOH $\xrightarrow{\text{Br}_2 \quad \text{CHCl}_3}$ PhCOCOOH   90%
|
OMe

Ber (1967) <u>100</u> 3777

i-PrCH$_2$CHCOOH $\xrightarrow[\text{2 NaOH} \quad \text{H}_2\text{O}]{\text{1 (CF}_3\text{CO)}_2\text{O}}$ i-PrCH$_2$COCOOH
|
NH$_2$

Annalen (1962) <u>658</u> 128

Me$_2$CHCOOH $\xrightarrow{\text{1 LiN(Pr-i)}_2 \quad \text{THF}}$ Me$_2$CCOOH
$\qquad\qquad$ 2 t-BuCOOMe $\qquad$ |
$\qquad\qquad$ 3 Me$_3$SiCl $\qquad$ t-BuCO
$\qquad\qquad$ 4 MeOH

JACS (1971) <u>93</u> 6321

$\xrightarrow[\text{2 CO}_2]{\text{1 Ph}_3\text{CK} \quad \text{Et}_2\text{O}}$

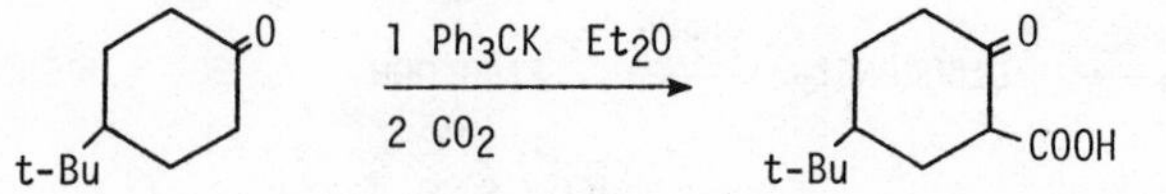

89%

Coll Czech (1961) <u>26</u> 847

PhCOCH$_3$ $\xrightarrow[\text{DMF}]{\text{CO}_2 \quad \text{PhOK}}$ PhCOCH$_2$COOH   60%

Chem Comm (1966) 618
JOC (1973) <u>38</u> 4086

PhCOCH$_3$ $\xrightarrow{\text{(MeOCOO)}_2\text{Mg} \quad \text{DMF}}$ PhCOCH$_2$COOH   68%

JACS (1959) <u>81</u> 2598

$$\text{1 Li NH}_3 \text{ Et}_2\text{O}$$
$$\text{2 CO}_2 \text{ Et}_2\text{O}$$

COOH

JACS (1965) 87 275
JOC (1968) 33 712

$$\text{C}_{10}\text{H}_{21}\text{COCl} \xrightarrow{\substack{\text{NaC(COOCH}_2\text{Ph)}_2 \\ \text{CH}_2\text{COOCH}_2\text{Ph}}} \text{C}_{10}\text{H}_{21}\text{COC(COOCH}_2\text{Ph)}_2 \xrightarrow[\text{2 170°}]{\text{1 H}_2 \text{ Pd-C}} \text{C}_{10}\text{H}_{21}\text{COCH}_2\text{CH}_2\text{COOH}$$

$$\text{CH}_2\text{COOCH}_2\text{Ph}$$

66%

JCS (1950) 325

$$\text{C}_5\text{H}_{11}\text{COOH} \xrightarrow{\substack{\text{1 SOCl}_2 \\ \text{2 i-PrCHCOOH} \\ \text{NH}_2 \\ \text{3 Ac}_2\text{O} \\ \text{4 CH}_2\text{=CHCN Et}_3\text{N}}}$$

$$\text{C}_5\text{H}_{11}\overset{\text{CH}_2\text{CH}_2\text{CN}}{\underset{\overset{\text{O CPr-i}}{\underset{\text{CO}}{\diagdown\diagup}}}{\text{C-N}}} \xrightarrow{\text{NaOH}} \text{C}_5\text{H}_{11}\text{COCH}_2\text{CH}_2\text{COOH}$$

Angew (1971) 83 727
(Internat Ed 10 655)

$$\text{EtCOCl} \xrightarrow[\text{2 CH}_2\text{=CO}]{\text{1 CH}_2\text{N}_2 \text{ Et}_2\text{O}} \text{EtC=CH} \xrightarrow{\text{NaOH H}_2\text{O}} \text{EtCOCH}_2\text{CH}_2\text{COOH}$$

29%

Annalen (1964) 678 113

$$\substack{\text{CH}_2\text{COCl} \\ | \\ \text{CH}_2\text{COCl}} \xrightarrow[\text{2 H}_2\text{SO}_4 \text{ H}_2\text{O}]{\text{1 Et}_3\text{Al}_2\text{Cl}_3} \text{EtCOCH}_2\text{CH}_2\text{COOH}$$

90%

Angew (1965) 77 810
(Internat Ed 4 785)

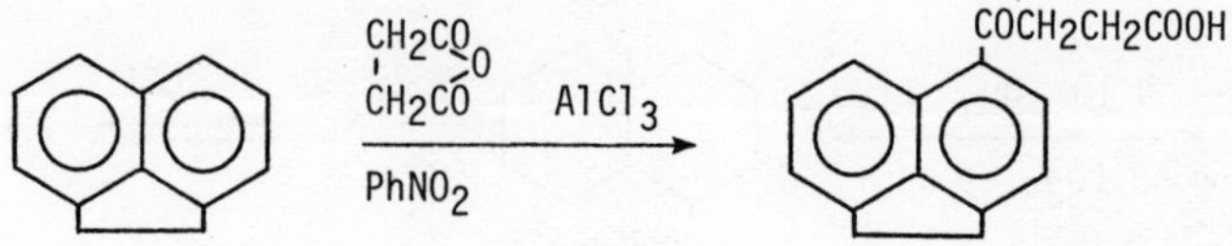

CH$_2$CO$>$O
CH$_2$CO   AlCl$_3$

PhNO$_2$

COCH$_2$CH$_2$COOH

Org Synth (1955) Coll Vol 3 6
(1943) Coll Vol 2 81
(1932) Coll Vol 1 517

C$_6$H$_{13}$CHO

CH$_2$COOEt
CH$_2$COOEt
——————
t-BuOK

C$_6$H$_{13}$CH=CCOOEt
CH$_2$COOEt

1 NaOH   H$_2$O
——————————
2 Br$_2$ CCl$_4$
3 NaOH   H$_2$O

C$_6$H$_{13}$COCH$_2$CH$_2$COOH   < 66%

JOC (1966) 31 616
Org Synth (1963) Coll Vol 4 430

PhI

HC≡CH   Ni(CO)$_4$ CO
————————————————
HCl   H$_2$O   30 atmos

PhCOCH$_2$CH$_2$COOH

Tetr Lett (1964) 2777

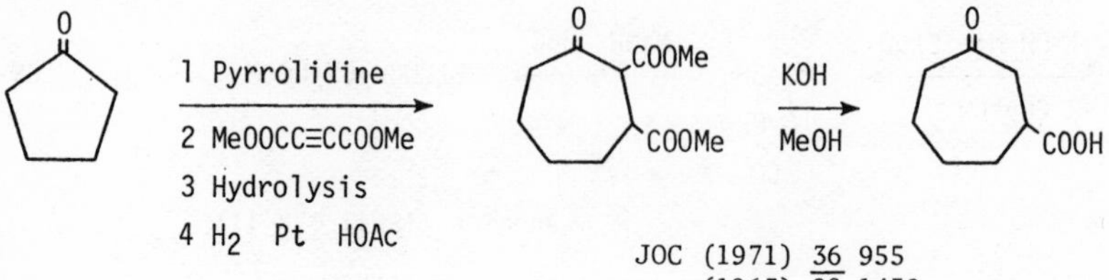

1 Pyrrolidine
2 MeOOCC≡CCOOMe
3 Hydrolysis
4 H$_2$ Pt HOAc

COOMe
COOMe

KOH
———
MeOH

COOH

JOC (1971) 36 955
(1963) 28 1459

PhCOCH$_3$

1 BrCH$_2$COOLi   LiNH$_2$ NH$_3$
————————————————————
2 BrCH$_2$COOH   Et$_2$O

PhCOCH$_2$CH$_2$COOH   61%

Chem Ind (1959) 255

CHO
Br—C₆H₄—COCH₃ COOH → Br—C₆H₄—COCH=CHCOOH → Br—C₆H₄—COCH₂CH₂COOH

base

Zn HOAc H₂O

JOC (1973) 38 4044
(1958) 23 1832

Me—(cyclohexanone, Me)—O

1 BrCH₂CH=CH₂ NaH
─────────────────
MeOCH₂CH₂OMe

2 RuO₄ NaIO₄ t-BuOH H₂O

→ Me—(cyclohexanone ring)—O, CH₂COOH, Me

Tetr Lett (1972) 1853

PhCOCH₂
 |
 Ph

HCHO Me₂NH·HCl
────────────────
Ac₂O

→ PhCOCHCH₂NMe₂
        |
        Ph

1 MeI
─────────
2 KCN

3 HCl H₂O

→ PhCOCHCH₂COOH
        |
        Ph

68%

JOC (1973) 38 4044

t-BuCOCH=CMe₂

1 KCN EtOH H₂O
─────────────────
2 KOH

→ t-BuCOCH₂CMe₂
              |
              COOH

8-12%

Annalen (1971) 751 168
JOC (1963) 28 1459

PhCOCH₂CH₂CHCH₂CH₂
            | |
            O ── CO

1 NaOH H₂O
──────────────
2 CrO₃ Pyr

→ PhCOCH₂CH₂COCH₂CH₂COOH

> 45%

JACS (1963) 85 2937

BuI
───
MeOH

MeOK
───→

(cyclohexanedione with Bu)

Ba(OH)₂ H₂O
───────────→

BuCH₂CO(CH₂)₃COOH

23%

Ber (1952) 85 61
JCS (1957) 2202
Newer Meth Prep Org Chem (1963) 2 51

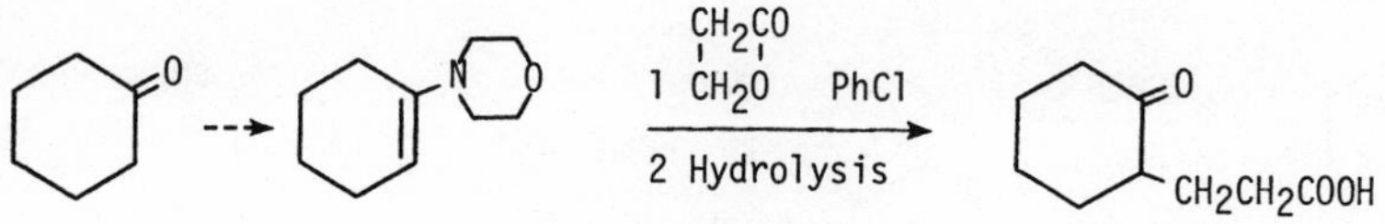
CH2CO
1 CH2O PhCl
2 Hydrolysis
CH2CH2COOH

81%

Acta Chem Scand (1964) 18 2201

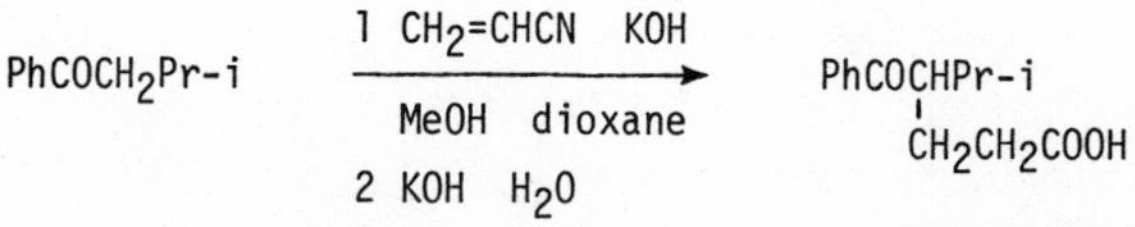
PhCOCH2Pr-i
1 CH2=CHCN KOH
MeOH dioxane
2 KOH H2O
PhCOCHPr-i
CH2CH2COOH

JCS (1948) 1741
Org React (1959) 10 179

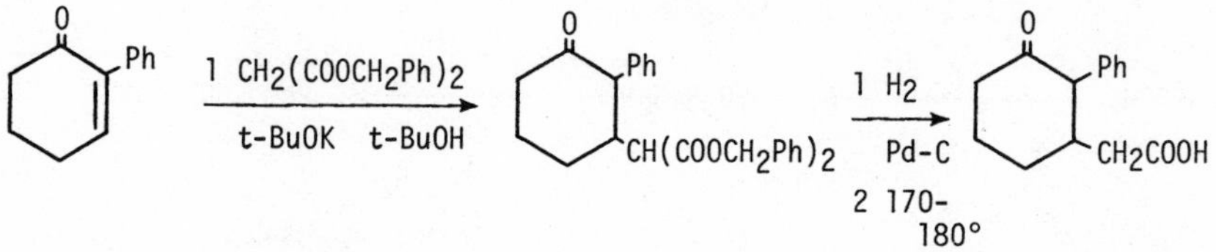
Ph
1 CH2(COOCH2Ph)2
t-BuOK t-BuOH
CH(COOCH2Ph)2
1 H2
Pd-C
2 170-180°
CH2COOH

Org React (1959) 10 179

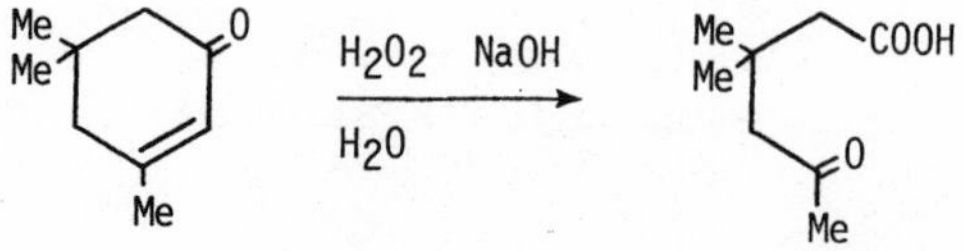
Me
Me
Me
H2O2 NaOH
H2O
COOH

JOC (1970) 35 1275

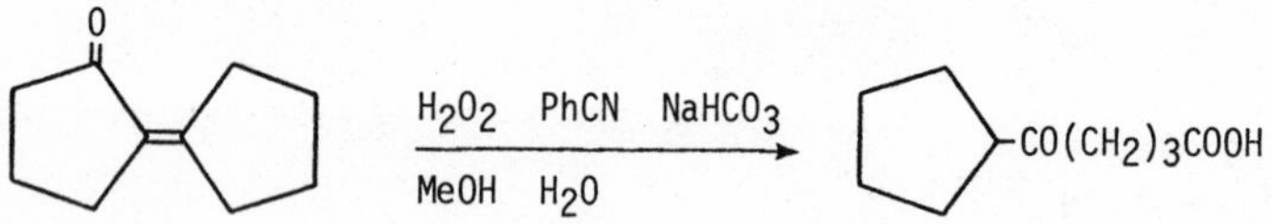
H2O2 PhCN NaHCO3
MeOH H2O
CO(CH2)3COOH

30%

Bull Soc Chim Fr (1969) 2871

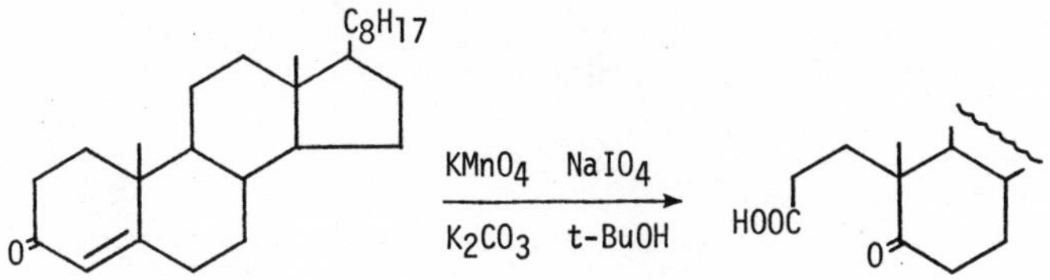
C_8H_{17}
$KMnO_4$ $NaIO_4$
K_2CO_3 t-BuOH
HOOC
O
O

79-88%

Can J Chem (1961) 39 599

O_3 Can J Chem (1962) 40 2153

RuO_2, $NaIO_4$ JOC (1969) 34 112

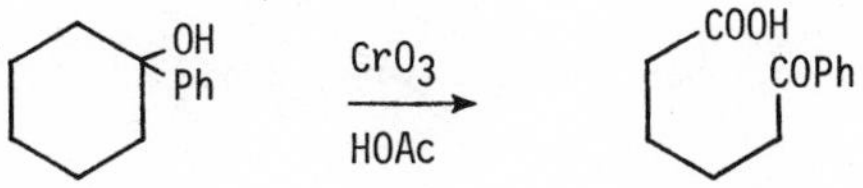
OH
Ph
CrO_3
HOAc
COOH
COPh

81%

JACS (1948) 70 3352

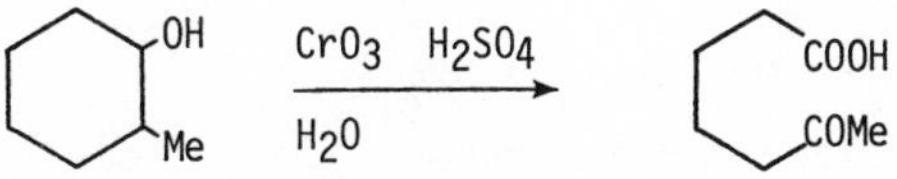
OH
Me
CrO_3 H_2SO_4
H_2O
COOH
COMe

46-55%

Org Synth (1963) Coll Vol 4 19

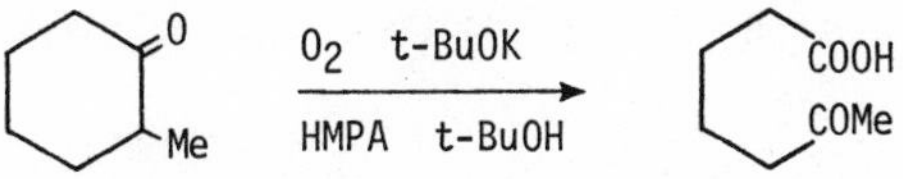
O
Me
O_2 t-BuOK
HMPA t-BuOH
COOH
COMe

Bull Soc Chim Fr (1967) 3742

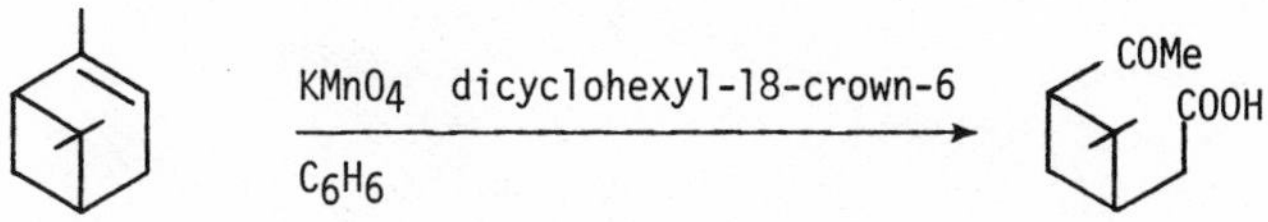
$KMnO_4$ dicyclohexyl-18-crown-6
C_6H_6
COMe
COOH

90%

JACS (1972) 94 4024

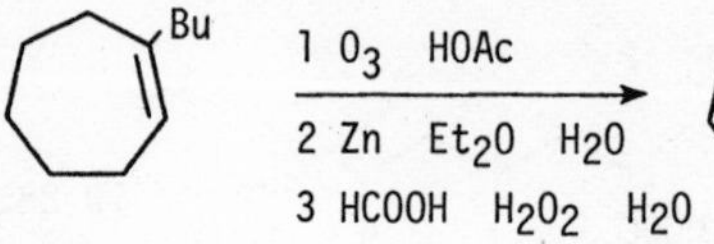

63%

Can J Chem (1955) 33 1720

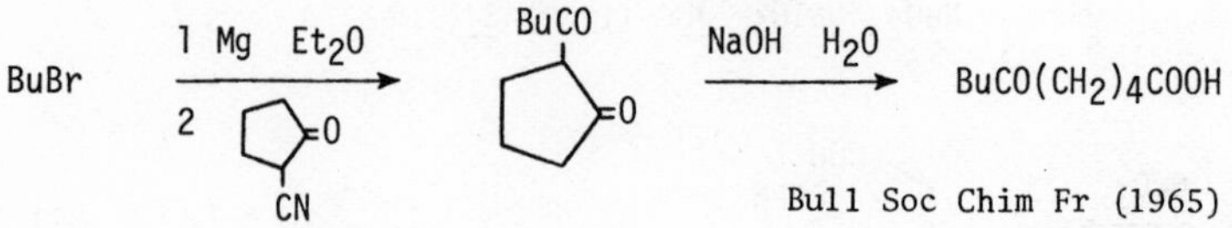

BuBr $\xrightarrow[\substack{2 \quad \overset{\displaystyle \bigcirc}{\underset{CN}{}} =O}]{1 \quad Mg \quad Et_2O}$ $\xrightarrow{NaOH \quad H_2O}$ $BuCO(CH_2)_4COOH$

Bull Soc Chim Fr (1965) 3041

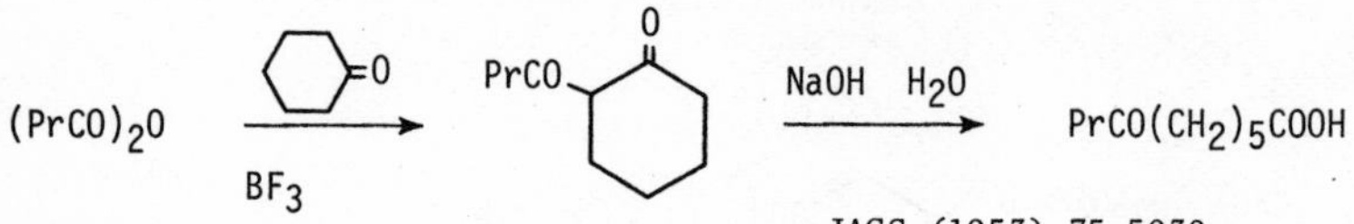

$(PrCO)_2O$ $\xrightarrow[BF_3]{\overset{\displaystyle \bigcirc}{=O}}$ $\xrightarrow{NaOH \quad H_2O}$ $PrCO(CH_2)_5COOH$

JACS (1953) 75 5030

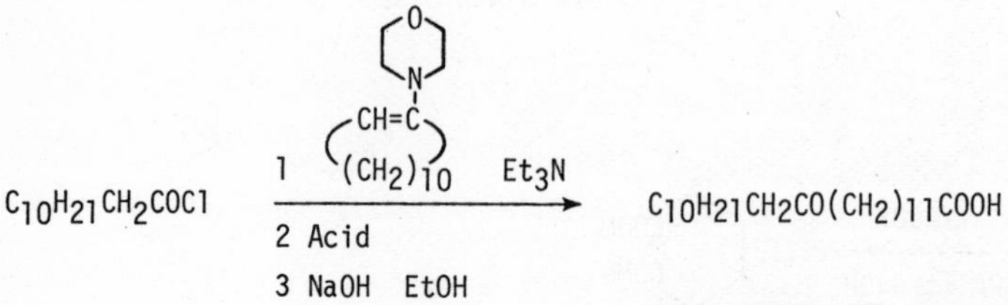

$C_{10}H_{21}CH_2COCl$ $\xrightarrow[\substack{2 \quad Acid \\ 3 \quad NaOH \quad EtOH}]{1 \quad \text{(morpholine enamine)} \quad Et_3N}$ $C_{10}H_{21}CH_2CO(CH_2)_{11}COOH$ 71%

Ber (1967) 100 4010 4017

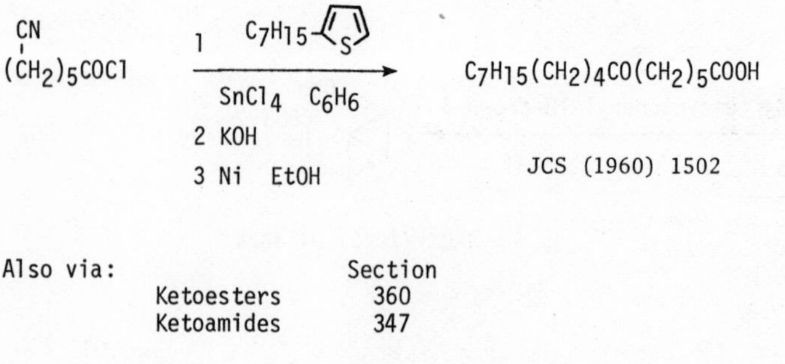

$\underset{(CH_2)_5COCl}{\overset{CN}{|}}$ $\xrightarrow[\substack{SnCl_4 \quad C_6H_6 \\ 2 \quad KOH \\ 3 \quad Ni \quad EtOH}]{1 \quad C_7H_{15}\text{-thiophene}}$ $C_7H_{15}(CH_2)_4CO(CH_2)_5COOH$

JCS (1960) 1502

Also via: Section
        Ketoesters   360
        Ketoamides   347

Section 321   Carboxylic Acid — Nitrile

Cyanoacids

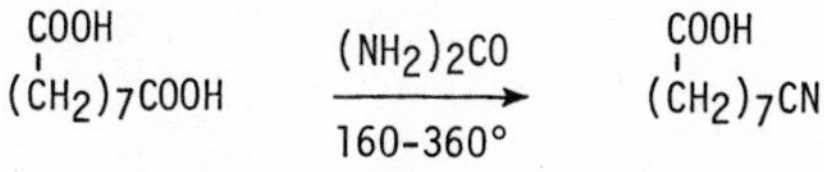

COOH
|
(CH₂)₇COOH

$\xrightarrow[160-360°]{(NH_2)_2CO}$

COOH
|
(CH₂)₇CN

<49%

JACS (1951) 73 343

KCN   280°

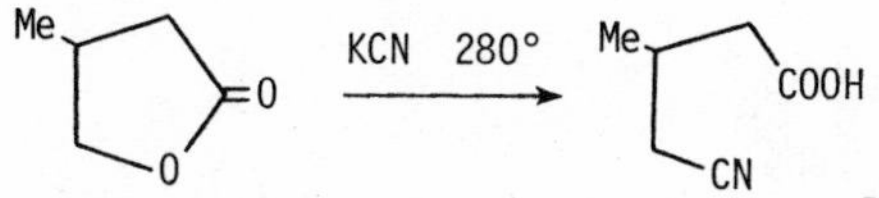

~94%

Bull Soc Chim Fr (1954) 88
Org Synth (1955) Coll Vol 3 174

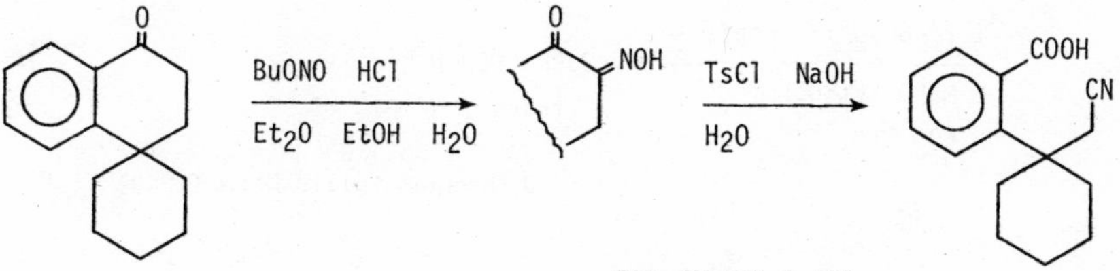

$\xrightarrow[Et_2O \quad EtOH \quad H_2O]{BuONO \quad HCl}$

$\xrightarrow[H_2O]{TsCl \quad NaOH}$

JOC (1941) 6 105
Tetr Lett (1968) 1127

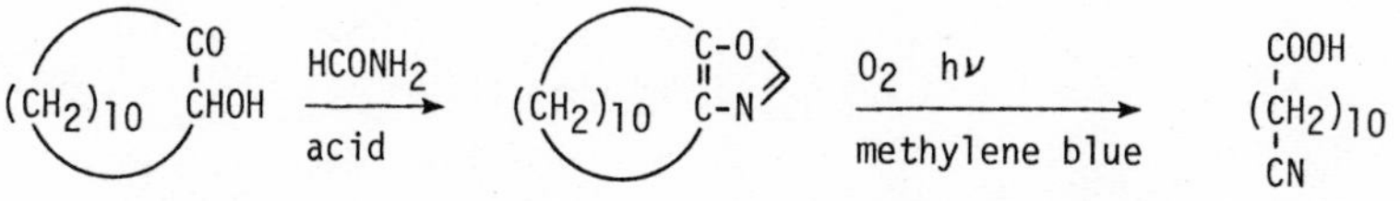

CO
|
(CH₂)₁₀   CHOH

$\xrightarrow[acid]{HCONH_2}$

C–O
(CH₂)₁₀   C–N

$\xrightarrow[methylene\ blue]{O_2 \quad h\nu}$

COOH
|
(CH₂)₁₀
|
CN

~80%

JACS (1968) 90 2440

$\xrightarrow[EtOH \quad H_2O]{NH_2OH \cdot HCl}$

$\xrightarrow[MeOH]{MeONa}$

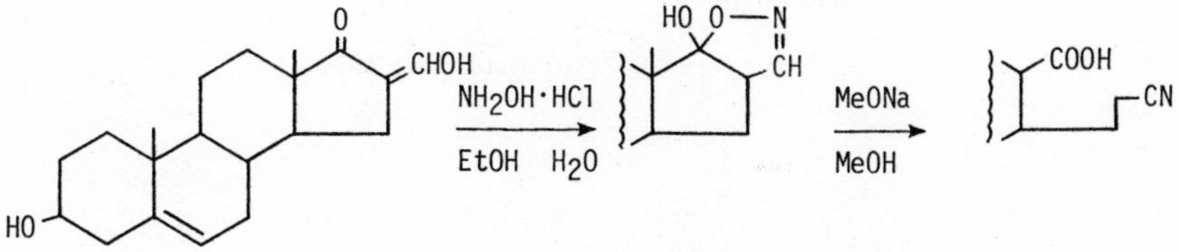

Ber (1961) 94 2897

Also via: Cyanoesters (Section 361)

Section 322   Carboxylic Acid — Olefin

Review: Recent Developments in the Synthesis of Fatty Acids

Chem Rev (1957) $\underline{57}$ 191

$$C_6H_{13}C\equiv CH \xrightarrow[\text{cyclohexane}]{\text{Ni(CO)}_4 \quad \text{HOAc}} \underset{\underset{COOH}{|}}{C_6H_{13}C=CH_2}$$

Chim Ind (Milan) (1964) $\underline{46}$ 1063
Bull Chem Soc Jap (1968) $\underline{41}$ 390
Annalen (1953) $\underline{582}$ 1
JCS (1951) 48

$$BuC\equiv CH \xrightarrow[\begin{array}{l}\text{2 MeLi   Et}_2\text{O}\\ \text{3 CO}_2\end{array}]{\text{1 (i-Bu)}_2\text{AlH   heptane}} \underset{\text{trans}}{BuCH=CHCOOH} \qquad 78\%$$

JACS (1967) $\underline{89}$ 2754
J Organometallic Chem (1968) $\underline{11}$ P7

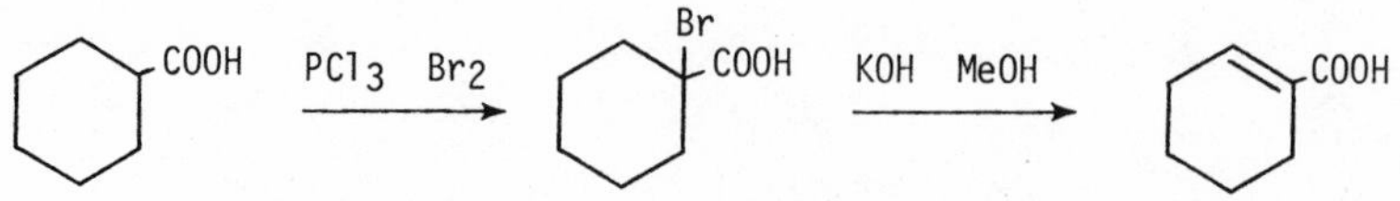

J Med Chem (1970) $\underline{13}$ 317

$$C_9H_{19}CH_2CH_2COONa \xrightarrow[\text{2 DDQ   C}_6\text{H}_6]{\text{1 (i-Pr)}_2\text{NLi   HMPA   THF}} C_9H_{19}CH=CHCOOH \qquad \sim 30\%$$

Chem Comm (1973) 94

$$C_{16}H_{33}CH_2COOH \xrightarrow[\text{2 } CH_2N_2]{\text{1 } SOCl_2} .C_{16}H_{33}CH_2COCHN_2 \xrightarrow[\text{2 } KOH]{\text{1 } Br_2 \quad CCl_4} C_{16}H_{33}CH=CHCOOH$$

35%

Chem Phys Lipids (1968) 2 213

$$C_7H_{15}CH_2COOH \xrightarrow[\text{2 } HCHO]{\text{1 } (i\text{-}Pr)_2NLi \quad HMPA \quad THF} \underset{\underset{CH_2}{\|}}{C_7H_{15}CCOOH}$$

90%

JOC (1972) 37 1256

$$PhCHO \xrightarrow[\text{EtCOONa}]{(MeCH_2CO)_2O} \underset{\underset{Me}{|}}{PhCH=CCOOH}$$

60-70%

Org React (1942) 1 210

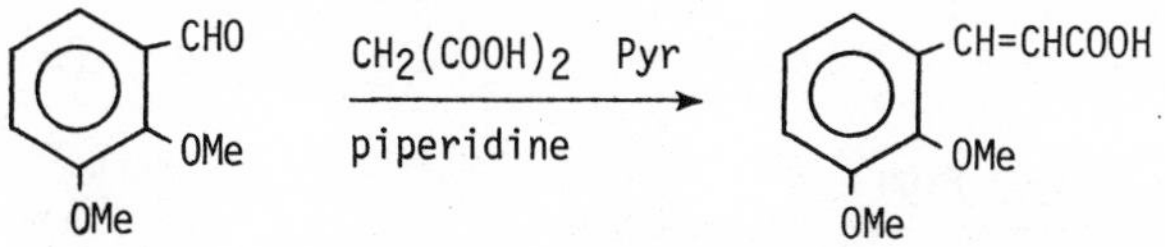

$$\xrightarrow[\text{piperidine}]{CH_2(COOH)_2 \quad Pyr}$$

87-98%

Org Synth (1963) Coll Vol 4 327
(1955) Coll Vol 3 425
Org React (1942) 1 210

$$t\text{-}BuCHO \xrightarrow[\substack{\text{2 Hydrolysis} \\ \text{3 } \triangle}]{\text{1 } CH_2(COOEt)_2 \quad ZnCl_2 \quad Ac_2O} t\text{-}BuCH=CHCOOH$$

JOC (1965) 30 917
JACS (1938) 60 2901

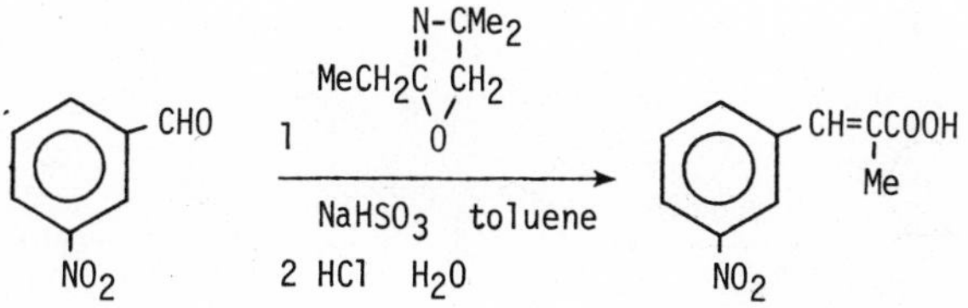

90%

JOC (1962) 27 4418

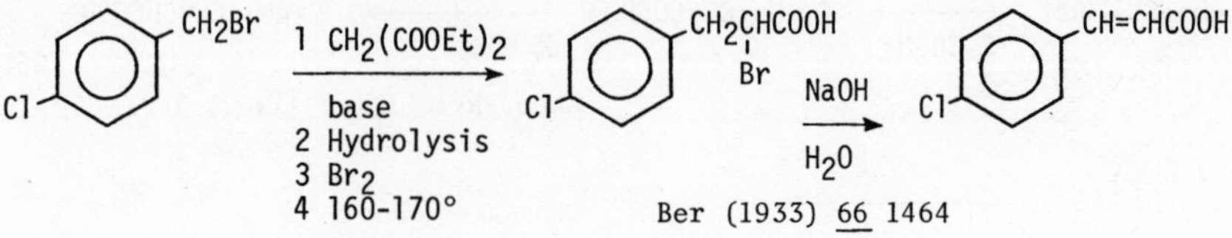

Cl—C₆H₄—CH₂Br →
1 CH₂(COOEt)₂
base
2 Hydrolysis
3 Br₂
4 160-170°

→ Cl—C₆H₄—CH₂CHCOOH (Br) →NaOH/H₂O→ Cl—C₆H₄—CH=CHCOOH

Ber (1933) 66 1464

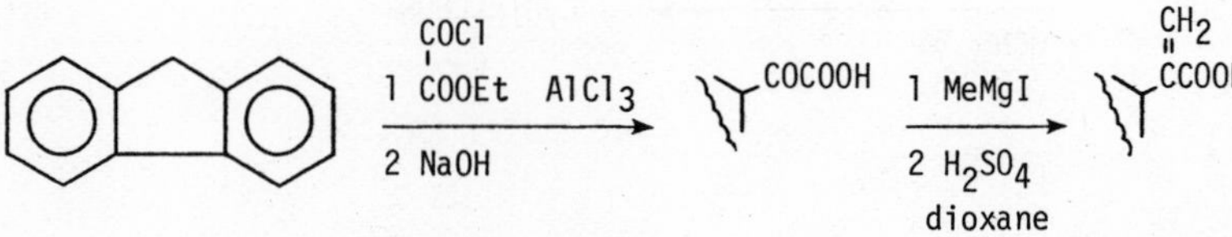

fluorene →
COCl
1 COOEt AlCl₃
2 NaOH

→ COCOOH →
1 MeMgI
2 H₂SO₄
dioxane
→ CH₂=CCOOH   76%

J Med Chem (1972) 15 1029

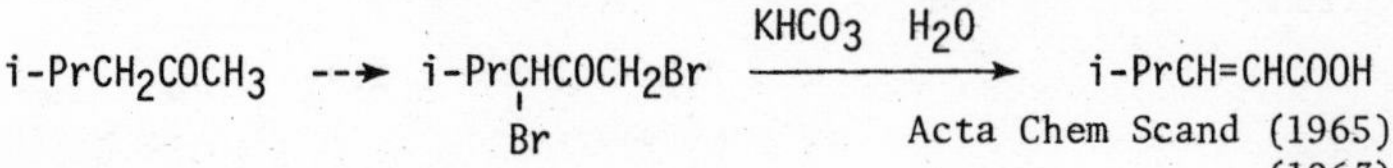

i-PrCH₂COCH₃ --→ i-PrCHCOCH₂Br (Br) →KHCO₃ H₂O→ i-PrCH=CHCOOH

Acta Chem Scand (1965) 19 383
                (1963) 17 2766
Chem Comm (1968) 306

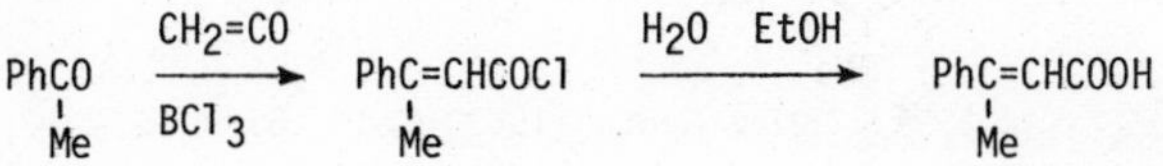

PhCO(Me) →CH₂=CO / BCl₃→ PhC=CHCOCl(Me) →H₂O EtOH→ PhC=CHCOOH(Me)   47%

Ber (1970) 103 2003

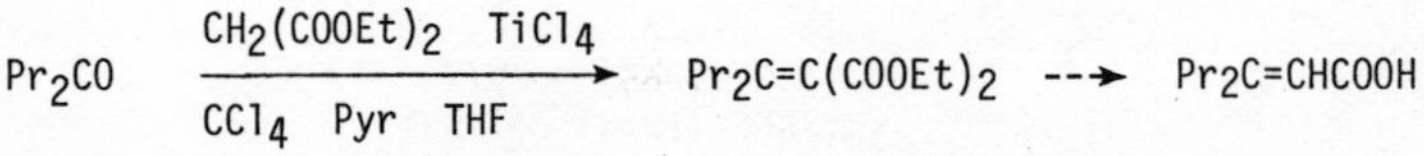

Pr₂CO →CH₂(COOEt)₂ TiCl₄ / CCl₄ Pyr THF→ Pr₂C=C(COOEt)₂ --→ Pr₂C=CHCOOH   <42%

Tetrahedron (1973) 29 635

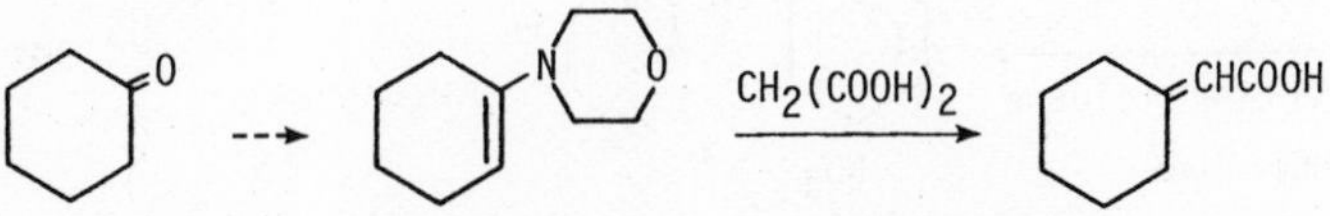

cyclohexanone --→ 1-morpholinocyclohexene →CH₂(COOH)₂→ cyclohexylidene-CHCOOH   <60%

JOC (1973) 38 399

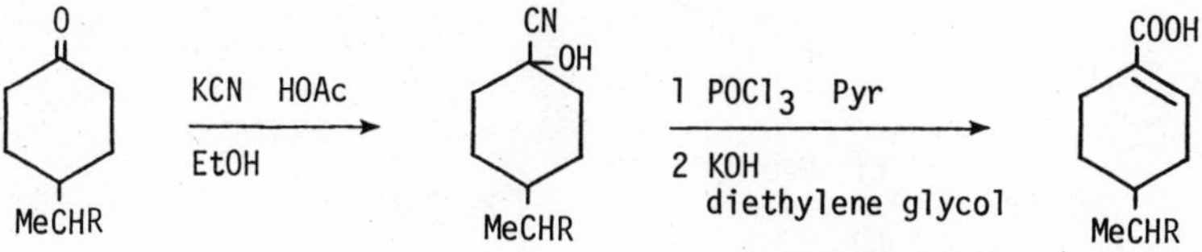

Tetrahedron (1968) 24 3127
JOC (1972) 37 2201

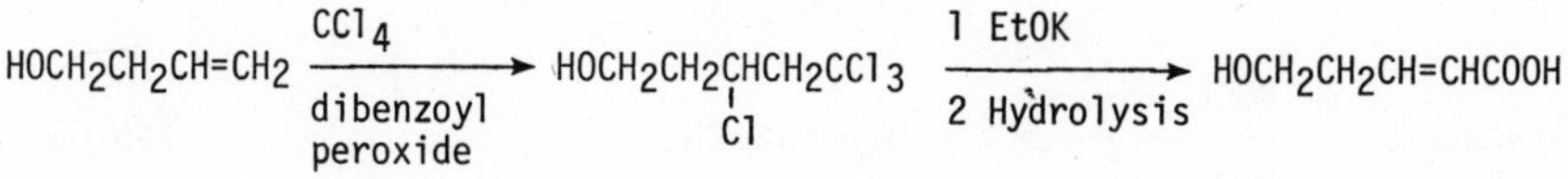

HOCH$_2$CH$_2$CH=CH$_2$ $\xrightarrow[\substack{\text{dibenzoyl} \\ \text{peroxide}}]{\text{CCl}_4}$ HOCH$_2$CH$_2$CHCH$_2$CCl$_3$ $\xrightarrow[\text{2 Hydrolysis}]{\text{1 EtOK}}$ HOCH$_2$CH$_2$CH=CHCOOH
$\hspace{9.5cm}$ Cl

Coll Czech (1966) 31 3765

PhC≡CCOOH $\xrightarrow{\text{N}_2\text{H}_4 \quad \text{O}_2}$ PhCH=CHCOOH

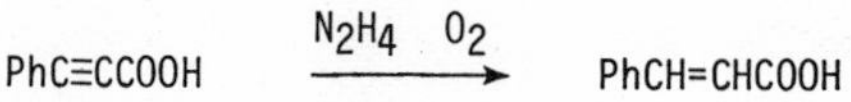

Tetr Lett (1961) 353
Catalytic and chemical reduction of acetylenic acids
Chem Rev (1957) 57 191

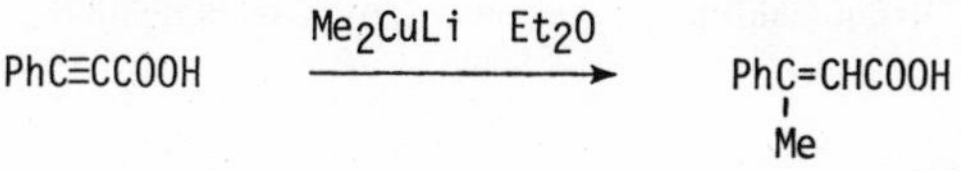

PhC≡CCOOH $\xrightarrow{\text{Me}_2\text{CuLi} \quad \text{Et}_2\text{O}}$ PhC=CHCOOH $\hspace{3cm}$ 93%
$\hspace{7.5cm}$ Me

JACS (1969) 91 6186

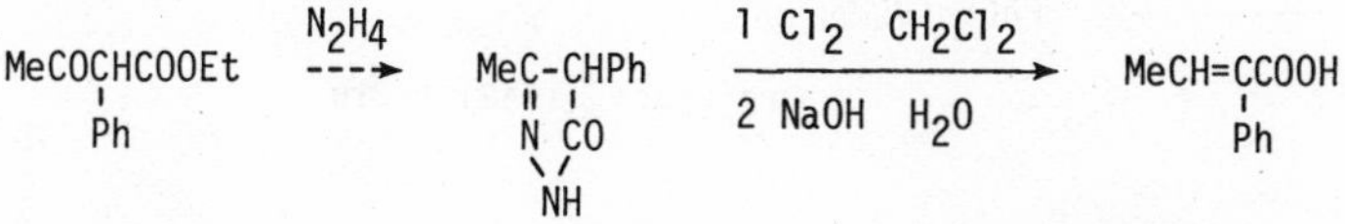

MeCOCHCOOEt $\xrightarrow{\text{N}_2\text{H}_4}$ MeC-CHPh $\xrightarrow[\text{2 NaOH} \quad \text{H}_2\text{O}]{\text{1 Cl}_2 \quad \text{CH}_2\text{Cl}_2}$ MeCH=CCOOH $\hspace{2cm}$ 31%
$\hspace{0.5cm}$ Ph $\hspace{3.5cm}$ N CO $\hspace{7cm}$ Ph
$\hspace{4.5cm}$ NH

JACS (1958) 80 601
JOC (1969) 34 1717
(1966) 31 2867

—CHBr $\xrightarrow[\text{2 CO}_2]{\text{1 Li MeOCH}_2\text{CH}_2\text{OMe}}$ $\rangle\!\rangle$=CHCOOH   60%

Angew (1965) <u>77</u> 173
(Internat Ed <u>4</u> 156)
Acta Chem Scand (1959) <u>13</u> 610

$Me_2C=CHCOMe$ $\xrightarrow[\text{H}_2\text{O}]{\text{KOCl dioxane}}$ $Me_2C=CHCOOH$ 49-53%

Org Synth (1955) Coll Vol 3 302

$BuC\equiv CH$ $\xrightarrow{\text{HOAc di-t-butyl peroxide}}$ $BuCH=CHCH_2COOH$ 24%

Can J Chem (1966) <u>44</u> 2241

$C_6H_{13}C\equiv CH$ $\xrightarrow[\text{butyl peroxide}]{\text{HC(COOEt)}_3 \text{ di-t-}}$ $C_6H_{13}CH=CHC(COOEt)_3$ $\xrightarrow[\substack{\text{EtOH} \\ \text{2 160°}}]{\text{1 NaOH}}$ $C_6H_{13}CH=CHCH_2COOH$

Synthesis (1969) 76

PhCHO $\xrightarrow[\text{2 } \Delta]{\substack{\text{CH}_2\text{COONa} \\ | \\ \text{1 CH}_2\text{COONa Ac}_2\text{O}}}$ $PhCH=CHCH_2COOH$

Org React (1942) <u>1</u> 210

$\xrightarrow[\text{NaH Me}_2\text{SO THF}]{\text{Cl}^- \text{ Ph}_3\overset{+}{P}\text{CH}_2\text{CH}_2\text{COOH}}$ CHCH$_2$COOH   66%

JACS (1964) <u>86</u> 1884
JOC (1962) <u>27</u> 3404

$C_7H_{15}CH_2CH=CHCOOH$ $\xrightarrow{h\nu \quad hexane}$ $C_7H_{15}CH=CHCH_2COOH$ 95%

JOC (1968) $\underline{33}$ 1671

$\xrightarrow[\substack{CH_2=CHPPh_3 \ Br \\ NaH \quad Et_2O \quad DMF}]{EtCH(COOEt)_2}$

CHCH$_2$C(COOEt)$_2$
                Et

$--\rightarrow$

CHCH$_2$CHCOOH
                Et

<14%

JOC (1966) $\underline{31}$ 467

$\xrightarrow[\substack{2 \ Ba(OH)_2 \\ 3 \ (COOH)_2 \ HCl \ Pyr \\ H_2O}]{\substack{CH_2COOEt \\ 1 \ CH_2COOEt \quad t\text{-}BuOK}}$

CH$_2$CH$_2$COOH Me   70%

JACS (1949) $\underline{71}$ 1384
Org React (1951) $\underline{6}$ 1

$BuCH_2CH=CH_2$ $\xrightarrow[\substack{Cu(OAc)_2 \quad HOAc \\ 2 \ Hydrolysis}]{1 \ CH_2(COOEt)_2 \quad Mn(OAc)_3}$ $BuCH=CHCH_2CH_2COOH$

Chem Comm (1973) 694
Synthesis (1970) 99

Br
Me$_2$CCOO
CH$_2$=CHCH$_2$

$\xrightarrow{Zn \quad toluene}$

Me$_2$CCOOH
CH$_2$CH=CH$_2$ 100%

Chem Comm (1973) 117

$\xrightarrow[\substack{NaH \quad Me_2SO}]{X \ Ph_3PCH_2(CH_2)_3COOH}$

Tetr Lett (1970) 311
JACS (1971) $\underline{93}$ 1490

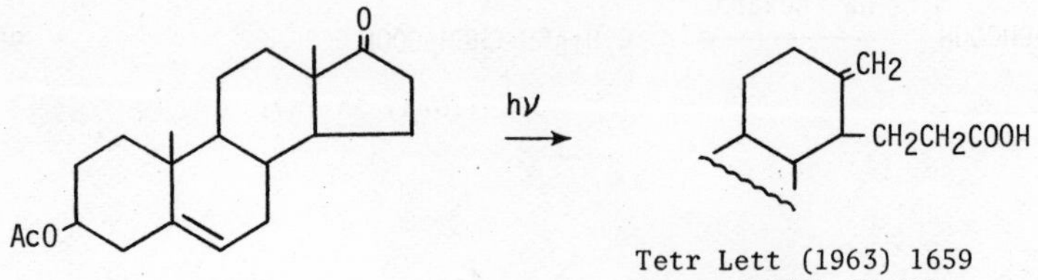

Tetr Lett (1963) 1659
Angew (1965) 77 229
(Internat Ed 4 216)

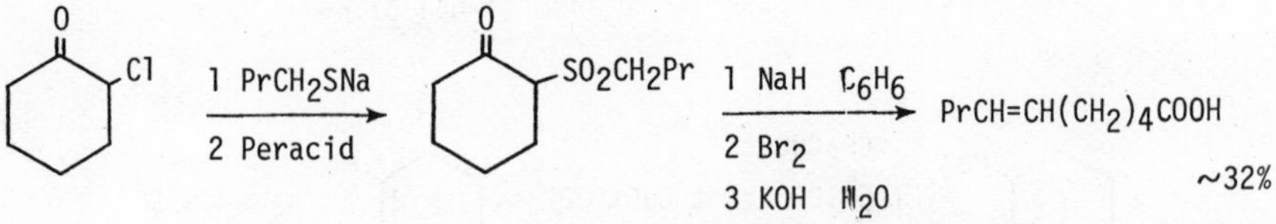

~32%

Bull Soc Chim Fr (1964) 723

		Section
Also via:	Olefinic esters	362
	Olefinic nitriles	376
	Olefinic amides	349
	Hydroxy acids	313
	Acetylenic acids	301

Section 323   Alcohol — Alcohol
              °°°°°°°°°°°°°°°°°°°

$$C_{10}H_{21}COCl \xrightarrow[\text{2 HOAc}]{\text{1 } CH_2N_2 \quad Et_2O} C_{10}H_{21}COCH_2OAc \xrightarrow[\text{i-PrOH}]{Al(OPr\text{-}i)_3} C_{10}H_{21}\underset{\overset{|}{OH}}{C}HCH_2OH \qquad 85\%$$

Org React (1954) 8 218 228

$$PhCOOH \xrightarrow[\text{MeOCH}_2CH_2OMe]{\text{MeLi} \quad TiCl_3} Ph\underset{\overset{|}{OH}}{\overset{\overset{Me}{|}}{C}}-\underset{\overset{|}{OH}}{\overset{\overset{Me}{|}}{C}}Ph$$

Chem Comm (1970) 451

PhCHO $\xrightarrow[\text{HMPA}]{\text{Mg Me}_3\text{SiCl}}$ PhCH-CHPh   > 90%
                                        OH OH

Tetr Lett (1972) 75

PhCOOEt $\xrightarrow[\text{t-BuOK}]{\text{Me}_2\text{SO}}$ PhCOCH$_2$SOMe $\xrightarrow[\substack{2 \text{ NaBH}_4 \text{ NaOH} \\ \text{EtOH H}_2\text{O}}]{1\text{HCl Me}_2\text{SO H}_2\text{O}}$ PhCHCH$_2$OH   75%
                                                                        OH

JACS (1966) __88__ 5498

COOMe   O   COOMe

CH=C(CH$_2$)$_2$CH=C(CH$_2$)$_2$CHCMe$_2$ $\xrightarrow[\text{H}_2\text{O}]{\text{HClO}_4 \text{ THF}}$ CH=C(CH$_2$)$_2$CH=C(CH$_2$)$_2$CH-CMe$_2$ 78%

    Me   Me   Me   Me   OH OH

JACS (1972) __94__ 5379

HIO$_4$ JACS (1949) __71__ 3938

2,4,6-Trinitrobenzenesulfonic acid, Me$_2$SO   Tetr Lett (1971) 411

$\xrightarrow[\text{KOH Me}_2\text{SO H}_2\text{O}]{\text{CCl}_3\text{COOH C}_6\text{H}_6}$ cis   trans

Tetr Lett (1965) 3421
JACS (1973) __95__ 2361

$\xrightarrow[\text{CH}_2\text{Cl}_2]{\text{Al-Hg}}$

Tetr Lett (1970) 4271

Na   Bull Soc Chim Fr (1967) 2011

Mg, MgI$_2$ JACS (1942) __64__ 30

Bu$_3$SnH hν JCS _C_ (1971) 1241

Electrolysis   Synthesis (1971) 285

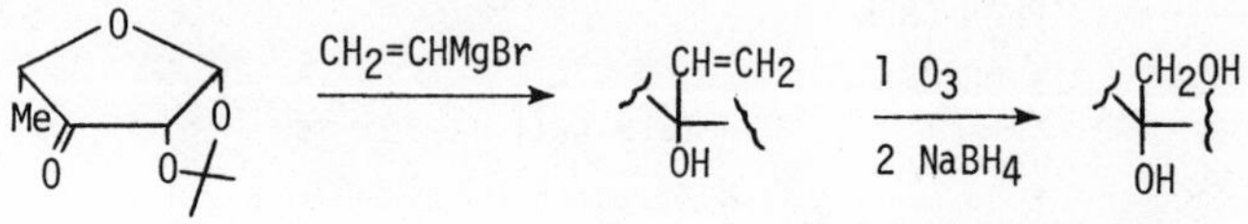
Me
O
O
O
O
$CH_2=CHMgBr$
CH=CH2
OH
1 O3
2 NaBH4
CH2OH
OH

JACS (1965) 87 654
Can J Chem (1967) 45 2921

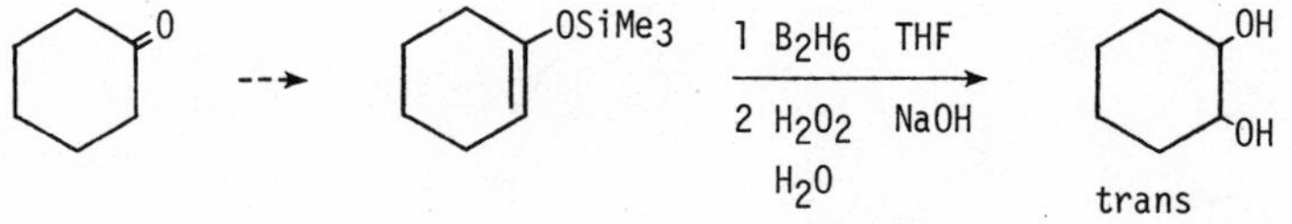
O
OSiMe3
1 B2H6 THF
2 H2O2 NaOH
H2O
OH
OH
trans

70%

Tetr Lett (1972) 2845

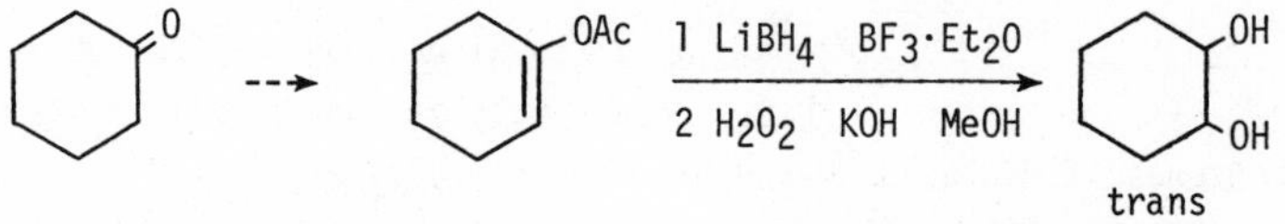
O
OAc
1 LiBH4 BF3·Et2O
2 H2O2 KOH MeOH
OH
OH
trans

<61%

JACS (1969) 91 2632
Chem Ind (1960) 720

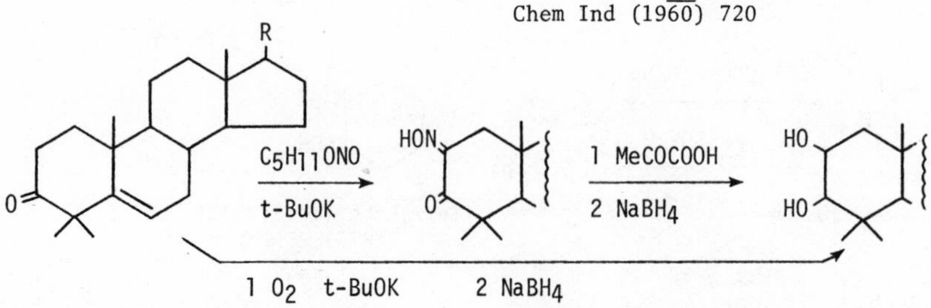
R
O
C5H11ONO
t-BuOK
HON
O
1 MeCOCOOH
2 NaBH4
HO
HO
1 O2 t-BuOK 2 NaBH4

Annalen (1972) 758 89

Review: Hydroxylation Methods — Advances in Org Chem (1960) 1 103
Epoxidation and Hydroxylation of Ethylenic Compounds with Organic Peracids — Org React (1953) 7 378

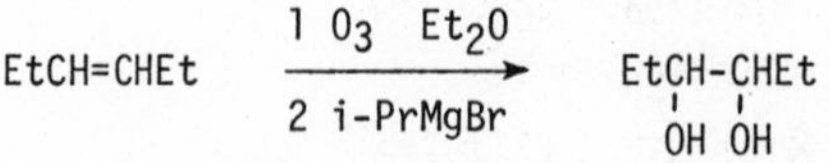
EtCH=CHEt
1 O3 Et2O
2 i-PrMgBr
EtCH-CHEt
OH OH

46%

JOC (1963) 28 1159

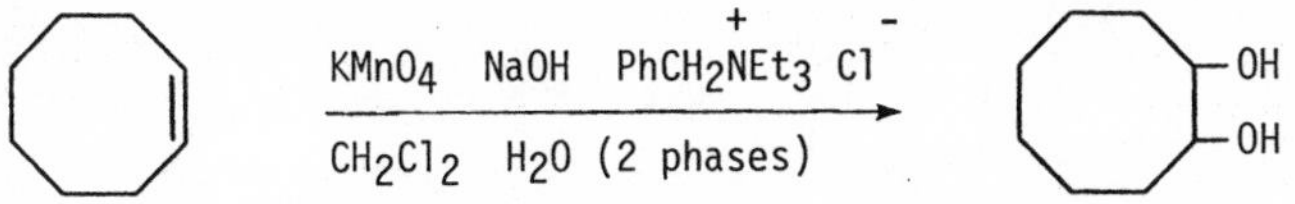

KMnO$_4$　NaOH　PhCH$_2$NEt$_3$ Cl$^-$ (with + over N)

CH$_2$Cl$_2$　H$_2$O (2 phases)

OH
OH

50%

Tetr Lett (1972) 4907

KMnO$_4$, MgSO$_4$ (neutral conditions)　J Heterocyclic Chem (1972) $\underline{9}$ 979

K$_2$MnO$_4$, NaOH　JCS (1956) 2452

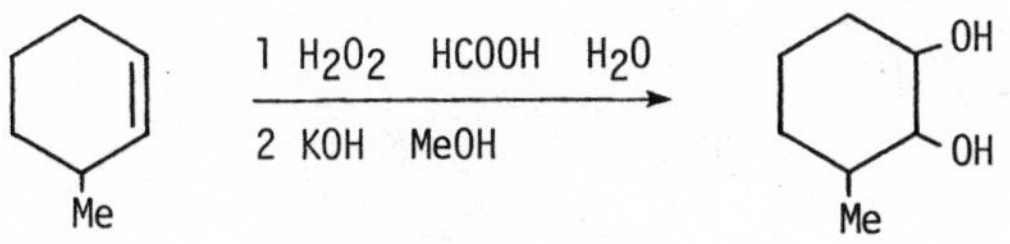

1 H$_2$O$_2$　HCOOH　H$_2$O

2 KOH　MeOH

Me

OH
OH
Me

Tetrahedron (1968) $\underline{24}$ 5701
Org React (1953) $\underline{7}$ 378

o-Sulfoperbenzoic acid　Tetr Lett (1971) 691

Disuccinoyl peroxide, detergent　Synthesis (1973) 156

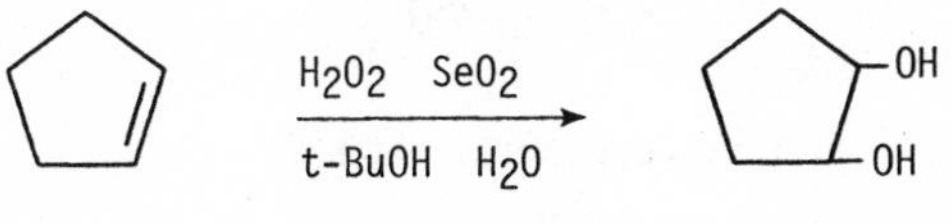

H$_2$O$_2$　SeO$_2$

t-BuOH　H$_2$O

OH
OH
trans

40%

Helv (1953) $\underline{36}$ 268

H$_2$O$_2$, H$_2$WO$_4$　JOC (1957) $\underline{22}$ 1682

H$_2$O$_2$, V$_2$O$_5$　JACS (1937) $\underline{59}$ 2342

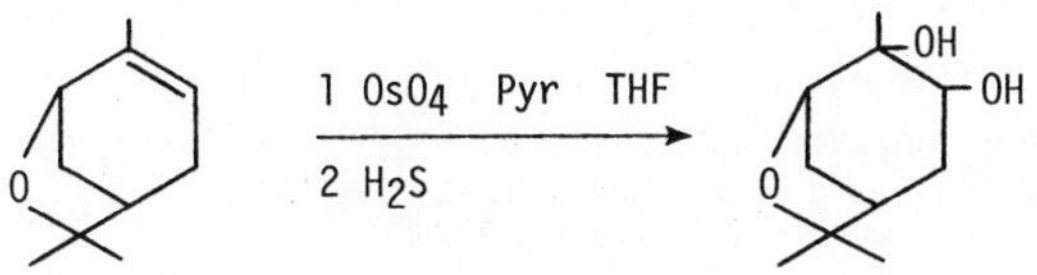

1 OsO$_4$　Pyr　THF

2 H$_2$S

OH
OH
cis

43%

Tetrahedron (1971) $\underline{27}$ 753
JOC (1960) $\underline{25}$ 257

OsO$_4$ (catalytic quantity), KClO$_4$　JACS (1968) $\underline{90}$ 5336

OsO$_4$ ("　　"), H$_2$O$_2$　JACS (1937) $\underline{59}$ 2345

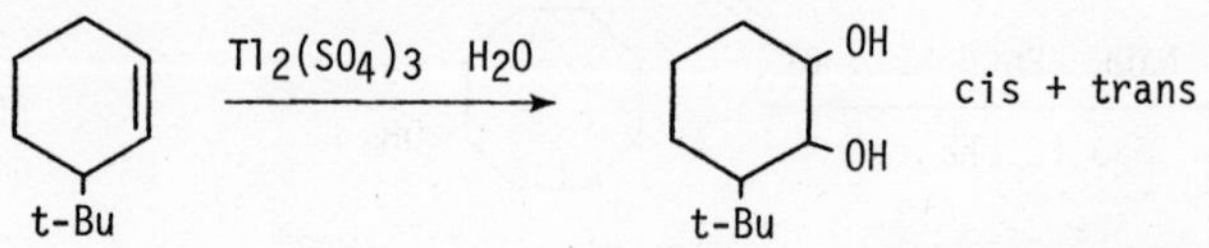

$Tl_2(SO_4)_3$ H_2O → OH, OH   cis + trans   <67%

t-Bu   t-Bu

Can J Chem (1971) 49 2586
Tetrahedron (1964) 20 1017

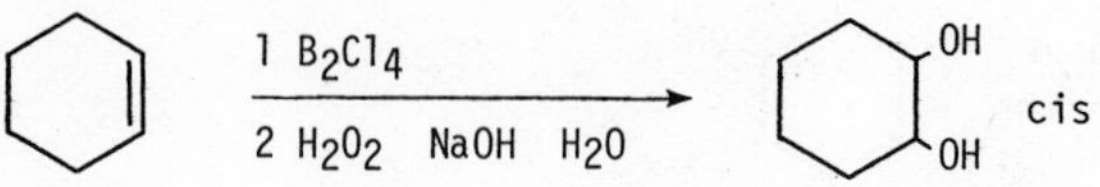

1 B_2Cl_4
2 H_2O_2 NaOH H_2O → OH, OH   cis

JACS (1967) 89 4217

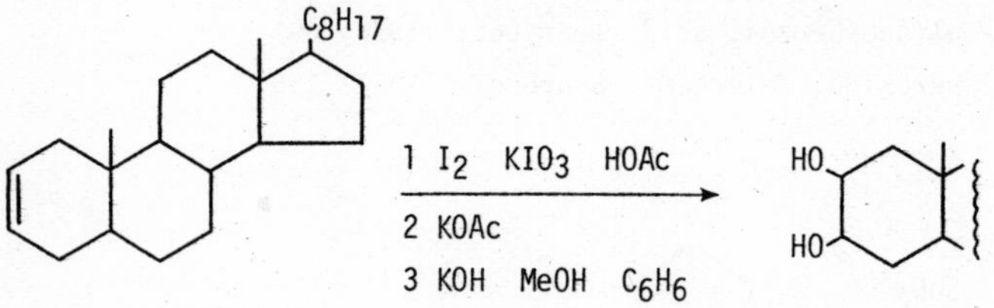

C_8H_{17}

1 I_2 KIO_3 HOAc
2 KOAc
3 KOH   MeOH C_6H_6 → HO, HO   70%

Tetr Lett (1973) 4485

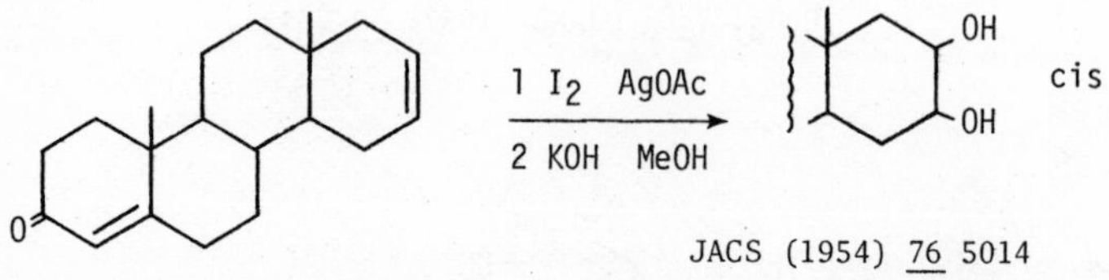

1 I_2 AgOAc
2 KOH   MeOH → OH, OH   cis   67%

JACS (1954) 76 5014

I_2, silver benzoate   JCS C (1966) 1327
                                Org React (1957) 9 332
                                JOC (1942) 7 227

I_2, silver trifluoroacetate   Chem Comm (1966) 202

$I(OCOCF_3)_3$ Angew (1973) 85 175
                        (Internat Ed) 12 163

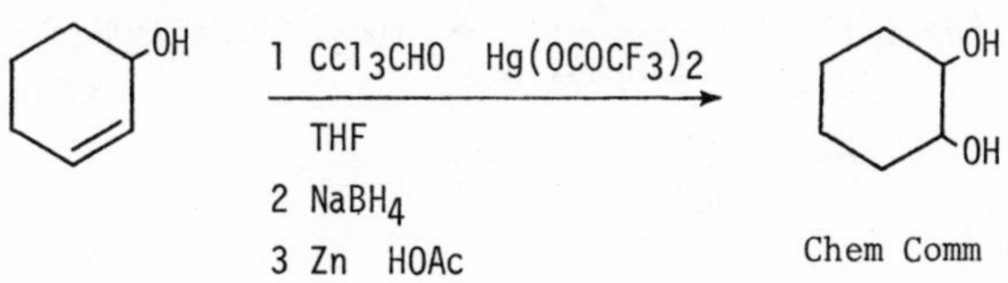
OH
1 CCl3CHO Hg(OCOCF3)2
THF
2 NaBH4
3 Zn HOAc
OH
OH

Chem Comm (1972) 1196

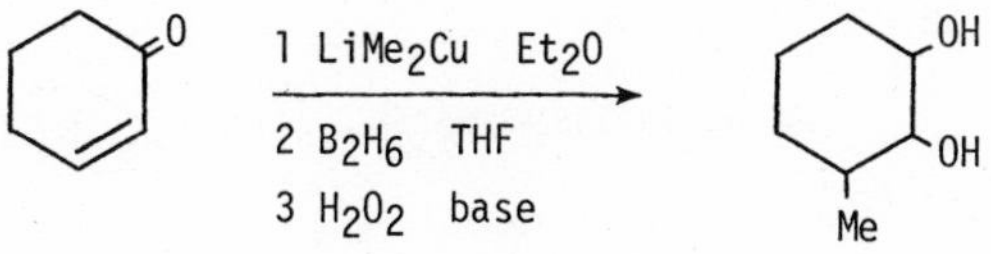
O
1 LiMe2Cu Et2O
2 B2H6 THF
3 H2O2 base
OH
OH
Me

59%

Tetr Lett (1972) 4031

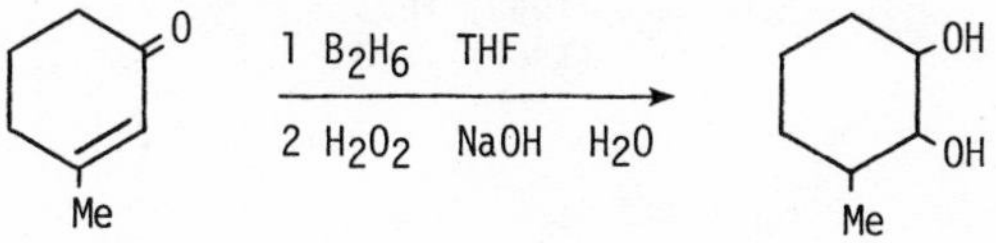
O
Me
1 B2H6 THF
2 H2O2 NaOH H2O
OH
OH
Me

trans

Tetrahedron (1968) 24 5701

$MeCOCH_2OH$ —Baker's yeast→ $MeCH(OH)CH_2OH$ 49-58%

Org Synth (1943) Coll Vol 2 545

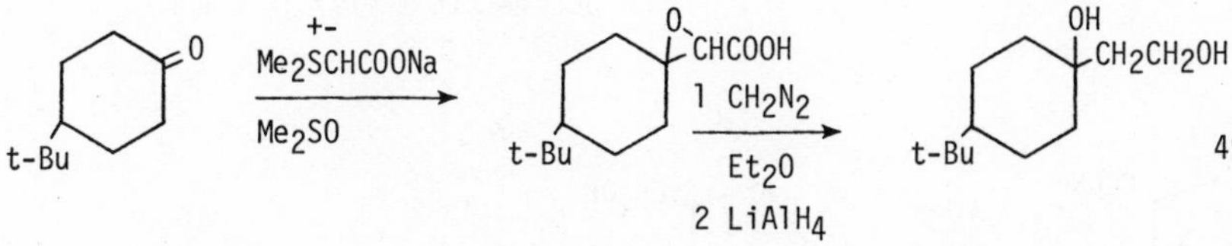
O
t-Bu
+-
Me2SCHCOONa
Me2SO
O
CHCOOH
t-Bu
1 CH2N2
Et2O
2 LiAlH4
OH
CH2CH2OH
t-Bu

49%

JOC (1970) 35 1600

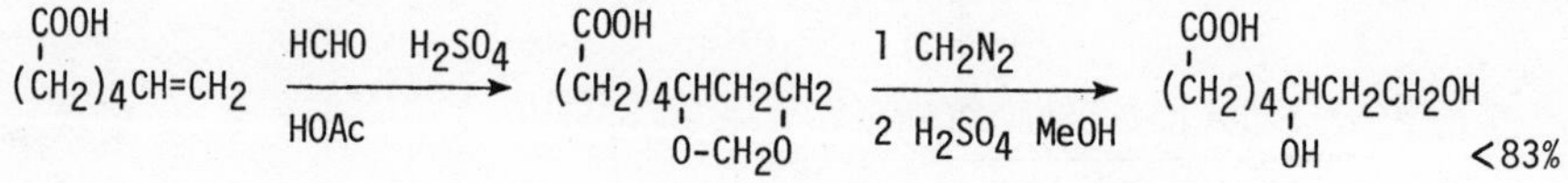

$$(CH_2)_4CH=CH_2 \xrightarrow[HOAc]{HCHO \quad H_2SO_4} (CH_2)_4CHCH_2CH_2 \xrightarrow[2 \; H_2SO_4 \; MeOH]{1 \; CH_2N_2} (CH_2)_4CHCH_2CH_2OH$$

with COOH groups; O–CH₂O ring; final product OH   <83%

JCS (1956) 3074

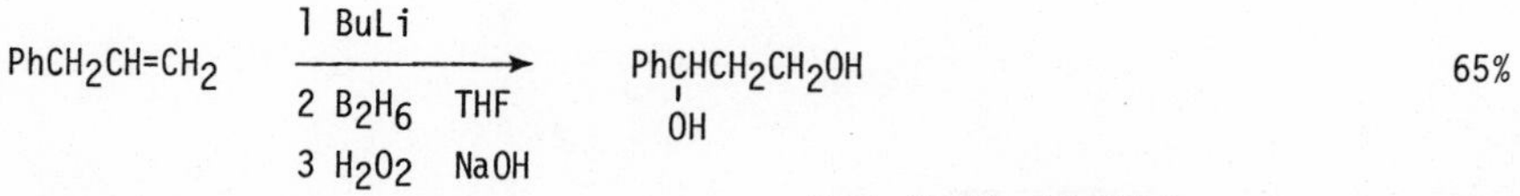

$$PhCH_2CH=CH_2 \xrightarrow[\substack{2 \; B_2H_6 \quad THF \\ 3 \; H_2O_2 \quad NaOH}]{1 \; BuLi} PhCHCH_2CH_2OH$$

OH   65%

JACS (1971) <u>93</u> 6313

$$PhCH=CHCH_2OH \xrightarrow[2 \; NaBH_4]{1 \; Hg(OAc)_2} PhCHCH_2CH_2OH$$

OH

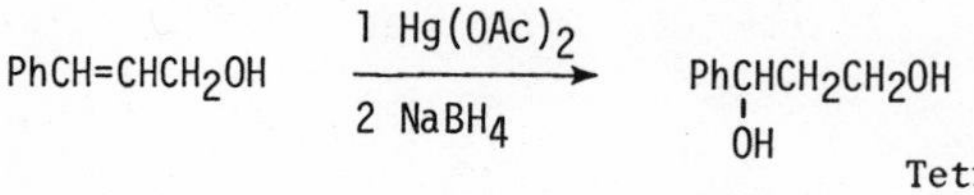

Tetr Lett (1968) 5105
Chem Comm (1968) 1073
(1967) 1283

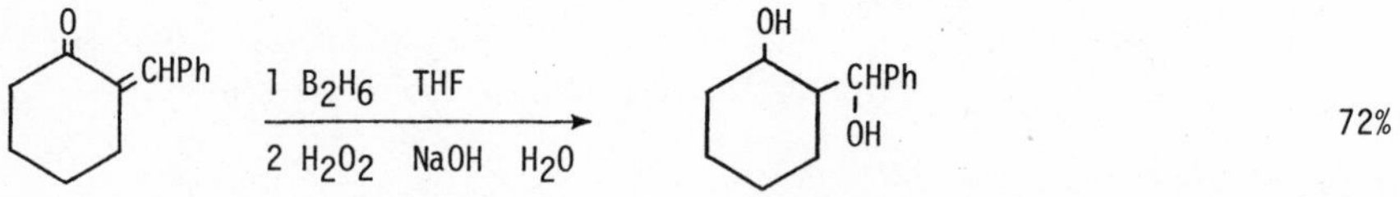

1 B₂H₆   THF

2 H₂O₂   NaOH   H₂O

72%

JCS Perkin I (1973) 1707

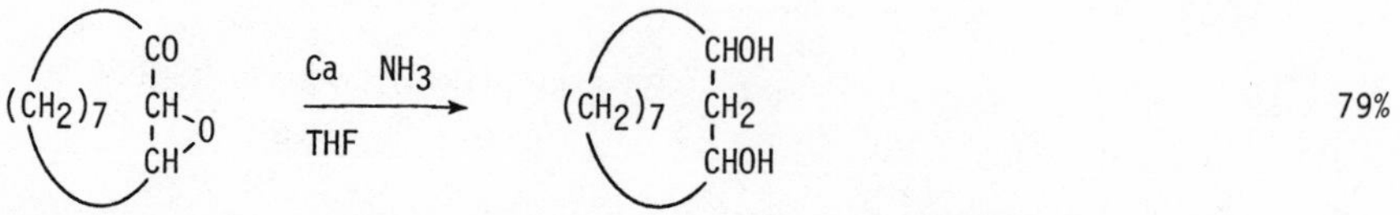

$$(CH_2)_7 \quad \xrightarrow{Ca \quad NH_3 \atop THF} \quad (CH_2)_7$$

79%

JCS Perkin I (1972) 1509

$$\xrightarrow[THF]{AlH_3}$$

82%

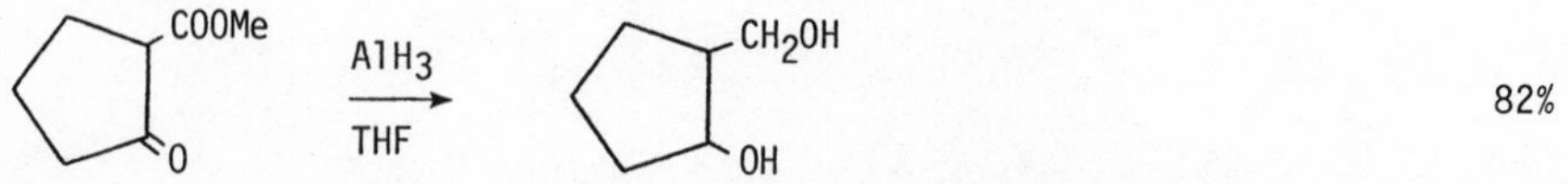

JACS (1968) <u>90</u> 2927

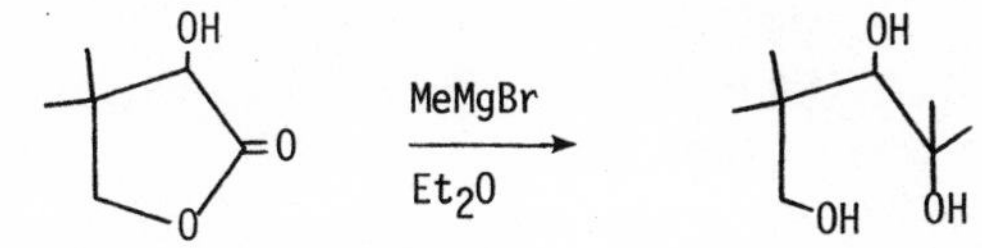
OH
O
O
MeMgBr
Et2O
OH
OH
OH

JACS (1940) 62 1779

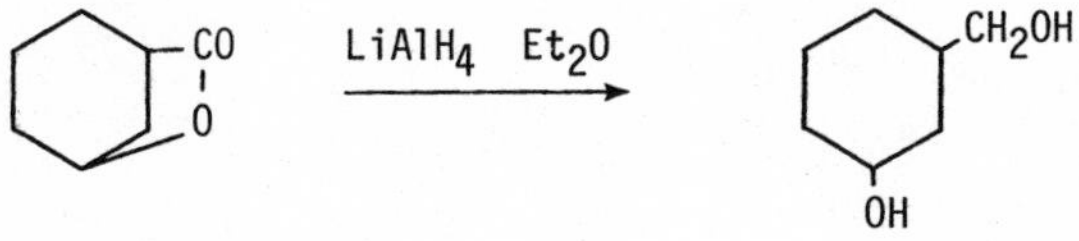
CO
O
LiAlH4 Et2O
CH2OH
OH

80%

JACS (1952) 74 5908 4336

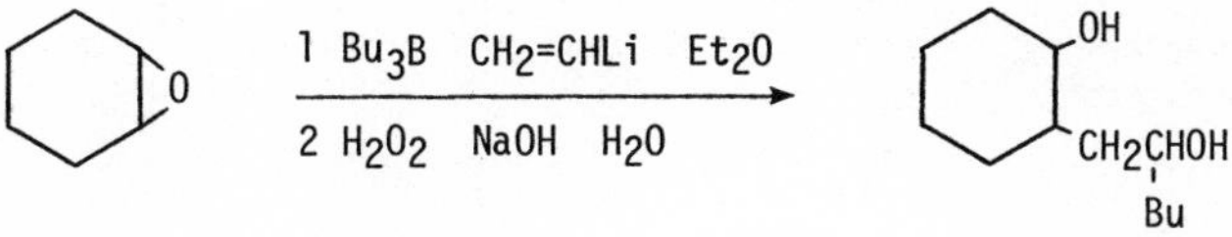
O
1 Bu3B CH2=CHLi Et2O
2 H2O2 NaOH H2O
OH
CH2CHOH
Bu

77%

Tetr Lett (1973) 4527

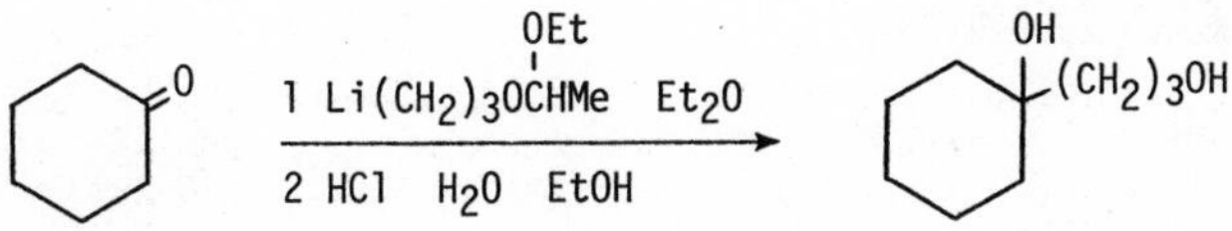
O
OEt
1 Li(CH2)3OCHMe Et2O
2 HCl H2O EtOH
OH
(CH2)3OH

86%

JOC (1972) 37 1947

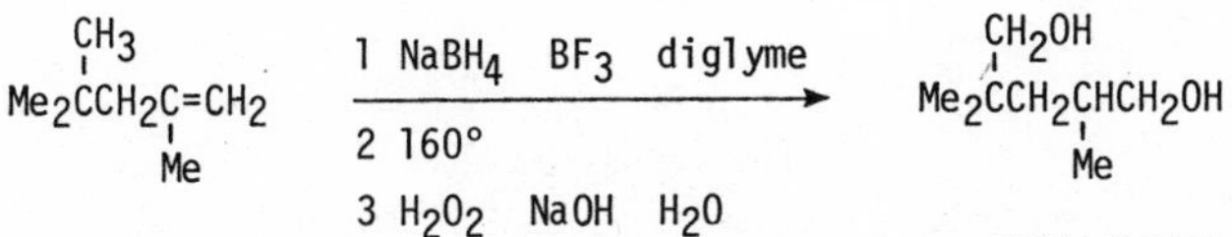
CH3
Me2CCH2C=CH2
Me
1 NaBH4 BF3 diglyme
2 160°
3 H2O2 NaOH H2O
CH2OH
Me2CCH2CHCH2OH
Me

JACS (1966) 88 1443

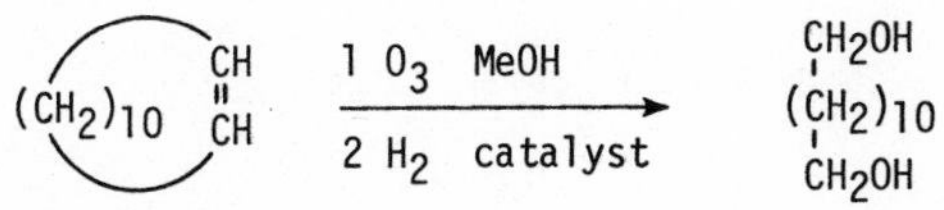
CH
(CH2)10
CH
1 O3 MeOH
2 H2 catalyst
CH2OH
(CH2)10
CH2OH

85%

Tetr Lett (1971) 3587
JACS (1957) 79 3165

$$
\begin{array}{c}
\text{CH=CH}_2 \\
| \\
\text{(CH}_2)_2 \\
| \\
\text{CH=CH}_2
\end{array}
\quad
\xrightarrow[\text{2 H}_2\text{O}_2 \quad \text{NaOH} \quad \text{H}_2\text{O}]{\overset{\displaystyle \text{Me}}{\overset{|}{\text{1 i-PrCBH}_2}}\quad \overset{|}{\underset{}{\text{Me}}}\quad \text{THF}}
\quad
\begin{array}{c}
\text{CH}_2\text{CH}_2\text{OH} \\
| \\
\text{(CH}_2)_2 \\
| \\
\text{CH}_2\text{CH}_2\text{OH}
\end{array}
\qquad \sim 90\%
$$

JACS (1972) <u>94</u> 3567
(1962) <u>84</u> 190

$$
\begin{array}{c}
\text{Me}_2\text{CCH}_2\text{OH} \\
| \\
\text{(CH}_2)_2 \\
| \\
\text{COOH}
\end{array}
\quad
\xrightarrow[\text{Na}_2\text{CO}_3 \quad \text{MeOH}]{\text{Electrolysis}}
\quad
\begin{array}{c}
\text{Me}_2\text{CCH}_2\text{OH} \\
| \\
\text{(CH}_2)_4 \\
| \\
\text{Me}_2\text{CCH}_2\text{OH}
\end{array}
\qquad 52\%
$$

Bull Soc Chim Fr (1970) 183

Also via: Section
                Hydroxyesters   327
                   Diesters   357

Section 324   Alcohol — Aldehyde
               ○○○○○○○○○○○○○○○○○○○○

$$
\text{PhCOOH} \dashrightarrow \text{PhCOCHN}_2
\quad
\xrightarrow[\text{2 MeSNa}]{\text{1 EtSCl} \quad \text{Et}_2\text{O}}
\quad
\text{PhCOCH(SEt)}_2
\quad
\xrightarrow[\substack{\text{2 Br}_2 \quad \text{HCl} \\ \text{H}_2\text{O} \quad \text{HOAc}}]{\text{1 LiBH}_4}
\quad
\begin{array}{c}
\text{PhCHCHO} \\
| \\
\text{OH}
\end{array}
$$

Ber (1958) <u>91</u> 1043
(1957) <u>90</u> 1230

$$
\text{i-PrCHO}
\quad
\xrightarrow[\substack{\text{BuLi} \quad \text{THF} \\ \text{2 Hydrolysis}}]{\text{1 MeSOCH}_2\text{SMe}}
\quad
\begin{array}{c}
\text{i-PrCHCHO} \\
| \\
\text{OH}
\end{array}
$$

Tetr Lett (1973) 4707

```
CHO                      CH2NO2                   CHO
|                        |                        |
CHOH                     CHOH                     CHOH
|           CH3NO2        |                        |
CHOH       ─────────►    CHOH      1 NaOH  H2O     CHOH
|           MeONa         |        ──────────►     |
CHOH                     CHOH      2 H2SO4  H2O    CHOH
|                        |                        |
CH2OH                    CHOH                     CHOH
                         |                        |
                         CH2OH                    CH2OH
```

JOC (1964) <u>29</u> 1790

```
                         CH=CH2                    CHO
CHO                      |                         |
|        CH2=CHMgCl       CHOH      1 O3  EtOAc     CHOH
CHO     ──────────►      CHO      ──────────►      CHO
|   ⤬     THF            |   ⤬     2 H2  PtO2      |   ⤬
CH2O                     CH2O                      CH2O
```

Can J Chem (1967) <u>45</u> 2921

```
            Me2SO                    1 SOCl2   CH2Cl2
PhCOOEt   ─────────►  PhCOCH2SOMe  ──────────────────►   PhCHCHO
            NaH                     2 MeSH   CH2Cl2            |
                                    3 NaBH4   H2O             OH
                                    4 I2   NaHCO3   H2O
```

JOC (1969) <u>34</u> 3618
JACS (1966) <u>88</u> 5498

```
          1 Mg                        Me
          ────────────────►          |
BuBr      2 MeCOCH=NOH              BuCCHO                              15%
          3 (COOH)2   H2O             |
                                     OH
```

Annales de Chimie (1939) <u>11</u> 453

```
                    ┌S┐
                    |  ⟩CHLi        ⎫ CHO                              62%
     [dioxolane]  1 └S┘            ⎬─⎧
                   ─────────►       ⎩  |
                    THF                OH
                    2 BF3  HgO
```

Ber (1972) <u>105</u> 1978
Angew (1965) <u>77</u> 1134
(Internat Ed <u>4</u> 1075)
Annalen (1972) <u>758</u> 89

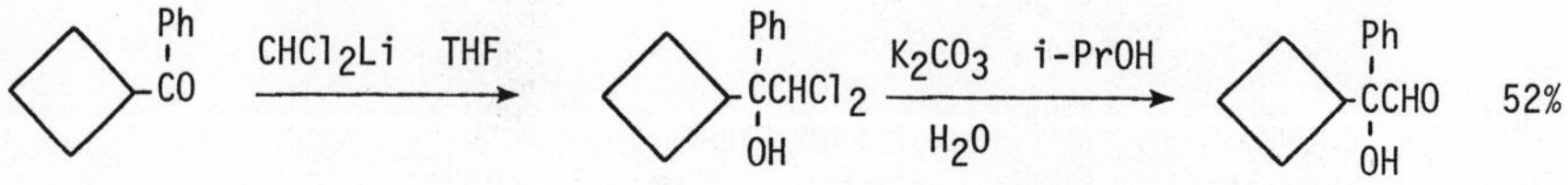

JOC (1972) <u>37</u> 1248
Ber (1973) <u>106</u> 2620

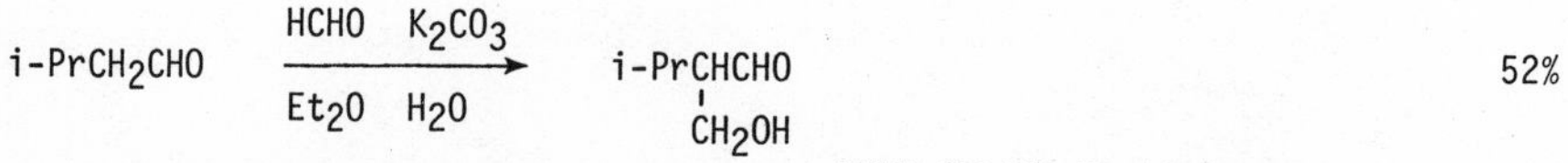

i-PrCH$_2$CHO $\xrightarrow[\text{Et}_2\text{O} \quad \text{H}_2\text{O}]{\text{HCHO} \quad \text{K}_2\text{CO}_3}$ i-PrCHCHO
                                                        |
                                                    CH$_2$OH   52%

JACS (1948) <u>70</u> 1694
Org React (1968) <u>16</u> 1

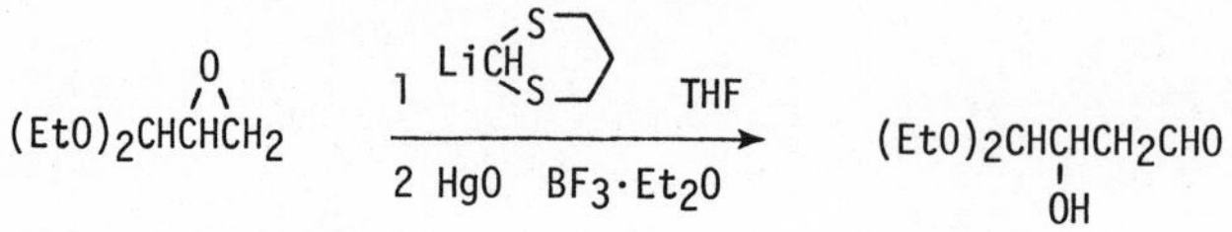

(EtO)$_2$CHCHCH$_2$ $\xrightarrow[\text{2 HgO} \quad \text{BF}_3\cdot\text{Et}_2\text{O}]{1 \quad \text{LiCH(S}\cdots\text{S)} \quad \text{THF}}$ (EtO)$_2$CHCHCH$_2$CHO
                                                                                |
                                                                               OH

JOC (1971) <u>36</u> 366

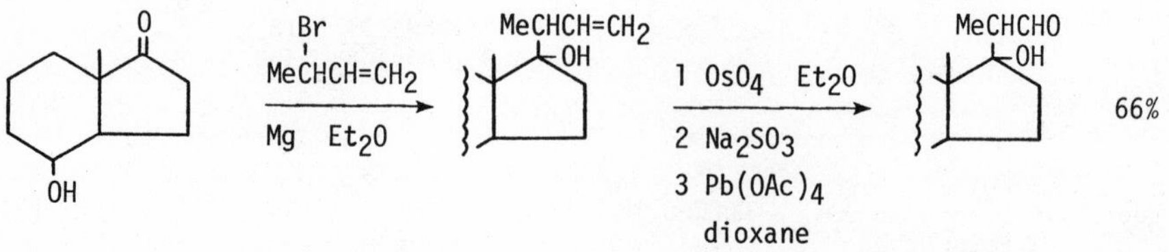

Ber (1959) <u>92</u> 1564

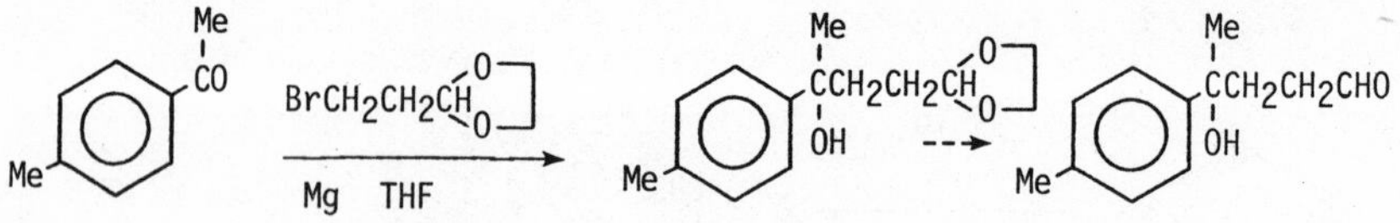

JOC (1969) <u>34</u> 1122

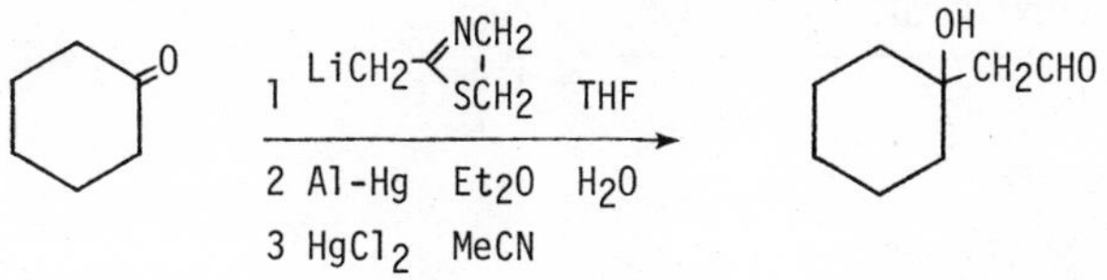

1 LiCH$_2$–C(=NCH$_2$)(SCH$_2$) THF

2 Al-Hg   Et$_2$O   H$_2$O

3 HgCl$_2$ MeCN

→ (cyclohexyl)OH–CH$_2$CHO

51%

Tetr Lett (1972) 3929
JOC (1973) <u>38</u> 36

PhCOO$_2$Bu-t

CuCl   C$_6$H$_6$

→ (tetrahydrofuranyl)-OCOPh --→ CHO / OH

Angew (1961) <u>73</u> 65
Acta Chem Scand (1961) <u>15</u> 249
Tetrahedron (1961) <u>13</u> 241

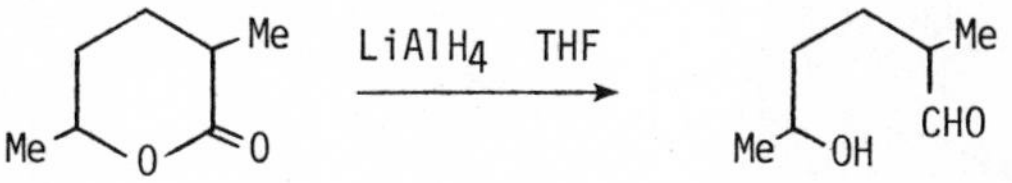

Me ... Me=O (lactone)   LiAlH$_4$ THF   →   Me–OH ... CHO Me   (lactol)

64%

JACS (1953) <u>75</u> 2413
(i-Bu)$_2$AlH   Helv (1963) <u>46</u> 2799

Also via: Acetoxyaldehydes (Section 336)

Section 325   Alcohol — Amide

(PhCH$_2$)$_2$CHCHO --→ (PhCH$_2$)$_2$CHCHCN / OH $\xrightarrow[\text{t-BuOH}]{\text{H}_2\text{SO}_4}$ (PhCH$_2$)$_2$CHCHCONHBu-t / OH

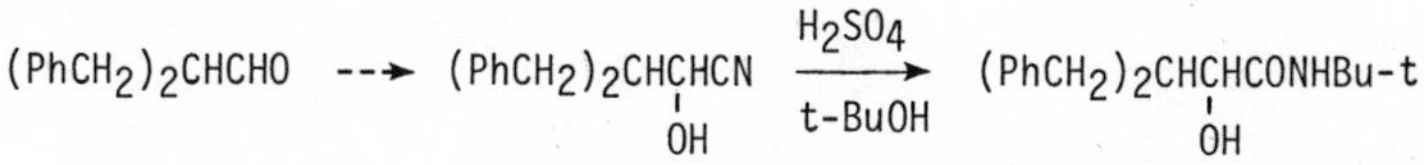

98%

Compt Rend <u>C</u> (1971) <u>272</u> 1157
HCl, HCOOH   Annalen (1971) <u>749</u> 198

$$\text{(CHO, Ur acetal)} \xrightarrow[\text{2 } H_2O_2 \quad H_2O]{\text{1 NaCN} \quad K_2CO_3 \quad MeOH \quad H_2O} \quad \begin{array}{c} CONH_2 \\ | \\ CHOH \end{array}$$ 48%

JACS (1971) <u>93</u> 3812

$$Me_2CO \dashrightarrow \begin{array}{c} Me_2CCN \\ | \\ OH \end{array} \xrightarrow{MnO_2 \quad CH_2Cl_2} \begin{array}{c} Me_2CCONH_2 \\ | \\ OH \end{array}$$ < 53%

Chem Comm (1966) 121

Ber (1965) <u>98</u> 936
Tetrahedron (1968) <u>24</u> 3795
Annalen (1968) <u>715</u> 47

$$\begin{array}{c} PhCO \\ | \\ Me \end{array} \xrightarrow[\text{THF}]{(Et_2NCO)_2Hg \quad BuLi} \begin{array}{c} OH \\ | \\ PhCCONEt_2 \\ | \\ Me \end{array}$$ 66%

Angew (1967) <u>79</u> 819
(Internat Ed <u>6</u> 805)

$$Ph_2CO \xrightarrow[\text{THF}]{Li[Me_2NCONi(CO)_3]} \begin{array}{c} Ph_2CCONMe_2 \\ | \\ OH \end{array}$$ < 30%

JOC (1971) <u>36</u> 2721

$$\begin{array}{c} PhCHCOOH \\ | \\ OH \end{array} \xrightarrow[\text{2 } NH_3]{\text{1 } Me_2CO \quad H_2SO_4} \begin{array}{c} PhCHCONH_2 \\ | \\ OH \end{array}$$ 62%

Org Synth (1955) Coll Vol 3 536

Me$_2$N-C$_6$H$_4$-COCONMe$_2$ $\xrightarrow[\text{C}_6\text{H}_6]{\text{MeMgBr \quad Et}_2\text{O}}$ Me$_2$N-C$_6$H$_4$-C(Me)(OH)CONMe$_2$ 79%

JOC (1959) 24 265

PhCHO $--\rightarrow$ PhCHC≡COEt (OH) $\xrightarrow{\text{PhNH}_2 \quad \text{EtOH}}$ PhCHCH$_2$CONHPh (OH)   <41%

Rec Trav Chim (1956) 75 1377

Ph$_2$CO $\xrightarrow{\text{LiCH}_2\text{CONMe}_2 \quad \text{THF}}$ Ph$_2$CCH$_2$CONMe$_2$ (OH)   90%

Ber (1968) 101 3113
JOC (1968) 33 4275

cyclohexanone $\xrightarrow[\text{toluene}]{\text{BrCHCONEt}_2\text{(Me)} \quad \text{Zn-Cu}}$ cyclohexane(OH)CHCONEt$_2$(Me)   30%

JACS (1948) 70 677
Organometallics in Chem Synth (1970) 1 57

EtCHCH$_2$ (O epoxide) $\xrightarrow[\text{NaNH}_2 \quad \text{NH}_3]{\text{CH}_3\text{CONMe}_2}$ EtCHCH$_2$CH$_2$CONMe$_2$ (OH)   44%

Ber (1972) 105 1621

C$_{10}$H$_{21}$-(lactone)=O $\xrightarrow[\text{Et}_2\text{O} \quad \text{C}_6\text{H}_6]{\text{PhNH}_2 \quad \text{EtMgBr}}$ C$_{10}$H$_{21}$-CH(OH)-CH$_2$CH$_2$CH$_2$-CONHPh

Rec Trav Chim (1965) 84 1177
Can J Chem (1969) 47 3671
Pharm Acta Helv (1945) 20 79

COOH
(CH$_2$)$_{10}$ ClCOOEt Et$_3$N $\overset{O\ O}{\overset{\|\ \|}{COCOEt}}$
CHOH   THF   →   (CH$_2$)$_{10}$
C$_6$H$_{13}$ CHOH
                                C$_6$H$_{13}$

NH$_2$—⬡ →

CONH—⬡
(CH$_2$)$_{10}$
CHOH
C$_6$H$_{13}$

J Am Oil Chem Soc (1963) <u>40</u> 101
(Chem Abs <u>59</u> 2641)

Me Et (cyclic imide structure, glutarimide) → NaBH$_4$ MeOH / H$_2$O →

Me Et
CH$_2$OH CONH$_2$

JOC (1968) <u>33</u> 206   40%

Section 326   Alcohol — Amine

PhCHO $\xrightarrow[\text{hexane}]{\text{Ph}_2\text{NCH}_2\text{SnBu}_3\ \ \text{BuLi}}$ Ph$_2$NCH$_2$CHPh
                                             OH   83%

JACS (1971) <u>93</u> 4027

EtCHCHO $\xrightarrow[\text{2 NaCN}]{\text{1 NaHSO}_3\ \ \text{H}_2\text{O}}$ $\overset{OH}{\text{EtCHCHCN}}$ $\xrightarrow[\text{Et}_2\text{O}]{\text{LiAlH}_4}$ $\overset{OH}{\text{EtCHCHCH}_2\text{NH}_2}$ 46%
Me   Me   Me

Ber (1973) <u>106</u> 1365

PhCOCl $\xrightarrow{\text{CuCN}}$ PhCOCN $\xrightarrow[\text{Et}_2\text{O}]{\text{LiAlH}_4}$ PhCHCH$_2$NH$_2$ 86%
                                                 OH

JACS (1952) <u>74</u> 5514

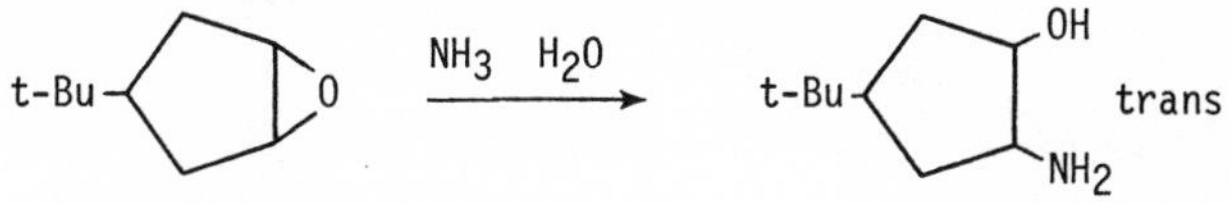

t-Bu ⟨epoxide⟩ → NH₃ H₂O → t-Bu ⟨cyclopentane OH, NH₂⟩ trans   94%

Tetrahedron (1972) 28 3475
Helv (1961) 44 1164
Synth Comm (1973) 3 177
JACS (1965) 87 1353
Chem Ind (1973) 1111

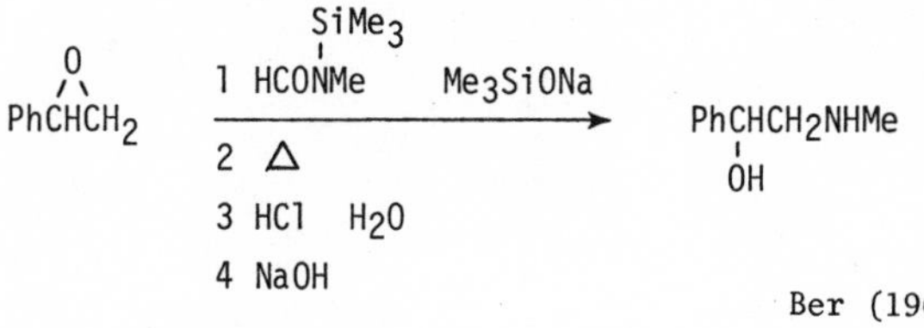

PhCHCH₂ ⟨epoxide, O⟩

1 HCONMe (SiMe₃)   Me₃SiONa
2 △
3 HCl H₂O
4 NaOH

→ PhCHCH₂NHMe
      |
      OH

Ber (1969) 102 14

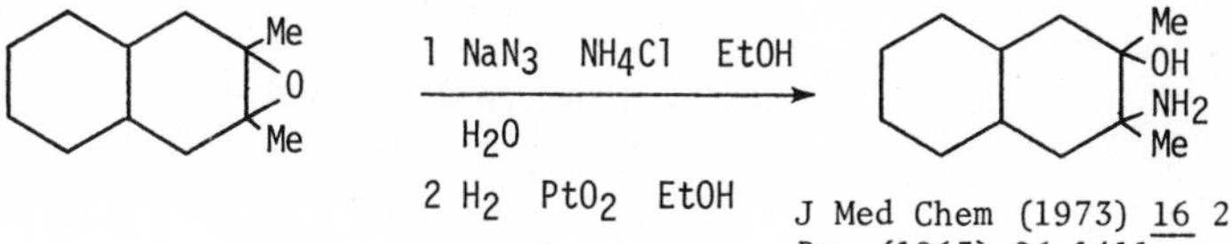

⟨decalin epoxide, Me, O, Me⟩

1 NaN₃ NH₄Cl EtOH
   H₂O
2 H₂ PtO₂ EtOH

→ ⟨decalin, Me, OH, NH₂, Me⟩   47%

J Med Chem (1973) 16 23
Ber (1963) 96 1411

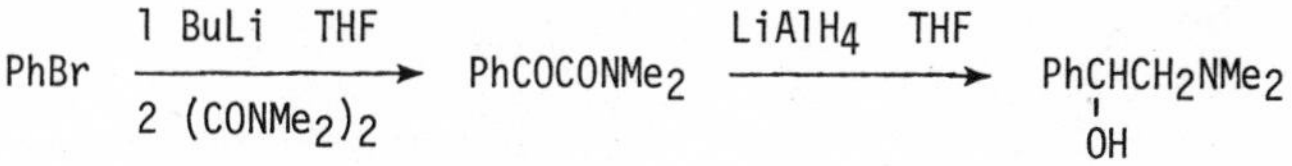

PhBr

1 BuLi THF
2 (CONMe₂)₂

→ PhCOCONMe₂

LiAlH₄ THF
→ PhCHCH₂NMe₂
      |
      OH   77%

Synth Comm (1973) 3 325

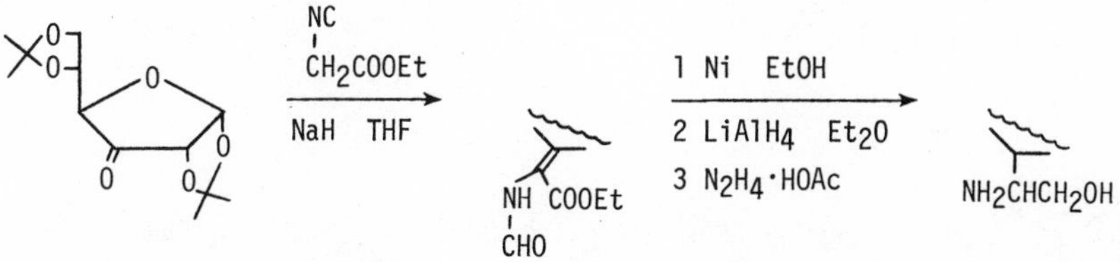

⟨sugar derivative⟩

NC
|
CH₂COOEt
NaH THF

→ ⟨NH COOEt, CHO⟩

1 Ni EtOH
2 LiAlH₄ Et₂O
3 N₂H₄·HOAc

→ NH₂CHCH₂OH

Tetr Lett (1972) 5353

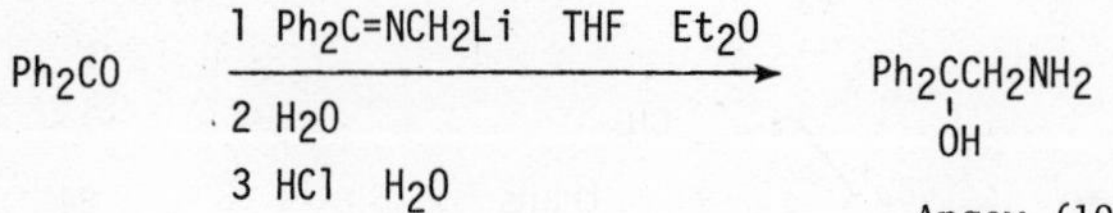

Ph_2CO
$\xrightarrow{\begin{array}{c}1\ Ph_2C=NCH_2Li\ \ THF\ \ Et_2O\\ \cdot 2\ H_2O\\ 3\ HCl\ \ H_2O\end{array}}$
$Ph_2CCH_2NH_2$
$\quad\quad\ \ \overset{|}{OH}$

Angew (1970) <u>82</u> 138
(Internat Ed <u>9</u> 163)

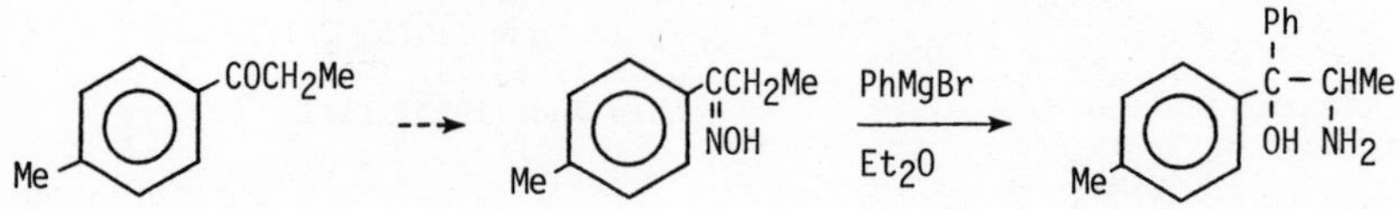

JOC (1943) <u>8</u> 99

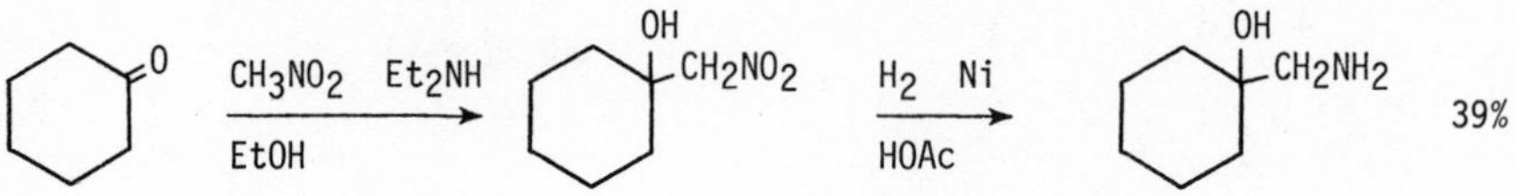

39%

JACS (1951) <u>73</u> 2359
(1952) <u>74</u> 2924
Org Synth (1963) Coll Vol 4 221

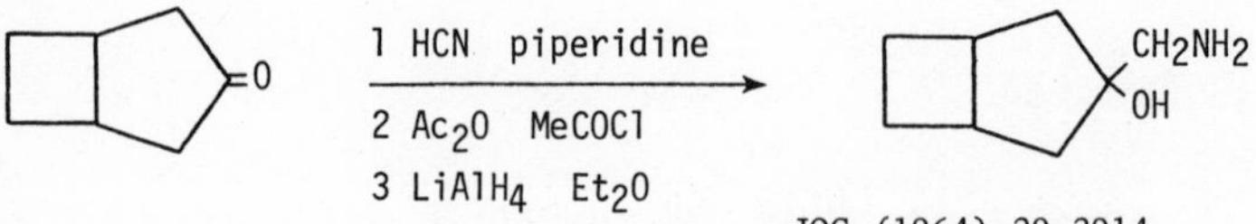

JOC (1964) <u>29</u> 2914
Helv (1950) <u>33</u> 1093
JACS (1952) <u>74</u> 1861

Catalytic reduction of cyanohydrins   Helv (1943) <u>26</u> 288

Reduction of cyanohydrin ethers   Compt Rend (1953) <u>237</u> 1006
<u>236</u> 387
Tetr Lett (1971) 923
Chem Comm (1973) 55

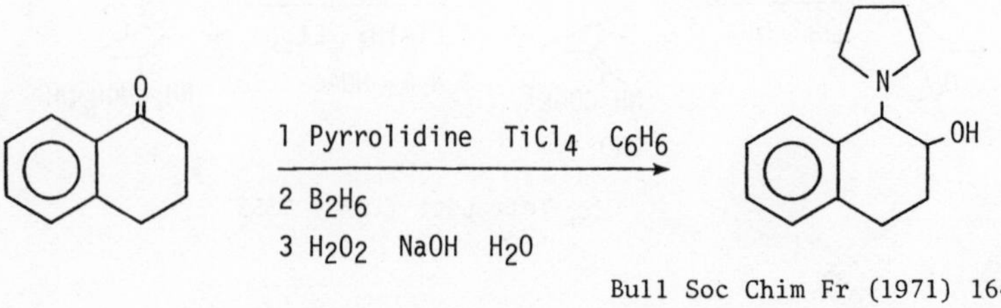

Bull Soc Chim Fr (1971) 1649 3978

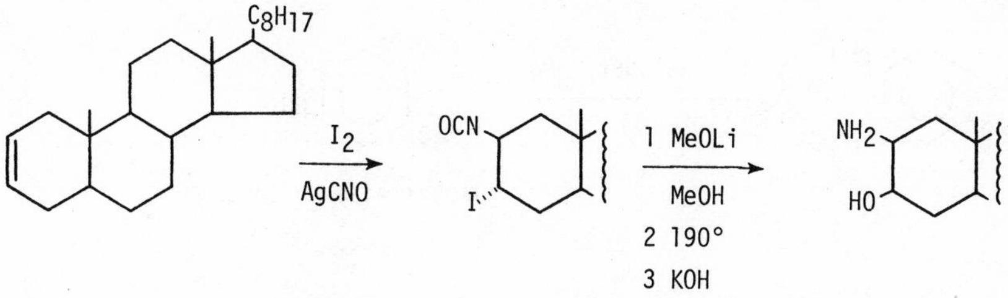

C_8H_{17}
I_2
AgCNO
OCN
I
1 MeOLi
MeOH
2 190°
3 KOH
NH_2
HO

JOC (1967) 32 540

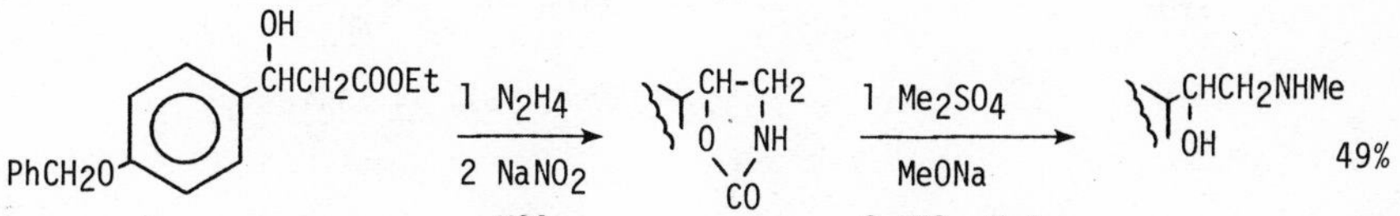

OH
$CHCH_2COOEt$
$PhCH_2O$
1 N_2H_4
2 $NaNO_2$
HOAc
$CH-CH_2$
O NH
CO
1 Me_2SO_4
MeONa
2 HCl H_2O
$CHCH_2NHMe$
OH
49%

JOC (1951) 16 84

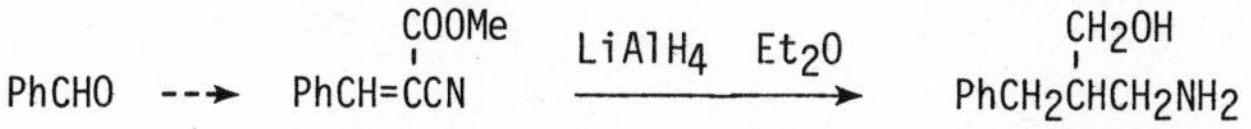

PhCHO
COOMe
PhCH=CCN
$LiAlH_4$ Et_2O
CH_2OH
$PhCH_2CHCH_2NH_2$

Ber (1950) 83 445

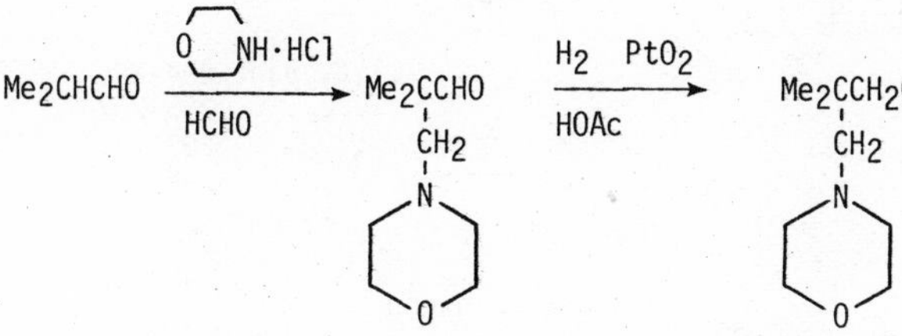

Me_2CHCHO
O NH·HCl
HCHO
Me_2CCHO
CH_2
N
O
H_2 PtO_2
HOAc
Me_2CCH_2OH
CH_2
N
O

JACS (1951) 73 685

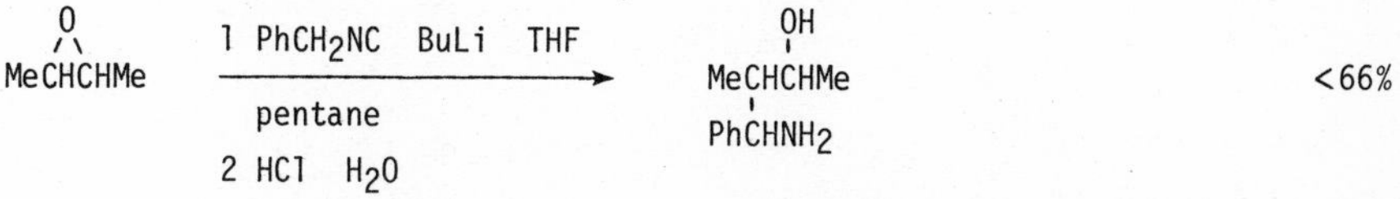

O
MeCHCHMe
1 $PhCH_2NC$ BuLi THF
pentane
2 HCl H_2O
OH
MeCHCHMe
$PhCHNH_2$
<66%

Angew (1973) 85 355
(Internat Ed 12 323)

JACS (1945) <u>67</u> 1472

$EtCH_2NO_2$ $\xrightarrow[\text{EtOH}]{\text{CH}_2\text{=CHCHO EtONa}}$ $EtCH(CH_2)_2CHO$ $\xrightarrow[\text{EtOH}]{\text{H}_2 \quad \text{PtO}_2}$ $EtCH(CH_2)_2CH_2OH$ 22%
                                                                          | |
                                                                         NO_2 NH_2

JACS (1952) <u>74</u> 1064

PrBr $\xrightarrow[\substack{2}]{\text{1 Mg Et}_2\text{O}}$ $HO(CH_2)_4CHNHPr$ 20%
                                                          |
                                                          Pr

Bull Soc Chim Fr (1963) 2468
                    (1954)   575

Section 327   Alcohol — Ester

$C_9H_{19}COOEt$ $\xrightarrow[\substack{\text{base} \\ 2 \text{ I}_2 \text{ acid}}]{1 \text{ Me}_2\text{SO}}$ $C_9H_{19}COCH(OMe)_2$ $\xrightarrow[\text{H}_2\text{O}]{\text{SnCl}_4 \quad \text{dioxane}}$ $C_9H_{19}CHCOOMe$
                                                                                                                                        |
                                                                                                                                       OH

JOC (1967) <u>32</u> 3947

~70%

JOC (1965) <u>30</u> 3804

$$\underset{\underset{Me}{|}}{PrCO} \xrightarrow[H_2O]{NaCN \quad NaHSO_3} \underset{\underset{Me}{|}}{Pr\overset{\overset{OH}{|}}{C}CN} \xrightarrow{HCl \quad EtOH} \underset{\underset{Me}{|}}{Pr\overset{\overset{OH}{|}}{C}COOEt} \qquad 74\%$$

JCS (1957) 3262

$$\underset{\underset{Ph}{|}}{\overset{\overset{OH}{|}}{C}COOEt}$$

JACS (1951) 73 2216
Tetr Lett (1972) 3011

$$\underset{\underset{\overset{\|}{CH_2}}{}}{\overset{\overset{COOMe}{|}}{CH_2CCOOMe}} \xrightarrow[CHCl_3]{CF_3COO_2H \quad Na_2HPO_4} \underset{\underset{\overset{|}{CH_2}}{O}}{\overset{\overset{COOMe}{|}}{CH_2C{-}COOMe}} \xrightarrow[Et_2O]{LiCu(Bu-i)_2} \underset{OH \quad CH_2Bu\text{-}i}{\overset{\overset{COOMe}{|}}{CH_2CCOOMe}}$$

Tetr Lett (1973) 4561

$$\xrightarrow[2 \; H_2SO_4 \quad EtOH]{1 \; EtCH_2C{=}N \quad BuLi}$$

JACS (1970) 92 6644

$$\underset{\underset{Et}{|}}{BuCHCHO} \xrightarrow[Zn \quad C_6H_6]{\overset{\overset{Me}{|}}{BrCHCOOEt}} \underset{Et \quad Me}{Bu\overset{\overset{OH}{|}}{C}HCHCOOEt} \qquad 87\%$$

Org Synth (1963) Coll Vol 4 444
(1955) Coll Vol 3 408
JOC (1970) 35 3966
Org React (1942) 1 1
JCS C (1969) 2799
Organometallics in Chem Synth (1970) 1 57

Continuous flow system   JOC (1974) 39 269

PhCHO $\xrightarrow{\text{LiCH}_2\text{COOEt} \quad \text{THF}}$ PhCHCH$_2$COOEt   80%
                                                    |
                                                    OH

JACS (1970) <u>92</u> 3222
           (1973) <u>95</u> 3050
JOC (1972) <u>37</u> 2468

PhCH$_2$COOEt $\xrightarrow[\text{DMSO}]{\text{HCHO} \quad \text{EtONa}}$ PhCHCOOEt   55%
                                                    |
                                                    CH$_2$OH

Ber (1966) <u>99</u> 2407

Me$_2$CHCOOEt $\xrightarrow[\text{2 PhCHO}]{\text{1 Ph}_3\text{CNa}}$
$\begin{array}{c}\text{COOEt}\\|\\\text{Me}_2\text{CCHPh}\\|\\\text{OH}\end{array}$

JACS (1939) <u>61</u> 793

$\overset{\text{O}}{\overset{\displaystyle /\backslash}{\text{MeCHCH}_2}}$ $\xrightarrow[\text{MeOH}]{\text{CO} \quad \text{Co}_2(\text{CO})_8}$ MeCHCH$_2$COOMe   40%
                                                    |
                                                    OH

JOC (1961) <u>26</u> 2102
Bull Chem Soc Jap (1964) <u>37</u> 672

PrI $\xrightarrow[\text{THF}]{\begin{array}{c}\text{CH}_2\text{OH}\\|\\\text{CH}_2\text{COOMe} \quad \text{LiN(Pr-i)}_2\end{array}}$
$\begin{array}{c}\text{CH}_2\text{OH}\\|\\\text{PrCHCOOMe}\end{array}$ 90%

Tetr Lett (1973) 2429

1 BrMgC≡COEt   Et$_2$O
$\xrightarrow{\hspace{3cm}}$
C$_6$H$_6$

2 H$_2$SO$_4$ THF

JACS (1954) <u>76</u> 1715
Advances in Org Chem (1960) <u>2</u> 117

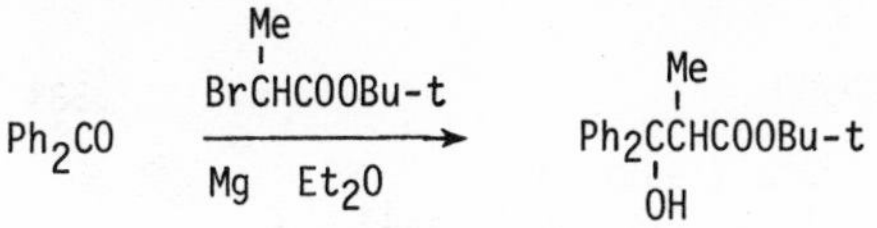

$$Ph_2CO \xrightarrow[\text{Mg } Et_2O]{\overset{\overset{\displaystyle Me}{|}}{BrCHCOOBu\text{-}t}} \underset{\underset{\displaystyle OH}{|}}{Ph_2C\overset{\overset{\displaystyle Me}{|}}{C}HCOOBu\text{-}t} \qquad 81\%$$

JOC (1966) <u>31</u> 983

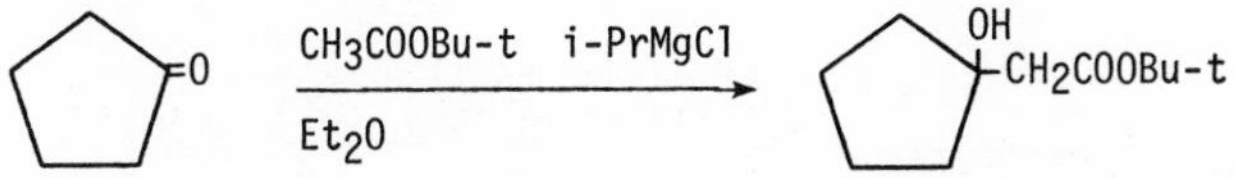

$$\xrightarrow[\text{Et}_2O]{CH_3COOBu\text{-}t \quad i\text{-}PrMgCl}$$

~75%

Bull Soc Chim Fr (1966) 3243 131 125

Et_2NMgBr, $CH_3COOBu\text{-}t$ JACS (1952) <u>74</u> 6254
Tetrahedron (1967) <u>23</u> 4271

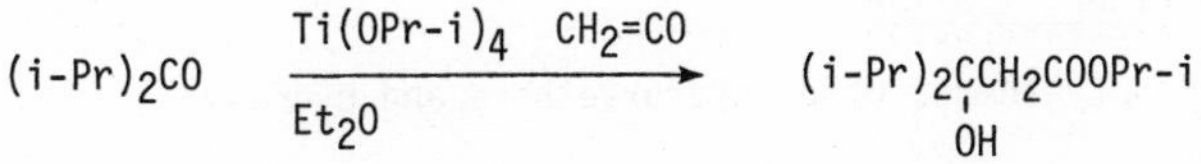

$$(i\text{-}Pr)_2CO \xrightarrow[\text{Et}_2O]{Ti(OPr\text{-}i)_4 \quad CH_2=CO} \underset{\underset{\displaystyle OH}{|}}{(i\text{-}Pr)_2CCH_2COOPr\text{-}i} \qquad 52\%$$

Synthesis (1972) 608

$$\xrightarrow[\text{(or NaBH}_4)]{H_2 \quad PtO_2 \quad EtOH}$$

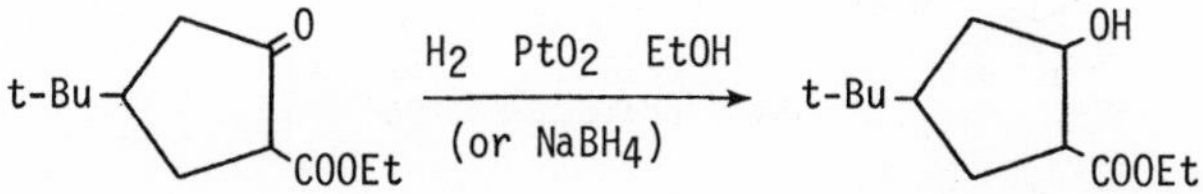

87-95%

Tetrahedron (1972) <u>28</u> 3475
KBH_4 JCS (1963) 2743

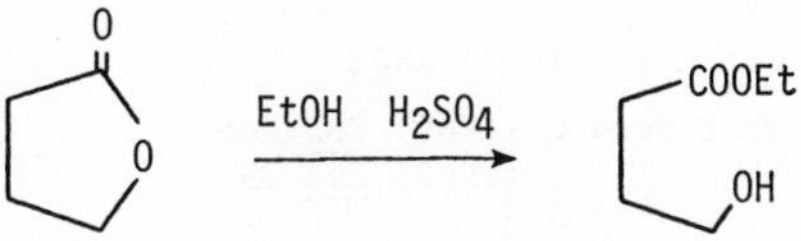

$$\xrightarrow{EtOH \quad H_2SO_4}$$

~53%

JOC (1966) <u>31</u> 485

$$\xrightarrow{H_2SO_5 \quad MeOH}$$

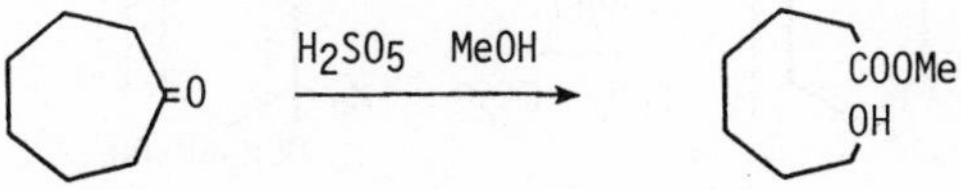

70%

Tetrahedron (1973) <u>29</u> 1447

$C_7H_{15}CN$ $\xrightarrow[\text{(reflux)}]{\text{HOCH}_2\text{CH}_2\text{OH}}$ $C_7H_{15}COOCH_2CH_2OH$ 58%

JCS (1963) 2417
Tetr Lett (1972) 4643

BuCOOH $\xrightarrow[\text{H}_2\text{O}]{\begin{array}{c}\text{O}\\ \diagup\diagdown\\ \text{CH}_2\text{-CH}_2 \quad \text{NaOH}\end{array}}$ $BuCOOCH_2CH_2OH$ 58-63%

JACS (1944) 66 1420
JOC (1961) 26 4563

Also via: Hydroxyacids (Section 313)

Section 328   Alcohol — Ether, Epoxide
°°°°°°°°°°°°°°°°°°°°°°°°°°°°°°°
1,2-Hydroxyethers, 1,3-hydroxyethers, other hydroxyethers and hydroxy-epoxides

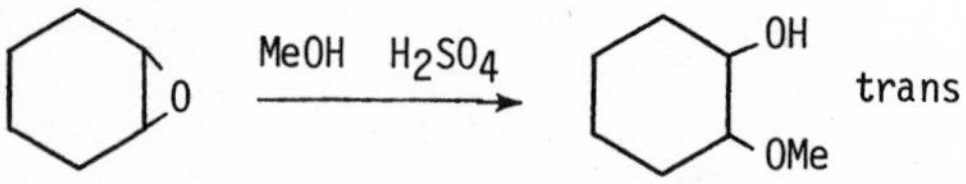

$\xrightarrow{\text{MeOH} \quad \text{H}_2\text{SO}_4}$ trans   82%

JACS (1943) 65 2196

Et_2CO $\xrightarrow[\text{CH}_2(\text{OMe})_2]{\text{LiCH}_2\text{OMe}}$ $Et_2\overset{\underset{\displaystyle OH}{|}}{C}CH_2OMe$ 75%

Tetr Lett (1964) 1503
EtOCH$_2$MgCl Compt Rend C (1966) 262 848
                              (1967) 264 399

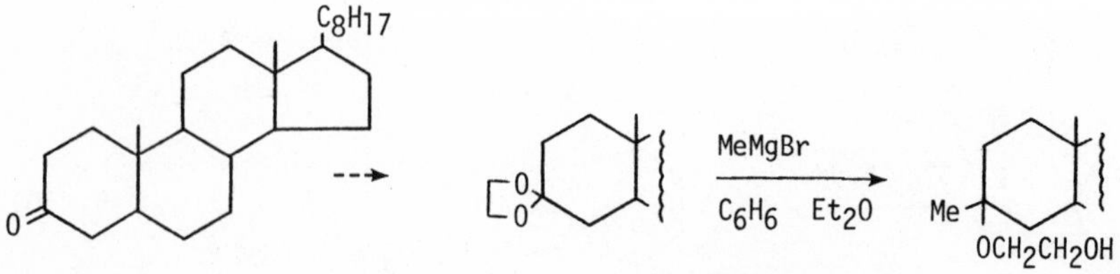

JOC (1967) 32 1417

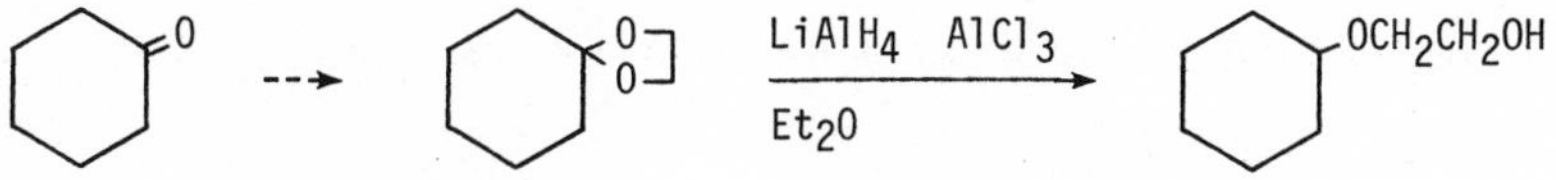

JACS (1962) 84 2371

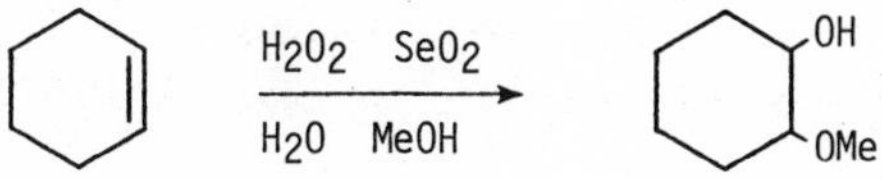

Bull Chem Soc Jap (1965) 38 958

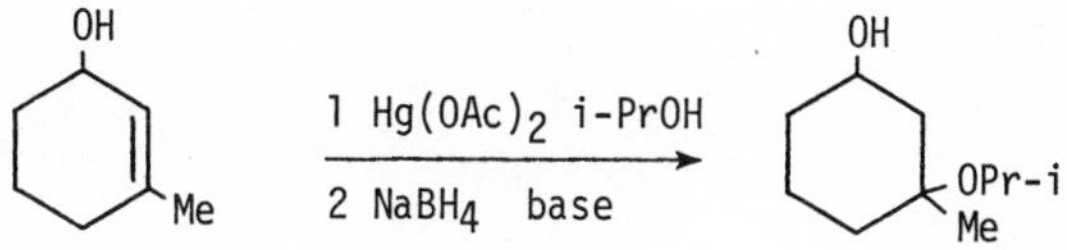

52%

JOC (1970) 35 1041

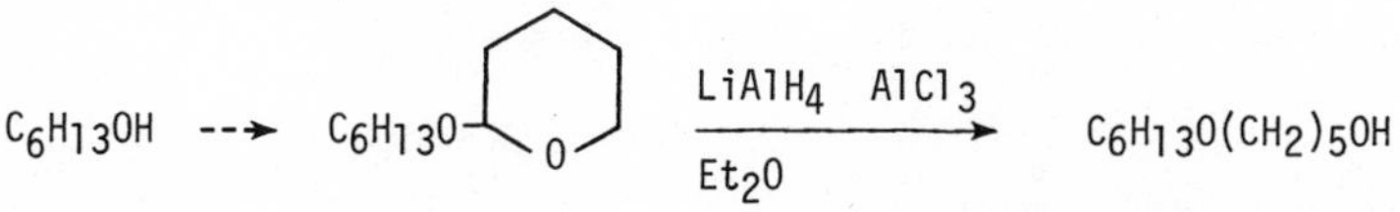

$C_6H_{13}OH$ $\dashrightarrow$ $C_6H_{13}O$⟨...⟩O $\xrightarrow[\text{Et}_2O]{\text{LiAlH}_4 \quad \text{AlCl}_3}$ $C_6H_{13}O(CH_2)_5OH$ <45%

Can J Chem (1967) 45 2547
JOC (1965) 30 2441

$HO(CH_2)_{10}OH$ $\xrightarrow{\text{Resin-COCl}}$ $Resin\text{-}COO(CH_2)_{10}OH$ $\xrightarrow[\substack{2 \text{ NH}_3 \quad \text{H}_2O \\ \text{dioxane}}]{1 \text{ Ph}_3CCl \quad \text{Pyr}}$ $HO(CH_2)_{10}OCPh_3$

Can J Chem (1973) 51 2452
(1972) 50 2892

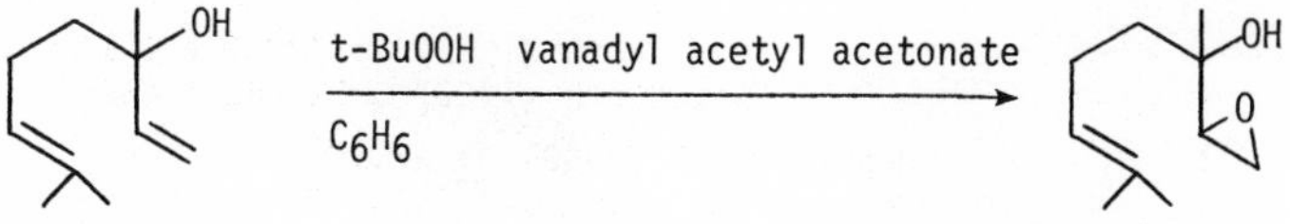

84%

JACS (1973) 95 6136

Section 329   Alcohol — Halide

Fluorohydrins, chlorohydrins, bromohydrins and iodohydrins

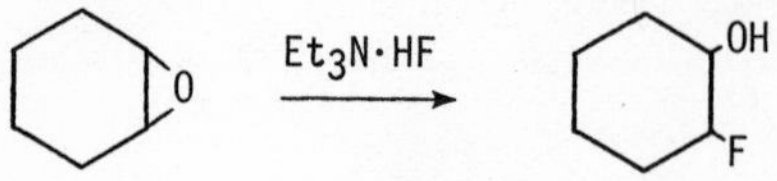

40%

Bull Soc Chim Fr (1969) 3647

HF   Can J Chem (1960) <u>38</u> 1495

KHF$_2$ J Med Chem (1972) <u>15</u> 1092

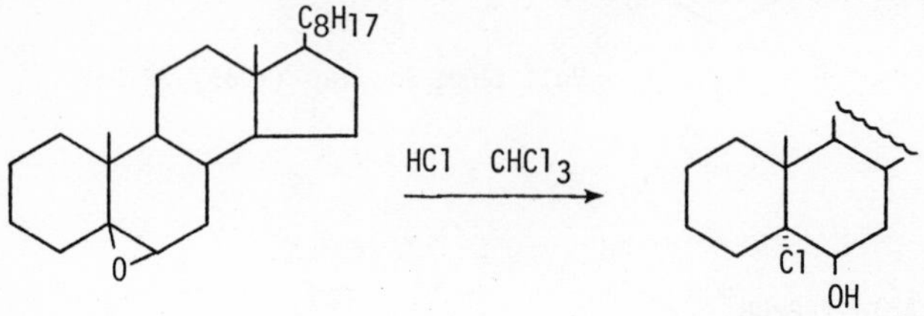

JCS (1958) 1657
JOC (1961) <u>26</u> 2403

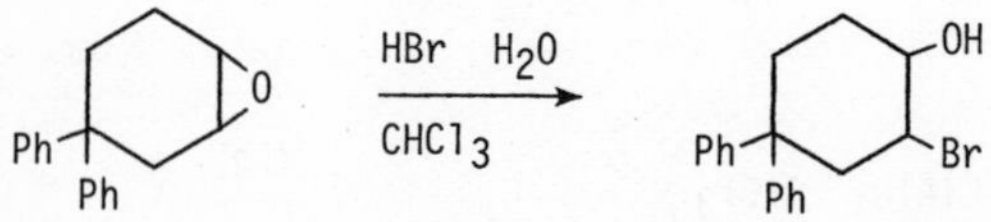

JOC (1968) <u>33</u> 3385
JCS (1951) <u>2598</u>

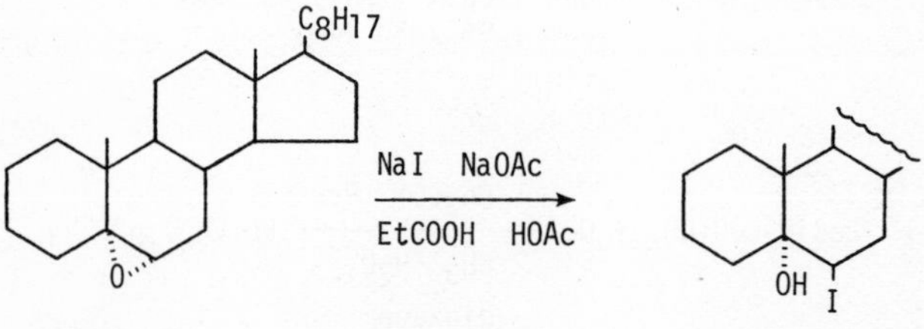

JCS <u>C</u> (1971) 1466
HI   JCS <u>(</u>1951) 2598

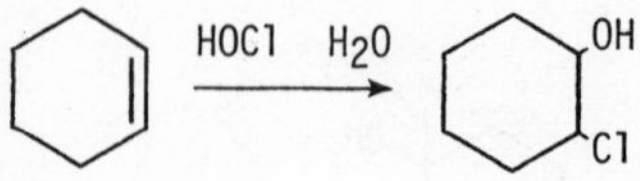

70-73%

Org Synth (1932) Coll Vol 1 158
NH$_2$CONHCl Org Synth (1963) Coll Vol 4 157
CrO$_2$Cl$_2$ JACS (1956) <u>78</u> 3749
        (1950) <u>72</u> 4353

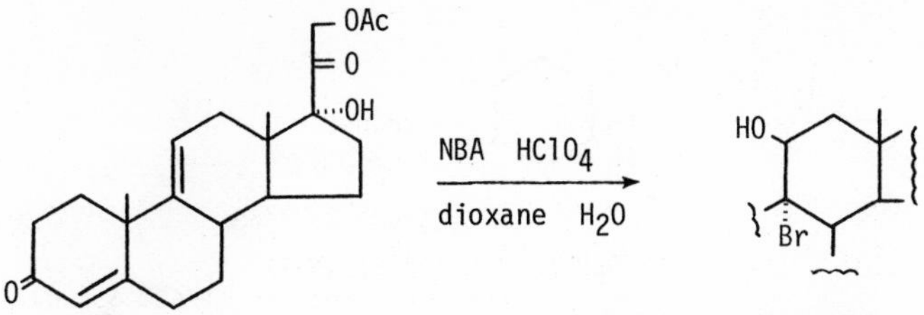

OAc
O
OH
NBA HClO4
dioxane H2O
HO
Br

>90%

JACS (1953) 75 2273
(1957) 79 1130

NBS, Me_2SO	JACS (1968) 90 5498
KBr, H_2O_2, enzyme	Tetr Lett (1968) 4057
BrN_3	Tetr Lett (1968) 3921

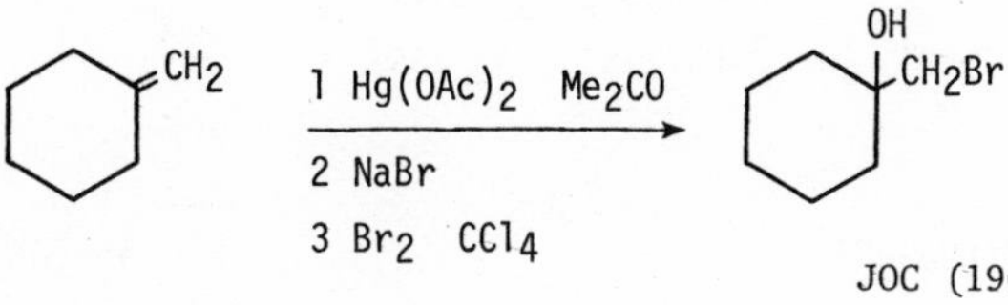

CH2
1 Hg(OAc)2 Me2CO
2 NaBr
3 Br2 CCl4
OH
CH2Br

JOC (1968) 33 3953

$PrCH=CH_2$ —(1 I_2 H_2O; 2 H_2O_2 Me_2CO H_2O)→ $PrCH(OH)CH_2I$ 85%

Can J Chem (1964) 42 2710

KIO_3, H_2SO_4 JCS C (1970) 846

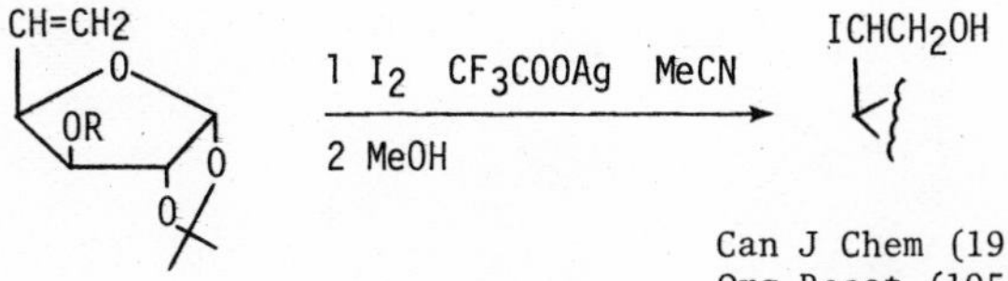

CH=CH2
O
OR
O
O
1 I2 CF3COOAg MeCN
2 MeOH
ICHCH2OH

Can J Chem (1972) 50 507
Org React (1957) 9 332

$PhCOCH_2Cl$ —(PhLi Et_2O)→ $Ph_2C(OH)CH_2Cl$ 78%

JACS (1951) 73 4030

$$\text{(cyclohexanone, Ph, Ph, Br)} \xrightarrow{\text{LiAlH}_4 \quad \text{Et}_2\text{O}} \text{(cyclohexanol, Ph, Ph, Br, OH)} \quad \text{trans} \qquad 93\%$$

JOC (1968) <u>33</u> 3385

NaBH$_4$ JCS (1958) 1657
Bull Chem Soc Jap (1962) <u>35</u> 2044

LiBH$_4$ Helv (1953) <u>36</u> 1241

Fluorohydrins   JCS Perkin I (1973) 1462

$$\text{MeCH}_2(\text{CH}_2)_3\text{OH} \xrightarrow[\text{2 h}\nu \quad \text{CHCl=CHCl}]{\text{1 NaOCl}} \underset{\overset{|}{\text{Cl}}}{\text{MeCH}(\text{CH}_2)_3\text{OH}} \qquad \sim 34\%$$

JOC (1972) <u>37</u> 3514

(i-Pr)$_2$NCl, hν JACS (1971) <u>93</u> 438

$$\text{HO}(\text{CH}_2)_6\text{OH} \xrightarrow[\text{toluene}]{\text{HCl} \quad \text{H}_2\text{O}} \text{HO}(\text{CH}_2)_6\text{Cl} \qquad 45\text{-}50\%$$

Org Synth (1955) Coll Vol 3 446

Section 330   Alcohol $\cdot$─ Ketone
~~○○○○○○○○○○○○○○○○○~~

$$\text{EtC}\equiv\text{CEt} \xrightarrow[\text{glyme H}_2\text{O}]{\text{Tl(NO}_3)_3 \quad \text{HClO}_4} \underset{\overset{|}{\text{OH}}}{\text{EtCOCHEt}} \qquad 70\text{-}90\%$$

JACS (1973) <u>95</u> 1296

$$\text{(furan)-COCl} \xrightarrow{CH_2N_2} \text{(furan)-COCHN}_2 \xrightarrow[\text{dioxane}]{H_2SO_4 \ H_2O} \text{(furan)-COCH}_2OH \quad <74\%$$

JACS (1948) <u>70</u> 142
(1950) <u>72</u> 5161

$$\text{(cyclohexane-1,2-diol)} \xrightarrow[C_6H_6]{Ag_2CO_3\text{-Celite}} \text{(2-hydroxycyclohexanone)} \quad 45\%$$

(Also applicable to 1,3 and 1,4-diols)

Chem Comm (1969) 1102
Tetrahedron (1973) <u>29</u> 2867
NCS, Me$_2$S   Tetr Lett (1974) 287

$$\xrightarrow{\text{Ph}_3\overset{+}{C} \ \overset{-}{BF}_4} \text{(product)}$$
CH$_2$Cl$_2$

JCS Perkin I (1972) 542

$$\text{(4-MeO-C}_6\text{H}_4\text{)-CHO} \xrightarrow[H_2O]{KCN \ EtOH} \text{(4-MeO-C}_6\text{H}_4\text{)-COCH(OH)-(4-MeO-C}_6\text{H}_4\text{)} \quad 52\%$$

Org React (1962) <u>4</u> 269
Org Synth (1941) Coll Vol 1 94
Bu$_4$N $\overset{+}{}$ CN $\overset{-}{}$ Tetr Lett (1971) 287

$$PhCH_2CHO \xrightarrow{KCN} PhCH_2\underset{OH}{CH}CN \xrightarrow{PhCH_2MgCl} PhCH_2\underset{OH}{CH}COCH_2Ph$$

Helv (1945) <u>28</u> 741
Bull Soc Chim Fr (1955) 784

$$\text{PhCHO} \xrightarrow[\substack{Me_2NCH_2CH_2NMe_2 \quad pentane \\ 2 \ HCl}]{1 \ CH_2=CHOEt \quad t\text{-}BuLi} \underset{\overset{|}{OH}}{PhCHCOMe} \qquad 57\%$$

Annalen (1972) <u>763</u> 208

$CH_2=CHSEt$ JACS (1973) <u>95</u> 2694

$$\begin{array}{l} \text{CHO} \\ \overset{|}{HOCH} \\ \overset{|}{HCOH} \\ \overset{|}{HCOH} \\ \overset{|}{CH_2OH} \end{array} \xrightarrow[\substack{2 \ H_2SO_4 \quad H_2O}]{1 \ HOCH_2CH_2NO_2 \quad base} \begin{array}{l} CH_2OH \\ \overset{|}{CO} \\ \overset{|}{HCOH} \\ \overset{|}{HOCH} \\ \overset{|}{HCOH} \\ \overset{|}{HCOH} \\ \overset{|}{CH_2OH} \end{array} \qquad 23\%$$

JACS (1950) <u>72</u> 3325

$$\text{PhCOOEt} \xrightarrow[t\text{-}BuOK]{Me_2SO} PhCOCH_2SOMe \xrightarrow[\substack{2 \ NaOSOCH_2OH \\ Cu(OAc)_2 \quad EtOH}]{1 \ HCl \quad Me_2SO} PhCOCH_2OH \qquad 60\text{-}65\%$$

JACS (1966) <u>88</u> 5498

$$C_{11}H_{23}COOMe \xrightarrow{Na \quad xylene} \underset{\overset{|}{OH}}{C_{11}H_{23}COCHC_{11}H_{23}} \qquad 80\text{-}90\%$$

Org React (1962) <u>4</u> 256
Org Synth (1963) Coll Vol 4 840

Review: The Acyloin Condensation as a Cyclization Method

Chem Rev (1964) <u>64</u> 573

$$\begin{array}{l} COOMe \\ \overset{|}{(CH_2)_8} \\ \overset{|}{COOMe} \end{array} \xrightarrow[\substack{2 \ HCl \quad H_2O \quad THF}]{1 \ Na \quad ClSiMe_3 \quad xylene} (CH_2)_8 \overset{\overset{CO}{|}}{\underset{CHOH}{}} \qquad 53\%$$

Ber (1964) <u>97</u> 1383
Can J Chem (1969) <u>47</u> 3266
Synthesis (1971) 236
JACS (1953) <u>75</u> 6231

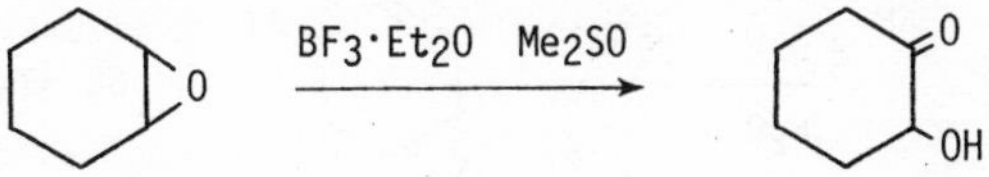
O
BF3·Et2O Me2SO
O
OH

76%

JOC (1961) 26 1681
Compt Rend (1965) 261 1990

$PhCH_2Br \xrightarrow{Li[BuCONi(CO)_3] \quad Et_2O} PhCH_2C(OH)(Bu)COBu$ 58%

JOC (1970) 35 4183

$BuBr$ → 1 Mg; 2 $MeCOC(Me){=}NOH$; 3 $(COOH)_2$ H_2O → $BuC(OH)(Me)COMe$

Annales de Chimie (1939) 11 453

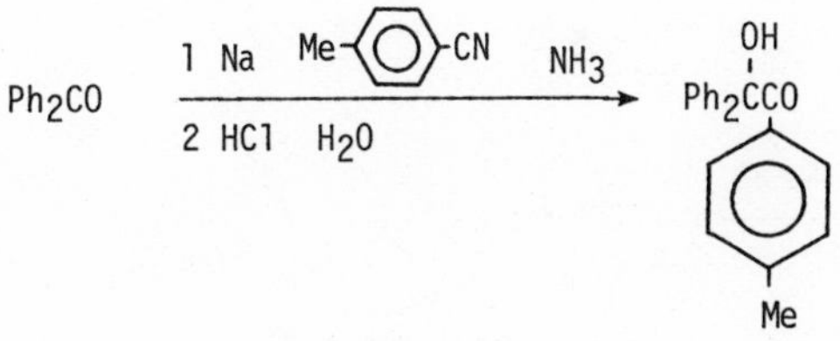
Ph2CO
1 Na Me-C6H4-CN NH3
2 HCl H2O
OH
Ph2CCO
Me

JOC (1965) 30 3804

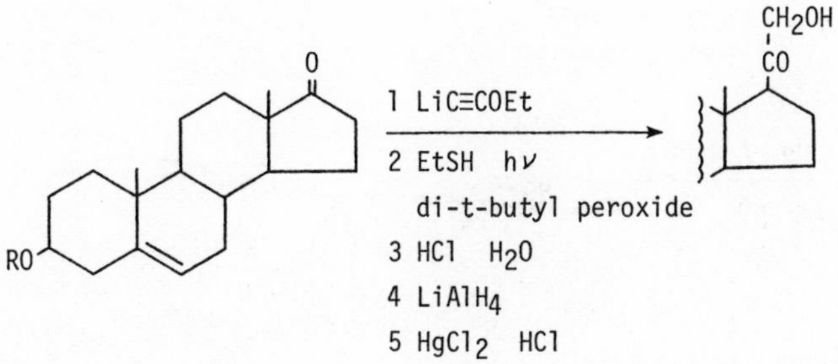
O
RO
1 LiC≡COEt
2 EtSH hν
di-t-butyl peroxide
3 HCl H2O
4 LiAlH4
5 HgCl2 HCl
CH2OH
CO

Rec Trav Chim (1960) 79 1293

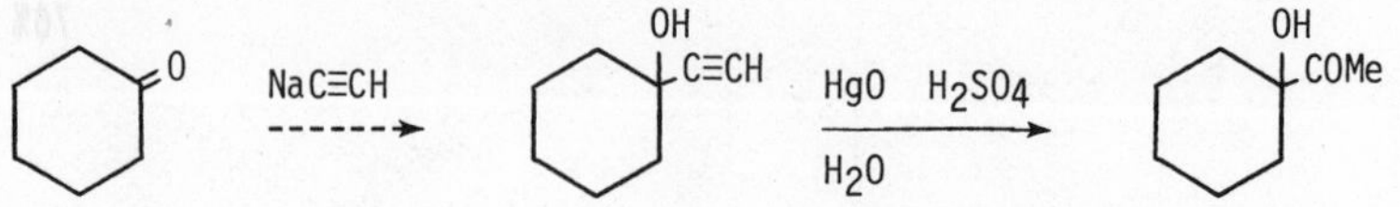

65-67%

Org Synth (1963) Coll Vol 4 13
Tetrahedron (1964) 20 1119

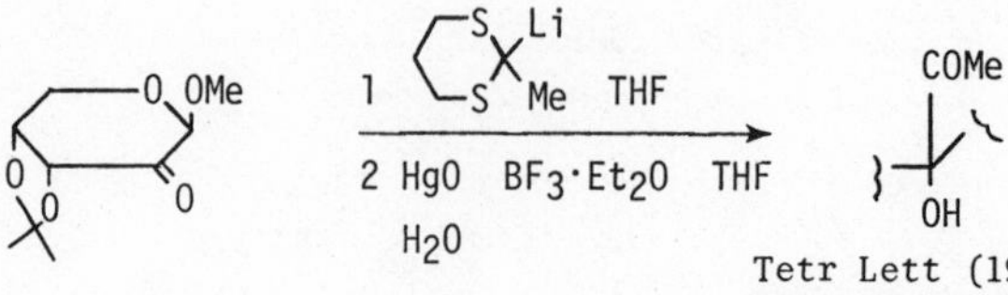

62%

Tetr Lett (1972) 3945
Annalen (1972) 758 89
JOC (1971) 36 2035

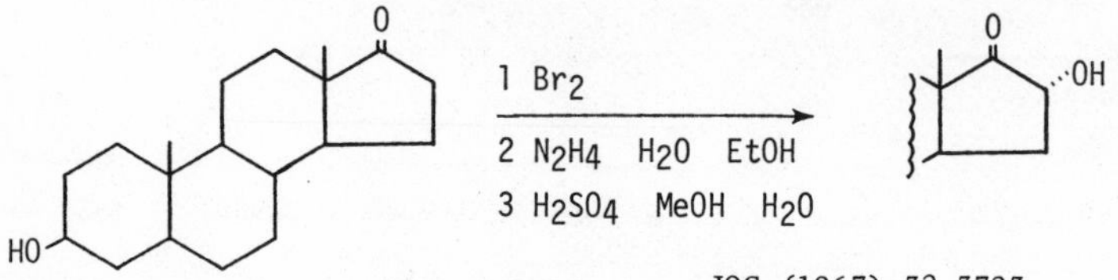

1 Br$_2$

2 N$_2$H$_4$ H$_2$O   EtOH

3 H$_2$SO$_4$ MeOH   H$_2$O

JOC (1967) 32 3723

PhCOCH$_2$Me --> PhCOCHMe $\underset{\text{Br}}{|}$ $\xrightarrow{\text{HCOOK   MeOH}}$ PhCOCHMe $\underset{\text{OH}}{|}$

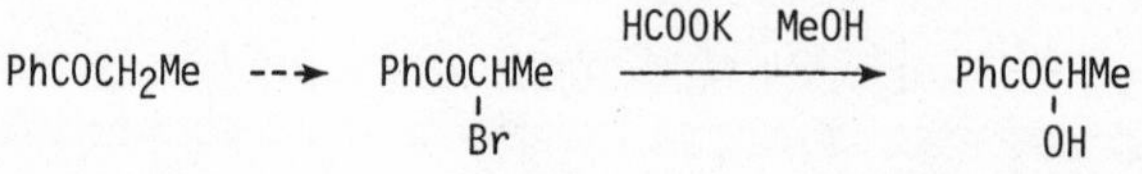

Annalen (1936) 526 143
Org Synth (1943) Coll Vol 2 5

NaOH   JCS (1954) 3257

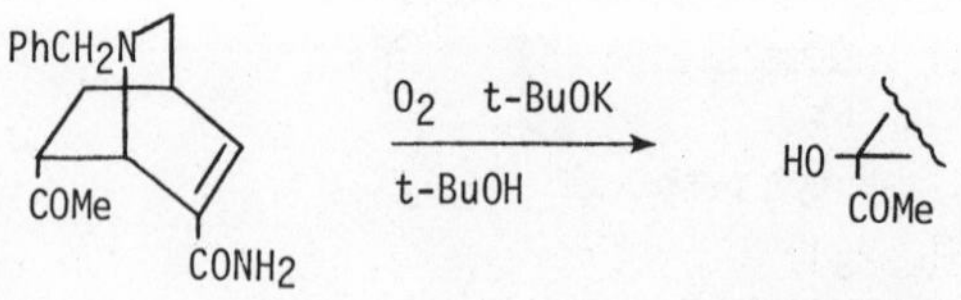

$\xrightarrow[\text{t-BuOH}]{\text{O}_2\quad\text{t-BuOK}}$

HO$-$⟨structure⟩COMe

JACS (1968) 90 2448
JOC (1968) 33 3294

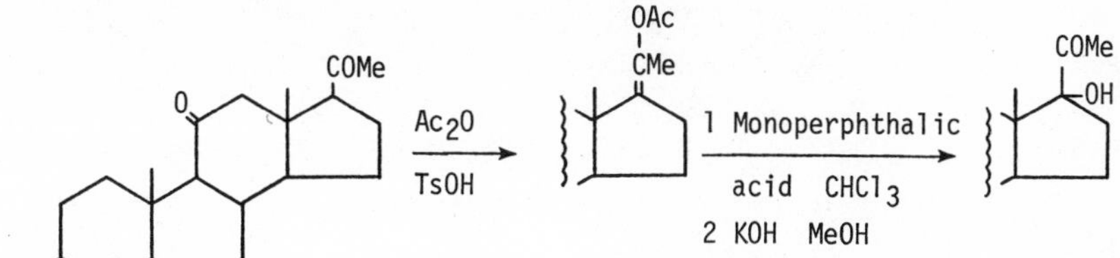

JCS (1954) 747
Helv (1959) 42 2043

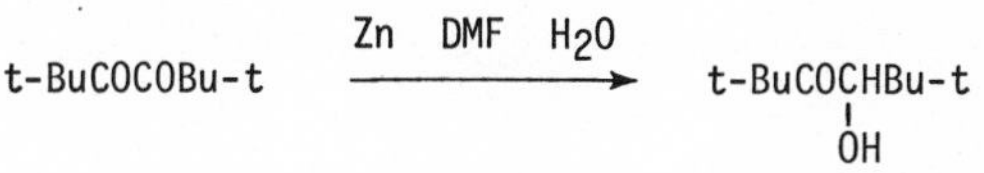

t-BuCOCOBu-t $\xrightarrow{\text{Zn \quad DMF \quad H}_2\text{O}}$ t-BuCOCHBu-t   87%
                                                |
                                              OH

Annalen (1971) 745 164

$(\text{EtO})_3\text{P}$ Can J Chem (1969) 47 3266

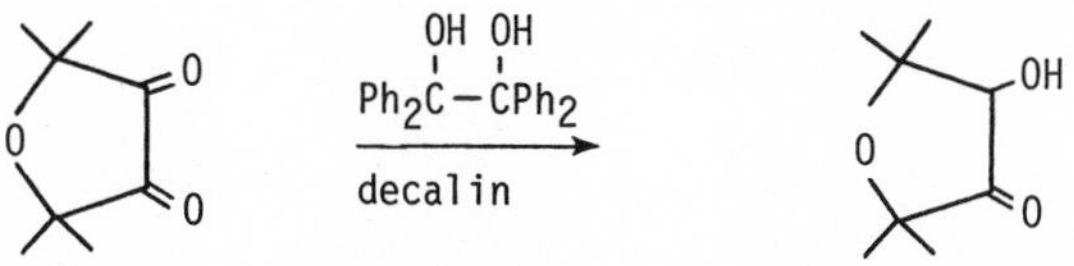

Tetr Lett (1971) 3403

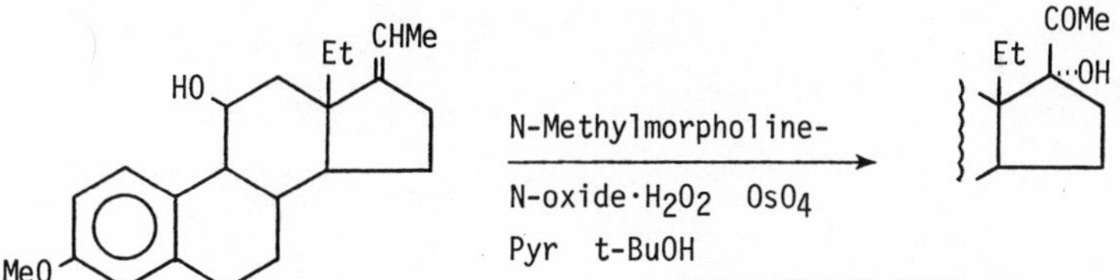

JCS C (1968) 2647
JOC (1960) 25 1968

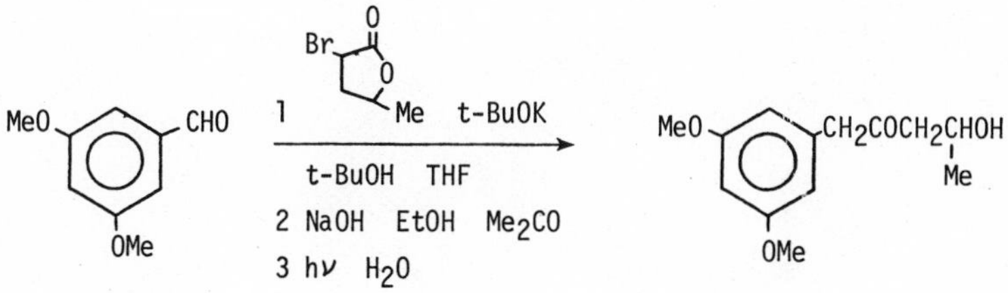

JACS (1971) 93 281

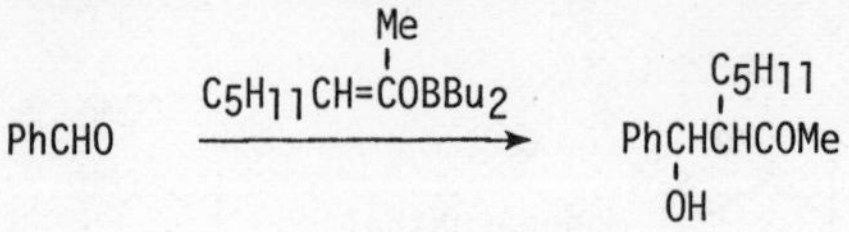

$$PhCHO \xrightarrow[]{C_5H_{11}\overset{Me}{\overset{|}{CH}}=COBBu_2} PhCHCHCOMe \overset{C_5H_{11}}{\underset{OH}{|}}$$ 91%

JACS (1973) <u>95</u> 967

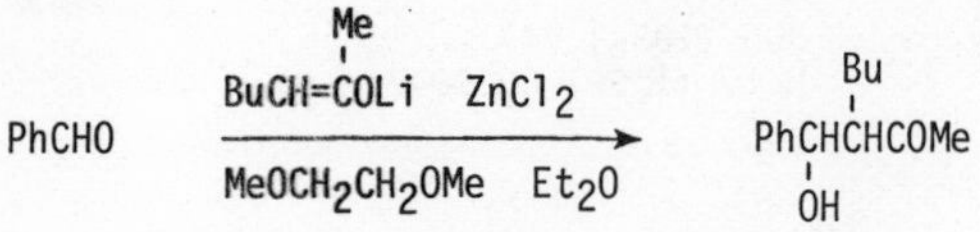

$$PhCHO \xrightarrow[MeOCH_2CH_2OMe \quad Et_2O]{BuCH=COLi \quad ZnCl_2} PhCHCHCOMe \overset{Bu}{\underset{OH}{|}}$$

JACS (1973) <u>95</u> 3310

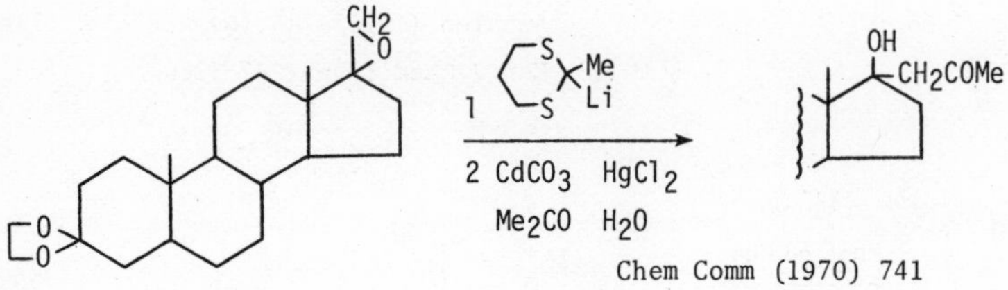

Chem Comm (1970) 741

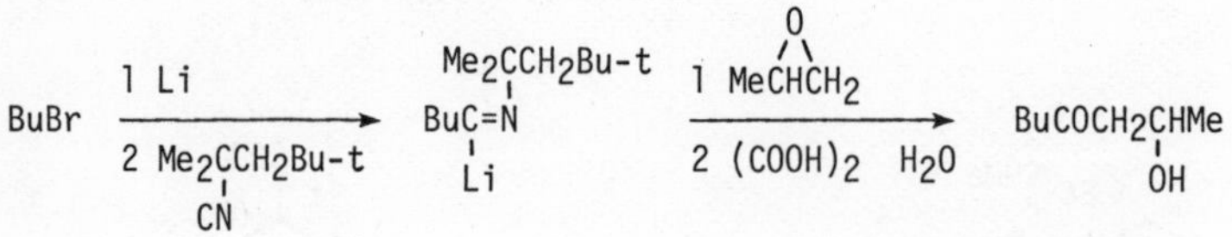

$$BuBr \xrightarrow[2 \ Me_2CCH_2Bu-t]{1 \ Li} \underset{Li}{\overset{Me_2CCH_2Bu-t}{BuC=N}} \xrightarrow[2 \ (COOH)_2 \quad H_2O]{1 \ MeCHCH_2} BuCOCH_2CHMe \atop OH$$ 90%

JACS (1970) <u>92</u> 6675

Review: The Aldol Condensation   Org React (1968) <u>16</u> 1

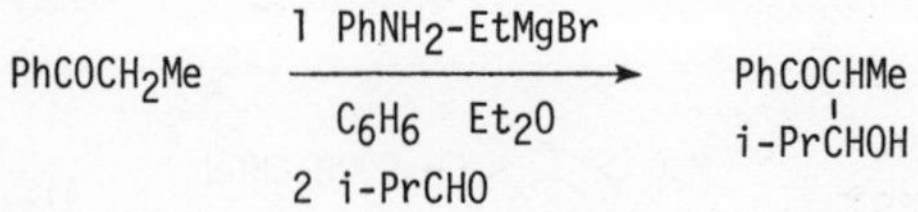

$$PhCOCH_2Me \xrightarrow[C_6H_6 \quad Et_2O \atop 2 \ i\text{-}PrCHO]{1 \ PhNH_2\text{-}EtMgBr} \underset{i\text{-}PrCHOH}{PhCOCHMe}$$ 83%

Org React (1968) <u>16</u> 1

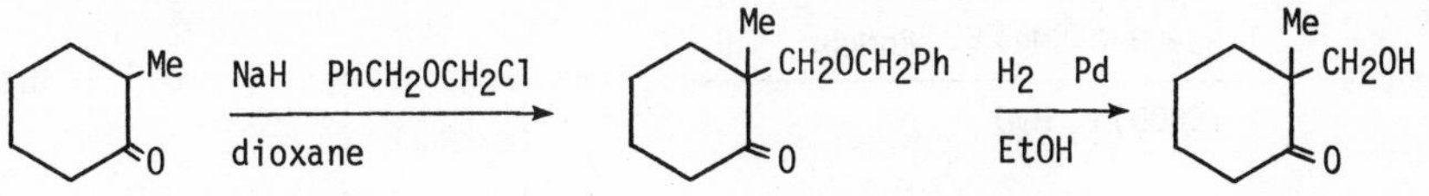

JCS (1963) 4634
     (1964) 4521

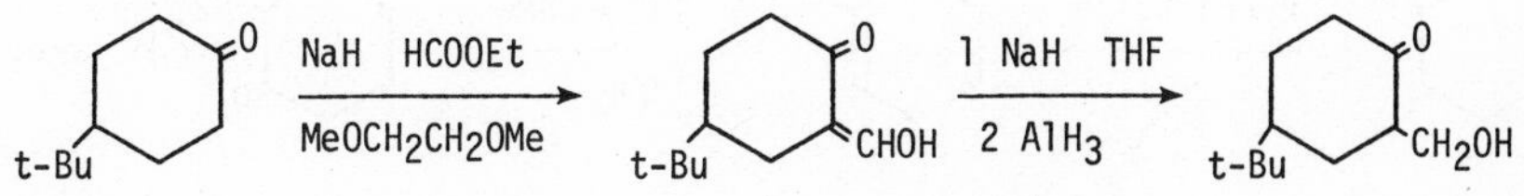

JOC (1971) <u>36</u> 3070

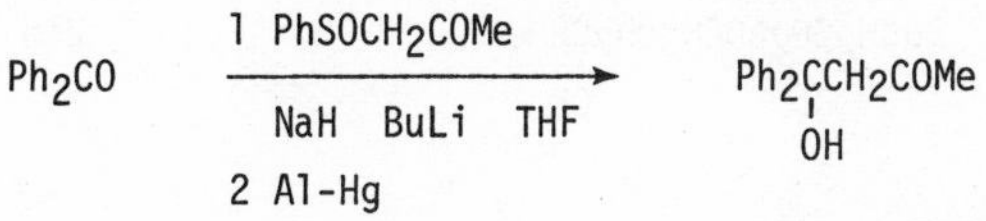

85%

Ph_2CO $\xrightarrow[\substack{NaH\ BuLi\ THF \\ 2\ Al-Hg}]{1\ PhSOCH_2COMe}$ Ph_2CCH_2COMe ~91%
                                                    $\overset{|}{OH}$

Tetr Lett (1974) 107

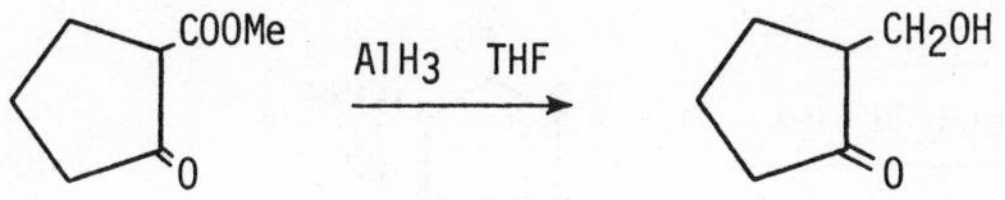

JACS (1968) <u>90</u> 2927

82%

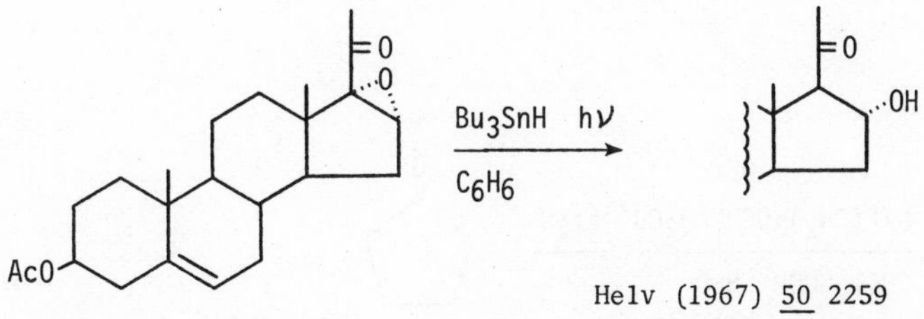

27%

Helv (1967) <u>50</u> 2259
$Cr(OAc)_2$ Coll Czech (1961) <u>26</u> 1207
Al-Hg JOC (1973) <u>38</u> 3187

$$\underset{CH_2-CH_2}{\overset{O}{\diagdown}} \quad \xrightarrow[\text{2 (COOH)}_2 \quad H_2O]{\begin{array}{c} Me \\ | \\ CH_2CHOSiMe_3 \\ 1 \; Me_2CN=C=CMe_2 \quad PhMgBr \quad THF \end{array}} \quad \begin{array}{c} PhCOCMe_2 \\ | \\ CH_2CH_2OH \end{array} \qquad 30\%$$

JOC (1973) <u>38</u> 2129

$$\xrightarrow{\text{Pyrrolidine}} \qquad \xrightarrow[\text{DMF}]{\underset{CH_2-CHMe}{\overset{O}{\diagdown}}} \qquad \sim 60\%$$

Tetrahedron (1969) <u>25</u> 3157

$$BuCH=CH_2 \quad \xrightarrow[\begin{array}{c} 3 \; \underset{CH_2-CHMe}{\overset{O}{\diagdown}} \\ 4 \; H_2O_2 \quad NaOH \quad H_2O \end{array}]{\begin{array}{c} 1 \; B_2H_6 \quad THF \\ 2 \; Me_3SiC{\equiv}CH \quad BuLi \end{array}} \quad BuCH_2CH_2COCH_2CH_2\underset{OH}{\overset{|}{C}HMe} \qquad 21\%$$

Tetr Lett (1973) 2741

$$\xrightarrow{MeCH_2OH \quad h\nu} \qquad 40\%$$

JCS (1967) 2032

$$\xrightarrow[\text{2 CHCl}_2COOH \quad H_2O]{\begin{array}{c} OEt \\ | \\ 1 \; Li[(CH_2)_3OCHMe]_2Cu \quad Et_2O \end{array}} \qquad (CH_2)_3OH$$

JOC (1972) <u>37</u> 1947

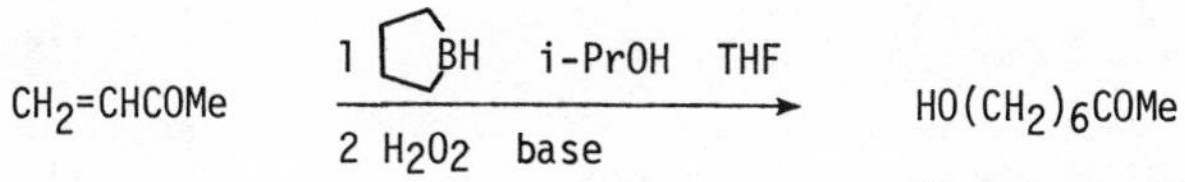

CH$_2$=CHCOMe $\xrightarrow[\text{2 H}_2\text{O}_2 \text{ base}]{\text{1 } \boxed{}\text{BH i-PrOH THF}}$ HO(CH$_2$)$_6$COMe

Chem Comm (1969) 1009

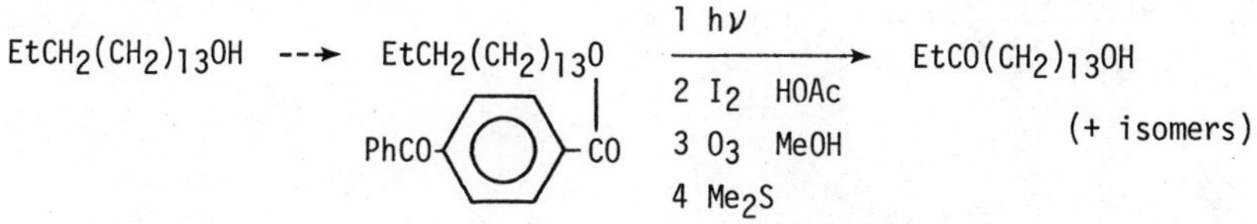

EtCH$_2$(CH$_2$)$_{13}$OH --→ EtCH$_2$(CH$_2$)$_{13}$O $\xrightarrow[\text{4 Me}_2\text{S}]{\substack{\text{1 h}\nu \\ \text{2 I}_2 \text{ HOAc} \\ \text{3 O}_3 \text{ MeOH}}}$ EtCO(CH$_2$)$_{13}$OH

PhCO—⬡—CO   (+ isomers)

JACS (1969) $\underline{91}$ 3083
     (1973) $\underline{95}$ 3251

Also via: Acyloxyketones (Section 360)

Section 331   Alcohol — Nitrile
oooooooooooooooooooo

α-Hydroxynitriles (cyanohydrins) and β-hydroxynitriles

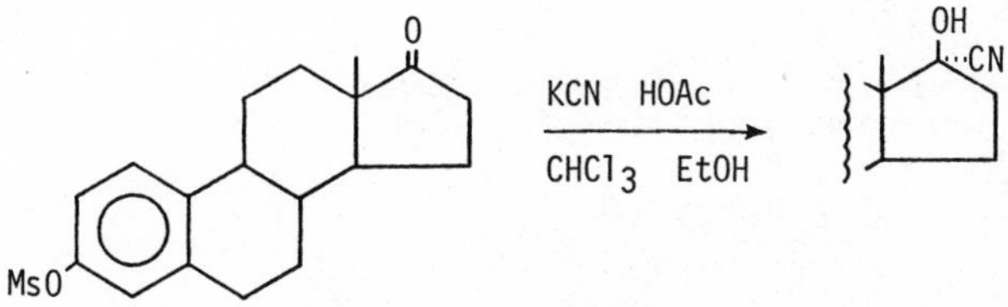

$\xrightarrow[\text{CHCl}_3 \text{ EtOH}]{\text{KCN HOAc}}$

JCS $\underline{C}$ (1968) 2283
NaCN, NaHSO$_3$ Org Synth (1941) Coll Vol 1 336
HCN, Et$_3$N JACS (1956) $\underline{78}$ 4100
Et$_2$AlCN Tetr Lett (1966) 1913
Me$_2$CCN JACS (1953) $\underline{75}$ 650
   |
   OH

Et
|
BrCHCN   Zn   THF
PrCHO ─────────────────→   −   Et
                              |
                              PrCHCHCN   66%
                              |
                              OH

Compt Rend C (1969) 269 861
Organometallics in Chem Synth (1970) 1 57

PhCHO $\xrightarrow[\text{THF}]{\text{CH}_3\text{CN} \quad \text{BuLi}}$ PhCHCH$_2$CN   83%
                                              |
                                              OH

Ber (1968) 101 3113

HCN   Et$_3$Al
─────────────────→
THF   56%

JCS C (1970) 2365

Ph$_2$CO $\xrightarrow{\text{CH}_3\text{CN} \quad \text{NaNH}_2 \quad \text{NH}_3}$ Ph$_2$CCH$_2$CN   93%
                                                      |
                                                      OH

JOC (1968) 33 3402
Ber (1968) 101 3113

Also via: Section
            Cyanohydrin trimethylsilyl ethers   366
                     Cyanohydrin esters   361

Section 332   Alcohol ── Olefin

Allylic and benzylic hydroxylation (C=C-CH ──→ C=C-C-OH, etc.) is listed
in section 41 (Alcohols and Phenols from Hydrides)

BuC≡CH $\xrightarrow{\begin{array}{l}1 \ (i\text{-}Bu)_2AlH \quad heptane \\ 2 \ MeLi \quad Et_2O \\ 3 \ RCHO\end{array}}$ BuCH=CHCHOH   73% (R=H)
                                       | 68% (R=Me)
                                       R

       trans

JACS (1967) 89 2754 5085
Tetr Lett (1971) 4571

$C_6H_{13}CHO$ $\xrightarrow{\begin{array}{l}1 \ Ph_3P=CHMe \quad THF \\ 2 \ BuLi \quad hexane\end{array}}$

3 HCHO →

H   Me
   C=C
C_6H_{13} CH_2OH 73%

3 MeCHO →

Me   Me
  C=C
H   CHOH
        |
       C_6H_{13} 67%

JACS (1970) 92 226 6636
      (1972) 94 4013
Tetr Lett (1970) 447

(cyclohexyl)-CH(COOEt)$_2$ $\xrightarrow{\begin{array}{l}NaH \quad LiAlH_4 \\ MeOCH_2CH_2OMe\end{array}}$ (cyclohexyl)- $\begin{array}{l}C=CH_2 \\ CH_2OH\end{array}$ 62%

JOC (1967) 32 113
Ber (1970) 103 3771

$EtCH_2CHCHPr$ (epoxide O) $\xrightarrow{\begin{array}{l}1 \ Ph_2Se_2\text{-}NaBH_4 \quad EtOH \\ 2 \ H_2O_2 \quad H_2O\end{array}}$ EtCH=CHCHPr   98%
                                                      OH

JACS (1973) 95 2697
Tetr Lett (1973) 1979

(cyclooctane with O-CHEt) $\xrightarrow{LiNEt_2}$ (cyclooctene with OH-CHEt)

JOC (1971) 36 1365
      (1969) 34 3583
LiNPr$_2$ JOC (1972) 37 2060
BuLi   JOC (1971) 36 3266
(i-Bu)$_2$AlH   Ber (1960) 93 2712
Li$_3$PO$_4$ (180°) or t-BuOK   Synthesis (1972) 194

C₆H₁₃I $\xrightarrow{\begin{array}{l}\text{1 PhSOCH}_2\text{CH=CMe}_2 \quad (\text{i-Pr})_2\text{NLi} \quad \text{THF} \\ \text{2 (MeO)}_3\text{P} \quad \text{MeOH}\end{array}}$ C₆H₁₃CH=CHCMe₂
$\qquad\qquad\qquad\qquad\qquad\qquad\qquad\qquad\qquad\qquad$ $\overset{|}{\text{OH}}$

$\qquad\qquad\qquad\qquad\qquad\qquad\qquad$ Tetr Lett (1973) 1385 1389
$\qquad\qquad\qquad\qquad\qquad\qquad\qquad$ JOC (1973) 38 2245
$\qquad\qquad\qquad\qquad\qquad\qquad\qquad$ Chem Comm (1972) 702

i-BuBr $\dashrightarrow$ i-BuCu·MgBr₂ $\xrightarrow{\begin{array}{l}\text{1 EtC≡CH} \\ \text{2 BrCH}_2\text{OCH}_2\text{CH}_2\text{Cl} \\ \text{3 BuLi}\end{array}}$ $\begin{array}{c}\text{Et} \quad\;\; \text{H} \\ \diagdown \;\;/ \\ \text{C=C} \\ /\;\;\diagdown \\ \text{i-Bu} \quad \text{CH}_2\text{OH}\end{array}$ 66%

$\qquad\qquad\qquad\qquad\qquad\qquad\qquad$ Tetr Lett (1973) 2407

$\qquad\qquad\qquad\qquad$ JCS (1952) 1610 642
$\qquad\qquad\qquad\qquad$ Org Synth (1955) Coll Vol 3 416
$\qquad\qquad\qquad\qquad$ JACS (1954) 76 4482

Me₂C=CH(CH₂)₂COMe $\xrightarrow{\text{CH}_2\text{=CHMgBr} \quad \text{THF}}$ Me₂C=CH(CH₂)₂$\overset{\overset{\displaystyle\text{OH}}{|}}{\underset{\underset{\displaystyle\text{Me}}{|}}{\text{C}}}$CH=CH₂ 83%

$\qquad\qquad\qquad\qquad\qquad$ Advances in Org Chem (1960) 2 1

37%

$\qquad\qquad\qquad\qquad$ Synthesis (1972) 575

Me₂CO $\xrightarrow{\begin{array}{l}\text{BuCH=CHAl(Bu-i)}_2 \\ \text{Et}_2\text{O}\end{array}}$ Me₂$\overset{}{\underset{\underset{\displaystyle\text{OH}}{|}}{\text{C}}}$CH=CHBu 42%

$\qquad\qquad\qquad\qquad\qquad$ Tetr Lett (1971) 4571

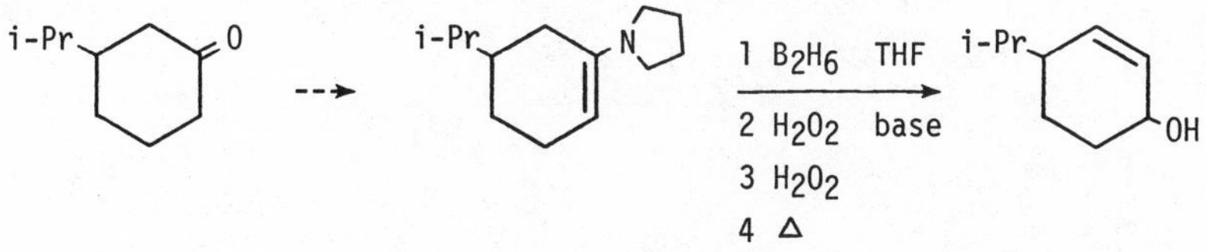

i-Pr
O
i-Pr
N
1 B_2H_6 THF
2 H_2O_2 base
3 H_2O_2
4 Δ
i-Pr
OH

Bull Soc Chim Fr (1971) 3978

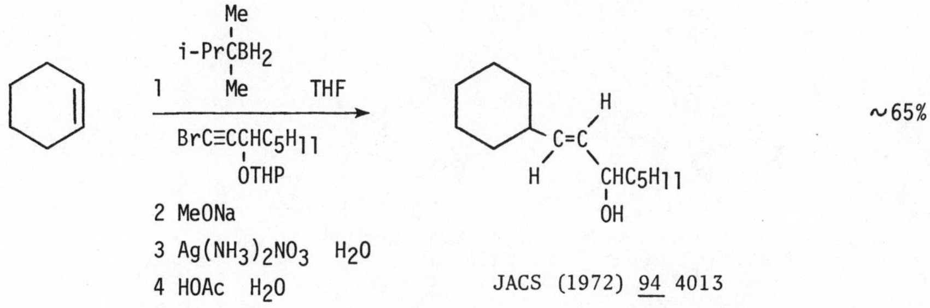

Me
i-PrCBH$_2$
1 Me THF
BrC≡CCHC$_5$H$_{11}$
OTHP
2 MeONa
3 $Ag(NH_3)_2NO_3$ H_2O
4 HOAc H_2O
H
C=C
H
CHC$_5$H$_{11}$
OH
~65%

JACS (1972) 94 4013

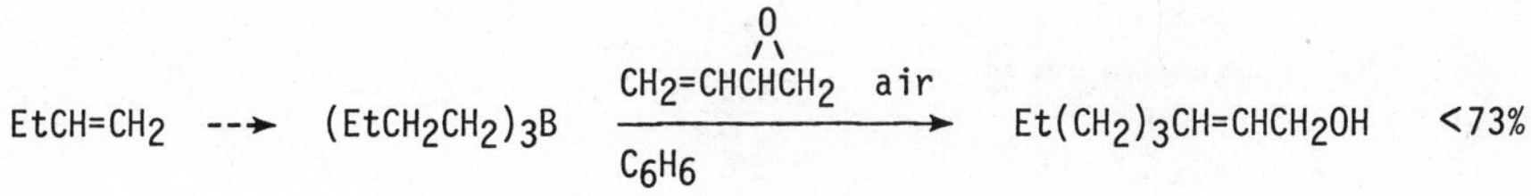

$EtCH=CH_2$
$(EtCH_2CH_2)_3B$
O
$CH_2=CHCHCH_2$ air
C_6H_6
$Et(CH_2)_3CH=CHCH_2OH$
<73%

JACS (1971) 93 2792

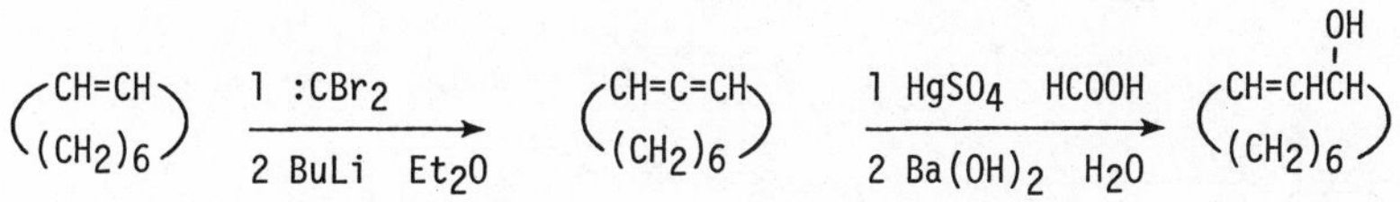

CH=CH
$(CH_2)_6$
1 :CBr_2
2 BuLi Et_2O
CH=C=CH
$(CH_2)_6$
1 $HgSO_4$ HCOOH
2 $Ba(OH)_2$ H_2O
OH
CH=CHCH
$(CH_2)_6$

Bull Soc Chim Fr (1964) 3273

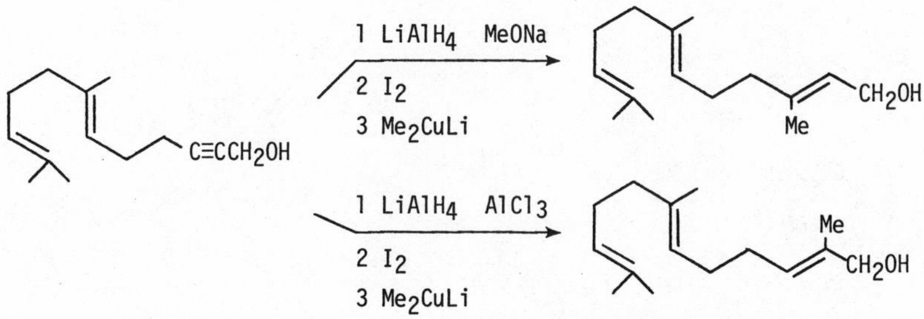

C≡CCH$_2$OH
1 $LiAlH_4$ MeONa
2 I_2
3 Me_2CuLi
CH$_2$OH
Me
1 $LiAlH_4$ $AlCl_3$
2 I_2
3 Me_2CuLi
Me
CH$_2$OH

JACS (1967) 89 4245
(1968) 90 5618

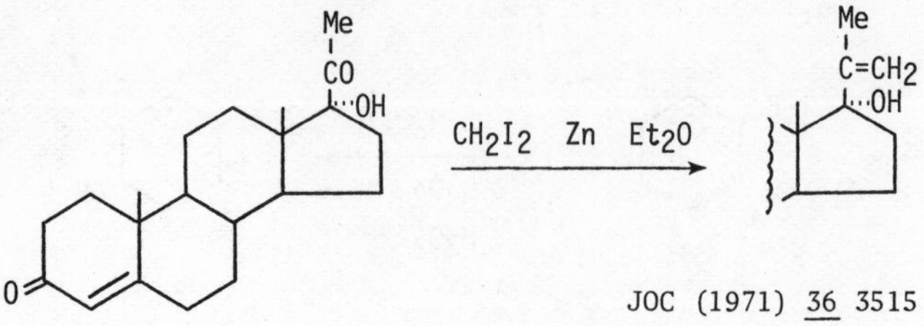
Me
CO
OH
O
CH_2I_2 Zn Et_2O
Me
$C=CH_2$
OH

93%

JOC (1971) 36 3515

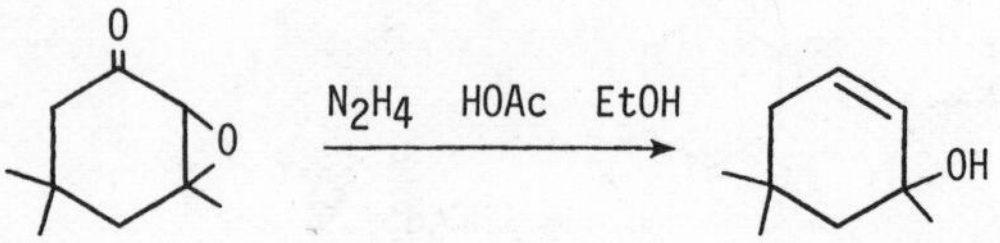
O
O
N_2H_4 HOAc EtOH
OH

66%

JOC (1961) 26 3615
(1968) 33 3347

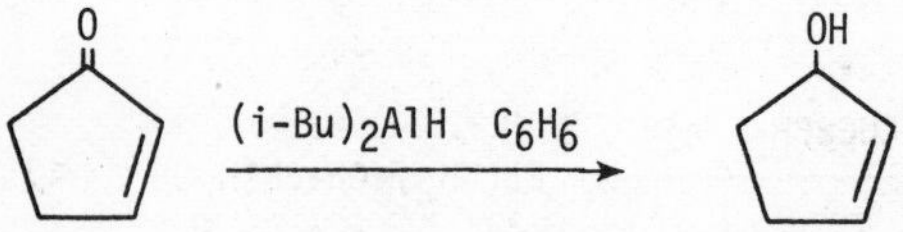
O
$(i-Bu)_2AlH$ C_6H_6
OH

83%

Chem Comm (1970) 213

AlH_3 JOC (1969) 34 2206

Et_2SiH, $(Ph_3P)_3RhCl$ Tetr Lett (1972) 5035

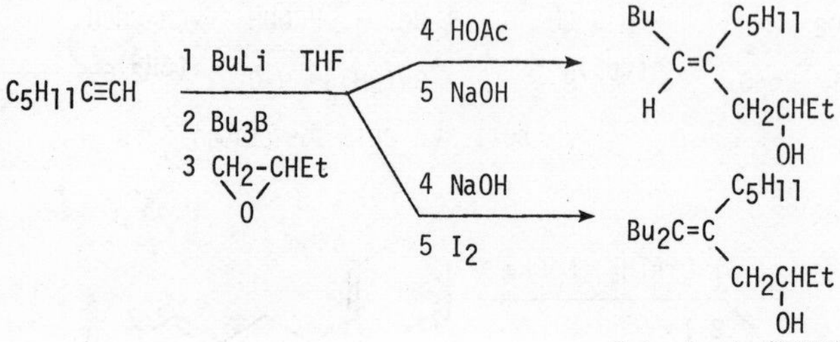
$C_5H_{11}C≡CH$
1 BuLi THF
2 Bu_3B
3 CH_2-CHEt
O
4 HOAc
5 NaOH
4 NaOH
5 I_2
Bu
C_5H_{11}
C=C
H
CH_2CHEt
OH
C_5H_{11}
$Bu_2C=C$
CH_2CHEt
OH

75%

34%

Tetr Lett (1973) 2741

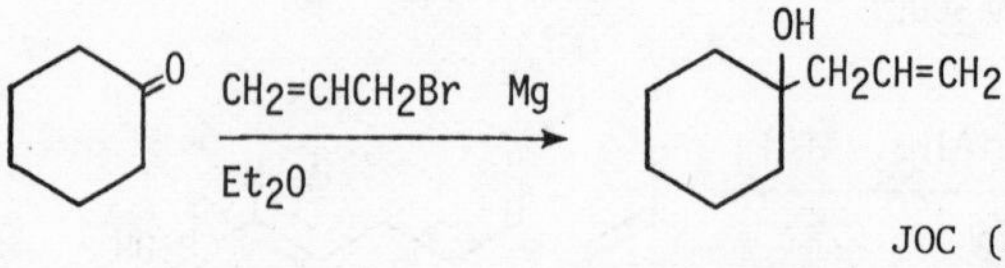
O
$CH_2=CHCH_2Br$ Mg
Et_2O
OH
$CH_2CH=CH_2$

(One-step procedure) 95%

JOC (1973) 38 326
(1963) 28 3269

$(CH_2=CHCH_2)_2Cd$ Bull Soc Chim Fr (1969) 4038

π-Methallylnickel bromide JACS (1967) 89 2755

Ph₃P with ring structure

furan-CHO → furan-CH=CH(CH₂)₂OH   52%

Ph_3P (cyclic)

furan-CHO $\longrightarrow$ furan-CH=CH(CH$_2$)$_2$OH   52%

JCS C (1968) 2448

BuBr $\xrightarrow[\;2\;Cl]{1\;Mg\;\;Et_2O}$ Cl / Bu O (pyran ring) $\xrightarrow{Na\;\;Et_2O}$ BuCH=CH(CH$_2$)$_3$OH   71%

trans

J Med Chem (1967) 10 533
(1971) 14 236

cyclohexanone =O $\xrightarrow[\substack{2\;SOCl_2\;\;Pyr\\3\;HCl\;\;H_2O\;\;EtOH}]{1\;Li(CH_2)_3OCHMe\;(OEt)}$ cyclohexene-(CH$_2$)$_3$OH   ~73%

JOC (1972) 37 1947

Pr$_2$C=CH(CH$_2$)$_2$CHO $\xrightarrow[THF]{1\;Ph_3P=CH(CH_2)_4OLi}$ Pr$_2$C=CH(CH$_2$)$_2$CH=CH(CH$_2$)$_4$OH   ~72%

Tetrahedron (1971) 27 5979

Also via: Acetylenic alcohols (Section 302)

Section 333   Aldehyde — Aldehyde
                 ○○○○○○○○○○○○○○○○○○○○○○

Dialdehydes

PhCH$_2$CH$_2$CHO $--\rightarrow$ PhCH$_2$CH$_2$CH(OEt)$_2$ $\xrightarrow[2\;NaOH\;\;H_2O]{1\;COCl_2\;\;HCONMe_2}$ PhCH$_2$CH(CHO)$_2$

Coll Czech (1958) 23 452
Synthesis (1973) 604
Tetr Lett (1973) 3979

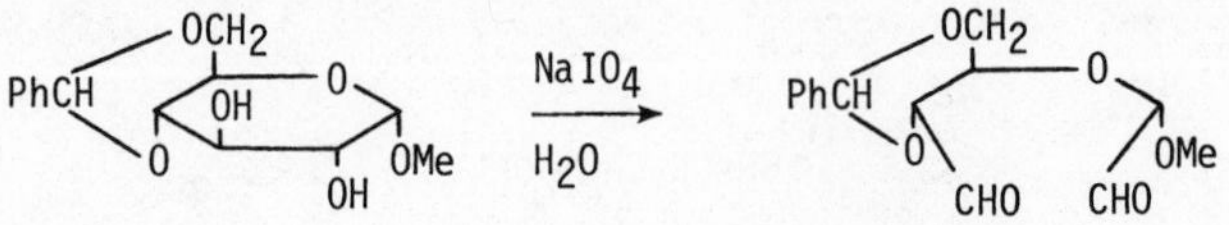

JCS (1959) 2441
Advances in Carbohydrate Chem (1961) 16 105

HIO$_4$ Ber (1956) 89 2224
Org React (1944) 2 341

Pb(OAc)$_4$ JACS (1949) 71 3310

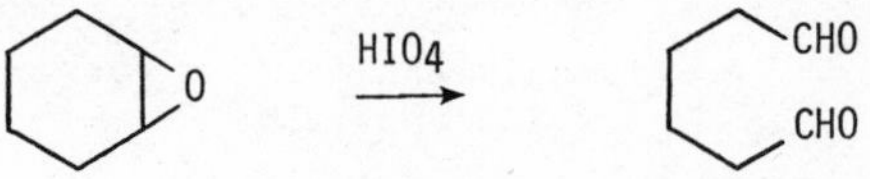

82%

Tetr Lett (1973) 4599

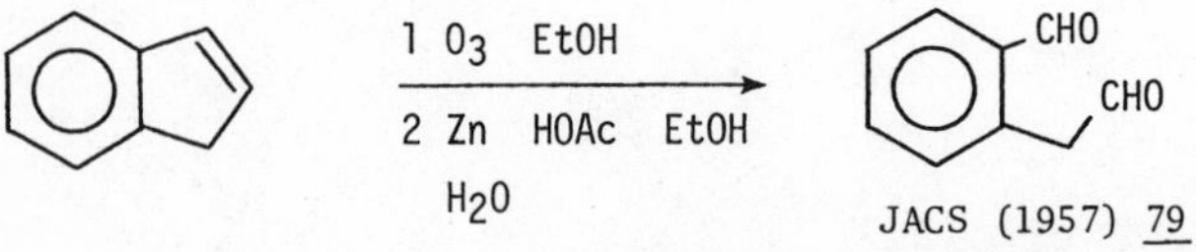

97%

JACS (1957) 79 3165
NaIO$_4$, OsO$_4$ JOC (1956) 21 478
RuO$_4$ JACS (1958) 80 6682

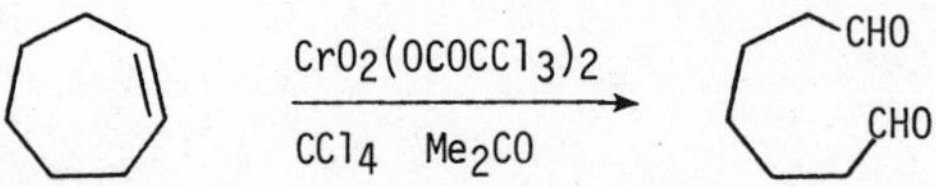

68%

Annalen (1962) 659 20

Me$_2$C(CH$_2$)$_2$COOH $\xrightarrow[\text{Na}_2\text{CO}_3 \quad \text{MeOH}]{\text{Electrolysis}}$ Me$_2$C(CH$_2$)$_4$CMe$_2$ 30%
   |CHO   CHO   CHO

Bull Soc Chim Fr (1970) 183

Section 334   Aldehyde — Amide

No examples

Section 335   Aldehyde — Amine

Aminoaldehydes

Me_2CHCHO $\xrightarrow{\begin{array}{c}O\!\!\diagdown\!\!NH\cdot HCl \quad HCHO\end{array}}$ $O\diagdown NCH_2\overset{\overset{\displaystyle Me}{|}}{\underset{\underset{\displaystyle Me}{|}}{C}}CHO$ 85%

JACS (1951) <u>73</u> 685
   (1948) <u>70</u> 2592
Ber (1932) <u>65</u> 378
Org React (1942) <u>1</u> 303

$PhCHO \dashrightarrow PhCH=N^+\diagdown \; Cl^-$ $\xrightarrow[\text{2 Chloramine-T}]{1 \; LiCH\diagup^S_S}$ $PhCHCHO$ <46%

Tetr Lett (1972) 2991

$Me_2C=CH_2 \dashrightarrow Me_2\overset{\overset{\displaystyle }{|}}{\underset{\underset{\displaystyle Cl}{|}}{C}}CH_2NO$ $\xrightarrow[\text{2 } NaNO_2 \quad HCl \quad H_2O]{1 \; EtNH_2 \quad EtOH}$ $Me_2\overset{\overset{\displaystyle }{|}}{\underset{\underset{\displaystyle EtNH}{|}}{C}}CHO$

Zh Obshch Khim (1960) <u>30</u> 805
(Chem Abs <u>55</u> 556)

$Me_2\overset{\overset{\displaystyle }{|}}{\underset{\underset{\displaystyle Br}{|}}{C}}CHO$ $\xrightarrow{Me_2NH \quad H_2O}$ $Me_2\overset{\overset{\displaystyle }{|}}{\underset{\underset{\displaystyle NMe_2}{|}}{C}}CHO$ 32%

JACS (1948) <u>70</u> 2592

Section 336   Aldehyde — Ester

Esters of hydroxyaldehydes and esters of carboxyaldehydes

$C_5H_{11}CH_2CHO$ $\xrightarrow[C_6H_6]{Pb(OAc)_4 \quad BF_3}$ $C_5H_{11}\underset{\underset{OAc}{|}}{C}HCHO$

Bull Soc Chim Fr (1968) 4083
Synthesis (1973) 567

$Tl(OAc)_3$ JOC (1968) 33 3359

RCH_2CHO - - -► $R\underset{\underset{I}{|}}{C}HCHO$ $\xrightarrow{AgOAc \quad DMF}$ $R\underset{\underset{OAc}{|}}{C}HCHO$

Bull Soc Chim Fr (1968) 4083

$Me_2\underset{\underset{Cl}{|}}{C}CHO$ $\xrightarrow[Et_2O]{MeONa}$ $Me_2\overset{\overset{O}{\diagup\diagdown}}{C}-CHOMe$ $\xrightarrow[2 \ Et_3N]{1 \ HOAc \quad Et_2O}$ $Me_2\underset{\underset{OAc}{|}}{C}CHO$ 10%

JACS (1957) 79 3448

$Ph\underset{\underset{OAc}{|}}{C}HCOCl$ $\xrightarrow[diglyme]{LiAlH_4 \quad t\text{-}BuOH}$ $Ph\underset{\underset{OAc}{|}}{C}HCHO$ 55%

Bull Soc Chim Fr (1968) 4083

Me_2CHCHO - - -► $Me_2C=CHN$⟨pyrrolidine⟩ $\xrightarrow[2 \ HCl \quad EtOH \quad H_2O]{1 \ N_2CHCOOEt \quad CuCl}$ $Me_2\underset{\underset{CH_2COOEt}{|}}{C}CHO$

Synth Comm (1973) 3 255

BuBr $\xrightarrow{\begin{array}{l} 1 \quad \text{(morpholine-type ring with } N=C-CH_2COOEt) \quad \text{t-BuOK} \quad \text{THF} \\ 2 \quad \text{HCl} \quad H_2O \\ 3 \quad NaBH_4 \quad \text{EtOH} \quad \text{THF} \quad \text{pH 5-6} \\ 4 \quad (COOH)_2 \end{array}}$ BuCHCHO
　　　　　　　　　　　　　　　　　　　　　　　　　　　　　　|
　　　　　　　　　　　　　　　　　　　　　　　　　　　　COOEt

67%

JOC (1973) <u>38</u> 36

$$\underset{CH_2=CCOOMe}{\overset{Me}{|}} \xrightarrow[\text{Rh}_2\text{O}_3 \quad Bu_3P \quad C_6H_6]{\text{CO} \quad H_2 \quad (600 \text{ atmos})} \underset{\underset{CHO}{|}}{\overset{\overset{Me}{|}}{MeCCOOMe}}$$

94%

Brennst-Chem (1967) <u>48</u> 46
(Chem Abs <u>66</u> 85430)

$$EtCH_2COOMe \xrightarrow[2 \quad HClO_4 \quad MeCN \quad H_2O]{1 \quad \overset{\overset{SOMe}{|}}{CH_2=CSMe} \quad LiN(Pr-i)_2 \quad THF} \underset{\underset{CH_2CHO}{|}}{EtCHCOOMe}$$

90%

Tetr Lett (1973) 4711 4715

$$MeCH=CHCOOMe \xrightarrow[2 \quad HClO_4]{1 \quad EtSCH_2SOEt \quad BuLi \quad THF} \underset{\underset{CHO}{|}}{MeCHCH_2COOMe}$$

Tetr Lett (1973) 3271

$$(CH_2)_{10}\overset{CO}{\underset{CH_2}{\big\rangle}} \xrightarrow[2 \quad TsOH]{1 \quad HC(OEt)_3 \quad H_2SO_4 \quad EtOH} (CH_2)_{10}\overset{COEt}{\underset{CH}{\big\rangle}} \xrightarrow[\begin{array}{c} 2 \quad H_2 \\ Pd-BaCO_3 \end{array}]{1 \quad O_3 \quad EtOH} \underset{\underset{CHO}{|}}{\overset{\overset{COOEt}{|}}{(CH_2)_{10}}}$$

Annalen (1962) <u>656</u> 97

$$(CH_2)_6\overset{CO}{\underset{CHOH}{\big\rangle}} \xrightarrow[2 \quad MeOH \quad H_2SO_4]{1 \quad Pb(OAc)_4 \quad MeOH} \underset{\underset{CH(OMe)_2}{|}}{\overset{\overset{COOMe}{|}}{(CH_2)_6}} \dashrightarrow \underset{\underset{CHO}{|}}{\overset{\overset{COOMe}{|}}{(CH_2)_6}}$$

~65%

JACS (1952) <u>74</u> 5324

Also via: Carboxyaldehydes (Section 314)

Section 337   Aldehyde — Ether, Epoxide

Aryloxyaldehydes, alkoxyaldehydes and epoxyaldehydes

PhOH $\xrightarrow{\begin{array}{l}1\ \text{EtONa}\\[2pt]2\ \text{BrCH}_2\text{CH(OEt)}_2\\[2pt]3\ \text{H}_2\text{SO}_4\ \ \text{H}_2\text{O}\end{array}}$ PhOCH$_2$CHO

JCS (1937) 1057

MeCH=CHCHO $\xrightarrow{\text{PrOH}\ \ \text{NaOH}}$ MeCHCH$_2$CHO   45%
                                                             |
                                                            OPr

JCS (1952) 4083

i-PrCH$\big\langle^{O}_{O}\big\rangle$X $\xrightarrow[\text{(vapor phase)}]{350°}$ i-PrCH$_2$ $\big\langle^{O}\big\rangle$X$_{CHO}$ 50%

JACS (1960) $\underline{82}$ 6419

PhCH=CHCHO $\xrightarrow[\text{MeOH}]{\text{t-BuOOH}}$ PhCHCHCHO   73%
                                                        \O/

JOC (1960) $\underline{25}$ 275

H$_2$O$_2$ JACS (1956) $\underline{78}$ 3087

Section 338   Aldehyde — Halide

Haloaldehydes

t-BuC≡CH $\xrightarrow[\text{H}_2\text{SO}_4\ \ \text{H}_2\text{O}]{\text{Et}_2\text{NCl}\ \ \text{h}\nu}$ t-BuCHCHO   60%
                                                              |
                                                             Cl

JOC (1967) $\underline{32}$ 3263

$$\text{cyclohexane-CHO} \xrightarrow{\text{Br}_2 \quad \text{CHCl}_3} \text{Br-cyclohexane-CHO}$$

80%

JCS (1949) 737

Br$_2$, CaCO$_3$ JACS (1957) $\underline{79}$ 456 3448

IBr   Chem Comm (1968) 849

CuBr$_2$ JOC (1965) $\underline{30}$ 587

$$\text{EtCH}_2\text{CHO} \xrightarrow{\text{CuCl}_2 \quad \text{DMF}} \text{EtCHCHO} \atop \qquad\qquad\quad \overset{|}{\text{Cl}}$$

JOC (1965) $\underline{30}$ 587
     (1967) $\underline{32}$ 4008

SO$_2$Cl$_2$ JACS (1957) $\underline{79}$ 3448
              (1954) $\underline{76}$ 2695

JACS (1957) $\underline{79}$ 1115

$$\text{C}_5\text{H}_{11}\text{CH}_2\text{CHO} \xrightarrow[\text{KOAc}]{\text{Ac}_2\text{O}} \text{C}_5\text{H}_{11}\text{CH=CHOAc} \xrightarrow[\substack{2\ \text{MeOH} \\ 3\ \text{HCl}\ \text{H}_2\text{O}}]{1\ \text{Br}_2\quad \text{CCl}_4} \text{C}_5\text{H}_{11}\underset{\overset{|}{\text{Br}}}{\text{CHCHO}}$$

~35%

Org Synth (1955) Coll Vol 3 127

Iodoaldehydes   JACS (1953) $\underline{75}$ 3493

$$\text{EtCH}_2\text{CHO} \dashrightarrow \text{EtCH=CHOEt} \xrightarrow[\text{MeOH}]{\text{N-Bromophthalimide}} \text{EtCH}\underset{\overset{|}{\text{Br}}}{\overset{\overset{|}{\text{OMe}}}{\text{CHOEt}}} \xrightarrow{\text{Acid}} \text{EtCH}\underset{\overset{|}{\text{Br}}}{\text{CHO}}$$

Tetr Lett (1972) 4055
Br$_2$; NaHCO$_3$, H$_2$O   JACS (1955) $\underline{77}$ 6365

$$Ph_2CO \xrightarrow[\text{2 LiCl}]{\text{1 LiCHCl}_2 \quad \text{THF}} Ph_2CCHO \atop | \atop Cl \qquad \qquad 74\%$$

Tetr Lett (1969) 2181
(1972) 4661
Annalen (1966) $\underline{691}$ 33

$$MeCH=CHMe \xrightarrow[\substack{\text{2 CH}_2\text{=CCHO} \\ | \\ Br}]{\text{1 B}_2\text{H}_6 \quad \text{THF}} MeCH_2CHCH_2CHCHO \atop \qquad | \qquad | \atop \qquad Me \quad Br \qquad \qquad 81\%$$

JACS (1968) $\underline{90}$ 4165

Section 339   Aldehyde — Ketone

Ketoaldehydes

Ber (1971) $\underline{104}$ 2475
Helv (1943) $\overline{26}$ 2050
Rec Trav Chim (1971) $\underline{90}$ 429

$$PhCH_2COCl \dashrightarrow PhCH_2COCHN_2 \xrightarrow[\substack{\text{2 EtSNa} \\ \text{3 Br}_2 \quad \text{HCl} \quad \text{HOAc} \\ \text{H}_2\text{O}}]{\text{1 EtSCl} \quad \text{Et}_2\text{O}} PhCH_2COCHO$$

Ber (1957) $\underline{90}$ 1230

$$PhCH_2CH_2COCl \dashrightarrow PhCH_2CH_2COCHN_2 \xrightarrow[\substack{\text{2 NaNO}_2 \quad \text{HCl} \\ \text{H}_2\text{O} \quad \text{THF}}]{\text{1 Ph}_3\text{P}} PhCH_2CH_2COCHO$$

Ber (1963) $\underline{96}$ 2259
(1959) $\underline{92}$ 1345

PhCOOEt $\xrightarrow[\text{t-BuOH}]{\text{Me}_2\text{SO}\quad \text{t-BuOK}}$ PhCOCH$_2$SOMe $\xrightarrow[\substack{2\ \text{Cu(OAc)}_2\\ \text{CHCl}_3}]{1\ \text{HCl}\quad \text{H}_2\text{O}}$ PhCOCHO　　64-73%

Org Synth (1968) 48 109
JOC (1967) 32 2786
JACS (1966) 88 5498

PrCOOEt $\xrightarrow[\text{BuLi}\quad \text{THF}]{\text{MeSOCH}_2\text{SMe}}$ PrCOCHSOMe $\xrightarrow{\text{Hydrolysis}}$ PrCOCHO
　　　　　　　　　　　　　　　 $\overset{|}{\text{SMe}}$

Tetr Lett (1973) 4707

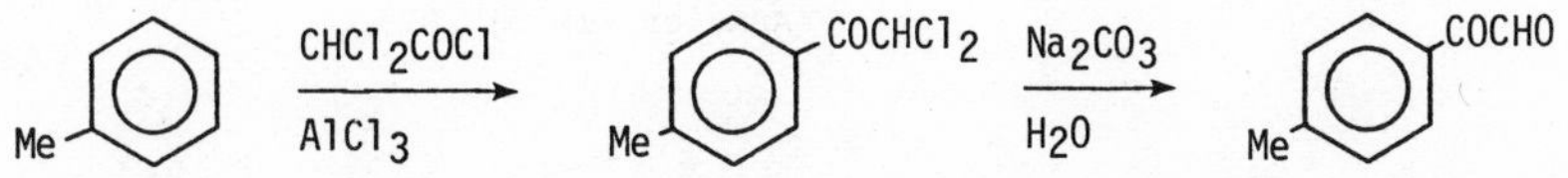

CHCl$_2$COCl / AlCl$_3$; Na$_2$CO$_3$ / H$_2$O

85%

Tetr Lett (1971) 199

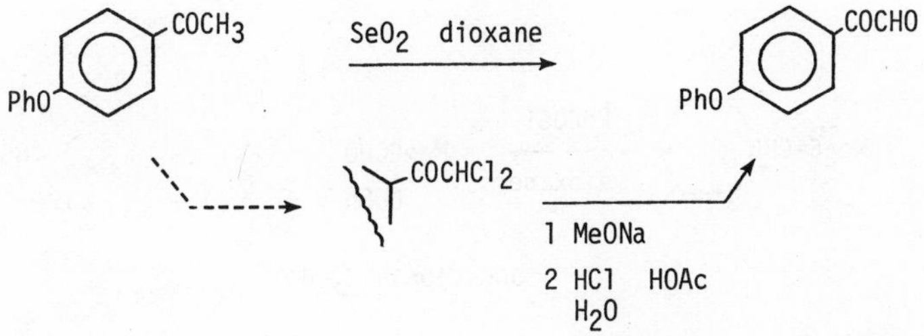

SeO$_2$　dioxane

~71%

1 MeONa
2 HCl　HOAc
　H$_2$O

J Med Chem (1964) 7 255
Org Synth (1943) Coll Vol 2 509
Org React (1949) 5 331

PhCOCH$_3$ $--\rightarrow$ PhCOCH$_2$Br $\xrightarrow{\text{Me}_2\text{SO}}$ PhCOCHO　　<71%

JACS (1957) 79 6562
α-picoline-l-oxide　Yakugaku Zasshi (1964) 84 287
(Chem Abs 61 638)

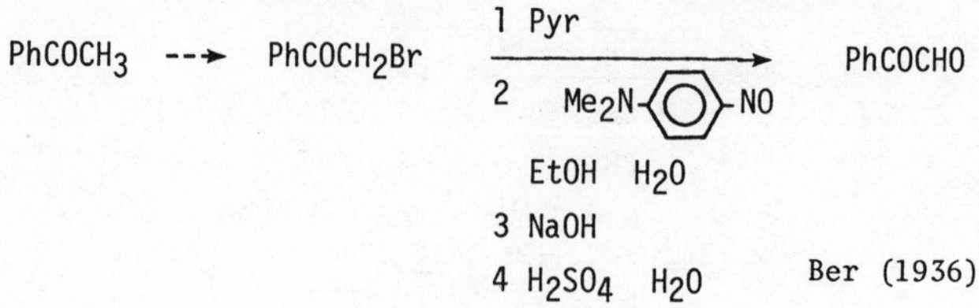

$PhCOCH_3$ --→ $PhCOCH_2Br$

1 Pyr

2 Me_2N-⟨○⟩-NO

EtOH H_2O

3 NaOH

4 H_2SO_4 H_2O Ber (1936) 69 2006

→ $PhCOCHO$ 87%

Air $Cu(OAc)_2$

MeOH

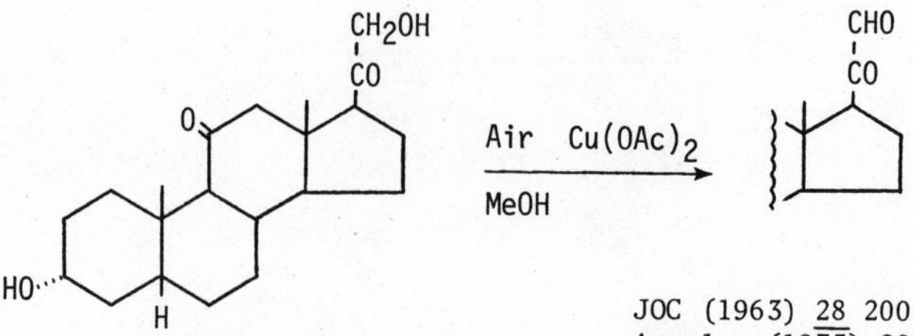

94%

JOC (1963) 28 2001
Annalen (1973) 2078

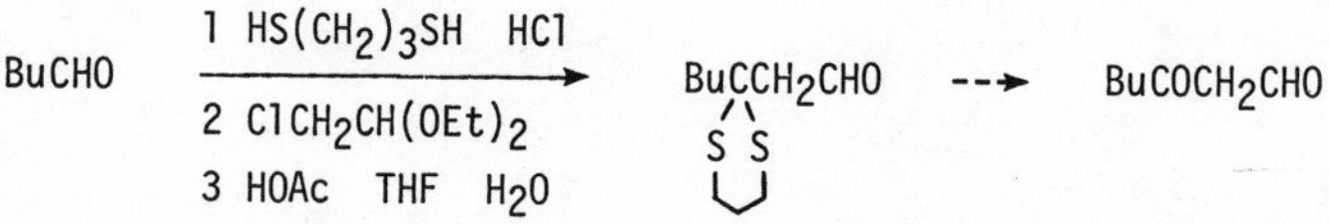

$BuCHO$

1 $HS(CH_2)_3SH$ HCl

2 $ClCH_2CH(OEt)_2$

3 HOAc THF H_2O

→ $BuCCH_2CHO$ (dithiane) --→ $BuCOCH_2CHO$ <47%

Tetr Lett (1972) 3735

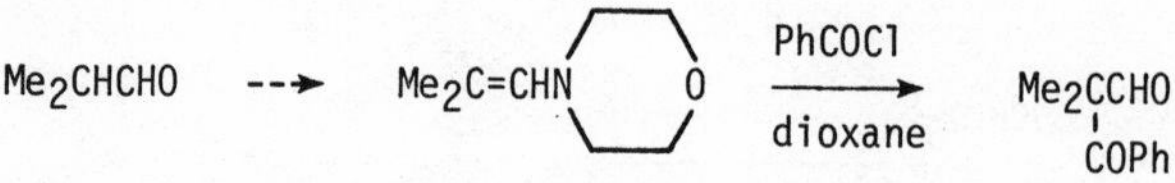

Me_2CHCHO --→ $Me_2C=CHN$⟨○⟩

$PhCOCl$

dioxane

→ Me_2CCHO / $COPh$ <86%

JOC (1967) 32 404

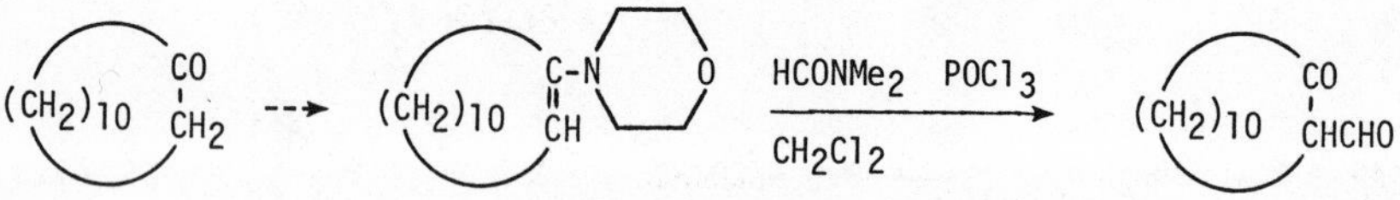

$(CH_2)_{10}$ CO/CH_2 --→ $(CH_2)_{10}$ $C-N$⟨○⟩ / CH

$HCONMe_2$ $POCl_3$

CH_2Cl_2

→ $(CH_2)_{10}$ CO/$CHCHO$ <59%

Angew (1965) 77 380
(Internat Ed 4 358)

90%

JOC (1971) $\underline{36}$ 3070
HCOOEt, MeONa   JCS Perkin I (1972) 1721
JOC (1965) $\underline{30}$ 2502

$C_{17}H_{35}COOCH=CH_2$ $\xrightarrow{\text{AlCl}_3 \quad \text{hexane}}$ $C_{17}H_{35}COCH_2CHO$ 60%

Tetr Lett (1969) 5205

87%

Tetr Lett (1973) 4711

i-PrCH$_2$CHO $\xrightarrow{\text{Piperidine}}$ i-PrCH=CHN $\xrightarrow[\text{2 (COOH)}_2 \quad \text{H}_2\text{O}]{\text{1 CH}_2\text{=CHCOMe}}$ i-PrCHCHO
                                                                       CH$_2$CH$_2$COMe

Chem Comm (1965) 340   60%

Section 340   Aldehyde — Nitrile

Cyanoaldehydes

BuCHCHO $\xrightarrow[\text{KOH} \quad \text{H}_2\text{O}]{\text{CH}_2\text{=CHCN}}$ BuCCHO   77%
  Et   Et   JACS (1944) $\underline{66}$ 56
                                                             JCS (1958) $\overline{3986}$
                                                             Org React (1949) $\underline{5}$ 79

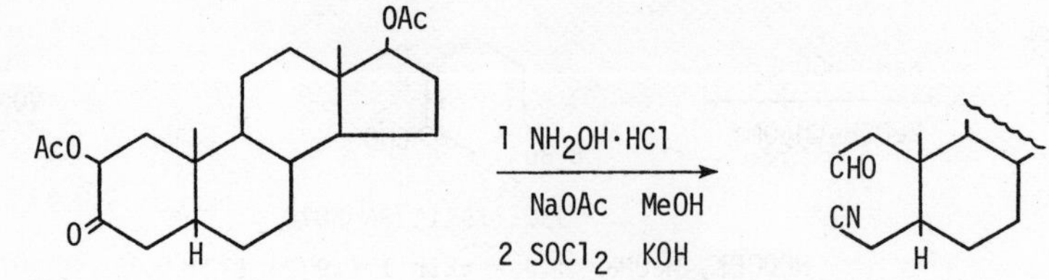

$$1 \quad NH_2OH \cdot HCl$$

NaOAc   MeOH

$$2 \quad SOCl_2 \quad KOH$$

49-56%

Tetr Lett (1972) 261

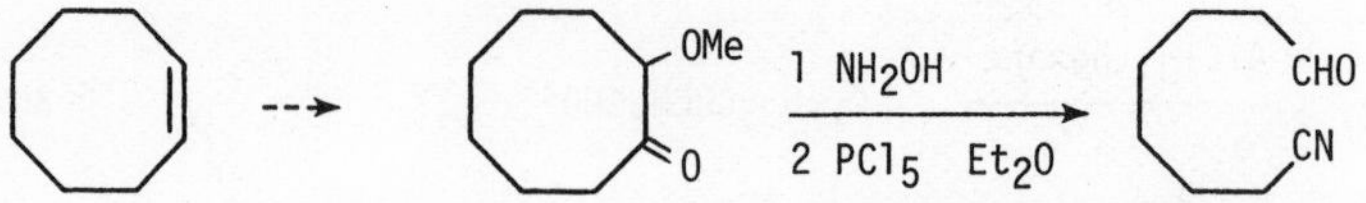

$$1 \quad NH_2OH$$

$$2 \quad PCl_5 \quad Et_2O$$

JACS (1966) 88 3168
JOC (1962) 27 29

Section 341   Aldehyde ── Olefin

αβ-Olefinic aldehydes . . . . page 328-333
Other olefinic aldehydes 334-335

For the oxidation of allylic alcohols to olefinic aldehydes see section 48
Vol 1 (Aldehydes from Alcohols)

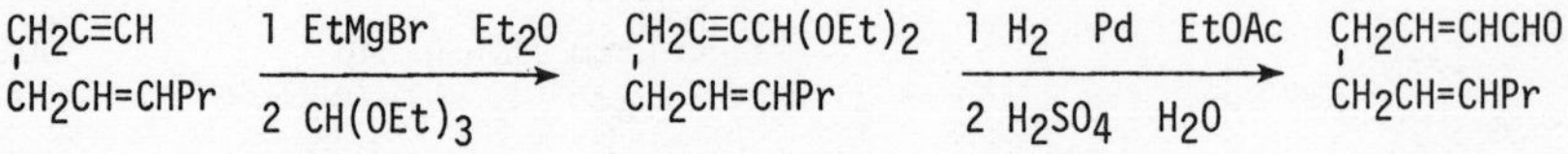

$$CH_2C\equiv CH \atop CH_2CH=CHPr$$

$$1 \quad EtMgBr \quad Et_2O$$

$$2 \quad CH(OEt)_3$$

$$CH_2C\equiv CCH(OEt)_2 \atop CH_2CH=CHPr$$

$$1 \quad H_2 \quad Pd \quad EtOAc$$

$$2 \quad H_2SO_4 \quad H_2O$$

$$CH_2CH=CHCHO \atop CH_2CH=CHPr$$

JCS (1955) 4244 1007

$$EtC\equiv CEt$$

$$1 \quad ClSO_2CNO \quad CH_2Cl_2$$

$$2 \quad LiAlH_4$$

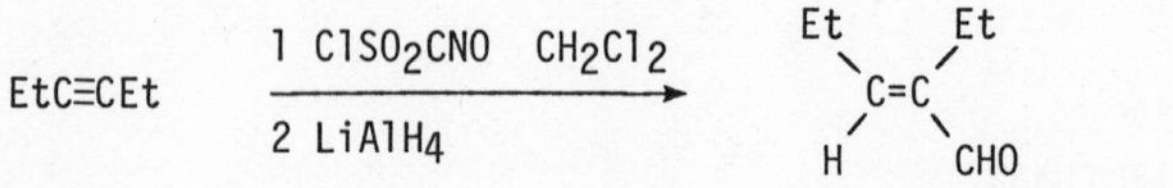

29%

Tetr Lett (1970) 27

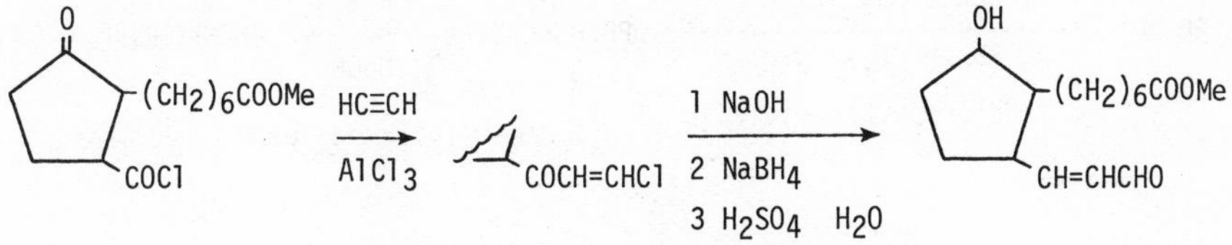

Tetr Lett (1966) 465

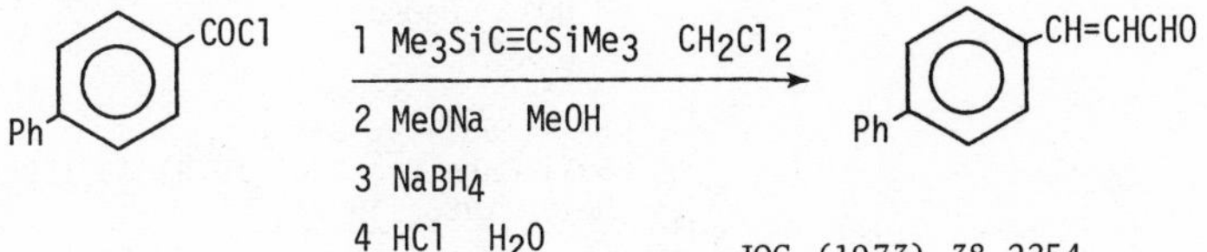

84%

JOC (1973) <u>38</u> 2254

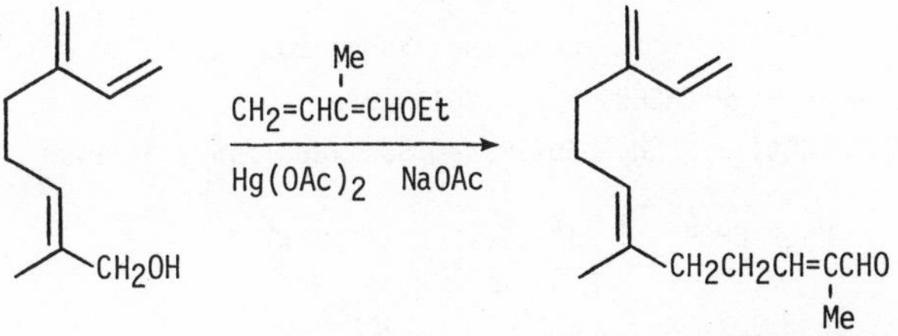

<43%

Chem Comm (1967) 947
JCS C (1970) 220
JCS Perkin I (1973) 2741

Review: The Aldol Condensation   Org React (1968) <u>16</u> 1

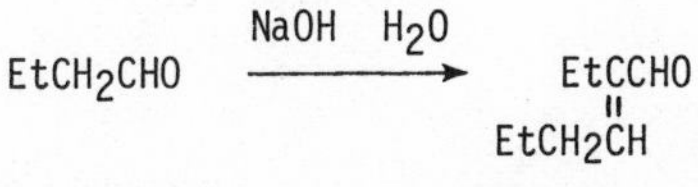

$EtCH_2CHO$ $\xrightarrow{\text{NaOH} \quad H_2O}$ EtCCHO   86%
                              ‖
                            $EtCH_2CH$ Org React (1968) <u>16</u> 1

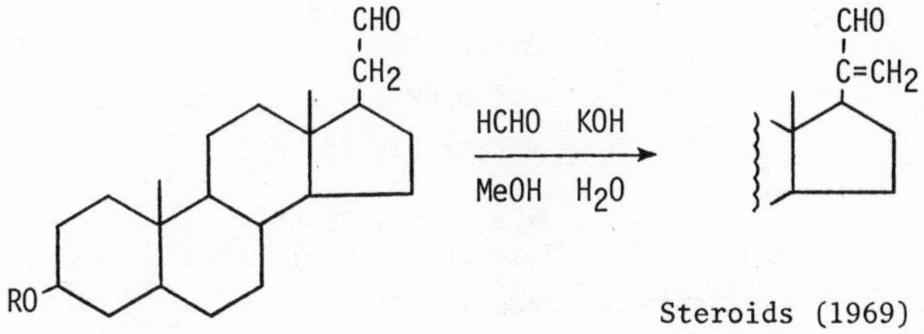

77%

Steroids (1969) <u>14</u> 637

PhCHO $\xrightarrow{\begin{array}{l}\text{1 ClMgC}\equiv\text{CSiMe}_3\ \text{Et}_2\text{O}\\ \text{2 H}_2\ \text{Pd-C Pyr EtOAc}\\ \text{3 m-Chloroperbenzoic acid}\end{array}}$ PhCHCHCHSiMe$_3$ $\xrightarrow[\text{MeOH}]{\text{H}_2\text{SO}_4}$ PhCH=CHCHO   41%

(with epoxide O and OH shown on intermediate)

JACS (1971) <u>93</u> 2080

1 CH$_2$=CHOEt
BF$_3$·Et$_2$O
2 HOAc   NaOAc
H$_2$O

<41%

Ber (1961) <u>94</u> 838
Bull Chem Soc Jap (1970) <u>43</u> 1586
Helv (1956) <u>39</u> 249

C$_6$H$_{13}$CHO $\xrightarrow{\text{Ph}_3\text{P=CCHO} \quad \text{C}_6\text{H}_6}$ C$_6$H$_{13}$CH=CCHO   72%

(Me substituents shown on ylide and product)

Chem Ind (1960) 202

Ph$_3$P=CHCHO   JCS (1961) 1266

(EtO)$_2$POCH$_2$CH(OEt)$_2$, NaNH$_2$ Bull Chem Soc Jap (1962) <u>35</u> 1498

Ph$_3$P=CHCH(O–CH$_2$CH$_2$–O)   JCS Perkin I (1974) 37

Br$_2$ / CHCl$_3$; PhNEt$_2$ 60%

JCS (1949) 737
Bull Soc Chim Fr (1971) 4012
(1968)   283

C$_8$H$_{17}$CH$_2$CH$_2$CHO $\xrightarrow[\text{KOAc}]{\text{Ac}_2\text{O}}$ C$_8$H$_{17}$CH$_2$CH=CHOAc $\xrightarrow{\begin{array}{l}\text{1 Br}_2\ \text{CCl}_4\\ \text{2 MeOH}\\ \text{3 KOH MeOH}\\ \quad\text{t-BuOH}\\ \text{4 Citric acid H}_2\text{O}\end{array}}$ C$_8$H$_{17}$CH=CHCHO   9%

JACS (1957) <u>79</u> 889
Tetr Lett (1969) 3139

Via enamine JACS (1957) <u>79</u> 1115

$C_9H_{19}CH_2CH_2CHO$ $\xrightarrow[\text{2 } H_2O_2 \quad H_2O \quad THF]{\text{1 PhSeCl} \quad HCl. \quad EtOAc}$ $C_9H_{19}CH=CHCHO$

JACS (1973) 95 6137

$C_5H_{11}Br$ $\xrightarrow[\text{2 } HgCl_2 \quad MeCN \quad H_2O]{\overset{\overset{\displaystyle OMe}{|}}{\text{1 MeSCH}_2CHCH_2SMe} \quad (i\text{-Pr})_2NLi}$ $C_5H_{11}CH=CHCHO$ 52%

JACS (1971) 93 1724

$C_8H_{17}Br$ $\xrightarrow[\text{2 } \triangle \quad CaCO_3 \quad MeOCH_2CH_2OMe]{\overset{\overset{\displaystyle Me}{|}}{\text{1 CH}_2=CHCH_2SCH=CH_2} \quad EtCHLi}$ $C_8H_{17}CH=CHCHO$ 57%

JACS (1973) 95 2693

PhBr $\xrightarrow[\substack{\text{2 } Me_2NCH=CHCH=NMe_2 \quad ClO_4 \\ + \qquad - \\ \text{3 } HCl \quad H_2O}]{\text{1 Mg} \quad Et_2O}$ PhCH=CHCHO   53%

Ber (1970) 103 222

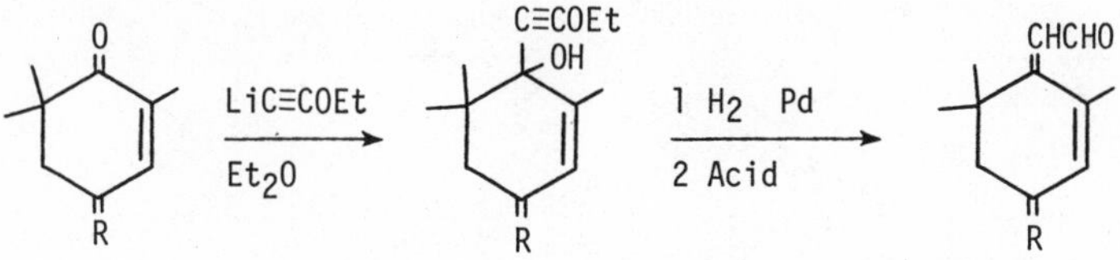

Helv (1967) 50 1606

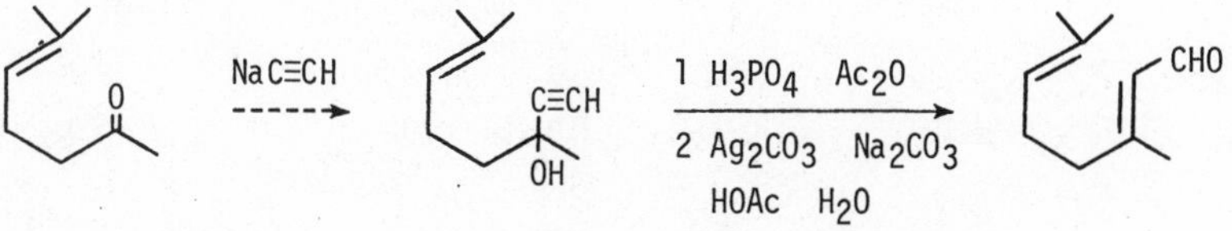

~61%

Helv (1959) 42 1945

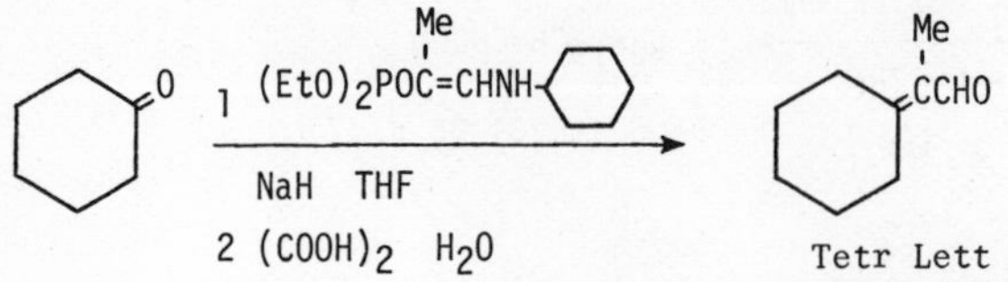

1 (EtO)$_2$POC=CHNH— (with Me)

NaH   THF

2 (COOH)$_2$ H$_2$O

→ (cyclohexane with CCHO, Me)

86%

Tetr Lett (1968) 4359
JCS C (1969) 460

(EtO)$_2$POCH$_2$CH=CHOEt   Bull Soc Chim Fr (1970) 1369

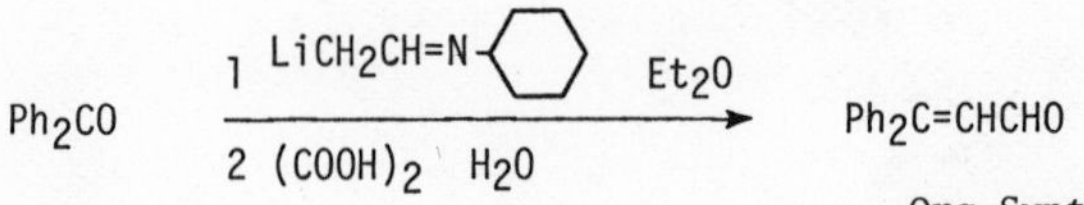

Ph$_2$CO

1 LiCH$_2$CH=N—◯   Et$_2$O

2 (COOH)$_2$ H$_2$O

→ Ph$_2$C=CHCHO   ~70%

Org Synth (1970) 50 66
Angew (1968) 80 8
(Internat Ed 7 7)
JOC (1969) 34 1122

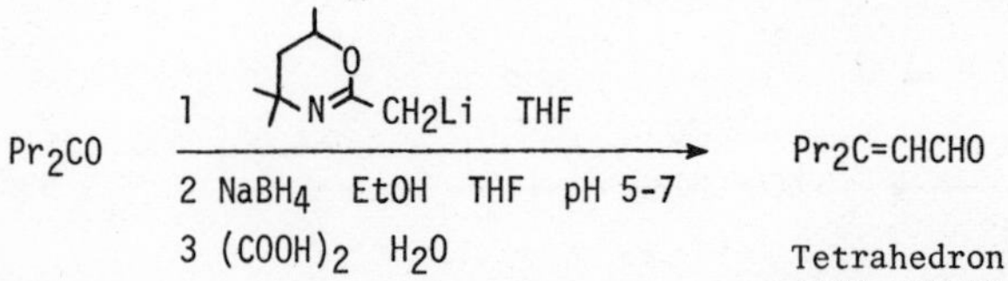

Pr$_2$CO

1 (oxazine with CH$_2$Li)   THF

2 NaBH$_4$ EtOH   THF   pH 5-7

3 (COOH)$_2$ H$_2$O

→ Pr$_2$C=CHCHO   ~65%

Tetrahedron (1971) 27 5979
JACS (1969) 91 764
JOC (1973) 38 36

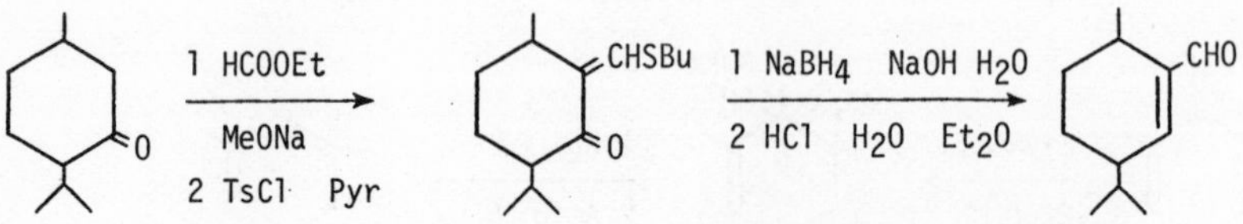

1 HCOOEt

MeONa

2 TsCl   Pyr

3 BuSH

→ (cyclohexanone with CHSBu)

1 NaBH$_4$ NaOH H$_2$O

2 HCl   H$_2$O   Et$_2$O

→ (cyclohexene with CHO)

JCS Perkin I (1972) 1721
Helv (1951) 34 728

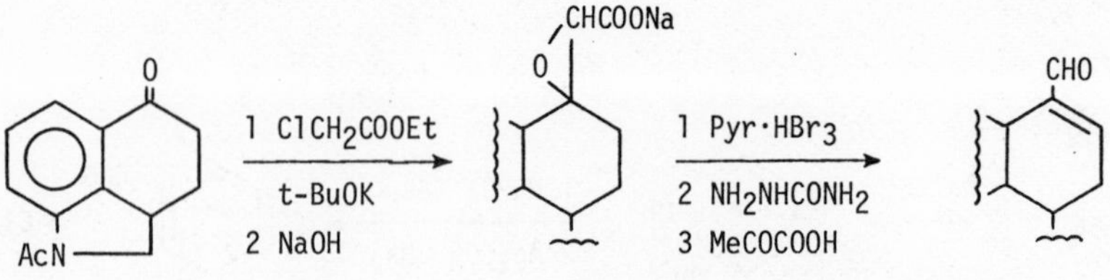

1 ClCH$_2$COOEt

t-BuOK

2 NaOH

→ (structure with CHCOONa, O)

1 Pyr·HBr$_3$

2 NH$_2$NHCONH$_2$

3 MeCOCOOH

→ (structure with CHO)   30%

JACS (1956) 78 3087

$C_5H_{11}CH_2COMe$ $\xrightarrow[\text{2 Zn EtOH}]{\text{1 HCONMe}_2\ \text{POCl}_3}$ $C_5H_{11}\underset{\overset{|}{CHO}}{C}=CHMe$

Chem Pharm Bull (1972) <u>20</u> 309

$C_5H_{11}\underset{\overset{|}{Me}}{CH_2CO}$ $\xrightarrow[\substack{\text{2 }\Delta \\ \text{3 LiClO}_4\ \text{CaCO}_3\ \text{HMPA}}]{\text{1 LiCHCl}_2\ \text{THF}}$ $C_5H_{11}CH=\underset{\overset{|}{Me}}{C}CHO$ 74%

Tetr Lett (1973) 2465

$\xrightarrow[\substack{\text{CH}_2\text{Cl}_2 \\ \text{2 (MeO)}_3\text{P}}]{\text{1 O}_3\ \text{MeOH}}$ Piperidine $\xrightarrow[\text{HOAc C}_6\text{H}_6]{}$ 54%

JOC (1960) <u>25</u> 1031

$\xrightarrow{\text{SeO}_2\ \text{EtOH}}$ 55%

JACS (1969) <u>91</u> 4933

$\underset{CH_2-CHCHO}{\overset{O}{\diagdown}}$ $\xrightarrow[\substack{\text{THF} \\ \text{2 HBF}_4 \\ \text{3 NaI Me}_2\text{CO}}]{\text{1 Sodium (cyclopentadienyl)-dicarbonylferrate}}$ $CH_2=CHCHO$

JACS (1972) <u>94</u> 7170

$C_5H_{11}C\equiv CCH_2O$ $\xrightarrow[\substack{\text{2 K}_2\text{CO}_3\ \text{MeOH}\ \text{H}_2\text{O} \\ \text{3 HOAc THF}}]{\text{1 BuLi THF}}$ $C_5H_{11}CH=CHCHO$

Tetr Lett (1972) 1815
(1971)   591

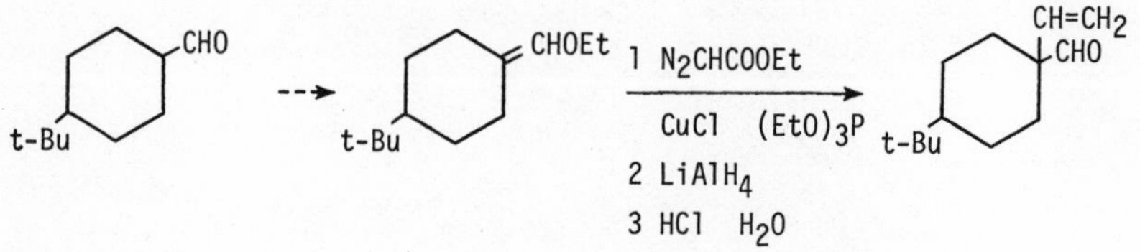
CHO
t-Bu
CHOEt
t-Bu
1 N2CHCOOEt
CuCl (EtO)3P
2 LiAlH4
3 HCl H2O
CH=CH2
CHO
t-Bu

Tetr Lett (1972) 5121

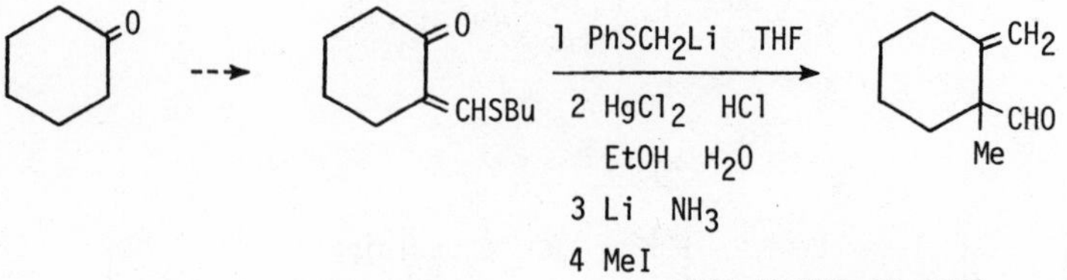
O
O
CHSBu
1 PhSCH2Li THF
2 HgCl2 HCl
EtOH H2O
3 Li NH3
4 MeI
CH2
CHO
Me

<41%

JACS (1972) 94 4758

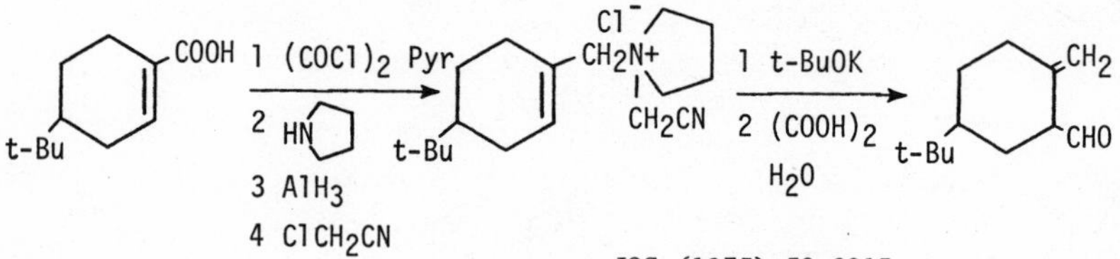
COOH
t-Bu
1 (COCl)2 Pyr
2 HN
3 AlH3
4 ClCH2CN
Cl⁻
CH2N+
CH2CN
t-Bu
1 t-BuOK
2 (COOH)2
H2O
CH2
CHO
t-Bu

JOC (1973) 38 2915

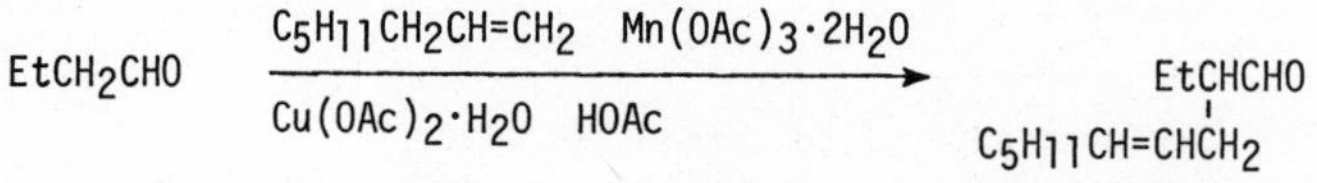
EtCH2CHO
C5H11CH2CH=CH2 Mn(OAc)3·2H2O
Cu(OAc)2·H2O HOAc
EtCHCHO
C5H11CH=CHCH2

25%

Synthesis (1972) 376

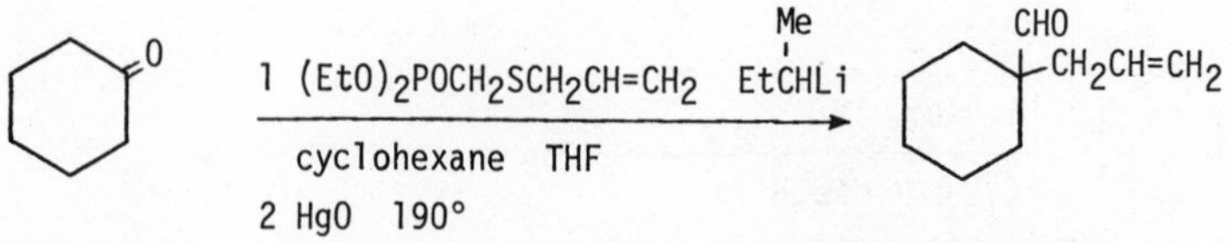
O
Me
1 (EtO)2POCH2SCH2CH=CH2 EtCHLi
cyclohexane THF
2 HgO 190°
CHO
CH2CH=CH2

82%

JACS (1970) 92 5522

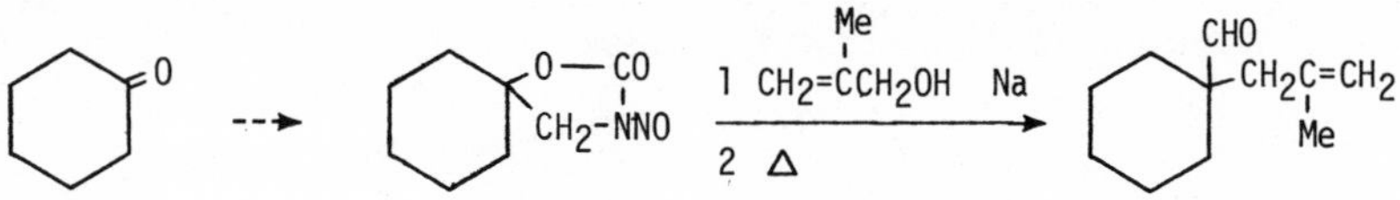

JOC (1969) 34 1220

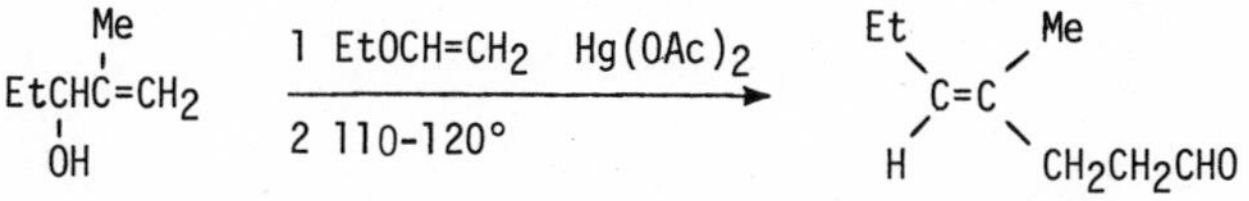

35%

JACS (1973) 95 553
JCS Perkin I (1973) 1791

Also via: β-Hydroxyaldehydes (Section 324)

Section 342　　Amide — Amide

Diamides

$$AcNH(CH_2)_5COOH \xrightarrow{\text{Electrolysis}} AcNH(CH_2)_{10}NHAc$$

Z Naturforsch (1947) 2b 182

Also via:　　　　　　　　　　Section
　　　　　　　Diamines　　　　350
　　　Dicarboxylic acids　　　312

Section 343　　Amide — Amine

Aminoamides

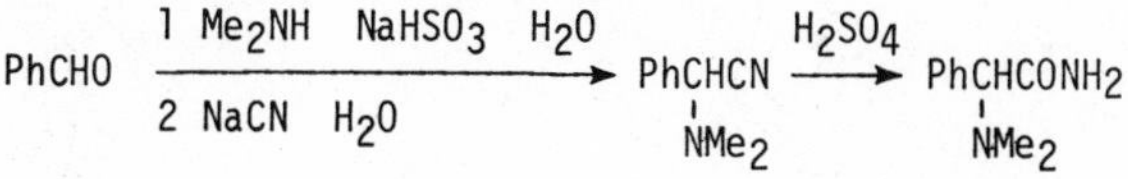

$$PhCHO \xrightarrow[\text{2 NaCN } H_2O]{\text{1 Me}_2NH \text{ NaHSO}_3 \text{ H}_2O} PhCHCN \xrightarrow{H_2SO_4} PhCHCONH_2$$

$$\underset{NMe_2}{|} \qquad \underset{NMe_2}{|}$$

JOC (1961) 26 4741
Annalen (1971) 749 198
JCS (1949) 2323

$$C_9H_{19}COCH_2CN \xrightarrow[\substack{NaOAc \quad EtOH \\ H_2O}]{NH_2OH \cdot HCl} C_9H_{19}\underset{\underset{O}{N}}{\overset{\overset{CH}{\parallel}}{C}}\underset{CNH_2}{} \xrightarrow[EtOH]{H_2 \quad Pt} C_9H_{19}\underset{NH_2}{CHCH_2CONH_2} \qquad 83\%$$

Can J Chem (1968) <u>46</u> 3617

$$CH_2=CHCONHPh \xrightarrow[EtOH]{PhCH_2NH_2} PhCH_2NHCH_2CH_2CONHPh \qquad 93\%$$

JOC (1960) <u>25</u> 1822

$$\underset{CN}{\overset{CN}{\underset{\displaystyle (CH_2)_4}{|}}} \xrightarrow[\substack{(basic) \quad H_2O}]{Ion\ exch\ resin} \underset{CN}{\overset{CONH_2}{\underset{\displaystyle (CH_2)_4}{|}}} \xrightarrow[NH_3]{H_2 \quad Ni} \underset{CH_2NH_2}{\overset{CONH_2}{\underset{\displaystyle (CH_2)_4}{|}}} \qquad 50\%$$

Ber (1959) <u>92</u> 2616

Section 344   Amide — Ester

Acylaminoesters

$$BuCH=CH_2 \xrightarrow[Me_2CO \quad t\text{-}BuOH]{\substack{NHAc \\ | \\ CH_2COOEt \quad h\nu}} BuCH_2CH_2\underset{NHAc}{CHCOOEt} \qquad 10\text{-}12\%$$

Chem Comm (1965) 471

Section 345   Amide — Epoxide

Epoxyamides

95%

JACS (1956) <u>78</u> 3087

Section 346   Amide — Halide
               °°°°°°°°°°°°°°°°

Haloamides

$$PhCHCOCl \xrightarrow{Et_2NH \quad CCl_4} PhCHCONEt_2 \qquad\qquad 49\%$$
  | |
  Br   Br

JACS (1949) <u>71</u> 3479
     (1955) <u>77</u> 4840

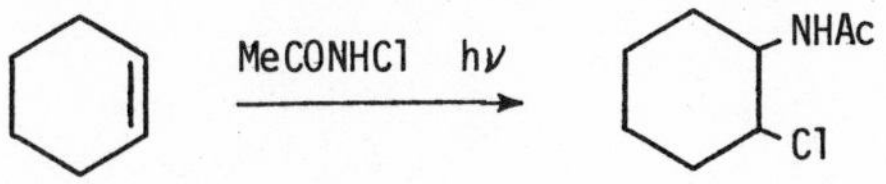

Tetr Lett (1971) 4425

Section 347   Amide — Ketone
               °°°°°°°°°°°°°°°°

Amides of α, β, γ and δ-ketoacids, acylaminoketones

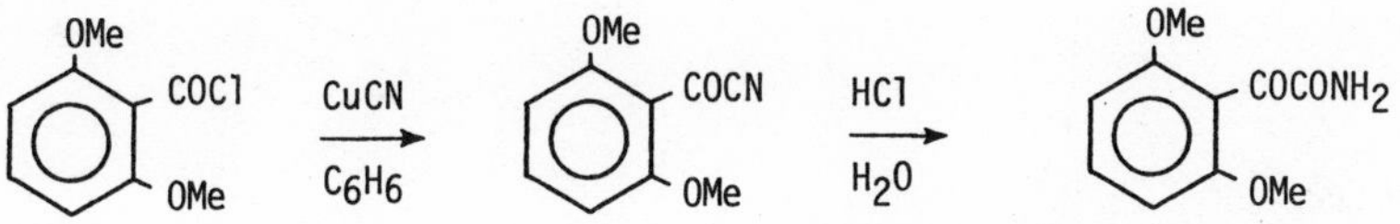

Can J Chem (1971) <u>49</u> 919   48%

$$PhCOCl \xrightarrow{BuNC} PhCOCONHBu \qquad\qquad 46\%$$

Ber (1961) <u>94</u> 1116

$$(PhCH_2)_2CHCHO \dashrightarrow (PhCH_2)_2CHCHCN \xrightarrow[\text{2 } CrO_3 \quad HOAc]{\text{1 } H_2SO_4 \quad t\text{-BuOH}} (PhCH_2)_2CHCOCONHBu\text{-}t$$
                                          |
                                          OH

Synthesis (1971) 538

PhBr $\xrightarrow[\text{2 (CONMe}_2)_2]{\text{1 BuLi   THF}}$ PhCOCONMe$_2$ 81%

Synth Comm (1973) 3 325
Compt Rend (1927) 184 825

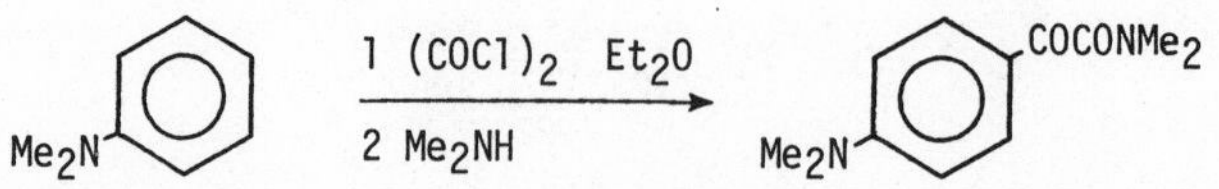

$\xrightarrow[\text{2 Me}_2\text{NH}]{\text{1 (COCl)}_2\text{   Et}_2\text{O}}$

68%

JOC (1959) 24 265

$\dashrightarrow$

$\xrightarrow[\text{2 KOH   EtOH}]{\substack{\text{1 H}_2\text{O}_2\text{   Na}_2\text{WO}_4\cdot2\text{H}_2\text{O} \\ \text{EtOH   H}_2\text{O}}}$

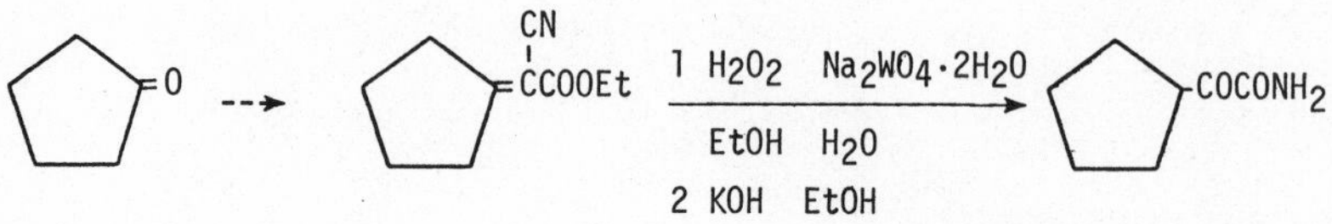

JOC (1963) 28 3088

BuCH$_2$CONMe$_2$ $\xrightarrow{\text{POCl}_3\text{   C}_6\text{H}_6}$

BuCHCONMe$_2$
|
BuCH$_2$CO   58%

Ber (1959) 92 1456

PhNH$_2$ $\xrightarrow[\text{Et}_2\text{O}]{\text{MeCOCH}_2\text{COF}}$ MeCOCH$_2$CONHPh   74%

JOC (1961) 26 225

PhNH$_2$ $\xrightarrow[\text{C}_6\text{H}_6]{\text{Diketene}}$ MeCOCH$_2$CONHPh   74%

Org Synth (1955) Coll Vol 3 10
Ber (1959) 92 1456
JACS (1964) 86 5654

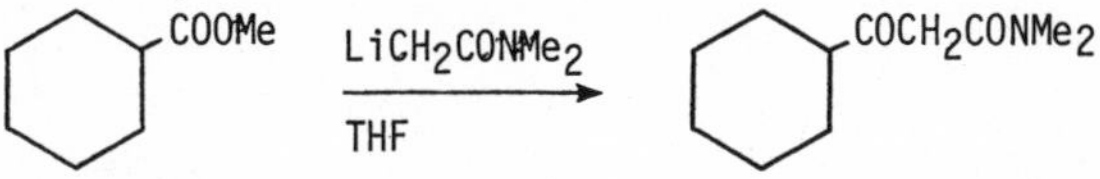

COOMe
LiCH2CONMe2
THF
COCH2CONMe2

Tetr Lett (1973) 1495
JOC (1959) 24 1551

O
1 Morpholine
2 PhNCO
3 H2SO4 H2O
O
CONHPh

Annalen (1964) 673 132
Ber (1962) 95 926
JOC (1961) 26 3043

Via silyl enol ether Tetr Lett (1973) 4271

$PhCOCH_2COOEt \xrightarrow[\text{xylene}]{PhNH_2} PhCOCH_2CONHPh$ 74-76%

Org Synth (1955) Coll Vol 3 108
(1963) Coll Vol 4 80

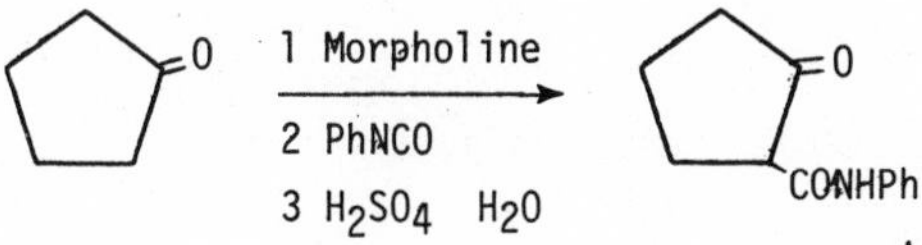

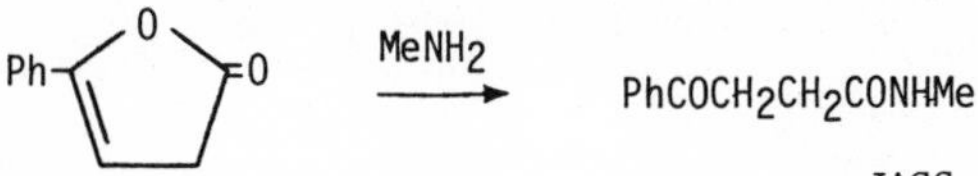

Ph
O
O
MeNH2
PhCOCH2CH2CONHMe

JACS (1958) 80 4573

$PhBr \xrightarrow[\text{2 } CH_2CO, CH_2CO > NH]{\text{1 Mg } Et_2O} PhCOCH_2CH_2CONH_2$ 65%

Chem Pharm Bull (1971) 19 391

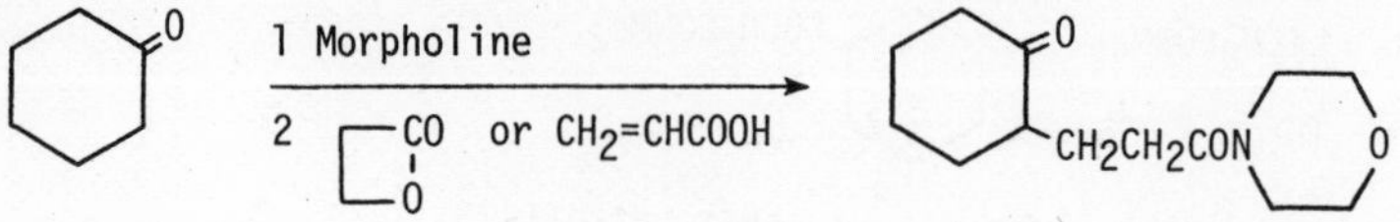

1 Morpholine

2 $\begin{array}{c}\text{—CO}\\\quad|\\\text{—O}\end{array}$ or $CH_2=CHCOOH$

64%

Acta Chem Scand (1964) 18 2201
Tetr Lett (1965) 2869

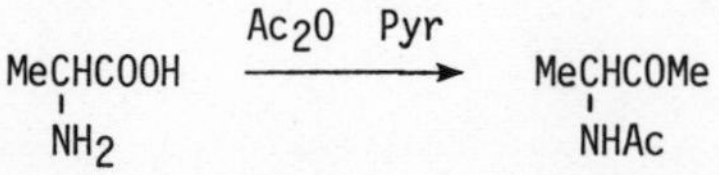

$\underset{NH_2}{MeCHCOOH} \xrightarrow{Ac_2O \quad Pyr} \underset{NHAc}{MeCHCOMe}$

81-88%

Org Synth (1963) Coll Vol 4 5

Also via: Ketoacids (Section 320)

Section 348   Amide — Nitrile

Cyanoamides

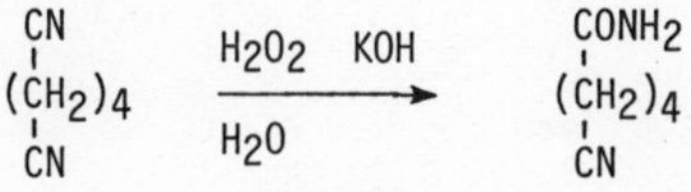

$\underset{CN}{\overset{CN}{(CH_2)_4}} \xrightarrow[H_2O]{H_2O_2 \quad KOH} \underset{CN}{\overset{CONH_2}{(CH_2)_4}}$

31%

JOC (1950) 15 800
Ber (1959) 92 2616

Also via: Cyanoacids (Section 321)

Section 349   Amide — Olefin

Olefinic amides

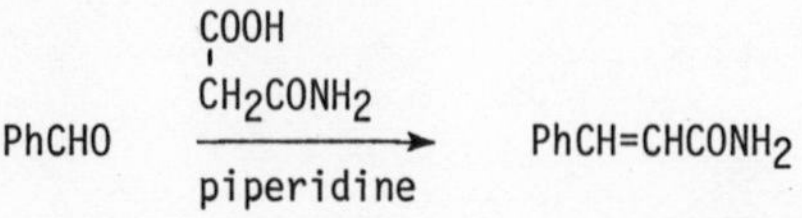

$PhCHO \xrightarrow[\text{piperidine}]{\overset{COOH}{\underset{}{\overset{|}{CH_2CONH_2}}}} PhCH=CHCONH_2$

JACS (1948) 70 2596
     (1949) 71 3562

PhCHO $\xrightarrow[\text{NaOAc}]{\text{MeCH}_2\text{CONHCOCH}_3}$ PhCH=CHCONH$_2$ + PhCH=CCONH$_2$

                                            46% $\overset{|}{\text{Me}}$ 37%

Rec Trav Chim (1951) <u>70</u> 146

PhCOMe $\xrightarrow[\text{BF}_3\cdot\text{Et}_2\text{O}]{\text{PhC}\equiv\text{CNMe}_2}$ $\overset{\text{Ph}}{\underset{\text{Me}}{\text{PhC}=\text{CCONMe}_2}}$ 84%

Ber (1970) <u>103</u> 564

EtCH=CHCH$_2$Br $\xrightarrow[\text{2 MeCH=CHCONEt}_2]{\text{1 Mg}}$ $\overset{\text{Et}}{\text{CH}_2\text{=CHCHCHCH}_2\text{CONEt}_2}$ 80%

                                            $\underset{\text{Me}}{}$

Tetr Lett (1971) 3251

Helv (1964) <u>47</u> 2425
Ber (1971) <u>104</u> 3689
JCS <u>C</u> (1971) 2950

Also via: Section
            Acetylenic amides   304
            Olefinic acids   322

Section 350   Amine — Amine
              ∘∘∘∘∘∘∘∘∘∘∘∘∘∘∘∘

1,1-Diamines, 1,2-diamines, 1,3-diamines and higher diamines

Et$_2$NH $\xrightarrow{\text{HCHO}}$ (Et$_2$N)$_2$CH$_2$

JACS (1932) <u>54</u> 4172

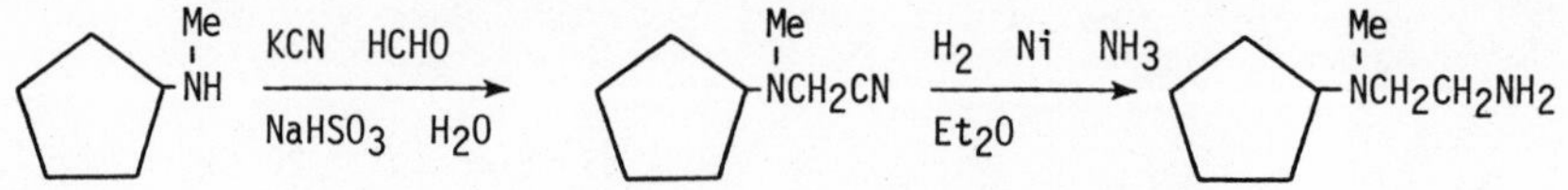

JACS (1946) $\underline{68}$ 1905

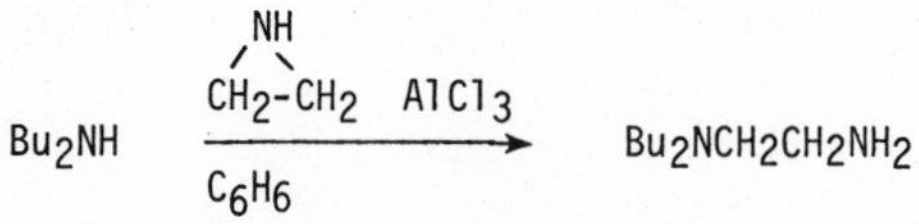

Bu_2NH $\xrightarrow[C_6H_6]{\overset{NH}{\overset{/ \backslash}{CH_2\text{-}CH_2}} \quad AlCl_3}$ $Bu_2NCH_2CH_2NH_2$

JACS (1946) $\underline{68}$ 2006
     (1948) $\underline{70}$ 184

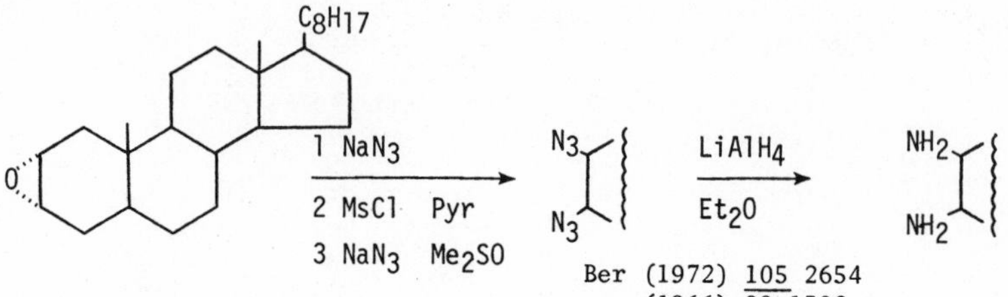

Ber (1972) $\underline{105}$ 2654
    (1966) $\underline{99}$ 1502
JOC (1967) $\underline{32}$ 511

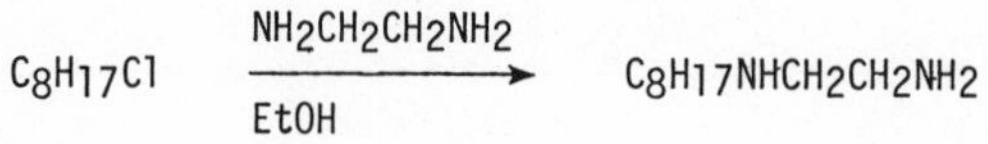

$C_8H_{17}Cl$ $\xrightarrow[EtOH]{NH_2CH_2CH_2NH_2}$ $C_8H_{17}NHCH_2CH_2NH_2$

JACS (1945) $\underline{67}$ 1581

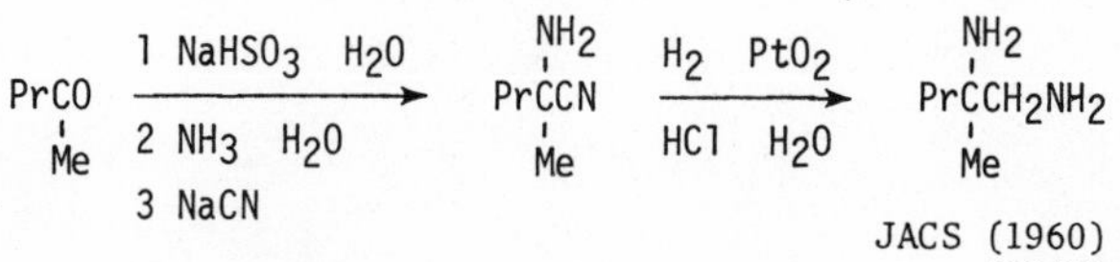

$\underset{Me}{PrCO}$ $\xrightarrow[\substack{2\ NH_3\ H_2O \\ 3\ NaCN}]{1\ NaHSO_3\ H_2O}$ $\underset{Me}{PrCCN}^{NH_2}$ $\xrightarrow[HCl\ H_2O]{H_2\ PtO_2}$ $\underset{Me}{PrCCH_2NH_2}^{NH_2}$ 28%

JACS (1960) $\underline{82}$ 696
    (1949) $\underline{71}$ 2530

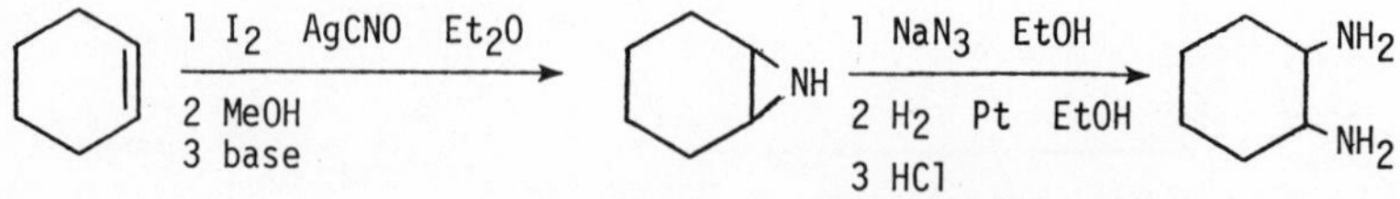

JOC (1967) $\underline{32}$ 511
Tetrahedron (1973) $\underline{29}$ 3137

$BuNH_2$ $\xrightarrow{\text{CH}_2=\text{CHCN}}$ $BuNH(CH_2)_2CN$ $\xrightarrow[\text{EtOH}]{\text{H}_2 \ \text{Ni} \ \ \text{NH}_3}$ $BuNH(CH_2)_2CH_2NH_2$ $\qquad <66\%$

JACS (1946) <u>68</u> 1217

$Br(CH_2)_3Br$ $\longrightarrow$ [phthalimide N-K] $\longrightarrow$ [phthalimide $N(CH_2)_3Br$] $\xrightarrow[\text{2 HCl} \ \ \text{H}_2\text{O}]{\text{1 Bu}_2\text{NH}}$ $NH_2(CH_2)_3NBu_2$ $\sim60\%$

Org Synth (1955) Coll Vol 3 256 254

$C_9H_{19}COCH_2CN$ $\xrightarrow[\text{2 H}_2 \ \ \text{Pt} \ \ \text{EtOH}]{\text{1 NH}_2\text{OH·HCl} \ \ \text{NaOAc}}$ $C_9H_{19}\underset{\underset{NH_2}{|}}{CH}CH_2CONH_2$ $\xrightarrow[\text{THF}]{\text{B}_2\text{H}_6}$ $C_9H_{19}\underset{\underset{NH_2}{|}}{CH}CH_2CH_2NH_2$

Can J Chem (1968) <u>46</u> 3617

[cyclooctene] $\xrightarrow[\begin{array}{l}\text{2 H}_2 \ \ \text{Pt}\\ \text{3 MeNH}_2 \ \ \text{H}_2 \ \ \text{Rh}\end{array}]{\text{1 O}_3 \ \ \text{ROH}}$ [cyclooctane with NHMe, NHMe] $\qquad 73\%$

Tetr Lett (1971) 3591 3587

Also via: Diamides (Section 342)

Section 351   Amine — Ester

Direct esterification of aminoacids and aminoacid halides. Other preparations of esters of aminoacids. Esters of aminoalcohols.

$\underset{\underset{NH_2}{|}}{Me}CHCOOH$ $\xrightarrow[\text{2 PhCH}_2\text{OH} \ \ \text{HCl} \ \ \text{Et}_2\text{O}]{\text{1 COCl}_2 \ \ \text{THF}}$ $\underset{\underset{NH_2}{|}}{Me}CHCOOCH_2Ph$ $\qquad 76\%$

Coll Czech (1958) <u>23</u> 1947

PhCH₂CHCOOH $\xrightarrow{\text{PhCH}_2\text{OH}\quad \text{PhSO}_3\text{H}}$ PhCH₂CHCOOCH₂Ph   75%
| |
NH₂   NH₂

JACS (1952) <u>74</u> 1092

TsOH   Chem Ind (1955) 16

SO₂Cl₂   Annalen (1961) <u>640</u> 139

SOCl₂   JOC (1965) <u>30</u> 3575

Polyphosphoric acid   JACS (1954) <u>76</u> 5781

i-PrCH₂CHCOOH $\xrightarrow{\text{MeOH}\quad \text{SOCl}_2}$ i-PrCH₂CHCOOMe   83-86%
| |
NH₂   NH₂

Helv (1953) <u>36</u> 1109
JACS (1956) <u>78</u> 381

$\underset{\text{H}}{\overset{\displaystyle\bigcap}{N}}\!\!-\!\text{COOH}$ $\xrightarrow[\text{TsOH}]{\text{Me}_2\text{SO}_3}$ $\underset{\text{H}}{\overset{\displaystyle\bigcap}{N}}\!\!-\!\text{COOMe}$ 91%

JCS (1963) 1927

PhCH₂CHCOOH $\xrightarrow[\text{dioxane}]{\text{Me}_2\text{C}=\text{CH}_2\quad \text{H}_2\text{SO}_4}$ PhCH₂CHCOOBu-t   ~70%
| |
NH₂   NH₂

Chem Ind (1959) 1121

t-BuOAc   HClO₄   Annalen (1961) <u>646</u> 134

HCl·NH₂CH₂COCl $\xrightarrow{\text{PhCH}_2\text{OH}}$ NH₂CH₂COOCH₂Ph

Helv (1929) <u>12</u> 332

i-PrCH₂CHCOOMe $\xrightarrow{\text{C}_6\text{H}_{13}\text{OH}\quad \text{MeONa}}$ i-PrCH₂CHCOOC₆H₁₃   93-96%
| |
NH₂   NH₂

Helv (1953) <u>36</u> 1109

$$PhCH_2Br \xrightarrow[Et_2O]{\substack{N(SiMe_3)_2 \\ | \\ CH_2COOEt}} \quad \xrightarrow{NaN(SiMe_3)_2} \quad PhCH_2\overset{\overset{\displaystyle N(SiMe_3)_2}{|}}{C}HCOOEt \quad \xrightarrow[H_2O]{HCl} \quad PhCH_2\overset{\overset{\displaystyle NH_2}{|}}{C}HCOOEt$$

Angew (1968) <u>80</u> 797
(Internat Ed <u>7</u> 809)

$$MeCH_2CN \xrightarrow[HCl]{MeOH} MeCH_2\overset{\overset{\displaystyle NH \cdot HCl}{\|}}{\underset{\underset{\displaystyle OMe}{|}}{C}} \xrightarrow[\substack{2 \ t\text{-BuOK} \\ 3 \ HCl \ H_2O}]{1 \ NaOCl \ \ H_2O} Me\overset{\overset{}{}}{C}HCOOMe \\ \overset{}{\underset{\displaystyle NH_2}{|}}$$

JACS (1960) <u>82</u> 4422

$$\underset{\underset{\displaystyle Br}{|}}{Me_2C}COCl \dashrightarrow \underset{\underset{\displaystyle Br}{|}}{Me_2C}CONHPr \xrightarrow[Et_2O]{t\text{-BuOK}} \underset{\underset{\displaystyle NHPr}{|}}{Me_2C}COOBu\text{-}t \qquad\qquad 41\%$$

JACS (1964) <u>86</u> 1356

JACS (1934) <u>56</u> 697

JACS (1945) <u>67</u> 1071

$$PhNH_2 \xrightarrow[HOAc]{CH_2=CHCOOMe} PhNHCH_2CH_2COOMe \qquad\qquad 69\%$$

JACS (1949) <u>71</u> 1901 2124 2532

$$Me_2C=CHCOOMe \xrightarrow[h\nu]{MeCH_2NEt_2} \underset{\underset{MeCHNEt_2}{|}}{Me_2CCH_2COOMe}$$

Chem Comm (1968) 180

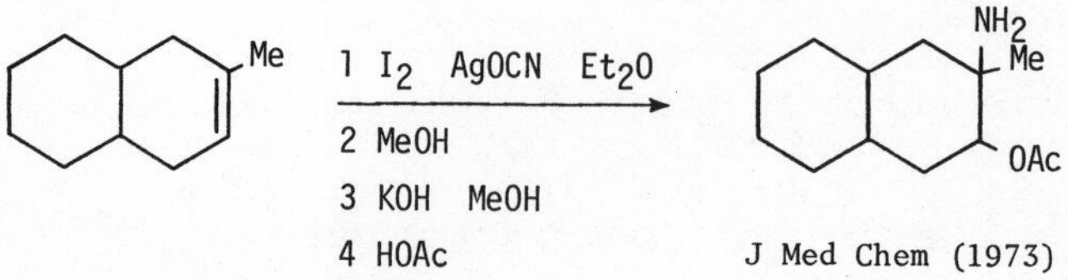

1 I_2 AgOCN Et_2O

2 MeOH

3 KOH   MeOH

4 HOAc

J Med Chem (1973) 16 23

Section 352   Amine ── Ether
∘∘∘∘∘∘∘∘∘∘∘∘∘∘

Aminoethers

$$C_7H_{15}OH \xrightarrow{Et_2NH \quad HCHO} C_7H_{15}OCH_2NEt_2$$

JACS (1932) 54 4172

$$PhCHO \xrightarrow[HgCl_2 \quad CaSO_4]{Et_2NH \quad BuOH} \underset{\underset{Et_2N}{|}}{PhCHOBu}$$

47%

JACS (1955) 77 1098

$$PhCH_2OH \xrightarrow[]{\overset{NH}{\overset{/ \ \backslash}{CH_2-CH_2}} \quad HBF_4} PhCH_2OCH_2CH_2NH_2$$

51%

Ber (1964) 97 510

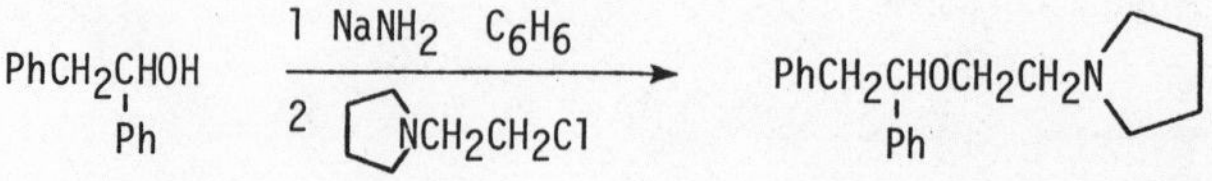

$$\underset{\underset{Ph}{|}}{PhCH_2CHOH} \xrightarrow[2 \quad \text{(NCH}_2\text{CH}_2\text{Cl)}]{1 \quad NaNH_2 \quad C_6H_6} \underset{\underset{Ph}{|}}{PhCH_2CHOCH_2CH_2N}\text{◯}$$

73%

JACS (1948) 70 3098

Section 353   Amine — Halide
              ○○○○○○○○○○○○○○○

Aminohalides

Me$_2$C=CMe$_2$ $\xrightarrow[\text{2 SnCl}_2 \quad \text{HCl} \quad \text{H}_2\text{O}]{\text{1 NOCl} \quad \text{MeOH}}$ Me$_2$C——CMe$_2$
                                                              | |
                                                            NH$_2$ Cl

                                        JACS (1960) <u>82</u> 6068

$\xrightarrow[\text{2 HCl} \quad \text{H}_2\text{O}]{\text{1 I}_2 \quad \text{AgOCN} \quad \text{Et}_2\text{O}}$

                                        J Med Chem (1973) <u>16</u> 23

$\xrightarrow[\text{2 HCl} \quad \text{H}_2\text{O}]{\text{1 N-Bromophthalimide}}$

                                        Bull Soc Chim Fr (1966) 1910

Me$_2$NCH$_2$CH$_2$OH $\xrightarrow{\text{SOCl}_2}$ Me$_2$NCH$_2$CH$_2$Cl   67-80%

                                        Org Synth (1963) Coll Vol 4 333

Et$_2$NH $\xrightarrow{\text{Br(CH}_2\text{)}_3\text{Cl}}$ Et$_2$N(CH$_2$)$_3$Cl   70%

                                        JACS (1945) <u>67</u> 1472

Also via: Haloamides (Section 346)

Section 354   Amine — Ketone
               ০০০০০০০০০০০০০০০০

α, β, γ and δ-aminoketones

$$\text{PrCOCl} \xrightarrow[\substack{\text{2 HCl}}]{\substack{\text{NC} \\ \text{1 CH}_2\text{COOEt} \\ \text{1,8-diazabicyclo[5.4.0]undecene}}} \text{PrCOCH}_2\text{NH}_2 \qquad 61\%$$

JOC (1973) 38 3571
Synth Comm (1972) 2 237

$$\text{PhCOCl} \xrightarrow{\text{CuCN}} \text{PhCOCN} \xrightarrow[\substack{\text{2 HCl} \quad \text{H}_2\text{O}}]{\substack{\text{1 SnCl}_2 \quad \text{HCl} \quad \text{Et}_2\text{O}}} \text{PhCOCH}_2\text{NH}_2 \qquad 67\%$$

JOC (1972) 37 318

$$\underset{\text{NH}_2}{\text{PhCHCH}_3} \xrightarrow[\substack{\text{2 MeONa} \quad \text{MeOH} \\ \text{3 HCl} \quad \text{H}_2\text{O}}]{\substack{\text{1 t-BuOCl} \quad \text{C}_6\text{H}_6}} \text{PhCOCH}_2\text{NH}_2 \qquad 55\text{-}72\%$$

Org Synth (1961) 41 82

$$\text{PhH} \xrightarrow[\substack{\text{AlCl}_3 \quad \text{CS}_2}]{\substack{\text{HCl·NH}_2\text{CH}_2\text{COCl}}} \text{PhCOCH}_2\text{NH}_2 \qquad 60\%$$

J Prakt Chem (1957) 5 91

1 NH$_2$OH·HCl K$_2$CO$_3$
   MeOH H$_2$O
2 TsCl Pyr
3 EtOK
4 HCl H$_2$O   <43%

JACS (1956) 78 3087

$$\text{PhCH}_2\text{COMe} \xrightarrow[\substack{\text{2 MeI} \quad \text{EtOH} \\ \text{3 Na}}]{\substack{\text{1 Me}_2\text{NNH}_2}} \underset{\text{NH}_2}{\text{PhCHCOMe}} \qquad 17\%$$

JOC (1957) 22 358

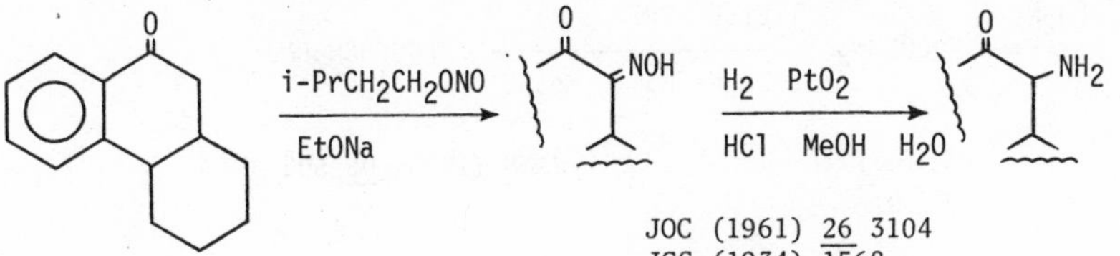

$$\xrightarrow[\text{EtONa}]{\text{i-PrCH}_2\text{CH}_2\text{ONO}}$$

$$\xrightarrow[\text{HCl} \quad \text{MeOH} \quad \text{H}_2\text{O}]{\text{H}_2 \quad \text{PtO}_2}$$

41%

JOC (1961) <u>26</u> 3104
JCS (1934) <u>1568</u>

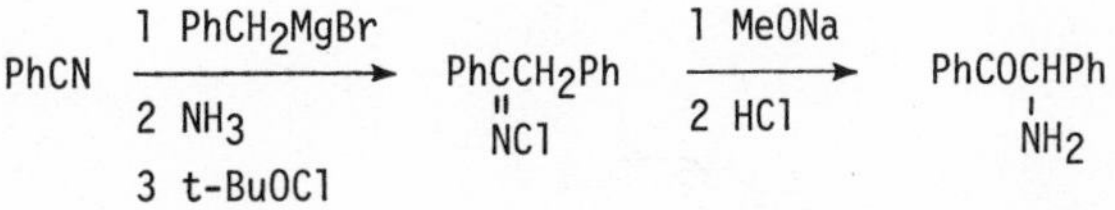

PhCN $\xrightarrow[\substack{2 \text{ NH}_3 \\ 3 \text{ t-BuOCl}}]{1 \text{ PhCH}_2\text{MgBr}}$ $\underset{\underset{\text{NCl}}{\|}}{\text{PhCCH}_2\text{Ph}}$ $\xrightarrow[2 \text{ HCl}]{1 \text{ MeONa}}$ $\underset{\underset{\text{NH}_2}{|}}{\text{PhCOCHPh}}$ 66%

JACS (1960) <u>82</u> 4422 459

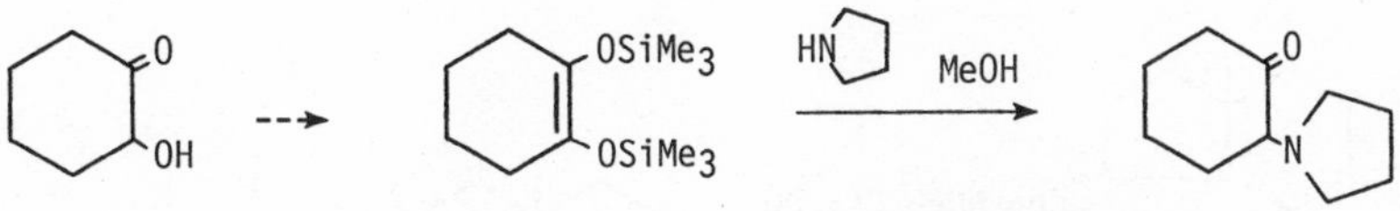

$$\xrightarrow[\text{MeOH}]{\text{HN}\diagdown}$$

Ber (1972) <u>105</u> 975

$$\xrightarrow{\text{i-PrNH}_2}$$

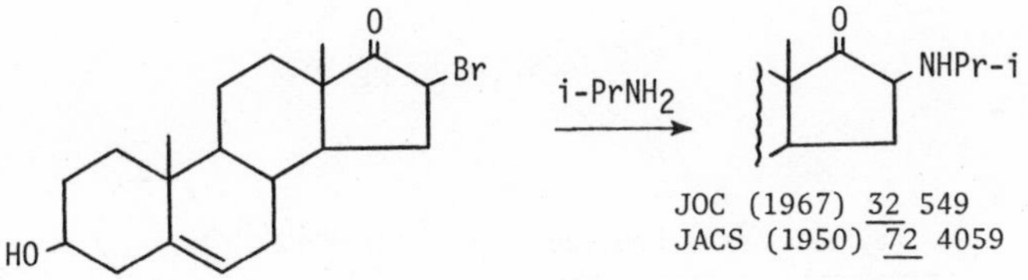

JOC (1967) <u>32</u> 549
JACS (1950) <u>72</u> 4059

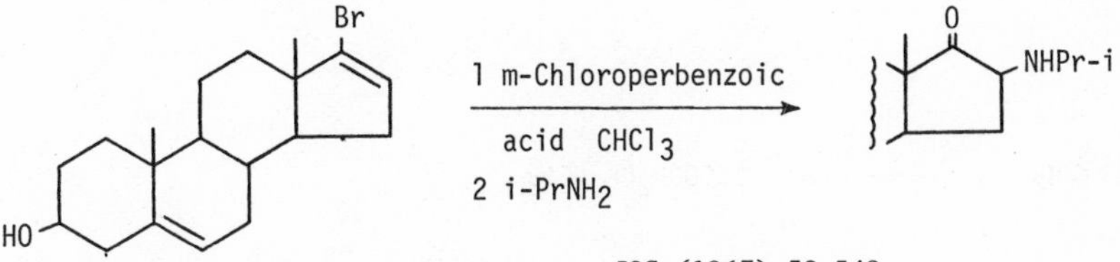

$$\xrightarrow[\substack{\text{acid} \quad \text{CHCl}_3 \\ 2 \text{ i-PrNH}_2}]{1 \text{ m-Chloroperbenzoic}}$$

JOC (1967) <u>32</u> 549

$$PrCOCl \xrightarrow{CH_2N_2} PrCOCHN_2 \xrightarrow[\text{2 } Me_2N=CH_2 \ I \ Me_2SO]{\text{1 } Et_3B \quad THF} PrCOCHCH_2NMe_2$$

with Et substituent:

PrCOCHCH₂NMe₂
          |
          Et

JACS (1973) 95 602

Reviews: The Mannich Reaction   Org React (1942) 1 303

            Advances in the Chemistry
            of Mannich Bases   Synthesis (1973) 703

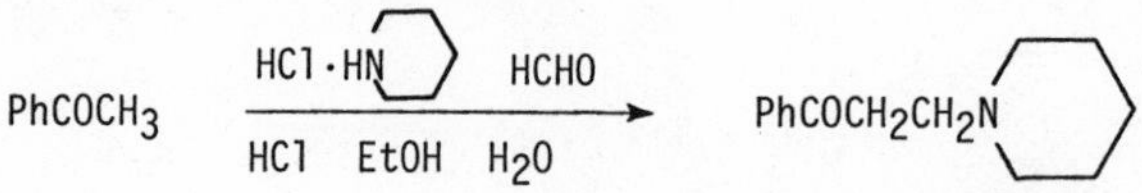

$$PhCOCH_3 \xrightarrow[\text{HCl \quad EtOH \quad H}_2O]{HCl \cdot HN \quad HCHO} PhCOCH_2CH_2N$$ 85%

Org React (1942) 1 303 329
Org Synth (1955) Coll Vol 3 305

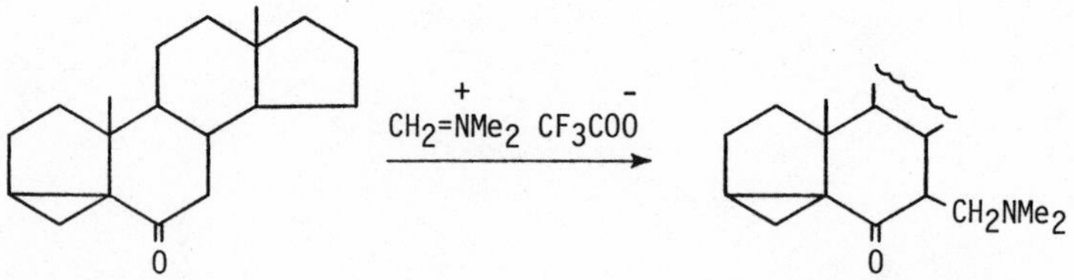

$$\xrightarrow{CH_2=NMe_2 \ CF_3COO} $$ 94%

with CH₂NMe₂ substituent

JACS (1968) 90 5622

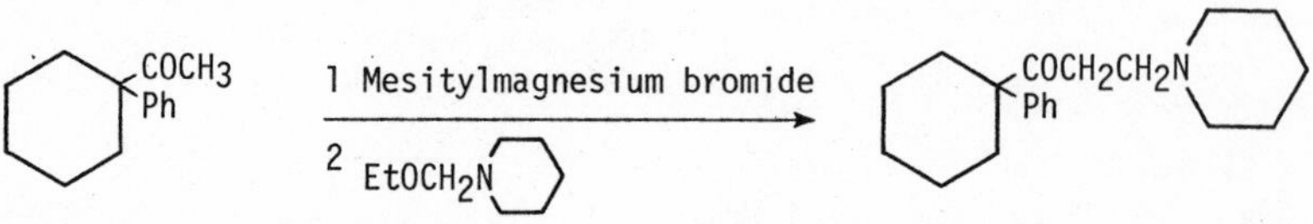

$$\xrightarrow[\text{2 } EtOCH_2N]{\text{1 Mesitylmagnesium bromide}}$$ 75%

Bull Soc Chim Fr (1962) 273

$$EtCOCH=CH_2 \xrightarrow{Et_2NH \quad Et_2O} EtCOCH_2CH_2NEt_2$$ >37%

JACS (1950) 72 4059
Org Synth (1941) Coll Vol 1 196
JOC (1964) 29 2346

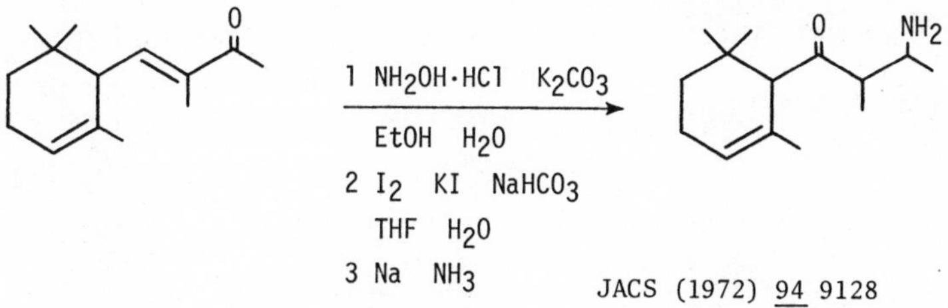

O
1 NH2OH·HCl K2CO3
EtOH H2O
2 I2 KI NaHCO3
THF H2O
3 Na NH3
O NH2

JACS (1972) 94 9128

1 CO NK CO DMF
2 HCl H2O
PrCOCH2CH2Cl
PrCOCH2CH2NH2 85%
PhCH2NHMe
PrCOCH2CH2NCH2Ph 82%
Me

JACS (1971) 93 2492

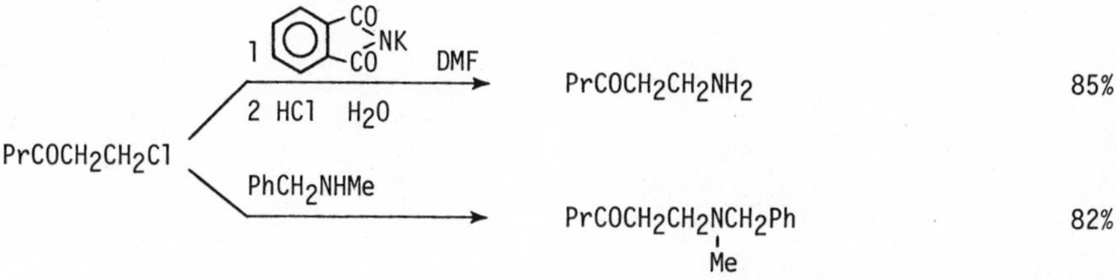

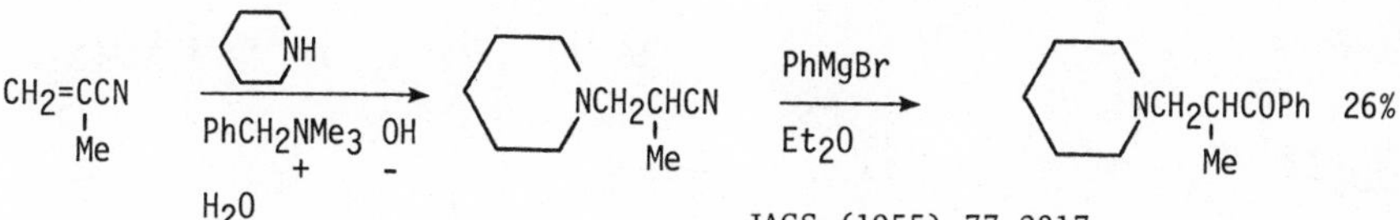

CH2=CCN
Me
NH
PhCH2NMe3 OH
+ -
H2O
NCH2CHCN
Me
PhMgBr
Et2O
NCH2CHCOPh 26%
Me

JACS (1955) 77 2817

OH
O
Et3N hν
O
CHMe
NEt2

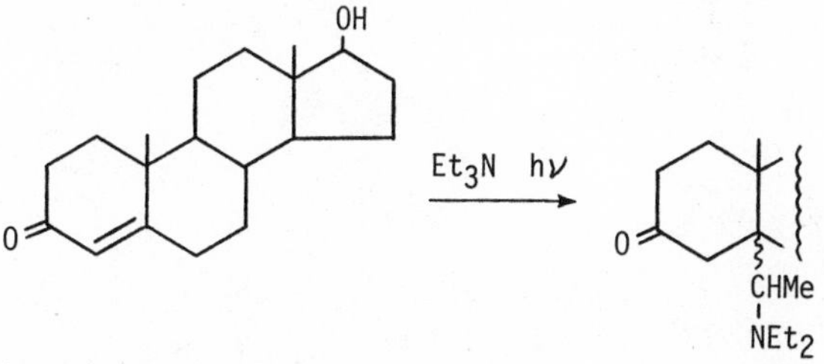

Chem Comm (1968) 180

$$Et_2N(CH_2)_3Cl \xrightarrow[\text{2 HCl } H_2O]{\text{1 } MeCOCH_2COOEt \text{ Na xylene}} Et_2N(CH_2)_3CH_2COMe$$

JACS (1947) 69 1258

Section 355   Amine — Nitrile

Cyanoamines, aminonitriles

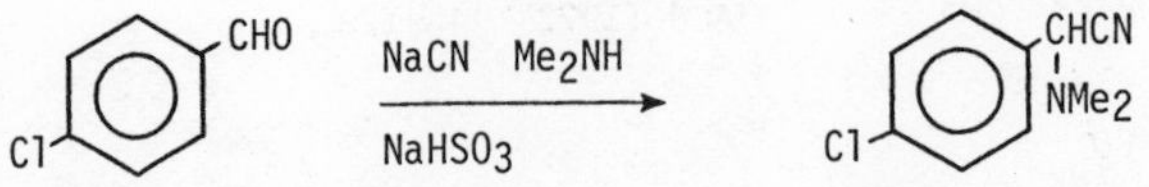

JOC (1961) 26 4741
JACS (1960) 82 696

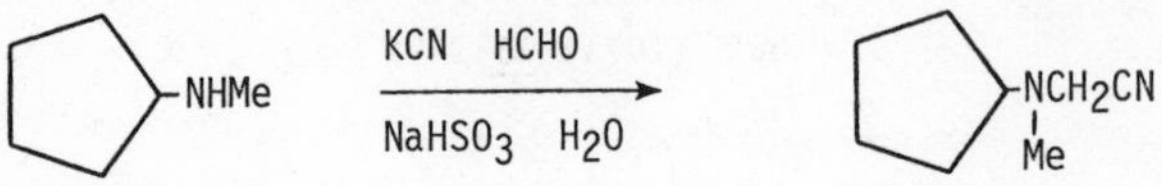

84%

JACS (1946) 68 1905
Org Synth (1955) Coll Vol 3 275

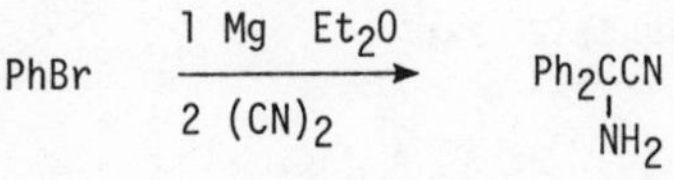

PhBr $\xrightarrow[\text{2 (CN)}_2]{\text{1 Mg Et}_2\text{O}}$ $\underset{\text{NH}_2}{\text{Ph}_2\text{CCN}}$

JOC (1954) 19 285

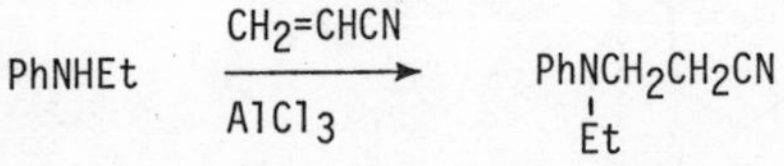

33%

PhNHEt $\xrightarrow[\text{AlCl}_3]{\text{CH}_2=\text{CHCN}}$ $\underset{\text{Et}}{\text{PhNCH}_2\text{CH}_2\text{CN}}$

Annalen (1972) 760 151
Org React (1949) 5 79
Org Synth (1963) Coll Vol 4 146

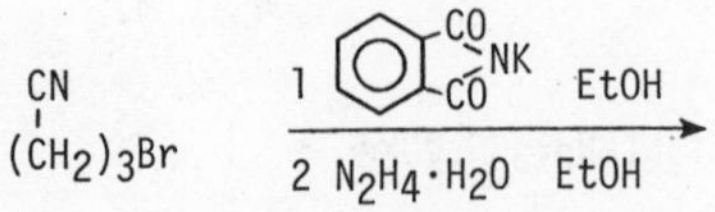

70%

$\underset{\text{(CH}_2)_3\text{Br}}{\text{CN}}$ $\xrightarrow[\text{2 N}_2\text{H}_4\cdot\text{H}_2\text{O   EtOH}]{1 \quad \text{NK   EtOH}}$ $\underset{\text{(CH}_2)_3\text{NH}_2}{\text{CN}}$

JCS (1947) 1369

Section 356   Amine — Olefin
            °°°°°°°°°°°°°°°°

Olefinic amines, enamines
For allylic amination see section 101 vol 1 (Amines from Hydrides)

Reviews: Enamines   Advances in Org Chem (1963) 4 1

              Enamines   Chem Ind (1970) 1188

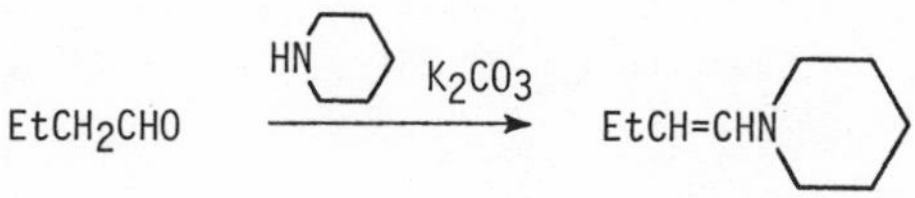

EtCH₂CHO $\xrightarrow{\text{HN} \quad K_2CO_3}$ EtCH=CHN

Ber (1936) 69 2106
JACS (1963) 85 207

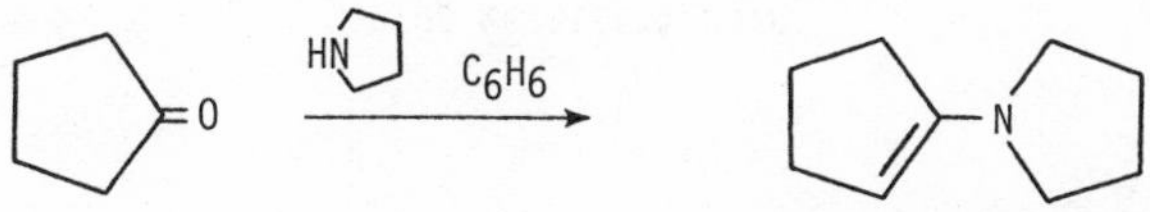

$\xrightarrow{\text{HN} \quad C_6H_6}$ 80-90%

JACS (1963) 85 207
TsOH xylene   Org Synth (1968) 48 56
Molecular sieves   JOC (1971) 36 1570
                   Rec Trav Chim (1972) 91 605
TiCl₄   JOC (1967) 32 213
Trisdialkylaminoboranes   JCS (1965) 5142

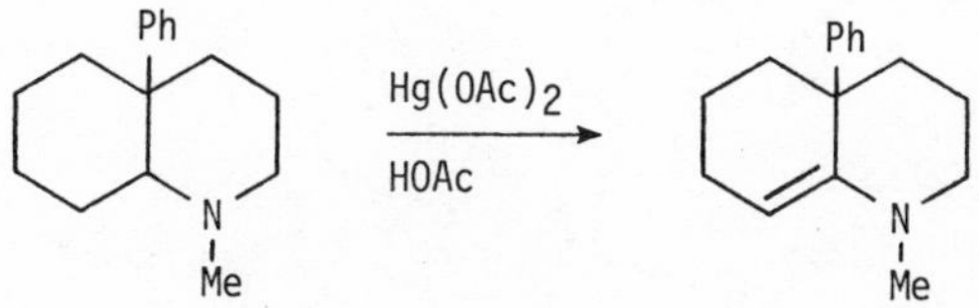

$\xrightarrow{\text{Hg(OAc)}_2 \quad \text{HOAc}}$

JOC (1961) 26 1104

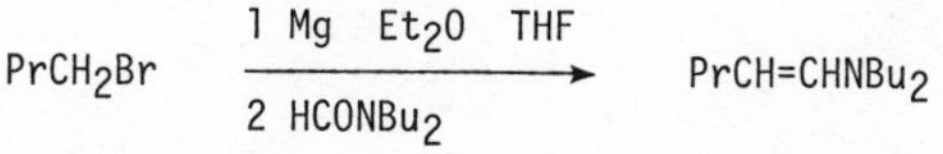

PrCH₂Br $\xrightarrow[\text{2 HCONBu}_2]{\text{1 Mg   Et}_2\text{O   THF}}$ PrCH=CHNBu₂   80%

JOC (1973) 38 3074

EtCHO $\xrightarrow[\text{2 CH}_2\text{=CHPPh}_3 \text{ Br DMF}]{\text{1 NaH Et}_2\text{NH}}$ EtCH=CHCH$_2$NEt$_2$ 17%

$\qquad\qquad\qquad\qquad\qquad\qquad\qquad\qquad$ JOC (1966) 31 467

$\qquad$ CH$_2$=CHC=CH$_2$ Sodium-naphthalene

$\qquad\qquad$ Me

$\xrightarrow{\text{THF}}$

Chem Ind (1973) 231

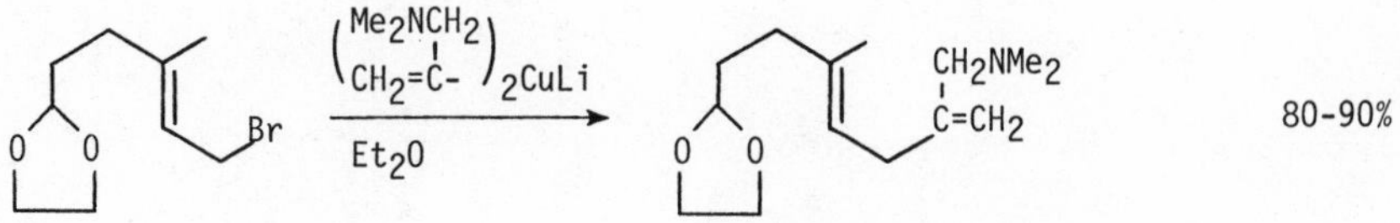

$\qquad$ $\begin{pmatrix} \text{Me}_2\text{NCH}_2 \\ \text{CH}_2\text{=C-} \end{pmatrix}_2$CuLi

$\xrightarrow{\text{Et}_2\text{O}}$ 80-90%

JACS (1971) 93 7016

PhCH=CCONMe$_2$ $\xrightarrow{\text{LiAlH}_4 \text{ Et}_2\text{O}}$ PhCH=CCH$_2$NMe$_2$ 70%
$\quad$ Ph $\qquad\qquad\qquad\qquad\qquad\qquad\quad$ Ph

$\qquad\qquad\qquad\qquad\qquad\qquad\qquad$ Ber (1970) 103 564

Also via: Acetylenic amines (Section 305)

Section 357   Ester — Ester
               ooooooooooooooo

PrCH$_2$COOEt $\xrightarrow[\text{EtONa}]{\text{CO(OEt)}_2}$ PrCH(COOEt)$_2$ 45%

                        JACS (1941) <u>63</u> 2056

           (COOEt)$_2$, EtONa   Org Synth (1963) Coll Vol 4 141
                                      (1943) Coll Vol 2 288

           ClCOOEt, Ph$_3$CNa   JACS (1941) <u>63</u> 3156

BuBr $\xrightarrow[\text{EtOH}]{\text{CH}_2\text{(COOEt)}_2 \quad \text{EtONa}}$ BuCH(COOEt)$_2$ 80-90%

                        Org Synth (1932) Coll Vol 1 250
                        JOC (1971) <u>36</u> 3944
                        Org React (1957) <u>9</u> 107

          EtOMgCH(COOEt)$_2$ JACS (1950) <u>72</u> 351

         CH$_2$(COOEt)$_2$, NaH   JOC (1961) <u>26</u> 644
                        Synth Comm (1973) <u>3</u> 359

CH$_2$(COOEt)$_2$, basic ion exch resin   JOC (1963) <u>28</u> 504

                Aromatic halides   JACS (1959) <u>81</u> 1627

                                JOC (1971) <u>36</u> 3260
                                Org Synth (1970) <u>50</u> 38

C$_5$H$_{11}$CH=CH$_2$ $\xrightarrow[\text{Mn(OAc)}_3 \cdot 2\text{H}_2\text{O}]{\text{CH}_2\text{(COOEt)}_2}$ C$_5$H$_{11}$CH$_2$CH$_2$CH(COOEt)$_2$ 50%

                        Chem Comm (1973) 693

MeCHCOOMe $\xrightarrow[\text{CO} \quad \text{MeOH}]{\text{NaCo(CO)}_4 \quad \text{dicyclohexylamine}}$ MeCH(COOMe)$_2$ 18%
|
Br

                        JACS (1963) <u>85</u> 2779

BuCH$_2$COOEt $\xrightarrow[\text{2 CuBr}_2]{\text{1 LiN}\hspace{-0.5em}\bigcirc\text{(i-Pr)}\quad\text{THF}}$ BuCHCOOEt
                                                              |
                                                          BuCHCOOEt   63%

JACS (1971) <u>93</u> 4605

Me$_2$CCOOEt $\xrightarrow[\text{THF}]{\text{Zn   CuCl}_2}$ Me$_2$CCOOEt
   | |
   Br   Me$_2$CCOOEt   38%

Synthesis (1971) 306

$\xrightarrow[\text{MeONa}]{\text{I}_2 \quad \text{O}_2}$ (structure) COOMe, COOMe   43%

J Med Chem (1969) <u>12</u> 192

Me$_2$C=CHCOOEt $\xrightarrow[\text{DMF}]{\text{Electrolysis   Et}_4\text{N}^+ \text{ TsO}^-}$ Me$_2$CCH$_2$COOEt
                                                                        |
                                                                    Me$_2$CCH$_2$COOEt

Compt Rend (1967) <u>265</u> 751
JOC (1964) <u>29</u> 1670

Br
|
(CH$_2$)$_{10}$ $\xrightarrow[\text{NH}_3 \quad \text{Et}_2\text{O}]{\text{CH}_3\text{COOBu-t   LiNH}_2}$ CH$_2$COOBu-t
| |
Br   (CH$_2$)$_{10}$ ~98%
                                                               |
                                                           CH$_2$COOBu-t

JACS (1959) <u>81</u> 5817

COOMe
|
(CH$_2$)$_8$COOH $\xrightarrow[\text{KOH   H}_2\text{O}]{\text{Electrolysis}}$ COOMe
                                                          |
                                                      (CH$_2$)$_{16}$COOMe   40-55%

Org Synth (1955) Coll Vol 3 401
JCS (1950) 3326
Advances in Org Chem (1960) <u>1</u> 1

PrCOOH $\xrightarrow[\text{MeONa \quad MeOH}]{\substack{\text{COOH \quad OAc} \\ \text{CH}_2\text{—CHCOOEt} \quad \text{electrolysis}}}$ $\text{PrCH}_2\overset{\text{OAc}}{\underset{}{\text{CHCOOEt}}}$

JCS (1954) 1460

$\xrightarrow[]{\substack{\text{OAc} \\ \text{CH}_2\text{=COEt}}}$

$\text{CHCH}_2\text{COOEt}$
OAc

70%

Chem Comm (1969) 1216

$\underset{\text{COOH}}{\text{BuCHCOOEt}}$ $\xrightarrow[\text{2 \quad 200°}]{\text{1 \quad Pb(OAc)}_4 \quad \text{C}_6\text{H}_6}$ $\underset{\text{OAc}}{\text{BuCHCOOEt}}$ 60%

Compt Rend (1960) 251 2544

$\xrightarrow[\text{HOAc}]{\text{AgOAc \quad Ac}_2\text{O}}$

62%

Coll Czech (1969) 34 340

PhCOCH_2Me $\overset{\substack{\text{Me} \\ \text{CH}_2\text{=COAc}}}{\underset{\text{Acid}}{\dashrightarrow}}$ $\underset{\text{OAc}}{\text{PhC=CHMe}}$ $\xrightarrow[\text{2 \quad Ac}_2\text{O}]{\text{1 \quad (i-Bu)}_2\text{AlH}}$ $\underset{\text{OAc}}{\overset{\text{MeCHOAc}}{\text{PhCHCHMe}}}$ ∼65%

Annalen (1972) 763 75
Synthesis (1972) 694

$\text{PhCH}_2\text{CH=CH}_2$ $\xrightarrow[\text{C}_6\text{H}_6]{\text{I}_2 \quad \text{PhCOOAg}}$ $\underset{\text{PhCOO}}{\text{PhCH}_2\text{CHCH}_2\text{OCOPh}}$ 85%

Org React (1957) 9 332
JOC (1972) 37 789

$C_6H_{13}CH=CH_2$ $\xrightarrow[\text{H}_2\text{SO}_4]{\text{SeO}_2 \quad \text{HOAc}}$ $C_6H_{13}\underset{\underset{\text{OAc}}{|}}{C}HCH_2OAc$ 32%

Tetr Lett (1969) 4439
Bull Chem Soc Jap (1969) $\underline{42}$ 2056

$AcO(CH_2)_5COOH$ $\xrightarrow{\text{Electrolysis}}$ $AcO(CH_2)_{10}OAc$ 83%

Chem Ing Tech (1970) $\underline{42}$ 170
Synthesis (1971) 285

Also via: Dicarboxylic acids   Section 312
                    Hydroxyesters   327
                              Diols   323

Section 358   <u>Ester — Ether, Epoxide</u>

Alkoxyesters and epoxyesters (glycidic esters)

$CH_2=CHCOOPr$ $\xrightarrow{\text{PrOH} \quad \text{PrONa}}$ $PrOCH_2CH_2COOPr$ 59%

JACS (1946) $\underline{68}$ 544
          (1947) $\underline{69}$ 2966

$BuCOONa$ $\xrightarrow{\text{ClCH}_2\text{OMe} \quad \text{H}_2\text{O}}$ $BuCOOCH_2OMe$

Suom Kemist (1969) $\underline{42}$ 6
(Chem Abs $\underline{70}$ 77263)
Synth Comm (1972) $\underline{2}$ 361
JCS (1965) 2127

$\underset{\underset{\text{Me}}{|}}{Ph}CO$ $\xrightarrow[\text{C}_6\text{H}_6]{\text{ClCH}_2\text{COOEt} \quad \text{NaNH}_2}$ $Ph\overset{\overset{\displaystyle O}{\diagup \backslash}}{C}-\underset{\underset{\text{Me}}{|}}{C}HCOOEt$ 62-64%

Org Synth (1955) Coll Vol 3 727
          (1963) Coll Vol 4 649
Org React (1949) $\underline{5}$ 413
continued

$$ClCH_2COOBu\text{-}t, \ t\text{-}BuOK \quad \text{JACS (1953) } \underline{75} \ 4995$$
Rec Trav Chim (1970) $\underline{89}$ 18

$$BrCH_2COOEt, \ LiN(SiMe_3)_2 \quad \text{Tetr Lett (1972) 3761}$$

$$Br_2CHCOOEt, \ BuLi \quad \text{Tetrahedron (1972) } \underline{28} \ 3009$$

$$CH_2=CHCOOMe \quad \xrightarrow[\text{radical inhibitor } ClCH_2CH_2Cl]{\text{m-Chloroperbenzoic acid}} \quad \overset{O}{\overset{/\backslash}{CH_2\text{-}CHCOOMe}}$$

Chem Comm (1972) 64
JOC (1966) $\underline{31}$ 2509
JACS (1959) $\underline{81}$ 680

$$CF_3COO_2H \quad \text{Tetr Lett (1973) 4561}$$

Also via: Section
          Alkoxyacids   318
          Epoxyacids   318

Section 359   Ester — Halide
            ○○○○○○○○○○○○○○○○

Esters of haloacids and esters of halohydrins

$$PhC\equiv CH \ \dashrightarrow \ PhC\equiv CBr \quad \xrightarrow[\text{MeOH}]{Tl(NO_3)_3} \quad \underset{Br}{PhCHCOOMe}$$

JACS (1973) $\underline{95}$ 1296

Proc Chem Soc (1962) 117
Org React (1960) $\underline{11}$ 189

BuCH$_2$COOEt $\xrightarrow[\text{2 } X_2]{\text{1 LiN}\langle\text{i-Pr}\rangle \quad \text{THF}}$ BuCHCOOEt$\underset{\text{X}}{|}$ 95% (X=I)
92% (X=Br)

Tetr Lett (1971) 3995

$\underset{\text{C}_{10}\text{H}_{21}\text{CHCOOMe}}{\overset{\text{Me}}{|}}$ $\xrightarrow[\text{CCl}_4]{\overset{\text{CH}_2\text{CO}}{\underset{\text{CH}_2\text{CO}}{|}}\text{NX} \quad \text{benzoyl peroxide}}$ $\underset{\text{C}_{10}\text{H}_{21}\text{CCOOMe}}{\overset{\text{Me}}{|}}\overset{}{\underset{\text{X}}{|}}$

X=Cl or Br

JOC (1953) <u>18</u> 842 649
Br$_2$ Ber (1960) <u>93</u> 2222 2549

BuBr $\dashrightarrow$ BuCH(COOEt)$_2$ $\xrightarrow[\text{EtOH}]{\text{KOH}}$ BuCHCOOEt$\underset{\text{COOH}}{|}$ $\xrightarrow[\text{CCl}_4]{\text{Br}_2}$ BuCHCOOEt$\underset{\text{Br}}{|}$ <67%

JACS (1949) <u>71</u> 3107

BuI $\xrightarrow[\text{NaH THF}]{\text{MeCOCH}_2\text{COOBu-t}}$ BuCHCOOBu-t$\underset{\text{COMe}}{|}$ $\xrightarrow[\substack{\text{2 Br}_2 \text{ CH}_2\text{Cl}_2 \\ \text{3 Ba(OH)}_2 \text{ EtOH}}]{\text{1 NaH}}$ BuCHCOOBu-t$\underset{\text{Br}}{|}$ 87%

Tetr Lett (1972) 4067

$\xrightarrow[\text{2 Br}_2\text{CHCOOEt} \quad \text{t-BuOK}]{\text{1 B}_2\text{H}_6 \quad \text{THF}}$ 47%

JACS (1968) <u>90</u> 1911
Can J Chem (1972) <u>50</u> 2387

$\underset{\text{OH}}{\overset{}{|}}$RCH(CH$_2$)$_N$COOEt $\xrightarrow{\text{Ph}_3\text{P} \quad \text{CCl}_4}$ $\underset{\text{Cl}}{\overset{}{|}}$RCH(CH$_2$)$_N$COOEt

N=0 or 5

Tetrahedron (1967) <u>23</u> 359
(PhO)$_3$P·MeI (iodoesters) JCS (1953) 2224

$CH_2=CHCOOMe$ $\xrightarrow{\text{HBr   Et}_2\text{O}}$ $BrCH_2CH_2COOMe$ 80-84%

Org Synth (1955) Coll Vol 3 576

$C_6H_{13}CH=CH_2$ $\xrightarrow[\text{diacetyl peroxide}]{\text{BrCH}_2\text{COOEt}}$ $C_6H_{13}\underset{\overset{|}{Br}}{C}HCH_2CH_2COOEt$

JACS (1948) <u>70</u> 1055

$\xrightarrow{\text{HBr   EtOH}}$ 77-84%

Org Synth (1965) <u>45</u> 42
Synthesis (1973) 538

$EtCH_2COOMe$ $\xrightarrow[\text{2   Br(CH}_2)_3\text{Br}]{\text{1   LiN(Pr-i)}_2\text{   THF}}$ $Et\underset{\overset{|}{(CH_2)_3Br}}{C}HCOOMe$ 90%

Tetr Lett (1973) 2425

$\underset{MeCHOH}{\overset{MeCHOH}{|}}$ $\xrightarrow{MeC(OMe)_3}$ $\underset{MeCHO}{\overset{MeCHO}{}}$✕$\underset{Me}{\overset{OMe}{}}$ $\xrightarrow{Me_3SiCl}$ $\underset{MeCHOAc}{\overset{MeCHCl}{|}}$

JOC (1973) <u>38</u> 4203 1173
JACS (1973) 95 278

$C_8H_{17}CH=CH_2$ $\xrightarrow{\text{NCS   DMF}}$ $\underset{HCOO}{\overset{}{}}C_8H_{17}CHCH_2Cl$ 90%

Ber (1973) <u>106</u> 606

1,2-Acyloxyhalides (acyl=HCO,
MeCO, EtCO, etc., halide=Br, Cl) JACS (1959) <u>81</u> 2195
                              Chem Pharm Bull (1972) <u>20</u> 2707

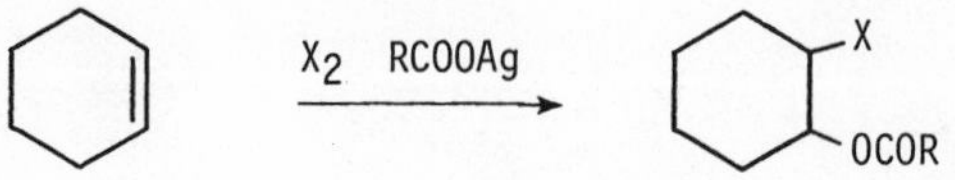

X=Br or I

R=Me or Ph

Org React (1957) 9 332
Tetrahedron (1972) 28 3475

I_2, KIO_3, HOAc   Tetr Lett (1973) 4485

Also via: Section
            Haloacids   319
            Halohydrins   329

Section 360   Ester — Ketone
              oooooooooooooooo

BuC≡CH --→ BuC≡CBr $\xrightarrow{O_3 \quad MeOH}$ BuCOCOOMe   40%

JOC (1973) 38 3653

PhCHO $\xrightarrow[\substack{Et_2O \quad EtOH \\ 2 \ HCl \quad H_2O}]{1 \ Me_2NCH_2COOEt \quad NaH}$ PhCH$_2$COCOOEt   72%

Annalen (1967) 703 37

PhCHO $\xrightarrow[HOAc]{PhSH \quad ZnCl_2}$ PhCH(SPh)$_2$ $\xrightarrow[\substack{2 \ ClCOOEt \\ 3 \ Hydrolysis}]{1 \ NaH \quad DMF}$ PhCOCOOEt

JOC (1963) 28 961

Ber (1960) 93 1387

PhCOOEt $\xrightarrow[\text{t-BuOK}]{\text{Me}_2\text{SO}}$ PhCOCH$_2$SOMe $\xrightarrow[\text{EtOH}]{\text{H}_2\text{SO}_4}$ PhCOCOOEt   55%

JACS (1966) 88 5498

BuBr $\xrightarrow[\text{2 NBS}]{1}$ BuCOCOOEt   56%

JOC (1972) 37 505

BuBr --→ Bu$_2$Cd $\xrightarrow[\text{CH}_2\text{Cl}_2]{\substack{\text{COCl} \\ \text{COOEt} \quad \text{LiBr} \quad \text{THF}}}$ BuCOCOOEt   70%

JCS A (1966) 456
JACS (1954) 76 1914

$\xrightarrow[\text{CH}_2\text{Cl}_2]{\substack{\text{COCl} \\ \text{COOEt} \quad \text{AlCl}_3}}$ 99%

J Med Chem (1972) 15 1029

PhCOCH$_3$ --→ PhCOCH$_2$Br $\xrightarrow[]{\text{SeO}_2 \quad \text{EtOH}}$ PhCOCOOEt   <70%

JOC (1959) 24 1825

BuC≡CCOOMe $\xrightarrow[\text{2 H}_2\text{O}_2 \quad \text{H}_2\text{O} \quad \text{pH 8}]{1 \; \left(\substack{\text{Me} \\ \text{i-PrCH}}\right)_2\text{BH}}$ BuCH$_2$COCOOMe   61%

Organometallics in Chem Synth (1971) 1 249

$$\overset{O}{\overset{/\backslash}{PhCHCHCOOEt}} \xrightarrow{BF_3 \quad C_6H_6} PhCH_2COCOOEt \qquad\qquad 80\%$$

JACS (1958) <u>80</u> 6386

$$EtC{\equiv}CEt \xrightarrow[\text{2 MeOH}]{\text{1 ClSO}_2\text{NCO} \quad \text{CH}_2\text{Cl}_2} \underset{\overset{|}{Et}}{EtCOCHCOOMe} \qquad 58\%$$

Tetr Lett (1970) 27

$$PhCOCl \xrightarrow[\substack{\text{2 Ph}_3\text{P} \quad \text{i-Pr}_2\text{O} \\ \text{3 MeOH} \quad \text{H}_2\text{O} \\ \text{4 ZnCl}_2 \quad \text{Et}_2\text{O} \\ \text{5 Pyr}}]{\text{1 N}_2\text{CHCOOEt}} PhCOCH_2COOEt \qquad 45\%$$

Ber (1963) <u>96</u> 1948

$$EtCH_2COCl \xrightarrow[\text{2 HCl} \quad \text{H}_2\text{O}]{\overset{\overset{\text{OSi(Me}_2)\text{Bu-t}}{|}}{\text{1 CH}_2{=}\text{COEt}} \quad \text{Et}_3\text{N} \quad \text{THF}} EtCH_2COCH_2COOEt \qquad \sim85\%$$

Tetr Lett (1973) 1297

$$EtCOCl \xrightarrow[\text{2 TsOH} \quad \text{C}_6\text{H}_6]{\overset{\overset{\text{COOBu-t}}{|}}{\text{1 EtOMg-CHCOOEt}} \quad \text{Et}_2\text{O}} EtCOCH_2COOEt \qquad 60\%$$

JACS (1944) <u>66</u> 1286
(1959) <u>81</u> 2907
JCS <u>C</u> (1971) 2821

$$PhCOCl \xrightarrow[\text{2 NH}_3 \quad \text{NH}_4\text{Cl} \quad \text{H}_2\text{O}]{\text{1 MeCOCH}_2\text{COOEt} \quad \text{Na} \quad \text{C}_6\text{H}_6} PhCOCH_2COOEt \qquad \sim50\%$$

Org Synth (1943) Coll Vol 2 266
(1963) Coll Vol 4 415

$$t\text{-BuOH} \xrightarrow{\text{Diketene}} MeCOCH_2COOBu\text{-}t \qquad 75\text{-}80\%$$

Org Synth (1962) <u>42</u> 28
MeCOCH_2COF   JOC (1961) <u>26</u> 225

Review: The Acetoacetic Ester Condensation

Org React (1942) $\underline{1}$ 266

$$\text{CH}_3\text{COOEt} \xrightarrow{\text{Na}} \text{MeCOCH}_2\text{COOEt}$$

Org Synth (1932) Coll Vol 1 235

NaH   Bull Soc Chim Fr (1954) 504

NaN(SiMe$_3$)$_2$ Angew (1963) $\underline{75}$ 793
(Internat Ed $\underline{2}$ 617)

Ba$_3$N$_4$ Z Naturforsch (1969) $\underline{24}$ 937

$$\text{Me}_2\text{CHCOOEt} \xrightarrow[\text{2 PhCOCl}]{\text{1 Ph}_3\text{CNa} \quad \text{Et}_2\text{O}} \underset{\text{Me}}{\overset{\text{Me}}{\text{PhCOCCOOEt}}}$$

50-55%

Org Synth (1943) Coll Vol 2 268

Lithium N-isopropylcyclohexylamide   Tetr Lett (1971) 2953

$$\text{PhCH}_2\text{COCH}_2\text{COOEt} \xrightarrow[\text{2 MeI}]{\text{1 NaH} \quad \text{DMF} \quad \text{C}_6\text{H}_6} \underset{\text{Me}}{\text{PhCH}_2\text{COCHCOOEt}}$$

77%

JOC (1956) $\underline{21}$ 245

RX, K$_2$CO$_3$ Synthesis ($\overline{1973}$) 316

γ-Alkylation of β-ketoesters   JACS (1970) $\underline{92}$ 6702

$$\xrightarrow[\text{ClCH}_2\text{CH}_2\text{Cl}]{\underset{\text{CH}_2\text{COOEt}}{\text{COCl}} \quad \text{AlCl}_3}$$

< 73%

JOC (1972) $\underline{37}$ 3687

$$\text{EtCOEt} \xrightarrow[\text{CH}_2\text{Cl}_2]{\text{N}_2\text{CHCOOEt} \quad \text{Et}_3\text{O}^+ \text{ BF}_4^-} \underset{\text{COOEt}}{\text{EtCOCHEt}}$$

86%

JACS (1970) $\underline{92}$ 5767

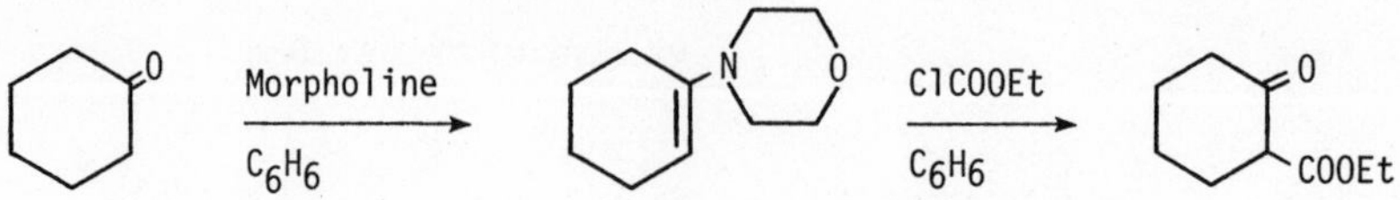

JACS (1963) **85** 207

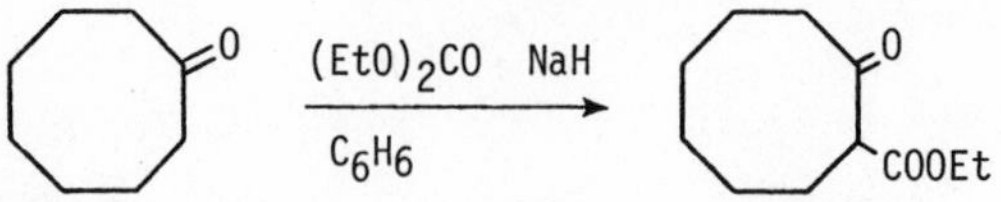

91-94%

Org Synth (1967) **47** 20
JOC (1973) **38** 3244
     (1972) **37** 2202

(COOEt)$_2$, EtONa; $\triangle$ Org Synth (1943) Coll Vol 2 531
(EtO)$_2$POCOOEt, NaH Tetr Lett (1966) 2201
Ph$_3$CNa; CO$_2$; CH$_2$N$_2$ Helv (1945) **28** 1677
     (COBr)$_2$; EtOH Ber (1960) **93** 551

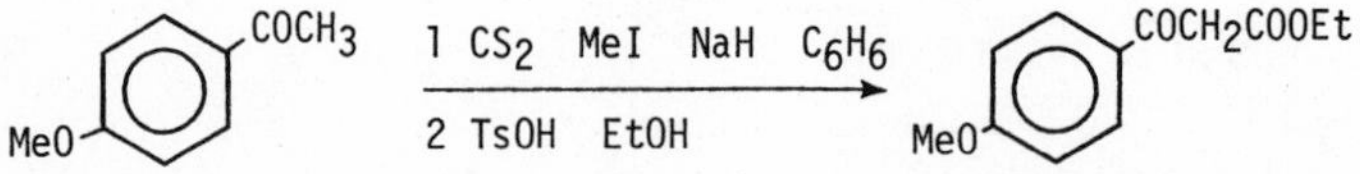

72%

Tetr Lett (1973) 4207

EtCN $\xrightarrow[\text{C}_6\text{H}_6]{\underset{\text{Me}_2\text{CCOOEt Zn}}{\overset{\text{Br}}{|}}}$ EtCOCCOOEt

80%

Bull Soc Chim Fr (1966) 1819
Org Synth (1963) Coll Vol 4 120

C$_6$H$_{13}$C≡CH $\xrightarrow{\begin{array}{l}\text{1 BuLi hexane}\\ \text{2 (C}_6\text{H}_{13}\text{)}_3\text{B diglyme}\\ \text{3 BrCH}_2\text{COOEt}\\ \text{4 NaOAc H}_2\text{O}_2\text{ H}_2\text{O}\end{array}}$ C$_6$H$_{13}$COCHCH$_2$COOEt

78%

Tetr Lett (1973) 4491

$$\text{PhCHO} \xrightarrow[\text{NaCN \quad DMF}]{\text{MeCH=CHCOOEt}} \underset{\underset{\text{Me}}{|}}{\text{PhCOCHCH}_2\text{COOEt}} \qquad 33\%$$

Angew (1973) <u>85</u> 89
(Internat Ed <u>12</u> 81)

$$\text{PhCOOEt} \xdashrightarrow[\text{t-BuOK}]{\text{Me}_2\text{SO}} \text{PhCOCH}_2\text{SOMe} \xrightarrow[\substack{\text{2 BrCH}_2\text{COOEt} \\ \text{3 Zn \quad HOAc}}]{\text{1 NaH \quad THF}} \text{PhCOCH}_2\text{CH}_2\text{COOEt} \qquad <52\%$$

JOC (1969) <u>34</u> 3624 3618

$$\text{EtCH}_2\text{COOEt} \xrightarrow[\substack{\text{2 MeI} \\ \text{3 HClO}_4 \quad \text{MeCN} \quad \text{H}_2\text{O}}]{\substack{\text{SOMe} \\ | \\ \text{1 CH}_2\text{=CSMe} \quad \text{LiN(Pr-i)}_2 \quad \text{THF}}} \underset{\underset{\text{CH}_2\text{COMe}}{|}}{\text{EtCHCOOEt}}$$

Tetr Lett (1973) 4715

$$\text{i-Pr(CH}_2)_2\text{Br} \xrightarrow[\substack{\text{2 CdCl}_2 \\ \text{3 (CH}_2)_2\text{COOMe} \quad \text{C}_6\text{H}_6 \\ | \\ \text{COCl}}]{\text{1 Mg \quad Et}_2\text{O}} \text{i-Pr(CH}_2)_2\text{CO(CH}_2)_2\text{COOMe} \qquad 59\%$$

JACS (1944) <u>66</u> 46
Org Synth (1955) Coll Vol 3 601

$$\text{PhI} \xrightarrow[\text{ROH \quad HCl}]{\text{HC≡CH \quad Ni(CO)}_4} \text{PhCOCH}_2\text{CH}_2\text{COOR}$$

Tetr Lett (1964) 2777

JACS (1963) <u>85</u> 207
Coll Czech (1966) <u>31</u> 602

(cyclohexanone) --→ (1-ethoxycyclohexene, OEt)

1 $N_2CHCOOEt$ copper bronze
cyclohexane
2 HCl EtOH H_2O

→ (2-substituted cyclohexanone, CH_2COOEt)

Synth Comm (1973) 3 255

PhCH=CHCOOMe

$\xrightarrow[\text{Et}_2O]{\text{BuLi} \quad \text{Ni(CO)}_4}$

$\underset{\text{BuCO}}{PhCHCH_2COOMe}$ 59%

JACS (1969) 91 4926

$C_5H_{11}C{\equiv}CCH_2COOEt$

$\xrightarrow[\text{EtOH}]{\text{HgSO}_4 \quad \text{H}_2SO_4}$

$C_5H_{11}COCH_2CH_2COOEt$ 85%

Can J Chem (1972) 50 1105

$PhCOCH_2Br$

$\xrightarrow[\text{2 K}_2CO_3 \quad \text{H}_2O]{\text{1 Pyr}}$

PhCOCHN⁻(ring)   +

$\xrightarrow[\substack{\text{NaH} \quad \text{DMF} \\ \text{2 Zn} \quad \text{HOAc}}]{\text{1 BrCH}_2COOEt}$

$PhCOCH_2CH_2COOEt$ 56%

Aust J Chem (1967) 20 2441

$PhCH_2COOEt$

$\xrightarrow[\text{LiNH}_2]{\text{CH}_2\text{=CHCOMe}}$

$\underset{CH_2CH_2COMe}{PhCHCOOEt}$ 29%

JOC (1964) 29 2346

(cyclohexanone)

$\xrightarrow[\text{C}_6H_6]{\text{HN(pyrrolidine)}}$

(1-pyrrolidinocyclohexene, N-ring)

$\xrightarrow[\text{DMF}]{\text{MeCH=CHCOOEt}}$

(cyclohexanone, $\underset{Me}{CHCH_2COOEt}$)   < 56%

JACS (1963) 85 207
Tetrahedron (1964) 20 1737

t-BuCOCH$_3$ $\xrightarrow[\text{NaNH}_2]{\text{PhCH=CHCOOEt}}$ t-BuCOCH$_2$CHCH$_2$COOEt   64%
                                                  |
                                                Ph

JOC (1949) <u>14</u> 261

Me$_2$C=CH$_2$ $\xrightarrow[\substack{\text{2 CH}_2\text{=CHCH}_2\text{COOEt} \\ \text{3 CO}}]{\substack{\text{Me} \\ | \\ \text{i-PrCBH}_2 \\ 1 \quad | \\ \text{Me}}}$ Me$_2$CHCH$_2$CO(CH$_2$)$_3$COOEt   84%

JACS (1967) <u>89</u> 5285

Br(CH$_2$)$_4$COOEt $\xrightarrow[\text{2 EtI}]{\text{1 Na}_2\text{Fe(CO)}_4}$ EtCO(CH$_2$)$_4$COOEt   74%

JACS (1972) <u>94</u> 1788

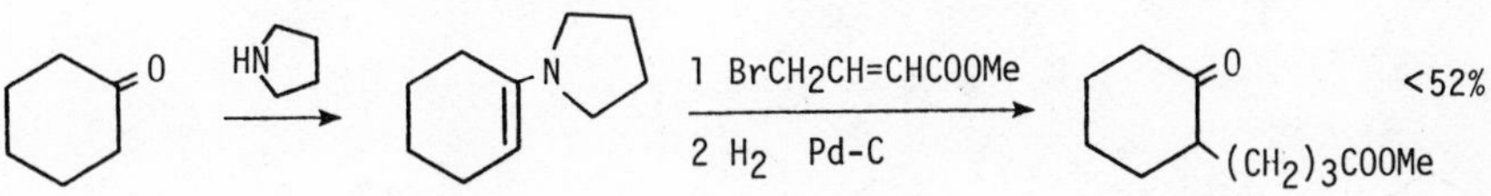

Tetr Lett (1965) 959

C$_6$H$_{13}$CHO $\xrightarrow[\text{benzoyl peroxide}]{\text{CH}_2\text{=CH(CH}_2)_8\text{COOEt}}$ C$_6$H$_{13}$CO(CH$_2$)$_{10}$COOEt

Rec Trav Chim (1953) <u>72</u> 84

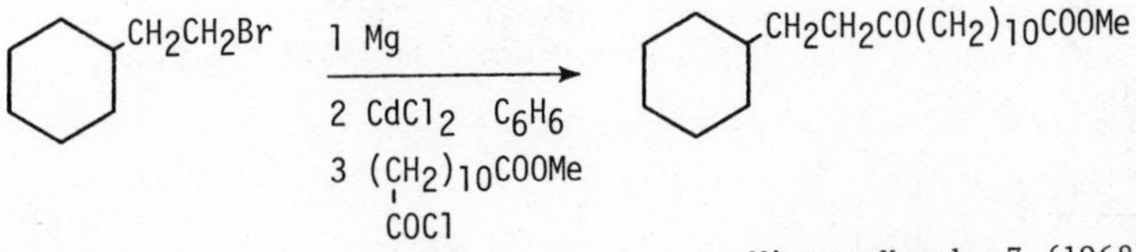

Nippon Kagaku Z (1968) <u>89</u> 516
(Chem Abs <u>69</u> 76716)
Org Synth (1955) Coll Vol 3 601

COOMe
|
(CH₂)₄COOH $\xrightarrow[\text{MeONa\quad MeOH}]{\text{MeCOCH}_2\text{CH}_2\text{COOH}\quad\text{electrolysis}}$ COOMe
|
(CH₂)₄CH₂CH₂COMe   20%

Nippon Kagaku Z (1956) 77 163
(Chem Abs 52 268)

C₁₀H₂₁COOH $\xrightarrow[\text{2 CH}_2\text{N}_2\quad\text{Et}_2\text{O}]{\text{1 SOCl}_2}$ C₁₀H₂₁COCHN₂ $\xrightarrow{\text{HOAc}}$ C₁₀H₂₁COCH₂OAc   >90%

Org React (1954) 8 228

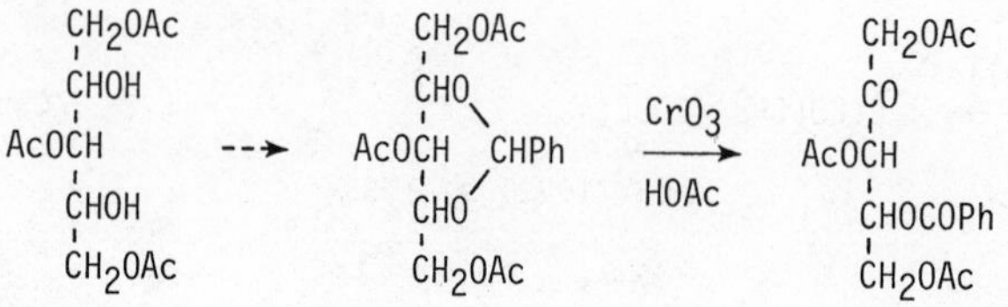

CH₂OAc
|
CHOH
|
AcOCH   - →
|
CHOH
|
CH₂OAc

CH₂OAc
|
CHO
|
AcOCH   CHPh
|
CHO
|
CH₂OAc

$\xrightarrow[\text{HOAc}]{\text{CrO}_3}$

CH₂OAc
|
CO
|
AcOCH
|
CHOCOPh
|
CH₂OAc

Aust J Chem (1971) 24 1219

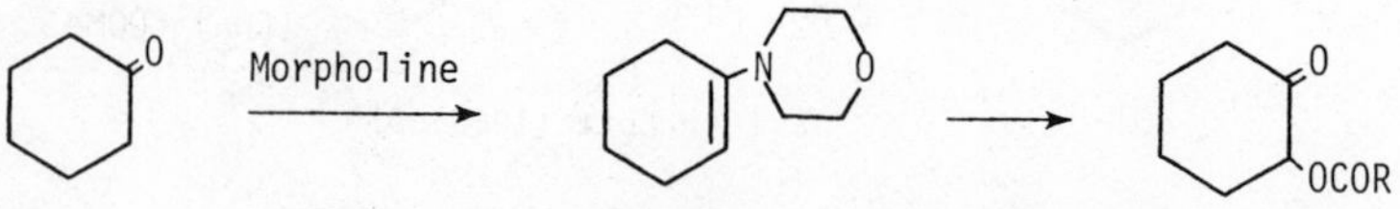

Tl(OAc)₃   JOC (1968) 33 3359
Dibenzoyl peroxide JOC (1963) 28 581

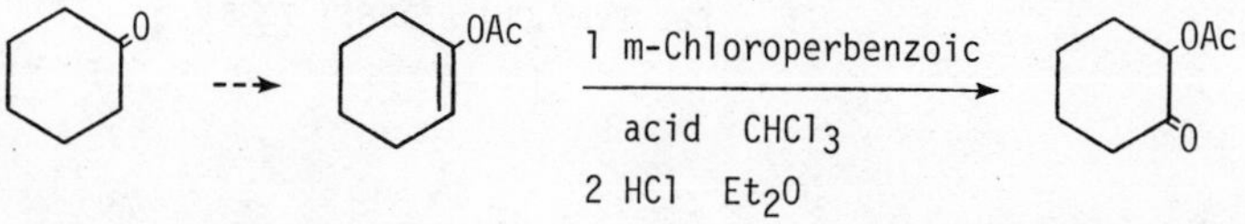

1 m-Chloroperbenzoic
acid CHCl₃
2 HCl Et₂O

JOC (1967) 32 3934
     (1962) 27 2131
Chem Pharm Bull (1972) 20 2156

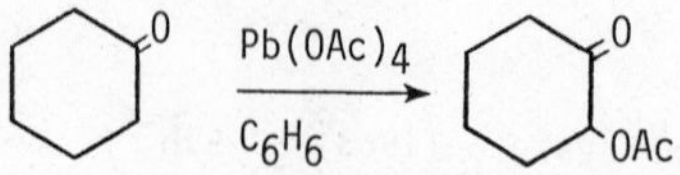

Pb(OAc)₄
C₆H₆

JCS (1955) 4426
Synthesis (1973) 567
JCS (1961) 4472
Chem Comm (1970) 406

61%

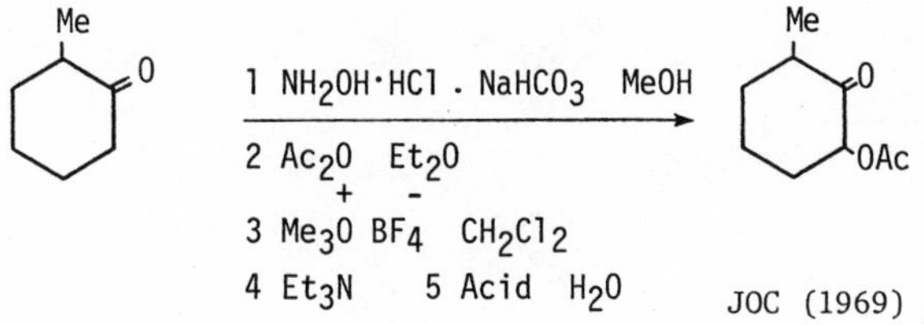

Me
O
1 NH2OH·HCl . NaHCO3 MeOH
2 Ac2O Et2O
3 Me3O BF4 CH2Cl2
4 Et3N 5 Acid H2O
Me
O
OAc

40-51%

JOC (1969) 34 1430

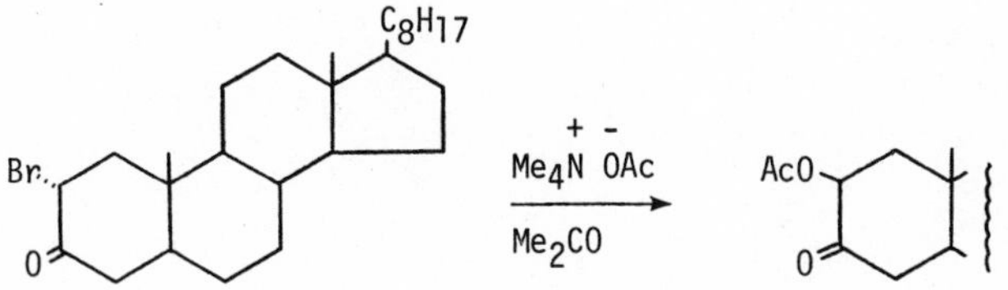

C8H17
Br
O
Me4N OAc
Me2CO
AcO
O

JOC (1961) 26 4563
JCS (1961) 2725
Tetrahedron (1963) 19 861

$C_6H_{13}CH{=}CH_2$ —[$Mn(OAc)_3$ Me_2CO / HOAc]→ $C_6H_{13}CH(OAc)CH_2CH_2COMe$ 24%

JACS (1971) 93 524

$C_8H_{17}CH{=}CH_2$ —[$KMnO_4$ Ac_2O]→ $C_8H_{17}COCH_2OAc$ 38%

JACS (1971) 93 3303

$C_5H_{11}Br$ —[$MeCOCH_2COOBu$-t / EtONa EtOH]→ $C_5H_{11}CH(COOBu$-t$)COMe$ —[1 NaH C_6H_6 / 2 Benzoyl peroxide / 3 TsOH]→ $C_5H_{11}CH(PhCOO)COMe$

Acta Chem Scand (1960) 14 1445

Tetrahedron (1967) <u>23</u> 2453
Bull Soc Chim Fr (1960) 1079

Also via: Section
                Ketoacids   320
        Hydroxyketones   330

Section 361   Ester — Nitrile
               ‍ooooooooooooooooo

α, β and higher cyanoesters. Esters of cyanohydrins

PrCHO → PrCH₂CHCOOEt   94-96%

$$\text{PrCHO} \xrightarrow[\text{HOAc}]{\overset{\text{CN}}{\underset{\mid}{\text{CH}_2\text{COOEt}}}\ \ \text{H}_2\ \ \text{Pd-C}} \text{PrCH}_2\overset{\overset{\text{CN}}{\mid}}{\text{CH}}\text{COOEt}$$

Org Synth (1955) Coll Vol 3 385

Review: The Alkylation of Esters and Nitriles

Org React (1957) <u>9</u> 107

$$\text{i-PrBr} \xrightarrow[\text{EtOH}]{\overset{\text{CN}}{\underset{\mid}{\text{BuCHCOOEt}}}\ \ \text{EtONa}} \overset{\overset{\text{CN}}{\mid}}{\underset{\underset{\text{i-Pr}}{\mid}}{\text{BuCCOOEt}}}$$ 87%

Org React (1957) <u>9</u> 161
Tetr Lett (1972) <u>1</u>279

JOC (1972) <u>37</u> 825

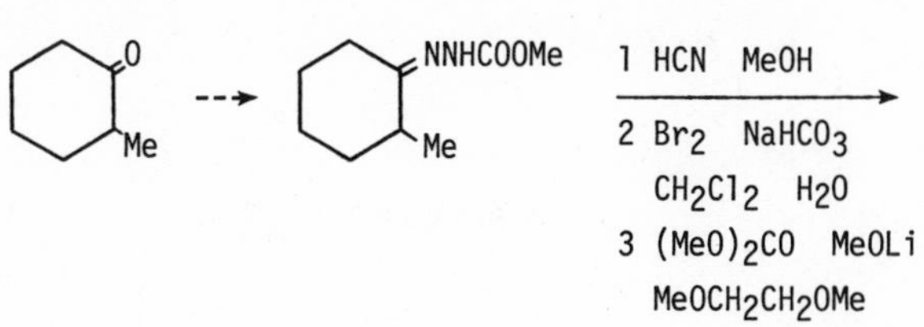

1 HCN   MeOH

2 Br$_2$ NaHCO$_3$
  CH$_2$Cl$_2$ H$_2$O

3 (MeO)$_2$CO   MeOLi
  MeOCH$_2$CH$_2$OMe

JACS (1971) 93 4318

<25%

PhCH$_2$CN $\xrightarrow[\text{toluene}]{\text{CO(OEt)}_2 \ \ \text{EtONa}}$ PhCHCN
                                                         |
                                                       COOEt

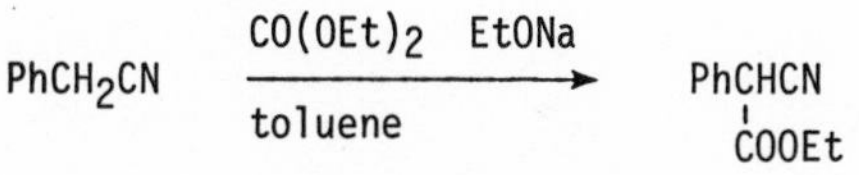

70-78%

Org Synth (1963) Coll Vol 4 461
JACS (1971) 93 4237

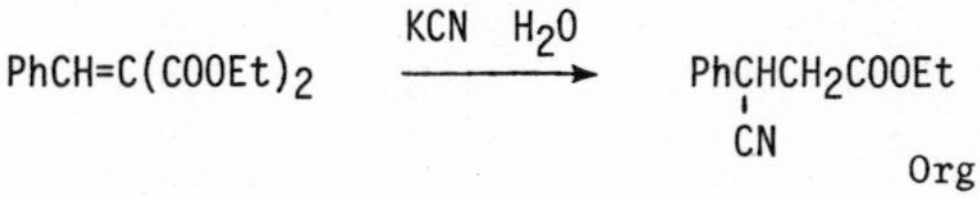

PhCH=C(COOEt)$_2$ $\xrightarrow{\text{KCN} \ \ \text{H}_2\text{O}}$ PhCHCH$_2$COOEt
                                                                      |
                                                                      CN

Org Synth (1963) Coll Vol 4 804
          (1955) Coll Vol 3 377
J Med Chem (1972) 15 1297

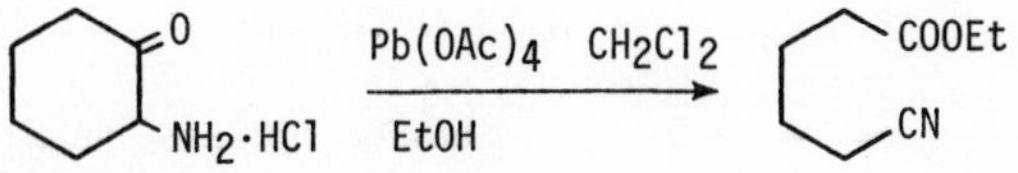

$\xrightarrow[\text{EtOH}]{\text{Pb(OAc)}_4 \ \ \text{CH}_2\text{Cl}_2}$

25%

JOC (1971) 36 3668

$\xrightarrow[\text{HOAc}]{\text{KCN} \ \ \text{Ac}_2\text{O}}$

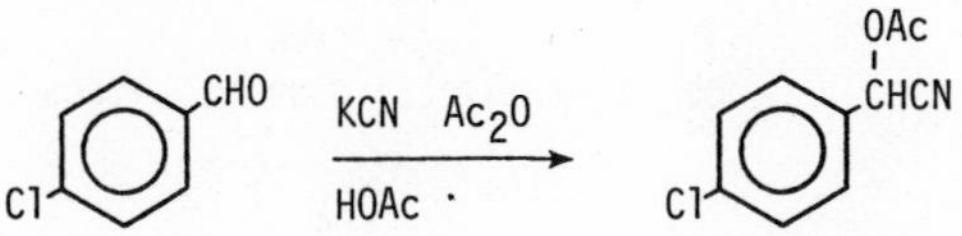

JOC (1959) 24 1650
    (1949) 14 1013

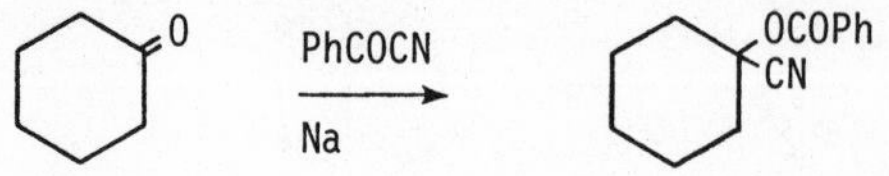

80%

Compt Rend (1971) <u>272</u> 1554

Also via: Section
              Hydroxynitriles   331
              Cyanoacids   321

Section 362   Ester — Olefin

For allylic acetoxylation see section 116 vol 1 and 2 (Esters from Hydrides)

$$C_5H_{11}CH_2C\equiv CH \xrightarrow[\text{HOAc}\quad H_2O]{\text{NBS}\quad \text{NaOAc}} C_5H_{11}CH_2COCHBr_2 \xrightarrow[\text{MeOH}]{\text{Et}_3N} C_5H_{11}CH=CHCOOMe \quad 46\%$$

cis

Proc Chem Soc (1964) 148

$$C_9H_{19}CH_2\underset{\underset{\text{Me}}{|}}{C}HCOOH \xrightarrow[\text{2 MeOH}]{\text{1 Br}_2\quad \text{PBr}_3} C_9H_{19}CH_2\overset{\overset{\text{Br}}{|}}{\underset{\underset{\text{Me}}{|}}{C}}COOMe \xrightarrow{\text{Quinoline}} C_9H_{19}CH=\underset{\underset{\text{Me}}{|}}{C}COOMe$$

Org Synth (1963) Coll Vol 4 608 616

$$PhCHO \xrightarrow[\text{EtOH}]{\text{CH}_3COOEt\quad \text{EtONa}} PhCH=CHCOOEt \qquad 68\text{-}74\%$$

Org Synth (1932) Coll Vol 1 252

BuCHO $\xrightarrow[\text{piperidine}]{\overset{\displaystyle\text{COOH}}{\underset{}{\text{CH}_2\text{COOEt}}}\ \text{Pyr}}$ BuCH=CHCOOEt   78%

JACS (1948) <u>70</u> 2601

PhCHO $\xrightarrow{\overset{\displaystyle 1\ \overset{\text{SCN}}{\text{CH}_2\text{COOEt}}\ \ \text{KF}\ \ \text{Et}_2\text{O}}{2\ \text{NaOH}\ \ \text{EtOH}}}$ PhCH=CHCOOEt

Bull Chem Soc Jap (1971) <u>44</u> 1357

$\overset{\text{CHO}}{\underset{}{(\text{CH}_2)_3\text{CHO}}}$ $\xrightarrow{\text{Ph}_3\text{P=CHCOOMe}}$ $\overset{\text{CHO}}{\underset{}{(\text{CH}_2)_3\text{CH=CHCOOMe}}}$

Annalen (1962) <u>658</u> 91
Org React (1965) <u>14</u> 270

$\overset{\text{Me}}{\underset{}{\text{Ph}_3\text{P=CCOOEt}}}$ JACS (1972) <u>94</u> 4298

(EtO)$_2$POCH$_2$COOEt, NaH   JOC (1971) <u>36</u> 1040

BrCH$_2$COOEt, (Me$_2$N)$_3$P, MeONa   Annalen (1965) <u>682</u> 58

CH$_2$=CHCOOEt, Ph$_3$P   JACS (1968) <u>90</u> 1647
Tetr Lett (1964) 1653

$\overset{}{\underset{\text{Et}}{\text{PrCHCH}_2\text{COOMe}}}$ $\xrightarrow{\overset{\displaystyle 1\ \bigcirc\!-\!\overset{\text{Pr-i}}{\text{NLi}}\ \ \text{THF}}{\begin{array}{l}2\ \text{Me}_2\text{S}_2\\3\ \text{NaIO}_4\ \ \text{MeOH}\ \ \text{H}_2\text{O}\\4\ 120°\end{array}}}$ $\overset{}{\underset{\text{Et}}{\text{PrC=CHCOOMe}}}$ 72%

JACS (1973) <u>95</u> 6840

C$_9$H$_{19}$CH$_2$CH$_2$COOEt $\xrightarrow{\overset{\displaystyle 1\ \text{R}_2\text{NLi}\ \ \text{THF}}{\begin{array}{l}2\ \text{PhSeCl}\\3\ \text{MeCOO}_2\text{H}\ \ \text{EtOAc}\end{array}}}$ C$_9$H$_{19}$CH=CHCOOEt

JACS (1973) <u>95</u> 6137

Cyclohexanone with CH_2, O group $\xrightarrow{Bu_3P=CHCOOEt}$ cyclohexane with $CH=CHCOOEt$

JACS (1967) <u>89</u> 5850
(1971) <u>93</u> 1693

$C_6H_{13}I \xrightarrow[\text{NaH THF}]{CH_2(COOMe)_2} C_6H_{13}CH(COOMe)_2 \xrightarrow[\substack{2\ MeI \\ 3\ 75°}]{\substack{1\ HCHO\ Me_2NH \\ Me_2NH\cdot HCl}} C_6H_{13}\underset{CH_2}{\overset{\|}{C}}COOMe$ <47%

Synth Comm (1973) <u>3</u> 359

$PrCH_2Br \dashrightarrow PrCH=PPh_3 \xrightarrow[C_6H_6]{\overset{I}{\underset{}{MeCHCOOEt}}} PrCH=\underset{Me}{C}COOEt$ 55%

Ber (1970) <u>103</u> 685

$PhI \xrightarrow[\text{N-methylpyrrolidone}]{CH_2=CHCOOMe\ \ Pr_3N\ \ Pd(OAc)_2} PhCH=CHCOOMe$ 81%

JOC (1972) <u>37</u> 2320

$\underset{CH=CMe_2}{CH_2Br} \xrightarrow[\substack{2\ CH_2=COAc\ TsOH \\ Me}]{\substack{1\ MeCOCH_2COOMe \\ NaH\ BuLi\ THF}} \underset{CH=CMe_2}{\overset{OAc}{CH_2CH_2C=CHCOOMe}} \xrightarrow[Et_2O]{Me_2CuLi} \underset{CH=CMe_2}{\overset{Me}{CH_2CH_2C=CHCOOMe}}$

Synth Comm (1973) <u>3</u> 321

Review: The Wittig Reaction   Org React (1965) <u>14</u> 270

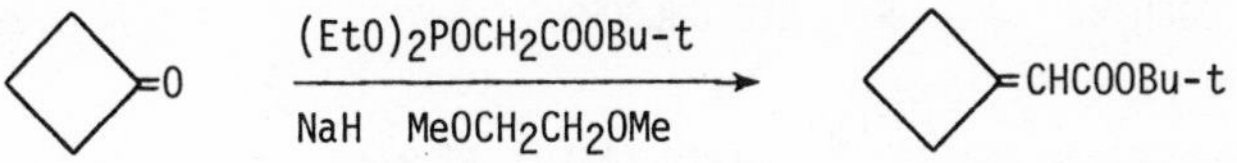

$(EtO)_2POCH_2COOBu$-t

NaH　$MeOCH_2CH_2OMe$

=CHCOOBu-t　　　83%

JOC (1971) 36 1024
Org Synth (1965) 45 44

Ph_3P=CHCOOMe 150°　Proc Chem Soc (1961) 454

Ph_3P=CHCOOEt, PhCOOH　Angew (1963) 75 858
(Internat Ed 2 619)

Review:　The Reformatsky Reaction　　Org React (1942) 1 1

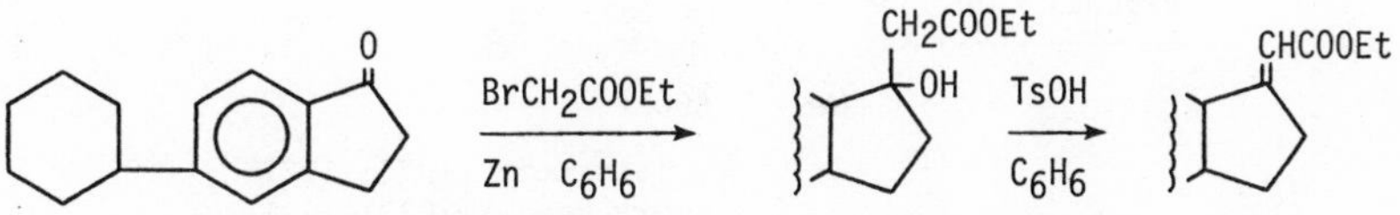

$BrCH_2COOEt$

Zn　C_6H_6

CH_2COOEt

OH　TsOH

C_6H_6

CHCOOEt

42%

J Med Chem (1972) 15 1297

$BrCH_2COOBu$-t, Mg　JOC (1966) 31 983

Dehydration of β-hydroxyesters with:
　　　　　　DCC, $CuCl_2$　JCS Perkin I (1973) 2738

AcCl, $PhNMe_2$; EtONa　Tetr Lett (1969) 5017

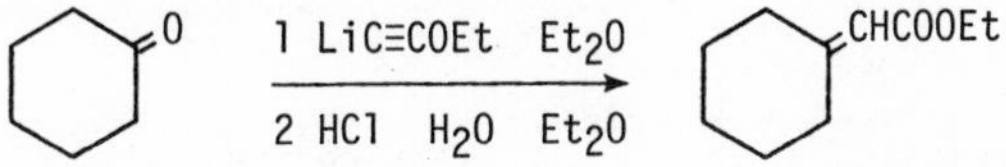

1 LiC≡COEt　Et_2O

2 HCl　H_2O　Et_2O

CHCOOEt

~50%

Advances in Org Chem (1960) 2 117 203 204
Rec Trav Chim (1956) 75 1377

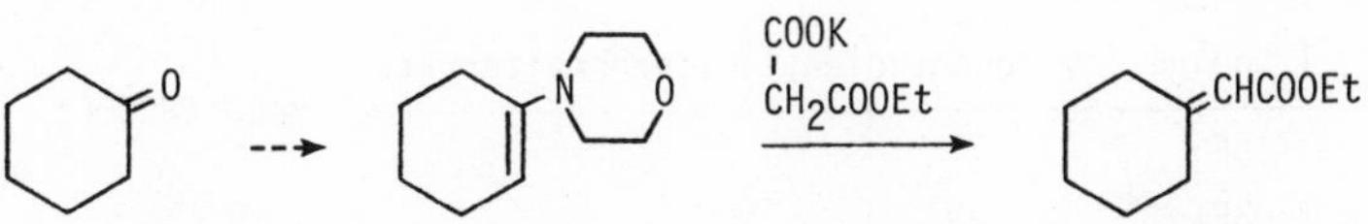

COOK

CH_2COOEt

CHCOOEt

<29%

JOC (1973) 38 399

EtCHCOCH$_3$ $\xrightarrow{Br_2}$ EtCCOCH$_2$Br $\xrightarrow{MeONa}$ EtC=CHCOOMe   29%
  | | |
  Me·   Me   Me

JACS (1950) $\underline{72}$ 974
JOC (1971) $\underline{36}$ 3266
Org React ($\overline{1960}$) $\underline{11}$ 261

(CH$_2$)$_9$ with CH$_2$/CO/CH$_2$ ring $\xrightarrow[Et_2O]{Br_2}$ (CH$_2$)$_9$ with CHBr/CO/CHBr ring $\xrightarrow[C_6H_6]{MeONa}$ (CH$_2$)$_9$ with CH=CCOOMe   ~90%

JOC (1968) $\underline{33}$ 2157

MeC≡CCOOEt $\xrightarrow{Pr_2CuLi\quad THF}$

          Pr   H
            \   /
             C=C
            /   \
          Me   COOEt   49%

Tetr Lett (1973) 1277
Org React (1972) $\underline{19}$ 1
JACS (1969) $\underline{91}$ 1851 1853 6186

BuC≡CCOOMe $\xrightarrow[\text{2 HOAc}]{1\ \left(\text{i-PrCH}_{\overset{|}{Me}}\right)_2BH\quad THF}$ BuCH=CHCOOMe

                                              cis

Organometallics in Chem Synth (1971) $\underline{1}$ 249

PhCH$_2$CHCOOMe $\dashrightarrow$ PhCH$_2$CCOOMe $\xrightarrow{MeONa}$ cis
          | ‖   PhCH=CHCOOMe
          NH$_2$ N$_2$ $\xrightarrow[DCC]{BF_3\cdot Et_2O}$ trans

Tetr Lett (1973) 4267

O
/\
MeCHCHCOOEt $\xrightarrow{\text{1 Sodium (cyclopentadienyl)dicarbonylferrate}}$ MeCH=CHCOOEt

          THF   96%
2 HBF$_4$
3 NaI Me$_2$CO

JACS (1972) $\underline{94}$ 7170
Ph$_2$PLi; MeI JACS (1973) $\underline{95}$ 822

PhCOCOOMe $\xrightarrow[\text{THF}]{\text{Ph}_3\text{P}=\text{CMe}_2}$ PhCCOOMe   72%
$\overset{\|}{\text{CMe}_2}$

Can J Chem (1971) <u>49</u> 2143

—COOMe $\xrightarrow[\text{octane}]{\text{Fe(CO)}_5}$ —COOMe

85%

JOC (1968) <u>33</u> 1550

$C_6H_{13}C\equiv CH$ $\xrightarrow{\begin{array}{l}\text{1 BuLi hexane}\\ \text{2 (C}_6\text{H}_{13}\text{)}_3\text{B diglyme}\\ \text{3 BrCH}_2\text{COOEt}\\ \text{4 HOAc}\end{array}}$

$$\underset{H}{\overset{C_6H_{13}}{\diagdown}}C=C\underset{CH_2COOEt}{\overset{C_6H_{13}}{\diagup}}$$

69%

Tetr Lett (1973) 4491

$EtCH=CH_2$ $\xrightarrow{\begin{array}{l}\text{1 B}_2\text{H}_6 \text{ THF}\\ \text{2 BrCH}_2\text{CH=CHCOOEt}\\ \text{potassium 2,6-di-t-}\\ \text{butylphenoxide}\end{array}}$ $EtCH_2CH_2CH=CHCH_2COOEt$ 72%

JACS (1970) <u>92</u> 1761

$MeCH=CHCOOEt$ $\xrightarrow[\text{2 EtI}]{\text{1 LiN(Pr-i)}_2 \text{ THF}}$ $CH_2=CHCHCOOEt$ 96%
$\overset{|}{Et}$

Tetr Lett (1973) 2433

$MeCH=CHCOOEt$ $\xrightarrow{h\nu \quad MeCN}$ $CH_2=CHCH_2COOEt$

Tetr Lett (1968) 4987
Chem Comm (1965) 137

$C_8H_{17}Br$

$\xrightarrow{\begin{array}{l} 1 \text{ MeCSSCH}_2\overset{\text{Me}}{\underset{}{C}}\text{=CH}_2 \quad \text{Et}\overset{\text{Me}}{\underset{}{C}}\text{HLi} \quad \text{THF} \\ 2 \text{ MeI} \\ 3 \text{ CuCl}_2 \quad \text{CuO} \quad \text{EtOH} \end{array}}$

$C_8H_{17}CH=\overset{\text{Me}}{\underset{}{C}}CH_2CH_2COOEt$

JACS (1973) $\underline{95}$ 5803

$CH_2=\overset{\text{Me}}{\underset{}{C}}CH_2CH_2\overset{\text{Me}}{\underset{\text{OH}}{C}}C=CH_2$ $\xrightarrow[\text{EtCOOH}]{\text{MeC(OEt)}_3}$ $CH_2=\overset{\text{Me}}{\underset{}{C}}CH_2CH_2CH=\overset{\text{Me}}{\underset{}{C}}CH_2CH_2COOEt$ 92%

JACS (1970) $\underline{92}$ 741
Synthesis (1971) 175

$C_8H_{17}CH=CH(CH_2)_7COOH$ $\xrightarrow[\text{MeOH}]{\overset{\begin{array}{c}\text{COOH} \\ | \\ (\text{CH}_2)_6\text{COOMe} \quad \text{electrolysis} \end{array}}{}}$ $C_8H_{17}CH=CH(CH_2)_{13}COOMe$

JCS (1954) 448   ~37%

$\overset{\text{COCl}}{\underset{(\text{CH}_2)_5\text{COOEt}}{|}}$ $\xrightarrow{\begin{array}{l} 1 \quad C_7H_{15}\text{-}\langle\!\langle_S\rangle\!\rangle \quad \text{SnCl}_4 \\ 2 \text{ KOH} \quad H_2O \\ 3 \text{ Wolff-Kishner} \\ 4 \text{ MeOH} \quad \text{acid} \\ 5 \text{ Nickel boride} \end{array}}$ $C_7H_{15}CH_2CH=CHCH_2(CH_2)_6COOMe$

Synth Comm (1972) $\underline{2}$ 415

RCHO $\xrightarrow{\text{Ph}_3\text{P=CH(CH}_2)_N\text{COOR'}}$ $RCH=CH(CH_2)_NCOOR'$

N=2   Acta Chem Scand (1966) $\underline{20}$ 992
N=4   Bull Soc Chim Fr (1965) 2988
N=9   Ber (1972) $\underline{105}$ 3591
N=10   Ivz (1960) 1900
        (Chem Abs $\underline{55}$ 14294)

$C_6H_{13}C\equiv CH$ $\xrightarrow{\begin{array}{l} 1 \text{ Hg(OAc)}_2 \quad \text{BF}_3\cdot\text{Et}_2\text{O} \quad \text{Ac}_2\text{O} \\ 2 \text{ KOH} \quad \text{Et}_2\text{O} \quad H_2O \end{array}}$ $C_6H_{13}\overset{}{\underset{\text{OAc}}{C}}=CH_2$ ~70%

JOC (1973) $\underline{38}$ 4254

$C_5H_{11}CH_2CHO$ $\xrightarrow{\text{Ac}_2\text{O} \quad \text{KOAc}}$ $C_5H_{11}CH=CHOAc$ 45-50%

Org Synth (1955) Coll Vol 3 127
JOC (1967) <u>32</u> 489
JCS (1953) 3864

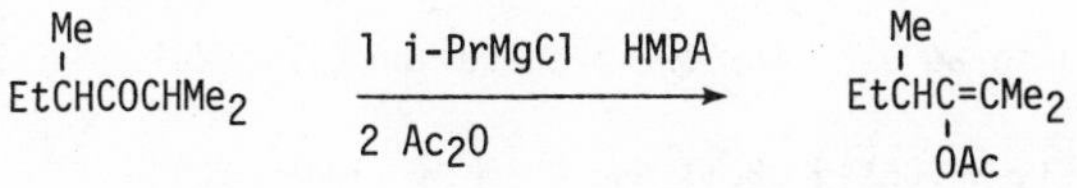

Me
|
$EtCHCOCHMe_2$ $\xrightarrow[\text{2 Ac}_2\text{O}]{\text{1 i-PrMgCl} \quad \text{HMPA}}$

Me
|
$EtCHC=CMe_2$
|
OAc

Tetrahedron (1973) <u>29</u> 479

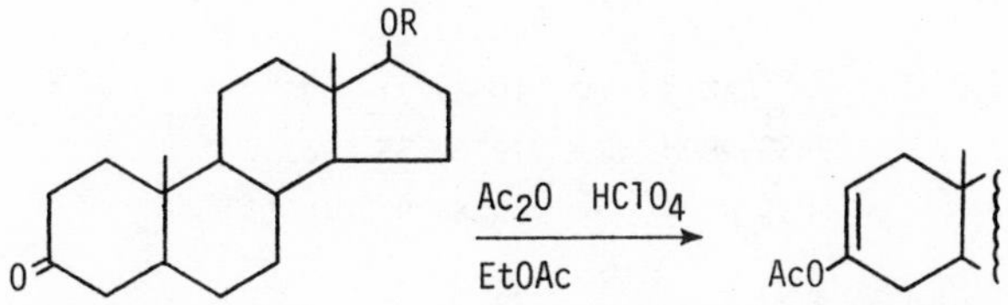

$\xrightarrow[\text{EtOAc}]{\text{Ac}_2\text{O} \quad \text{HClO}_4}$

96%

JOC (1966) <u>31</u> 324
Ac$_2$O, HClO$_4$, CCl$_4$ Org Synth (1972) <u>52</u> 39
(PhCO)$_2$O, HClO$_4$ (enol benzoate)   JOC (1969) <u>34</u> 1962
Ac$_2$O, BF$_3$ JOC (1969) <u>34</u> 1425
CH$_2$=COAc, acid   Tetrahedron (1971) <u>27</u> 5987
            | JCS (1962) 1323
            Me

$\xrightarrow{\text{Zn} \quad \text{Ac}_2\text{O}}$

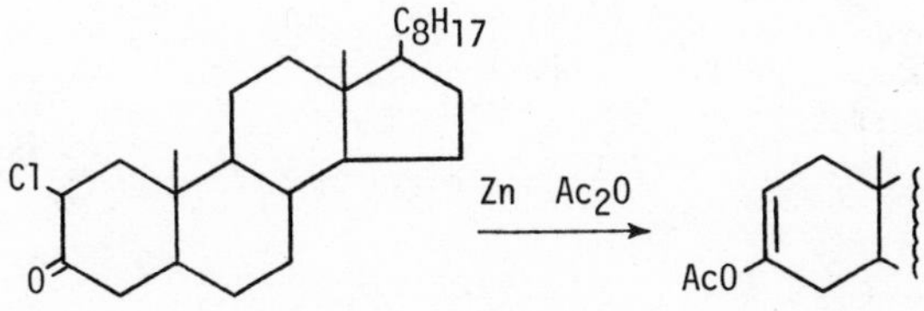

JCS (1953) 3869

Also via: Section
        Acetylenic esters   306
        β-Hydroxyesters   327
        Olefinic acids   322

Section 363   Ether —— Ether
                ∞∞∞∞∞∞∞∞∞∞∞∞

1,1-Diethers (acetals and ketals)

$$C_6H_{13}CHO \xrightarrow[\text{HC(OMe)}_3]{\text{MeOH TsOH}} C_6H_{13}CH(OMe)_2$$

JCS Perkin I (1973) 2738

The following reagents may also be used for the preparation of acetals
from aldehydes and alcohols:

NH_4Cl, $HC(OEt)_3$ Bull Soc Chim Fr (1965) 1007

NH_4NO_3, $HC(OEt)_3$ Org Synth (1963) Coll Vol 4 21

TsOH, molecular sieves   Synthesis (1972) 419

HCl   JCS (1953) 3864

$CaCl_2$ JOC (1966) 31 3406

$PhSO_2NHOH$ JOC (1970) 35 1962

Me_2SO_4, NaOH   Ber (1958) 91 410

>74%

JOC (1941) 6 489
Org Synth (1955) Coll Vol 3 701

82%

Synthesis (1972) 419

The following reagents may also be used for the preparation of ketals
from ketones and alcohols:

Ion exch resin (acid)   JOC (1959) 24 1731

$BF_3 \cdot Et_2O$ JACS (1954) 76 1728

TsOH, $HC(OMe)_3$ JACS (1968) 90 2448
                    Annalen (1962) 656 97

TsOH, $Me_2C(OMe)_2$ JOC (1960) 25 521 525
                       continued

$$\text{SeO}_2 \quad \text{JACS (1954)} \ \underline{76} \ 6113$$
$$(\text{Ph}_3\text{P})_3\text{RhCl} \quad \text{Ber (1968)} \ \underline{101} \ 1154$$

Section 364   Ether — Halide

α-Haloethers, β-haloethers, γ-haloethers and higher haloethers

Review: The α-Haloalkyl Ethers   Chem Rev (1955) $\underline{55}$ 301

$$\text{HO(CH}_2)_{10}\text{OH} \quad \xrightarrow[\text{C}_6\text{H}_6]{\text{HCHO} \ \ \text{HCl}} \quad \text{Cl(CH}_2)_{10}\text{OCH}_2\text{Cl}$$

JOC (1949) $\underline{14}$ 754 696

α-Bromoethers   JACS (1949) $\underline{71}$ 258

$$\text{PhCH}_2\text{CHO} \quad \xrightarrow[\text{Na}_2\text{SO}_4]{\text{EtOH} \ \ \text{HCl}} \quad \underset{\underset{\text{Cl}}{|}}{\text{PhCH}_2\text{CHOEt}}$$ < 94%

JACS (1949) $\underline{71}$ 4007

$$\text{EtOCH}_2\text{Me} \quad \xrightarrow{\text{PhICl}_2 \ \ h\nu} \quad \underset{\underset{\text{Cl}}{|}}{\text{EtOCHMe}}$$ 79%

Annalen (1969) $\underline{728}$ 12

56%

Synthesis (1972) 693

75%

JOC (1952) $\underline{17}$ 68
JACS (1955) $\underline{77}$ 1738

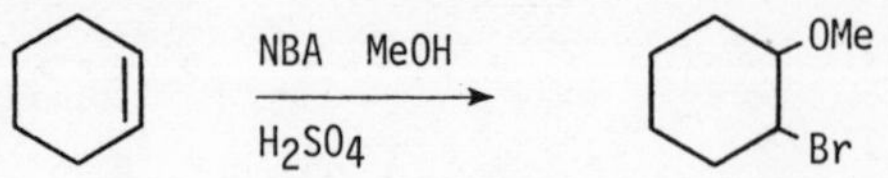

JACS (1943) 65 2196
t-BuOBr Bull Soc Chim Fr (1961) 2360
BrN$_3$ Tetr Lett (1968) 3921

Me$_2$C=CH$_2$ $\xrightarrow[\text{ZnCl}_2]{\text{ClCH}_2\text{OMe}}$ Me$_2$CCH$_2$CH$_2$OMe   60%
                          Cl

Ber (1936) 69 2706

MeCH$_2$(CH$_2$)$_4$OMe $\xrightarrow[\text{H}_2\text{SO}_4\ \text{H}_2\text{O}]{\text{(i-Pr)}_2\text{NCl}\ h\nu}$ MeCH(CH$_2$)$_4$OMe
              Cl

JACS (1971) 93 438

Section 365   Ether, Epoxide — Ketone

α-Alkoxyketones, β-alkoxyketones and epoxyketones

EtC≡CEt $\xrightarrow{\text{Tl(OAc)}_3\ \text{MeOH}}$ EtCOCHEt   85%
               OMe

JACS (1973) 95 1296

PhCOCl --→ PhCOCHN$_2$ $\xrightarrow{\text{i-PrOH}\ \text{BF}_3\cdot\text{Et}_2\text{O}}$ PhCOCH$_2$OPr-i   ~69%

JACS (1950) 72 5161

MeCOCl $\xrightarrow{\text{Et}_2\text{O}\ h\nu}$ MeCHOEt
             COMe

Angew (1965) 77 216
(Internat Ed 4 239)

PhBr $\xrightarrow[\text{2 MeOCH}_2\text{CN}]{\text{1 Mg Et}_2\text{O}}$ PhCOCH$_2$OMe   71-78%

Org Synth (1955) Coll Vol 3 562

BuC≡CCH(OEt)$_2$ $\xrightarrow[\text{2 H}_2\text{O}_2 \quad \text{NaOH} \quad \text{H}_2\text{O}}{\text{1 (i-PrCH)}_2\text{BH} \quad \text{THF}}$ BuCH$_2$COCH$_2$OEt   61%

(with Me substituent on the borane reagent)

JACS (1974) 96 316

PhBr $\xrightarrow[\begin{array}{c}\text{2 CdCl}_2 \quad \text{C}_6\text{H}_6 \\ \text{3 EtOCH}_2\text{CH}_2\text{COCl}\end{array}]{\text{1 Mg Et}_2\text{O}}$ PhCOCH$_2$CH$_2$OEt   82%

JACS (1949) 71 3480

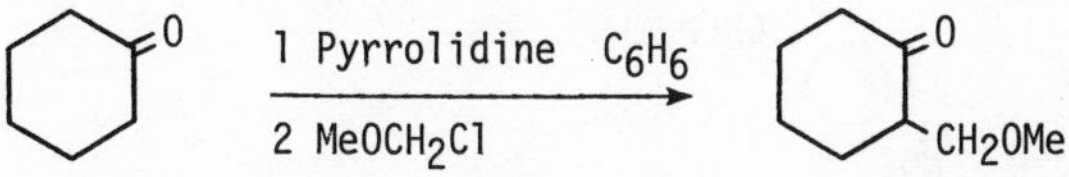

33%

JACS (1963) 85 207

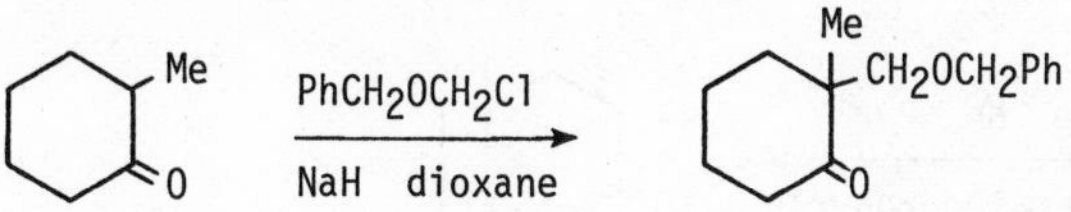

JCS (1963) 4634
(1964) 4521

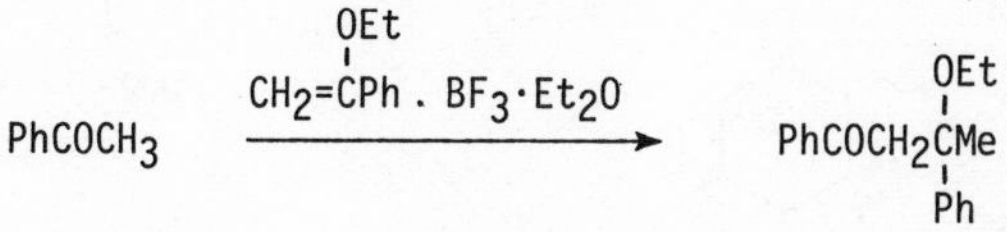

PhCOCH$_3$ $\xrightarrow{\begin{array}{c}\text{OEt} \\ \text{CH}_2=\text{CPh} . \text{BF}_3\cdot\text{Et}_2\text{O}\end{array}}$ PhCOCH$_2$CMe (with OEt and Ph substituents)   37%

Synthesis (1972) 378

$Me_2C=CHCOMe$ $\xrightarrow{\text{MeOH   ion exch resin (acid)}}$ Me_2CCH_2COMe 57%
                                                                  |
                                                                 OMe

JOC (1958) 23 937

EtOH hν

                                                                         62%

JACS (1967) 89 3949
Tetr Lett (1968) 2025
Bull Chem Soc Jap (1967) 40 945

Review: The α-Epoxyketones   Annales de Chimie (1966) 11 159

$\xrightarrow[\text{MeONa   EtOH}]{BrCH_2COPh}$

                                                                         88%

JACS (1955) 77 590
JCS C (1970) 1154

$\xrightarrow[\text{hematoporphyrin}]{O_2 \quad hν \quad Pyr}$

                                                                         75%

JACS (1963) 85 1894

$\xrightarrow[\text{MeOH   H}_2\text{O}]{H_2O_2 \quad NaOH}$

                                                                         70-72%

                            Org Synth (1963) Coll Vol 4 552
                     +   - JCS Perkin I (1972) 2051
t-BuOOH, PhCH₂NMe₃ OH JACS (1958) 80 5845
                                (1968) 90 6495

continued

H_2O_2, DMF (neutral)   Chem Pharm Bull (1969) <u>17</u> 1206
NaOCl, Pyr   JOC (1963) <u>28</u> 250
Tetrahedron (1968) <u>24</u> 6583
$PhCOO_2H$ JACS (1950) <u>72</u> 367

Section 366   <u>Ether, Epoxide — Nitrile</u>

α-Cyanoethers, β-cyanoethers and epoxynitriles

$$C_6H_{13}\overset{\overset{\displaystyle OSiMe_3}{|}}{\underset{\underset{\displaystyle Me}{|}}{C}}O \quad \xrightarrow[\text{18-crown-6}]{Me_2CCN \quad KCN} \quad C_6H_{13}\overset{\overset{\displaystyle OSiMe_3}{|}}{\underset{\underset{\displaystyle Me}{|}}{C}}CN \qquad 95\%$$

Tetr Lett (1973) 4929
Ber (1973) <u>106</u> 587

$$BuOH \quad \xrightarrow{CH_2=CHCN \quad PhCH_2\overset{+}{N}Me_3 \ \overset{-}{O}H} \quad BuOCH_2CH_2CN \qquad 86\%$$

Org React (1949) <u>5</u> 79 110
JCS (1945) 535

$$\overset{\displaystyle O}{\overset{/\backslash}{Me_2C-CH_2}} \quad \xrightarrow{Me_3SiCN \quad AlCl_3} \quad Me_2\overset{|}{\underset{\underset{\displaystyle CN}{|}}{C}}CH_2OSiMe_3 \qquad 93\%$$

Tetr Lett (1973) 1449

$$\xrightarrow[\text{NaOH} \quad H_2O]{ClCH_2CN \quad PhCH_2\overset{+}{N}Et_3 \ \overset{-}{C}l} \qquad 79\%$$

Tetr Lett (1972) 2395
J Indian Chem Soc (1963) <u>40</u> 114

EtCHCHO —NaCN H₂O→ EtCHCHCN (O epoxide)   30%
|Cl   Bull Soc Chim Fr (1967) 1587

Section 367   Ether ─── Olefin

Vinyl ethers (enol ethers) and allyl ethers

C₆H₁₃C≡CH —1 Hg(OAc)₂ MeOH / 2 NaBH₄ NaOH→ C₆H₁₃C=CH₂   59%
                                              |OMe
                                         JOC (1973) 38 4254

CH₂=CCHOH (Et, Me) —EtOCH=CH₂ / Hg(OAc)₂→ CH₂=CCHOCH=CH₂ (Et, Me)   77%
                                         JACS (1973) 95 553
                                         JCS Perkin I (1973) 1791
                                 CH₂=CHOAc JACS (1953) 75 2678

—1 NaH CH₂=CHCH₂Br C₆H₆ / 2 t-BuOK Me₂SO→ CH₂OCH=CHMe
                                         JCS (1965) 2205
                                         JCS Perkin I (1972) 1535
                                         JOC (1973) 38 3224

NO₂-C₆H₄-OH —1 BrCH₂CH₂Br NaOH H₂O / 2 t-BuOK C₆H₆→ NO₂-C₆H₄-OCH=CH₂
                                         Synthesis (1972) 693

Me_2CHCHO $\xrightarrow[\text{EtOH}]{\text{HC(OEt)}_3 \quad \text{NH}_4\text{Cl}}$ $Me_2CHCH(OEt)_2$ $\xrightarrow{\text{H}_3\text{PO}_4}$ $Me_2C=CHOEt$ 55%

Bull Soc Chim Fr (1965) 1007

$PrCH_2CHO$ $\xrightarrow[\text{HCl}]{\text{EtOH}}$ $PrCH_2\overset{\underset{|}{Cl}}{C}HOEt$ $\xrightarrow{\text{NaOAc}}$ $PrCH=CHOEt$ 37%

JOC (1958) 23 670
Ber (1956) 89 1468

$C_{11}H_{23}CH_2Br$ $\xrightarrow[\text{2 MeONa \quad THF}]{\text{1 Ph}_3\text{P}}$ $C_{11}H_{23}CH=PPh_3$ $\xrightarrow{\text{HCOOEt}}$ $C_{11}H_{23}CH=CHOEt$ 6%

Zh Org Khim (1966) 2 2181
(Chem Abs 66 85406)

$\xrightarrow[\text{BuOH}]{\text{BuONa}}$

<85%

JOC (1969) 34 1220

$Ph_3\overset{+}{P}CH_2OMe \ \overset{-}{Cl} \quad PhLi$

Et_2O

71%

Org React (1965) 14 270 399
$(Me_2N)_2POCH_2OR$ Compt Rend (1971) 272 100

$(CH_2)_{10}\overset{\text{CO}}{\underset{\overset{|}{C}H_2}{}}$ $\xrightarrow[\text{2 TsOH}]{\text{1 HC(OEt)}_3 \quad \text{H}_2\text{SO}_4 \quad \text{EtOH}}$ $(CH_2)_{10}\overset{\text{COEt}}{\underset{\text{CH}}{\overset{\|}{}}}$

Annalen (1962) 656 97

PhCOCH$_3$ $\xrightarrow[\text{2 Me}_2\text{SO}_4]{\text{1 t-BuOK   Me}_2\text{SO}}$ PhC=CH$_2$ 68%
                                              |
                                             OMe   Chem Comm (1966) 51

PhCHO $\xrightarrow[\text{EtOH}]{\text{CH}_2\text{=CHPPh}_3 \overset{+}{} \text{ Br} \overset{-}{} \text{ EtONa}}$ PhCH=CHCH$_2$OEt   54%

                                              JOC (1966) 31 467

$\xrightarrow[\text{2 ROH}]{\text{1 PhSeBr}}$

                                              JOC (1974) 39 429

BuC≡CCH(OEt)$_2$ $\xrightarrow[\substack{\text{2 HOAc} \\ \text{3 H}_2\text{O}_2 \text{   NaOH   H}_2\text{O}}]{\substack{\text{Me} \\ | \\ \text{1 (i-PrCH)}_2\text{BH   THF}}}$ BuCH=CHCH$_2$OEt   65%
                                                       cis
                                              JACS (1974) 96 316

Section 368   Halide — Halide
              ∘∘∘∘∘∘∘∘∘∘∘∘∘∘∘∘∘

1,1-Dihalides and 1,2-dihalides

BuC≡CH $\xrightarrow{\text{HF   Pyr   THF}}$ BuCF$_2$ 70%
                                              |
                                             Me   Synthesis (1973) 779

C$_6$H$_{13}$CHO $\xrightarrow{\text{R}_2\text{NSF}_3 \text{   CCl}_4}$ C$_6$H$_{13}$CHF$_2$ 75%

                                              Synthesis (1973) 787
                                    MoF$_6$ Tetrahedron (1971) 27 3965

PhCHO $\xrightarrow{\text{Ph}_3\text{P·Cl}_2}$ PhCHCl$_2$ 　　　　　　　　　59%

　　　　　　　　　　　　　　　　Annalen (1959) <u>626</u> 26

　　　　　　　　　　　PCl$_5$　Org Synth (1943) Coll Vol 2 549

　　　　　　　　MeOCHCl$_2$　Ber (1959) <u>92</u> 83

　　　　　　　NH$_2$OH; Cl$_2$　JOC (1971) <u>36</u> 2146

i-PrCHO $\xrightarrow[\text{2 I}_2 \ \text{Et}_3\text{N} \ \text{Et}_2\text{O}]{\text{1 N}_2\text{H}_4}$ i-PrCHI$_2$ 　　　　　　　62%

　　　　　　　　　　　　　　　　Aust J Chem (1970) <u>23</u> 989
　　　　　　　　　　　　　　　　JCS (1962) 470

　　　　　　　　　　　1 NaNO$_2$　HBr　H$_2$O
　　　　　　　　　　　2 CH$_2$=CHBr　CuBr$_2$
　　　　　　　　　　　　NaOAc　H$_2$O

　　　　　　　　　　　　　　　< 77%

　　　　　　　　　　　　　　JACS (1954) <u>76</u> 3005
　　　　　　　　　　　　　　Org React (1960) <u>11</u> 189

BuCHMe$_2$ $\xrightarrow{\text{PhHgCCl}_2\text{Br}}$ BuCMe$_2$ 　　　　　　　20%
　　　　　　　　　　　　　　　　　|
　　　　　　　　　　　　　　　CHCl$_2$

　　　　　　　　　　　　　　JOC (1970) <u>35</u> 1989
　　　　　　　　　　　　　　Annalen (1969) <u>729</u> 33

$(CH_2)_{10}$CO $\xrightarrow[\text{CH}_2\text{Cl}_2]{\text{SF}_4 \ \text{HF}}$ $(CH_2)_{10}$CF$_2$ 　　　　　23%

　　　　　　　　　　　　　　JOC (1971) <u>36</u> 818
　　　　　　　　　　　　　　　　(1961) <u>26</u> 2436
　　　　　　　　　　　　　　JACS (1960) <u>82</u> 543

　　　　　　　　　　R$_2$NSF$_3$　Synthesis (1973) 787
　　　　　　　　　　　COF$_2$　JACS (1962) <u>84</u> 4275
　　　　　　　　　　　MoF$_6$　Tetrahedron (1971) <u>27</u> 3965

Ph$_2$CO $\xrightarrow{\text{PCl}_3}$ Ph$_2$CCl$_2$ 　　　　　　　86%

　　　　　　　　　　　　　　Ber (1963) <u>96</u> 1387

i-PrCOMe $\xrightarrow{\begin{array}{c} 1 \ N_2H_4 \\ \hline 2 \ I_2 \ \ Et_3N \ \ Et_2O \end{array}}$ i-PrCI$_2$ 50%
                                              |
                                             Me

Aust J Chem (1970) 23 989

$\xrightarrow[CCl_4]{Br_2 \ h\nu}$ ~50%

Rec Trav Chim (1964) 83 67

$\xrightarrow{Ph_3P \cdot X_2}$ X=Cl or Br

Tetrahedron (1971) 27 2617
Tetr Lett (1972) 3869
COCl$_2$ Ber (1962) 95 2976
SO$_2$Cl$_2$ Can J Chem (1966) 44 2339
Me$_2$N=CHCl Cl   Chem Comm (1967) 1152
   +   -

MeCH$_2$CMe$_2$ $\xrightarrow{NBS \ \ CCl_4}$ MeCH-CMe$_2$ 20%
        | | |
        Br   Br Br
                                              JOC (1953) 18 649

Me$_2$C=CMe$_2$ $\xrightarrow[2 \ AgF]{1 \ HF \ \ NBS \ \ Pyr \ \ Et_2O}$ Me$_2$C-CMe$_2$
                                                         | |
                                                         F F

Synthesis (1973) 780
Pb(OAc)$_4$, HF   JACS (1962) 84 1050

C$_5$H$_{11}$CH=CH$_2$ $\xrightarrow{Cl_2}$ C$_5$H$_{11}$CHCH$_2$Cl   86%
                                          |
                                          Cl   Synth Comm (1972) 2 129
                        Cl$_2$, Pyr or Cl$_2$, SbCl$_3$ JOC (1959) 24 1621

Continued

Cl_2, $SbCl_5$ Chem Comm (1971) 1064
JACS (1951) $\underline{73}$ 3329

$PhICl_2$ JOC (1959) $\underline{24}$ 1621
(1971) $\underline{36}$ 1024

NCS, HCl   JOC (1959) $\underline{24}$ 1621
JACS (1959) $\underline{81}$ 2191

SO_2Cl_2, Pyr   Rec Trav Chim (1971) $\underline{90}$ 549
JOC (1971) $\underline{36}$ 3566

NCl_3 Synthesis (1969) 135
JOC (1971) $\underline{36}$ 3566

PCl_5 JOC (1971) $\underline{36}$ 3566

$CuCl_2$ Bull Chem Soc Jap (1971) $\underline{44}$ 1973
(1970) $\underline{43}$ 1439

$TlCl_3$ Bull Chem Soc Jap (1972) $\underline{45}$ 1482

$Pb(OAc)_4$, MeCOCl   Tetrahedron (1969) $\underline{25}$ 1545

50%

Rec Trav Chim (1964) $\underline{83}$ 67
Org Synth (1943) Coll Vol 2 171

Pyr·HBr_3 Org Synth (1966) $\underline{46}$ 46
JACS (1972) $\underline{94}$ 7118

$CuBr_2$ JOC (1971) $\underline{36}$ 3324

$PrCH=CH_2$ $\xrightarrow{I_2}$ $PrCHCH_2I$
                                          |
                                          I

Can J Chem (1964) $\underline{42}$ 2710

$BuCH=CH_2$ $\xrightarrow[\text{sulfolane}]{\text{NCS   HF   Pyr}}$ $BuCHCH_2Cl$
                                                                            |
                                                                            F

40%

Synthesis (1973) 780

JACS (1959) 81 4107 2191
Org Synth (1966) 46 10

NBS, HF   Synthesis (1973) 780

Br$_2$, AgF   Can J Chem (1969) 47 361
Chem Ind (1967) 1787

BuCH=CH$_2$ $\xrightarrow[\text{sulfolane}]{\text{HF I}_2\text{ AgNO}_3\text{ Pyr}}$ BuCHCH$_2$I   80%
                                                                              |
                                                                              F

Synthesis (1973) 780

I$_2$, AgF or NIS, HF   Can J Chem (1969) 47 361
JACS (1960) 82 4007

PhCH=CH$_2$ $\xrightarrow[\text{H}_2\text{O}]{\text{NBA HCl}}$ PhCHCH$_2$Br   36-44%
                                                              |
                                                              Cl

JACS (1951) 73 998
(1959) 81 2191
(1952) 74 4891

t-BuCH=CH$_2$ $\xrightarrow[\text{MeCN}]{\text{I}_2\text{ CuCl}_2}$ t-BuCHCH$_2$I   73%
                                                                  |
                                                                  Cl

JOC (1971) 36 3324 2088

I$_2$, HgCl$_2$ JACS (1948) 70 828

Section 369   Halide — Ketone

PrC≡CMe $\xrightarrow[\text{HOAc}]{\text{Et}_2\text{NCl H}_2\text{SO}_4}$ PrCHCOMe   33%
                                                                  |
                                                                  Cl

JOC (1967) 32 3263

$C_5H_{11}C{\equiv}CH$ $\xrightarrow[\text{2 } Br_2]{\text{1 EtMgBr } Et_2O}$ $C_5H_{11}C{\equiv}CBr$ $\xrightarrow[\text{2 HCl } H_2O]{\text{1 HgO } BF_3}$ $C_5H_{11}COCH_2Br$

JACS (1937) 59 1307

▷—COCl $\xrightarrow[\substack{\text{2 } PhICl_2 \text{ } C_6H_6 \\ \text{3 } Na_2CO_3 \text{ } H_2O}]{\text{1 } Ph_3P{=}CHPr}$ ▷—COCHPr
$\quad\quad\quad\quad\quad\quad\quad\quad\quad\quad\quad\quad\quad$ $\overset{|}{Cl}$

Tetrahedron (1969) 25 1871

$\xrightarrow[Et_2O]{CH_2N_2}$ ⋎⋏—$CH_2CH_2COCHN_2$ $\xrightarrow[H_2O \text{ } Et_2O]{HCl}$ ⋎⋏—$CH_2CH_2COCH_2Cl$

J Med Chem (1971) 14 641
Org Synth (1955) Coll Vol 3 119

Bromoketones　JOC (1970) 35 1381

Iodoketones via bromoketones　JACS (1943) 65 1516

Me　　　　　　　　　　　　OMe　Me　　　　　　　　　　　Me
$\overset{|}{CH_2}C{=}CHCOOMe$ $\xrightarrow[THF]{Ph_3P{=}CEt}$ $\overset{|}{CH_2}C{=}CHCOOMe$ $\xrightarrow[NaOAc]{NCS}$ $\overset{|}{CH_2}C{=}CHCOOMe$
$\overset{|}{CH_2}CH{=}C(CH_2)_2CHO$ $\quad$ $\overset{|}{CH_2}CH{=}C(CH_2)_2CH{=}CEt$ $\quad$ $\overset{|}{CH_2}CH{=}C(CH_2)_2CHCOEt$
$\overset{|}{Me}$ $\quad\quad\quad\quad\quad\quad\quad\quad\quad$ $\overset{|}{Me}\quad\quad\overset{|}{OMe}$ $\quad\quad\quad\quad\quad$ $\overset{|}{Me}\quad\quad\overset{|}{Cl}$

60%

JACS (1972) 94 5374 5379

$\xrightarrow[\text{polyphosphoric acid}]{ClCH_2COOH}$ 40%

Synth Comm (1972) 2 97

$ClCH_2COCl$, $AlCl_3$　Org Synth (1955) Coll Vol 3 183

PhBr $\xrightarrow[\substack{\text{2 } EtCHCOOEt \\ \overset{|}{F}}]{\text{1 Mg } Et_2O}$ $PhCOCHEt$
$\quad\quad\quad\quad\quad\quad\quad\quad\quad\quad$ $\overset{|}{F}$

50%

Bull Soc Chim Fr (1970) 991

BuBr $\xrightarrow{\begin{array}{c}1 \text{ Mg} \quad Et_2O\\ \hline 2 \text{ ClCH}_2\text{COCl}\end{array}}$ BuCOCH$_2$Cl   51%

Org React (1954) <u>8</u> 28
JACS (1945) <u>67</u> 1944

PhBr $\xrightarrow{\quad\quad\quad\quad}$

1 Mg THF

2 (structure: OSiMe$_3$ / N=C=CMe$_2$)

3 Br$_2$

4 (COOH)$_2$ H$_2$O

PhCOCMe$_2$
         |
         Br   40%

JOC (1973) <u>38</u> 2129

Me
|
CH$_2$C=CHCOOMe
|
CH$_2$CH=C(CH$_2$)$_2$Br
|
Et

$\xrightarrow{\begin{array}{c}1 \text{ EtCOCHCOEt (Li)}\\ \hline 2 \text{ CuCl}_2 \quad \text{LiCl} \quad \text{DMF}\\ 3 \text{ Ba(OH)}_2 \quad \text{EtOH}\end{array}}$

Me
|
CH$_2$C=CHCOOMe
|
CH$_2$CH=C(CH$_2$)$_2$CHCOEt
            | |
            Et   Cl

JACS (1968) <u>90</u> 6225
      (1944) <u>66</u> 1132

(CF$_3$)$_3$COF
or SF$_5$OF
or CF$_3$OF

40-62%

Chem Comm (1972) 122
FClO$_3$ JCS Perkin I (1973) 2365
Via ethoxalyl derivative JACS (1960) <u>82</u> 2312
Via enamine JOC (1958) <u>23</u> 1406

PhCO (cyclopropyl) $\xrightarrow{\begin{array}{c}1 \text{ LiCHCl}_2\\ \hline 2 \text{ Lithium piperidide}\\ \text{THF}\end{array}}$

Cl
|
PhCHCO (cyclopropyl)   <70%

Tetr Lett (1972) 4117 4661
Ber (1973) <u>106</u> 2626

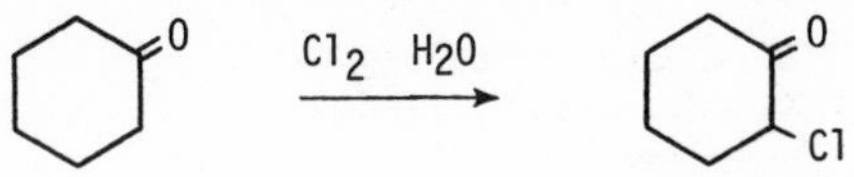

61-66%

	Org Synth (1955) Coll Vol 3 188
Cl$_2$, HOAc	JCS (1953) 3869
Pyr·HCl$_3$	JACS (1953) 75 3500
CuCl$_2$	JOC (1963) 28 630
SO$_2$Cl$_2$	Org Synth (1963) Coll Vol 4 162
	JOC (1972) 37 2436
SeOCl$_2$	JOC (1963) 28 1128
PCl$_5$	Compt Rend (1963) 256 1996
t-BuOCl	JACS (1953) 75 3500
Pb(OAc)$_4$, MeCOCl	Tetrahedron (1969) 25 1545
Chlorination of imines	Tetrahedron (1970) 26 5191

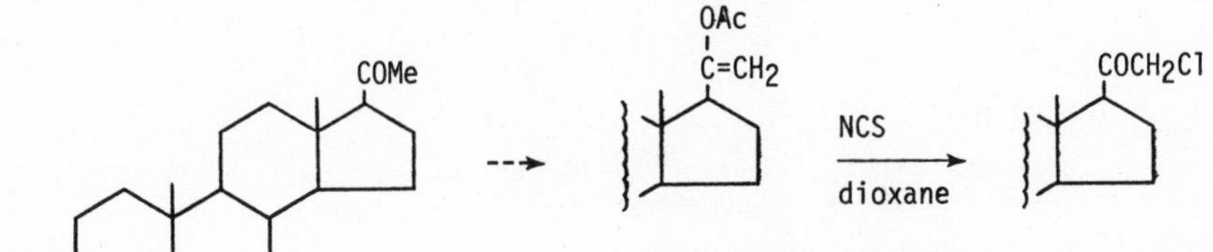

JACS (1959) 81 2383
Cl$_2$ JOC (1961) 26 4569

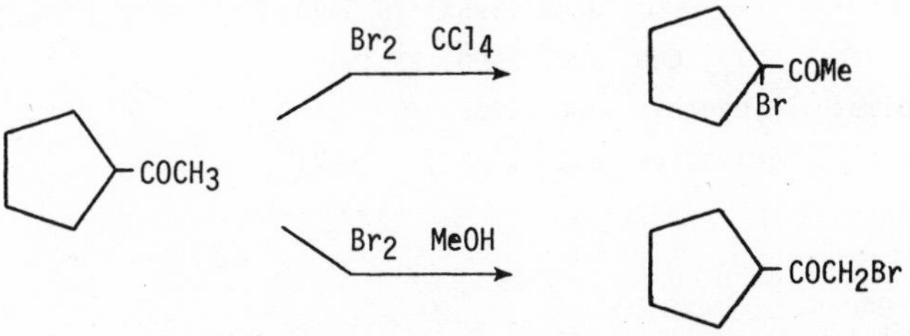

	Tetrahedron (1970) 26 5611
Br$_2$, HBr, HOAc or dioxane·Br$_2$	JCS (1954) 3257
Pyrrolidone·HBr$_3$	Can J Chem (1969) 47 706
CuBr$_2$	JOC (1964) 29 3459
	JACS (1968) 90 6218
NBS, benzoyl peroxide	JCS Perkin I (1972) 50
	Angew (1959) 71 349

continued

$$PhNMe_3 \overset{+}{} \overset{-}{Br_3}$$

JOC (1971) 36 4124
Bull Soc Chim Fr (1962) 90

Bromination of Ketals   Tetrahedron (1963) 19 861
Bull Soc Chim Fr (1962) 90
JACS (1965) 87 817

Bromination of enol acetates   Chem Pharm Bull (1969) 17 1585
(1972) 20 2156

Bromination of lithium enolates   JOC (1973) 38 2576

Bromination of enol ethers   JOC (1959) 24 1564
Tetr Lett (1972) 4055

Bromination of imines   Tetrahedron (1970) 26 5191

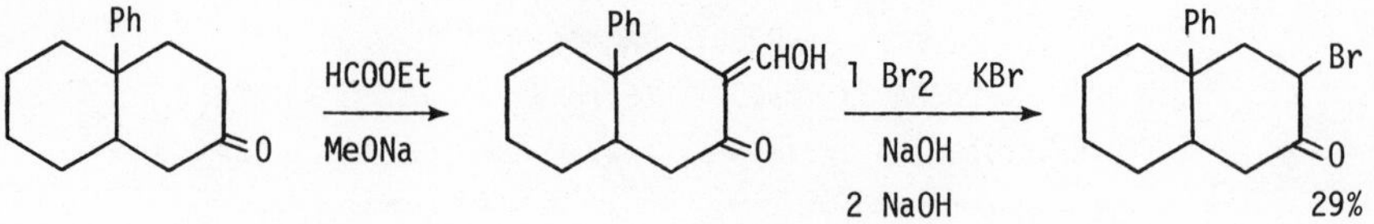

JOC (1970) 35 753

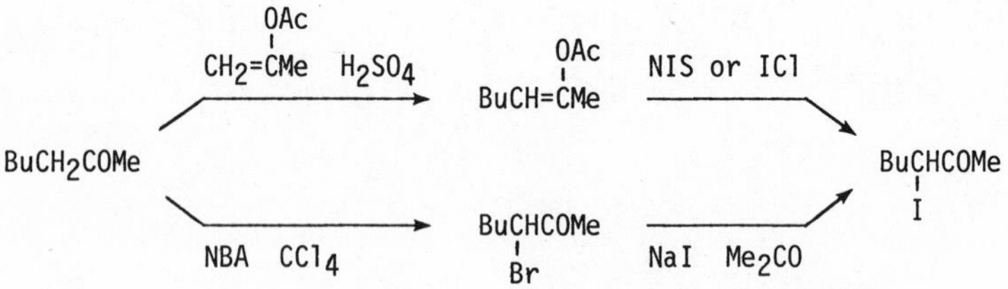

JACS (1953) 75 3493
I$_2$, CaO   JOC (1960) 25 1966
1,3-Diiodo-5,5-dimethylhydantoin   JOC (1965) 30 1101
Via oxalyl derivative   JACS (1952) 74 4974

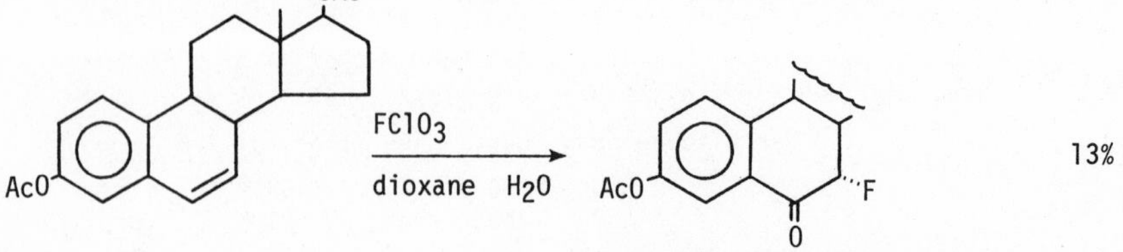

13%

JCS Perkin I (1973) 1462
NOF   Chem Ind (1965) 1929

$$\text{cycloheptene} \xrightarrow[\text{CCl}_4]{\text{NOCl}} \text{(NO, Cl substituted cycloheptane)} \xrightarrow[\text{HCl } H_2O]{\text{MeCOCH}_2\text{CH}_2\text{COOH}} \text{(2-chlorocycloheptanone)} \quad 93\%$$

JOC (1967) <u>32</u> 4136

$$(CH_2)_{10}\begin{matrix}CH \\ \| \\ CH\end{matrix} \xrightarrow[\text{Me}_2\text{CO}]{\text{CrO}_2\text{Cl}_2} (CH_2)_{10}\begin{matrix}CO \\ | \\ CHCl\end{matrix} \quad 79\%$$

JOC (1973) <u>38</u> 185

BuCH=CH$_2$

1 B$_2$H$_6$ THF

2 i-PrCOCHN$_2$

3 NBS

4 NaOH H$_2$O

BuCH$_2$CH$_2$CHCOPr-i
$\quad\quad\quad$ |
$\quad\quad\quad$ Br

Can J Chem (1972) <u>50</u> 2387

CH$_2$CH=CHPh
OAc (bicyclic)

$\xrightarrow[\text{dioxane } H_2O]{\text{NBA HClO}_4}$

Br OH
CH$_2$CH-CHPh
OAc (bicyclic)

$\xrightarrow[\text{Me}_2\text{CO }H_2O]{\text{CrO}_3 H_2SO_4}$

Br
CH$_2$CHCOPh
OAc (bicyclic)

JACS (1972) <u>94</u> 5386
$\quad\quad$(1956) <u>78</u> 3749
Helv (1944) <u>27</u> 821

MeCHCOCl
$\,$|
$\,$Cl

1 C$_6$H$_{13}$CH(COOTHP)$_2$ Na C$_6$H$_6$

2 C$_6$H$_6$ (reflux)

$\longrightarrow$

MeCHCOCH$_2$C$_6$H$_{13}$
$\,$|
$\,$Cl

JCS (1952) 3945

PhH $\xrightarrow[\text{AlCl}_3\text{ CS}_2]{\text{BrCH}_2\text{CH}_2\text{COCl}}$ BrCH$_2$CH$_2$COPh $\quad\quad$ <93%

JACS (1940) <u>62</u> 1435

$$I(CH_2)_{10}COCl \xrightarrow[\text{Et}_2O]{\text{Bu}_2CuLi} I(CH_2)_{10}COBu \qquad 93\%$$

JACS (1972) <u>94</u> 5106

Section 370   <u>Halide — Nitrile</u>

Halonitriles

$$PhNH_2 \dashrightarrow \overset{+}{PhN_2}\overset{-}{Cl} \xrightarrow{CH_2=CHCN \ Cu^{2+}} \underset{\underset{Cl}{|}}{PhCH_2CHCN} \qquad \sim 60\%$$

Proc Chem Soc (1962) 117
Org React (1960) <u>11</u> 189

$$\underset{\underset{OH}{|}}{Me_2CCN} \xrightarrow{PCl_5 \ C_6H_6} \underset{\underset{Cl}{|}}{Me_2CCN} \qquad 38\%$$

JACS (1945) <u>67</u> 690
$POCl_3$, lutidine   JACS (1950) <u>72</u> 1753

$$EtCH=CH_2 \xrightarrow[\substack{2 \ Cl_2CHCN \ \ \text{potassium 2,6-di-}\\ \text{t-butylphenoxide} \ \ THF}]{1 \ 9\text{-BBN}} \underset{\underset{Cl}{|}}{EtCH_2CH_2CHCN} \qquad 75\%$$

JACS (1970) <u>92</u> 5790

Section 371   <u>Halide — Olefin</u>

Vinylic halides, allylic halides, homoallylic halides and other olefinic halides.

For the conversion of allylic alcohols to allylic halides see section 138 vol 1 and vol 2 (Halides from Alcohols)
For allylic halogenation with NBS etc. see section 146 vol 1 and vol 2 (Halides from Hydrides)

$$C_5H_{11}CHO \xrightarrow[\text{THF}]{Ph_3P=CHF} C_5H_{11}CH=CHF \qquad\qquad 35\%$$

Synthesis (1969) 75

$$\xrightarrow[\text{CH}_2\text{Cl}_2]{SF_4 \quad HF} \qquad \xrightarrow{Al_2O_3} \qquad\qquad 44\%$$

JOC (1971) 36 818

$$PhC{\equiv}CH \xrightarrow[\text{CH}_2\text{Cl}_2]{PhCH_2Cl \quad ZnCl_2} \underset{\underset{Cl}{|}}{PhC}=CHCH_2Ph \qquad 95\%$$

JCS Perkin I (1973) 2491

$$BuC{\equiv}CH \xrightarrow[\text{2 Me}_3\text{SiCl}]{1 \text{ EtMgBr} \quad Et_2O} BuC{\equiv}CSiMe_3 \xrightarrow{\substack{1 \text{ Dicyclohexylborane} \\ 2 \text{ HOAc} \\ 3 \text{ H}_2\text{O}_2 \quad \text{base} \\ 4 \text{ X}_2 \\ 5 \text{ MeONa}}} \underset{\text{trans}}{BuCH=CHX} \quad \substack{X=Cl \\ \text{or Br}}$$

Tetr Lett (1974) 543

$$\xrightarrow[\text{piperidine} \quad Et_2O]{\overset{+}{Ph_3P}CH_2Cl \quad \overset{-}{Cl} \quad BuLi}$$

81%

Ber (1972) 105 1683
Org React (1965) 14 270

PhHgCHClBr, Ph3P   J Organometallic Chem (1965) 3 337

$$\xrightarrow[\text{LiX}]{MeOCH_2CH_2OLi} \qquad\qquad X=Cl \text{ or } I$$

JACS (1970) 92 4309

$(CH_2)_7$ CO / CH$_2$ $\xrightarrow[\text{2 KOH EtOH}]{\text{1 PCl}_5 \ \ C_6H_6}$ $(CH_2)_7$ CCl ‖ CH   32%

JACS (1952) <u>74</u> 3643

Ph$_3$PCl$_2$ Annalen (1959) <u>626</u> 26

Cl$_2$CHOMe, ZnCl$_2$ Ber (1959) <u>92</u> 83

Ph$_3$P, CCl$_4$ Chem Comm (1972) 443

[catechol]PCl$_3$ Ber (1963) <u>96</u> 1387

[cyclobutylidene]=CH$_2$ $\xrightarrow[\text{CH}_2\text{Cl}_2]{\text{PhICl}_2}$ [cyclobutane]—CH$_2$Cl / Cl $\xrightarrow[\text{EtOH}]{\text{EtONa}}$ [cyclobutylidene]=CHCl

JOC (1971) <u>36</u> 1024

PrC≡CH $\xrightarrow[\text{2 Br}_2 \ \ \text{CCl}_4]{\text{1 (EtCH)}_2\text{BH (Me)} \ \ \text{THF}}$

3 NaOH → cis PrCH=CHBr

3 Δ → trans

JACS (1967) <u>89</u> 4531

(1973) <u>95</u> 6456

MeO—[benzene]—C≡CH $\xrightarrow{\text{HBr} \ \ \text{DMF}}$ MeO—[benzene]—C=CH$_2$ / Br   78%

Helv (1964) <u>47</u> 194

BuC≡CH $\xrightarrow{\text{HBr} \ \ \text{peroxides}}$ BuCH=CHBr   74%

Org React (1963) <u>13</u> 150 234

[cyclohexane]—Br $\xrightarrow[\text{2 BrCH}_2\text{C=CH}_2 / \text{Br}]{\text{1 Mg} \ \ \text{Et}_2\text{O}}$ [cyclohexane]—CH$_2$C=CH$_2$ / Br

Org Synth (1932) Coll Vol 1 186

$$\text{(cyclohexanone)} \xrightarrow[\text{Et}_2\text{O}]{\overset{+}{\text{Ph}_3\text{PCH}_2\text{Br}} \ \overset{-}{\text{Br}} \ \text{PhLi}} \text{(=CHBr cyclohexylidene)} \quad 89\%$$

JOC (1965) 30 2208
Org React (1965) 14 270

$$\xrightarrow[\text{2 KOH EtOH}]{\text{1 Br}_2 \ \text{CCl}_4} \text{CH=CHBr}$$

J Med Chem (1972) 15 1262

t-BuOK   JOC (1965) 30 2208

NaNH$_2$, t-BuOK   Chem Comm (1972) 1289

Bu$_4$N Br, CH$_2$CH$_2$ Ber (1973) 106 1648
$\overset{+}{} \overset{-}{}$ \O/

$$\text{(cyclopentylidene)=CHCOOH} \xrightarrow[\text{2 Na}_2\text{CO}_3 \ \text{H}_2\text{O}]{\text{1 Br}_2 \ \text{CH}_2\text{Cl}_2} \text{(cyclopentylidene)=CHBr} \quad 85\%$$

JOC (1965) 30 2208
JCS C (1971) 2352

$$\text{BuC}\equiv\text{CX} \xrightarrow[\text{2 HOAc}]{\text{1 Dicyclohexylborane THF}} \begin{array}{l} \text{BuCH=CHX} \\ \quad \text{cis} \end{array} \begin{array}{l} <95\% \ (\text{X=Br}) \\ <85\% \ (\text{X=I}) \end{array}$$

JACS (1967) 89 5086

$$\text{BuC}\equiv\text{CH} \xrightarrow[\substack{\text{2 X}_2 \ \text{THF} \\ \text{3 H}_2\text{SO}_4 \ \text{H}_2\text{O}}]{\text{1 (i-Bu)}_2\text{AlH heptane}} \begin{array}{l} \text{BuCH=CHX} \\ \quad \text{trans} \end{array} \begin{array}{l} 72\% \ (\text{X=Br}) \\ 74\% \ (\text{X=I}) \end{array}$$

JACS (1967) 89 2753

$$\text{C}_6\text{H}_{13}\text{C}\equiv\text{CH} \xrightarrow[\substack{\text{2 H}_2\text{O} \\ \text{3 NaOH H}_2\text{O Et}_2\text{O} \\ \text{4 I}_2}]{\text{1 Catecholborane}} \text{C}_6\text{H}_{13}\text{CH=CHI} \quad 71\%$$

JACS (1973) 95 5786
Tetr Lett (1974) 543

$C_5H_{11}\underset{\overset{|}{OTHP}}{CH}C\equiv CH$ $\xrightarrow[\text{2 } I_2]{\text{1 BuLi Et}_2O}$ $C_5H_{11}\underset{\overset{|}{OTHP}}{CH}C\equiv CI$ $\xrightarrow[\text{HOAc MeOH}]{\overset{\displaystyle \underset{\|}{NCOOK}}{NCOOK}}$ $C_5H_{11}\underset{\overset{|}{OTHP}}{CH}CH=CHI$ ~65%

cis

JACS (1972) <u>94</u> 9256

$PhCOCH_2Me$ $\xrightarrow[\text{2 } I_2 \text{ Et}_3N \text{ Et}_2O]{\text{1 } N_2H_4 \cdot H_2O \text{ EtOH}}$ $PhC\underset{\overset{|}{I}}{=}CHMe$ 60-65%

Aust J Chem (1971) <u>24</u> 1425
(1970) <u>23</u> 989
JCS (1962) 470
JOC (1969) <u>34</u> 3502

$\square$=CH_2 $\xrightarrow[]{\text{ICl CH}_2Cl_2}$ $\underset{-CH_2I}{\overset{Cl}{\square}}$ $\xrightarrow[]{\text{KOH EtOH}}$ $\square$=CHI   27%

JOC (1971) <u>36</u> 1024

$C_{10}H_{21}Br$ $\xrightarrow[\text{2 MeI NaHCO}_3 \text{ DMF}]{\text{1 } CH_2=CHCHS \overset{\overset{\displaystyle Li}{|}}{\underset{}{\diagdown}} \text{ THF HMPA}}$ $C_{10}H_{21}CH=CHCH_2I$ 50%

Tetr Lett (1972) 2743

$CH_2=CHCH_2Br$ $\xrightarrow[\substack{NaH \text{ THF} \\ \text{2 Ba(OH)}_2 \\ \text{3 LiAlH}_4}]{\text{1 } \triangleright\!\!-\!COCH_2COOEt}$ $CH_2=CH(CH_2)_2CH\underset{\overset{|}{OH}}{-}\triangleright$ $\xrightarrow[\text{2 ZnBr}_2]{\text{1 PBr}_3 \text{ LiBr collidine}}$

$CH_2=CH(CH_2)_2\underset{H}{\overset{Me}{\diagup}}C=C\underset{(CH_2)_2Br}{\overset{Me}{\diagdown}}$

JACS (1968) <u>90</u> 2882 6225

$I(CH_2)_3Cl$ $\xrightarrow[]{\overset{\overset{\displaystyle Me}{|}}{EtC=CHCu}}$ $Et\underset{\overset{|}{Me}}{C}=CH(CH_2)_3Cl$ 46%

Tetr Lett (1973) 2407

Also via: Acetylenic halides (Section 308)

Section 372   <u>Ketone — Ketone</u>

$$BuC\equiv CBu \xrightarrow[\substack{CCl_4 \quad H_2O}]{\substack{RuO_2 \quad NaIO_4}} BuCOCOBu \qquad\qquad 70\%$$

Tetr Lett (1971) 2941

NBS   Can J Chem (1971) <u>49</u> 1099

HNO₃   JACS (1956) <u>78</u> 2518 2522

KMnO₄   JOC (1952) <u>17</u> 1063

Tl(NO₃)₃   JACS (1973) <u>95</u> 1296

$$\text{◇–COCl} \xrightarrow[\substack{2\ KMnO_4\quad MgSO_4\quad C_6H_6\quad H_2O}]{\substack{1\ PrCH_2\overset{+}{P}Ph_3\ \overset{-}{Br}\quad PhLi\quad Et_2O}} \text{◇–COCOPr}$$

Tetrahedron (1968) <u>24</u> 2419
Ber (1969) <u>102</u> 2259

$$\underset{\substack{OH\ OH}}{C_6H_{13}CH\text{-}CHC_7H_{15}} \xrightarrow[\substack{H_2O}]{\substack{NBS\quad EtOAc}} C_6H_{13}COCOC_7H_{15} \qquad\qquad 35\%$$

JCS <u>C</u> (1968) 2617

(Ph₃P)₃RuCl₂, PhCH=CHCOMe   JOC (1972) <u>37</u> 1832

$$PhCHO \xrightarrow[\substack{ZnCl_2}]{\substack{PhSH}} PhCH(SPh)_2 \xrightarrow[\substack{2\ Hydrolysis}]{\substack{1\ NaH\quad PhCOCl\quad DMF}} PhCOCOPh$$

JOC (1963) <u>28</u> 961

$$PhCHO \dashrightarrow PhCH=NPh \xrightarrow[\substack{2\ HCl\quad H_2O}]{\substack{1\ NaCN\quad DMF}} PhCOCOPh \qquad\qquad 71\%$$

JOC (1972) <u>37</u> 135

$(CH_2)_9 \begin{array}{c} CHCOOMe \\ | \\ CHCOOMe \end{array}$ $\xrightarrow[Me_3SiCl]{Na}$ $(CH_2)_9 \begin{array}{c} CH-COSiMe_3 \\ \| \\ CH-COSiMe_3 \end{array}$ $\xrightarrow[H_2O]{HCl}$ $(CH_2)_9 \begin{array}{c} CH_2 \\ | \\ CO \\ | \\ CO \\ | \\ CH_2 \end{array}$ 71%

Can J Chem (1969) <u>47</u> 3266

BuBr $\dashrightarrow$ Bu$_2$Cd $\xrightarrow[THF]{(COCl)_2 \quad LiBr}$ BuCOCOBu   37%

JCS <u>A</u> (1966) 456

$\xrightarrow{SeO_2 \quad dioxane \quad H_2O}$

Org Synth (1963) Coll Vol 4 229
Annalen (1962) <u>659</u> 64
Synthesis (1971) 215

$\xrightarrow[HMPA \quad t-BuOH]{O_2 \quad t-BuOK}$

Bull Soc Chim Fr (1967) 3742
Tetr Lett (1961) 554

BuCH$_2$COMe $\xrightarrow[Et_2O \quad H_2O]{MeONO \quad HCl}$ $\underset{NOH}{BuCCOMe}$ $\dashrightarrow^{H_2SO_4 \quad H_2O}$ BuCOCOMe

JOC (1959) <u>24</u> 1726
Org Synth (1955) Coll Vol 3 20

$\dashrightarrow$ $\xrightarrow[Me_2CO]{KMnO_4 \quad MgSO_4}$

JACS (1964) <u>86</u> 3068
Via hydroxymethylene derivative   Chem Comm (1968) 1055

1 Pyr

2 p-Nitrosodi-
methylaniline

3 HCl H_2O

4 KOH H_2O

Helv (1944) 27 524
JOC (1964) 29 1677

$$\begin{pmatrix} CH=CH \\ (CH_2)_{10} \end{pmatrix} \xrightarrow{\text{KMnO}_4 \quad \text{Ac}_2\text{O}} \begin{pmatrix} COCO \\ (CH_2)_{10} \end{pmatrix}$$ 48%

JACS (1971) 93 3303

$$\begin{pmatrix} \overset{\text{OH}}{\underset{|}{\text{COCH}}} \\ (CH_2)_8 \end{pmatrix} \xrightarrow[\text{MeOH} \quad H_2O]{\text{Cu(OAc)}_2 \quad \text{HOAc}} \begin{pmatrix} COCO \\ (CH_2)_8 \end{pmatrix}$$ 88-89%

Org Synth (1963) Coll Vol 4 838
JOC (1964) 29 1677

Air, $CuSO_4$, Pyr   Org Synth (1932) Coll Vol 1 87

CrO_3, H_2SO_4, Me_2CO Chem Ind (1972) 807

Ac_2O, Me_2SO JOC (1967) 32 3204

MnO_2 JOC (1966) 31 615

Bi_2O_3 JACS (1956) 78 2518
JCS (1951) 793

TlOEt   JCS (1952) 2771

$HgCl_2$ JCS B (1966) 366

$Na[Sb(OAc)_6]$ Ber (1964) 97 124

$Ph_3P \cdot Br_2$ Synthesis (1972) 697

$(Ph_3P)_3RuCl_2$, PhCH=CHCOMe   JOC (1972) 37 1832

NBS   JCS C (1968) 2617

HCl H_2O

65%

JOC (1959) 24 719
Helv (1962) 45 2575

$$BuC\equiv CH \xrightarrow[\text{2 MeOH}]{\text{1 MeCOOH   }(CF_3CO)_2O} BuCOCH_2COMe$$

20%

$$\begin{array}{c}+ \ - \\ RCO \ BF_4\end{array}$$
JCS (1953) 3628
Tetr Lett (1972) 4935
(1971) 3101

$$(EtCH_2CO)_2O \xrightarrow[\text{2 }(BuCO)_2O]{\text{1 }BF_3} \begin{array}{c}EtCH_2COCHEt \\ | \\ COBu\end{array}$$

80%

Annalen (1963) <u>668</u> 15

$$\begin{array}{c}CHOH \\ | \\ (CH_2)_7 \ CH_2 \\ | \\ CHOH\end{array} \xrightarrow[\text{Me}_2CO]{\text{CrO}_3 \ H_2SO_4} \begin{array}{c}CO \\ | \\ (CH_2)_7 \ CH_2 \\ | \\ CO\end{array}$$

JCS Perkin I (1972) 1509

Review: The Acylation of Ketones to Form β-Diketones or β-Keto Aldehydes

Org React (1954) <u>8</u> 59

$$\begin{array}{c}CO \\ (CH_2)_{11} \\ CH_2\end{array} \dashrightarrow \begin{array}{c}C-N \quad O \\ (CH_2)_{11} \quad \| \\ CH\end{array} \xrightarrow[\text{2 HCl   }H_2O]{\text{1 }CH_3COCl \ Et_3N \quad Et_2O} \begin{array}{c}CO \\ | \\ (CH_2)_{11} \ CH_2 \\ | \\ CO \\ | \\ CH_2\end{array}$$

<36%

Ber (1972) <u>105</u> 2216

$$\text{Morpholine} \qquad \xrightarrow[\text{C}_6H_6]{\text{C}_5H_{11}COCl}$$

JACS (1963) <u>85</u> 207

$$BuCH_2COMe \xrightarrow[]{Ac_2O \quad TsOH} BuCH=CMe \atop \underset{OAc}{|} \xrightarrow[Ac_2O]{AcOH \cdot BF_3} \underset{COMe}{\overset{|}{BuCHCOMe}} \qquad 64\text{-}77\%$$

Org Synth (1971) <u>51</u> 90
JACS (1953) <u>75</u> 50$\overline{30}$
Tetrahedron $\overline{(1966)}$ <u>22</u> 2039
JOC (1969) <u>34</u> 1425

Acylation of ketals   JACS (1963) <u>85</u> 3901

JOC (1962) <u>27</u> 2742
Org Synth $(\overline{1955)}$ Coll Vol 3 251 291

Annalen (1962) <u>659</u> 64

$$C_6H_{13}CH=CH_2 \xrightarrow[peroxides]{MeCOCH_2COMe} \underset{COMe}{\overset{|}{C_6H_{13}CH_2CH_2CHCOMe}}$$

Proc Chem Soc (1964) 142

$$C_{17}H_{35}COOC=CH_2 \atop \underset{CH_3}{|} \xrightarrow[]{AlCl_3 \quad Hexane} C_{17}H_{35}COCH_2COC_{17}H_{35} \qquad 65\text{-}70\%$$

Tetr Lett (1969) 2553
JOC (1971) <u>36</u> 1447

$$EtCOCH_2Br \xrightarrow[Et_2O]{PrCOSH \quad Et_3N} EtCOCH_2SCOPr \xrightarrow[\substack{propyl]\text{-phenylphosphine} \\ LiBr \quad MeCN}]{\text{Bis-[3-dimethylamino-}} EtCOCH_2COPr \qquad \sim86\%$$

Helv (1971) <u>54</u> 710

$$RCOCH_2COR \xrightarrow[\text{base}]{RX} RCOCHCOR$$
$$\overset{|}{R}$$

Base:

K$_2$CO$_3$ Org Synth (1973) Coll Vol 5 785

TlOEt   JACS (1968) $\underline{90}$ 2421

NaH   JOC (1961) $\underline{26}$ 4112

NaNH$_2$ Org Synth (1967) $\underline{47}$ 92

$$C_6H_{13}C \equiv CH \xrightarrow{\substack{1 \ BuLi \ hexane \\ 2 \ (C_6H_{13})_3B \ diglyme \\ 3 \ BrCH_2COMe \\ 4 \ H_2O_2 \ NaOAc \ H_2O}} C_6H_{13}COCHCH_2COMe \quad 75\%$$
$$\overset{|}{C_6H_{13}}$$

Tetr Lett (1973) 4491

$$EtCOCl \dashrightarrow EtCOCHN_2 \xrightarrow{\substack{1 \ CH_2=COAc \overset{Me}{|} \\ \text{cuprous acetylacetonate} \\ 2 \ Base}} EtCOCH_2CH_2COMe$$

Tetr Lett (1971) 2575

$$PhCHO \xrightarrow[\text{DMF}]{PhCH=CHCOMe \ KCN} PhCOCHCH_2COMe \quad 67\%$$
$$\overset{|}{Ph}$$

Angew (1973) $\underline{85}$ 89
(Internat Ed $\overline{12}$ 81)
Tetr Lett (1973) 1461

$$C_6H_{13}CHO \dashrightarrow C_6H_{13}CH(SPh)_2 \xrightarrow{\substack{1 \ BuLi \ THF \ hexane \\ 2 \ CuI \\ 3 \ CH_2=CHCOMe \\ 4 \ CuCl \ CuO}} C_6H_{13}COCH_2CH_2COMe \quad 72\%$$

Me$_2$CO   H$_2$O   JACS (1972) $\underline{94}$ 8641

$$C_6H_{13}OTs \xrightarrow{\substack{1 \ EtSOCH_2SMe \ BuLi \ HMPA \\ 2 \ BuLi \\ 3 \ CH_2=CHCOMe \\ 4 \ HgCl_2 \ HCl}} C_6H_{13}COCH_2CH_2COMe \quad 76\%$$

Tetr Lett (1973) 3275 3271

CONMe$_2$
|
(CH$_2$)$_N$CONMe$_2$ Et$_2$O
PhBr --→ PhLi $\xrightarrow{\hspace{3cm}}$ PhCO(CH$_2$)$_N$COPh   N=2 or 3

JOC (1973) <u>38</u> 901

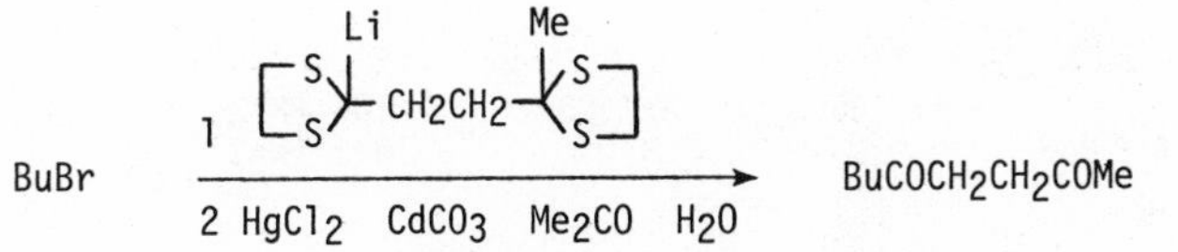

BuBr $\xrightarrow[\text{2 HgCl}_2 \quad \text{CdCO}_3 \quad \text{Me}_2\text{CO} \quad \text{H}_2\text{O}]{\text{1}}$ BuCOCH$_2$CH$_2$COMe

Chem Comm (1972) 529

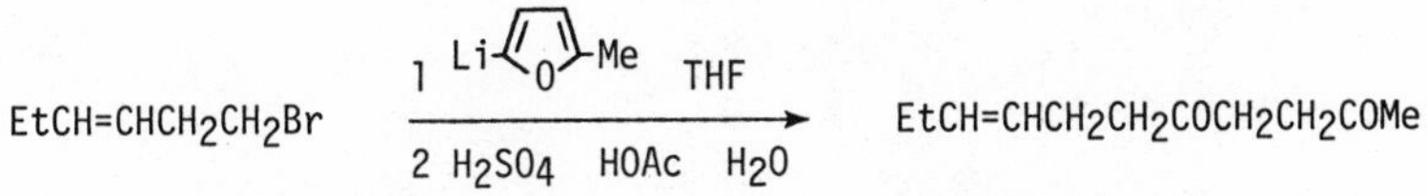

EtCH=CHCH$_2$CH$_2$Br $\xrightarrow[\text{2 H}_2\text{SO}_4 \quad \text{HOAc} \quad \text{H}_2\text{O}]{\text{1 Li} \quad \text{Me} \quad \text{THF}}$ EtCH=CHCH$_2$CH$_2$COCH$_2$CH$_2$COMe

JOC (1966) <u>31</u> 977

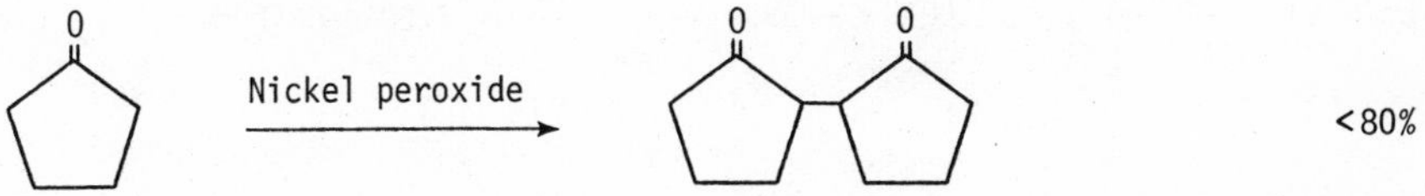

Nickel peroxide $\longrightarrow$ <80%

JCS Perkin I (1974) 280

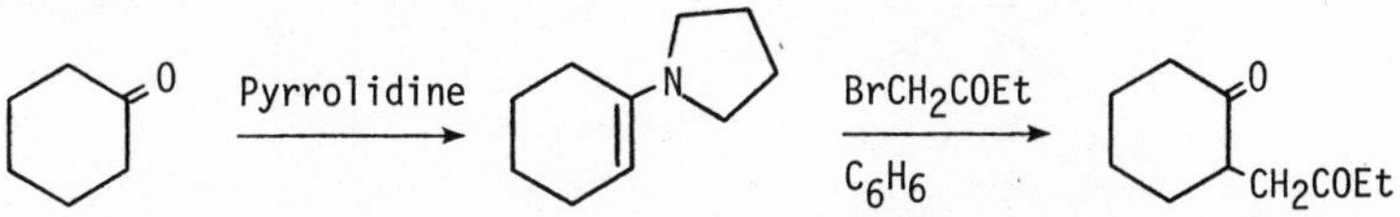

Pyrrolidine → $\xrightarrow[\text{C}_6\text{H}_6]{\text{BrCH}_2\text{COEt}}$

JACS (1963) <u>85</u> 207

--→ $\xrightarrow[\text{2 HCl} \quad \text{H}_2\text{O} \quad \text{Et}_2\text{O}]{\text{1 N}_2\text{CHCOMe} \quad \text{copper} \quad \text{bronze}}$

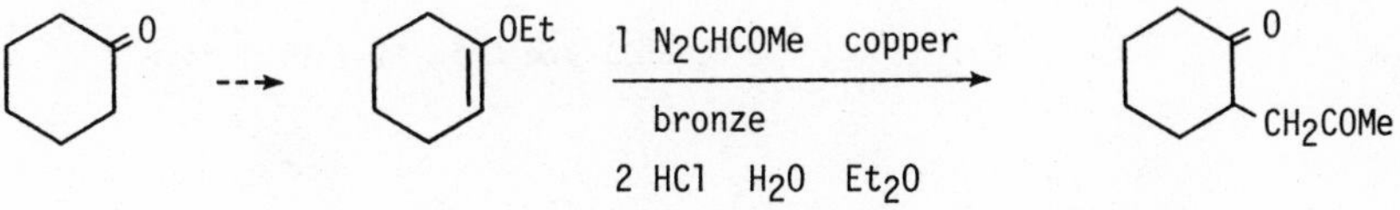

Synth Comm (1973) <u>3</u> 255

PhCOCHO $\xrightarrow{\text{Ph}_3\text{P=CHCOPh}}$ PhCOCH=CHCOPh $\xrightarrow[\text{HOAc\quad H}_2\text{O}]{\text{SnCl}_2\quad\text{HCl}}$ PhCOCH$_2$CH$_2$COPh   <85%

Aust J Chem (1971) <u>24</u> 2137

BuC≡C(CH$_2$)$_N$COMe $\xrightarrow[\text{MeOH\quad H}_2\text{O}]{\text{HgSO}_4\quad\text{H}_2\text{SO}_4}$ BuCH$_2$CO(CH$_2$)$_N$COMe   85% (N=2)
                                                                                                              93% (N=3)

JACS (1964) <u>86</u> 935

Me$_2$C=CHCOMe $\xrightarrow[\text{pentane}]{\text{Ni(CO)}_4\quad\text{BuLi}}$
$$\text{BuCOCCH}_2\text{COMe}$$
with Me above and Me below the central C   89%

Synthesis (1971) 55

EtCH$_2$NO$_2$ $\xrightarrow[\text{(i-Pr)}_2\text{NH}]{\text{CH}_2\text{=CHCOMe}}$ EtCHCH$_2$CH$_2$COMe (NO$_2$) $\xrightarrow[\text{glyme}]{\text{TiCl}_3\quad\text{H}_2\text{O}}$ EtCOCH$_2$CH$_2$COMe   64%

JACS (1971) <u>93</u> 5309
Tetr Lett (1972) 1331
JOC (1974) <u>39</u> 259

Ph furanone structure $\xrightarrow[\text{2 HCl\quad H}_2\text{O\quad HOAc}]{\text{1 Et}_2\text{NC≡CMe\quad MeCN}}$ PhCOCH$_2$CH$_2$COCH$_2$Me   56%

Tetr Lett (1971) 1565 1569

MeI $\xrightarrow[\text{2 \quad Me\quad OEt/OEt}]{\text{1 Mg}}$ MeCO(CH$_2$)$_3$C(OEt)$_2$ (Me) - -→ MeCO(CH$_2$)$_3$COMe   <20%

Bull Soc Chim Fr (1970) 4429

PhBr $\xrightarrow[\displaystyle 2 \text{ Ph} \overset{O}{\diagdown} O]{1 \text{ Mg} \quad Et_2O}$ PhCO(CH$_2$)$_3$COPh

Zh Org Khim (1969) 5 83
(Chem Abs 70 87453)

$\xrightarrow{\text{Pyrrolidine}}$ $\xrightarrow[\substack{\text{dioxane} \\ 2 \text{ HOAc} \quad \text{NaOAc} \\ H_2O}]{1 \quad CH_2=CHCOEt}$ (CH$_2$)$_2$COEt

65%

JACS (1963) 85 207

BuCOOH $\xrightarrow[\text{THF}]{CH_2=CHMgCl \quad CuCl}$ BuCO(CH$_2$)$_4$COBu   53%

Can J Chem (1972) 50 2786

$\begin{array}{c} COCl \\ | \\ (CH_2)_4 \\ | \\ COCl \end{array}$ $\xrightarrow{R_2CuLi \quad Et_2O}$ $\begin{array}{c} COR \\ | \\ (CH_2)_4 \\ | \\ COR \end{array}$ 92% (R=Me)
90% (R=Bu)

JACS (1972) 94 5106

$\begin{array}{c} COCl \\ | \\ (CH_2)_4 \\ | \\ COCl \end{array}$ $\xrightarrow{PhH \quad AlCl_3}$ $\begin{array}{c} COPh \\ | \\ (CH_2)_4 \\ | \\ COPh \end{array}$ 75-81%

Org Synth (1943) Coll Vol 2 169

MeCO(CH$_2$)$_2$COOH $\xrightarrow[\text{MeONa} \quad \text{MeOH}]{\text{Electrolysis}}$ MeCO(CH$_2$)$_4$COMe   21-26%

Annalen (1953) 580 125

Section 373   Ketone — Nitrile

α, β, γ, δ and ε-ketonitriles

$$\text{BuC} \equiv \text{CH} \quad \dashrightarrow \quad (\text{BuC} \equiv \text{C})_2\text{Hg} \quad \xrightarrow{\text{NOCl}} \quad \text{BuCOCN}$$

Proc Chem Soc (1963) 13

$$\text{PhCOCl} \quad \xrightarrow{\text{CuCN}} \quad \text{PhCOCN}$$

60-65%

Org Synth (1955) Coll Vol 3 112
Can J Chem (1971) 49 919

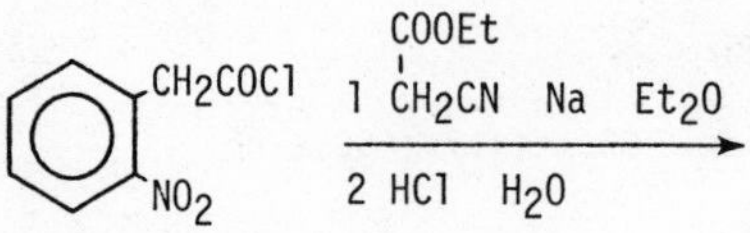

JOC (1960) 25 736

40%

$$\text{PrCOOEt} \quad \xrightarrow[\text{Et}_2\text{O}]{\text{CH}_3\text{CN} \quad \text{Ph}_3\text{CNa}} \quad \text{PrCOCH}_2\text{CN}$$

JACS (1942) 64 2720

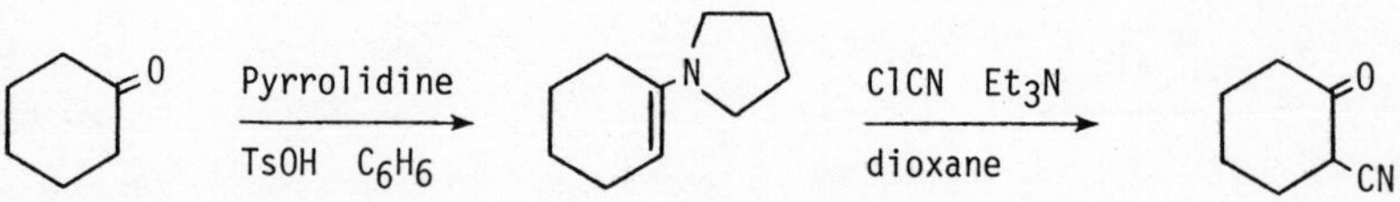

JACS (1959) 81 5400

<93%

Synthesis (1973) 682

69%

PhCH$_2$CN $\xrightarrow[\text{EtOH}]{\text{MeCOOEt EtONa}}$ PhCHCN   66-73%
                                      |
                                    COMe

Org Synth (1943) Coll Vol 2 487
(1963) Coll Vol 4 174

PhCN $\xrightarrow[\text{2 HCl EtOH H}_2\text{O}]{\text{1 CH}_3\text{CN NaNH}_2\text{ Et}_2\text{O}}$ PhCOCH$_2$CN

Ber (1949) __82__ 254

$\xrightarrow[\substack{\text{2 HCl H}_2\text{SO}_4\text{ H}_2\text{O} \\ \text{toluene}}]{\text{1 NaH dioxane}}$

           46%

Can J Chem (1965) __43__ 2512

PhCOCH$_2$Br $\xrightarrow{\text{LiCN DMF}}$ PhCOCH$_2$CN

JOC (1964) __29__ 1970
JACS (1947) __69__ 990

$\xrightarrow[\text{HOAc}]{\text{NH}_2\text{OH·HCl}}$ $\xrightarrow[\text{Et}_2\text{O}]{\text{MeONa MeOH}}$

Tetrahedron (1968) __24__ 5959

$\xrightarrow[\substack{\text{2 i-PrCHCOOH} \\ \quad\quad | \\ \quad\quad \text{NH}_2 \\ \text{3 Ac}_2\text{O}}]{\text{1 SOCl}_2}$ $\xrightarrow[\text{2 NaOH}]{\substack{\text{1 CH}_2\text{=CHCN} \\ \text{Et}_3\text{N}}}$

Angew (1971) __83__ 727
(Internat Ed __10__ 655)

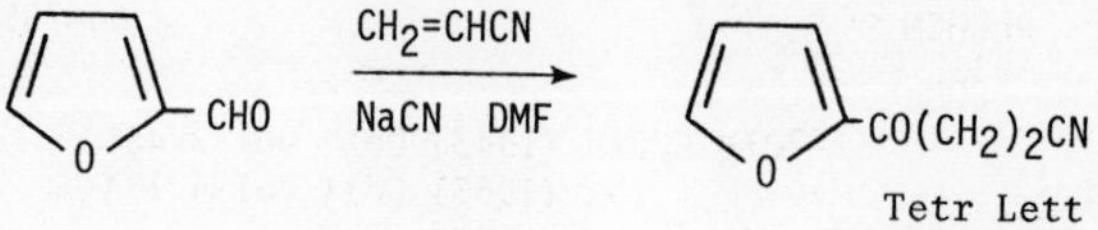

Tetr Lett (1973) 1461
Angew (1973) 85 89
(Internat Ed 12 81)

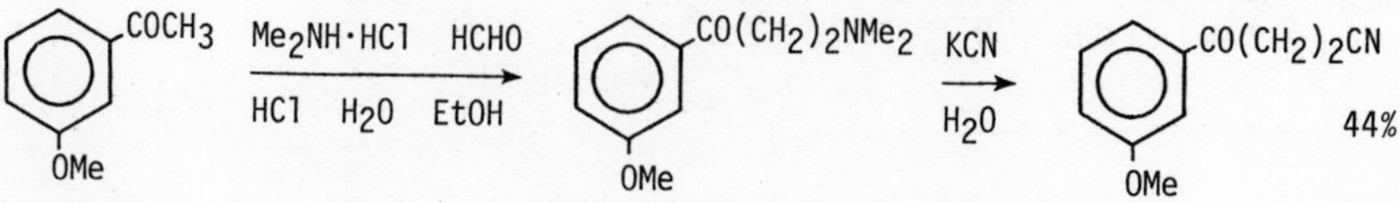

32%

JCS (1947) 1190

44%

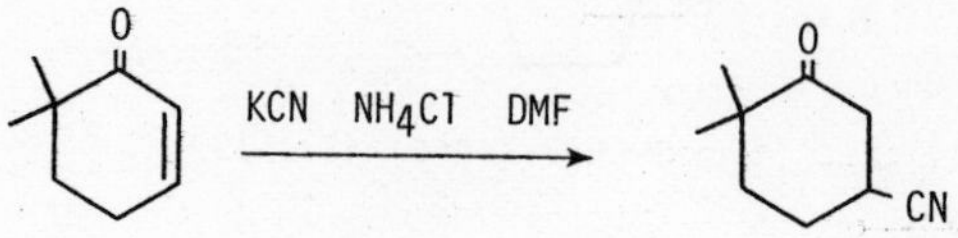

JOC (1972) 37 268
Org Synth (1943) Coll Vol 2 498
JOC (1959) 24 879
JCS Perkin I (1972) 2765

HCN, Et₃Al   Org Synth (1972) 52 100
                 JACS (1972) 94 4635 4654

Me₂CCN   JCS (1958) 4193
     |
     OH   Bull Soc Chim Fr (1963) 2471
                 Tetr Lett (1973) 141

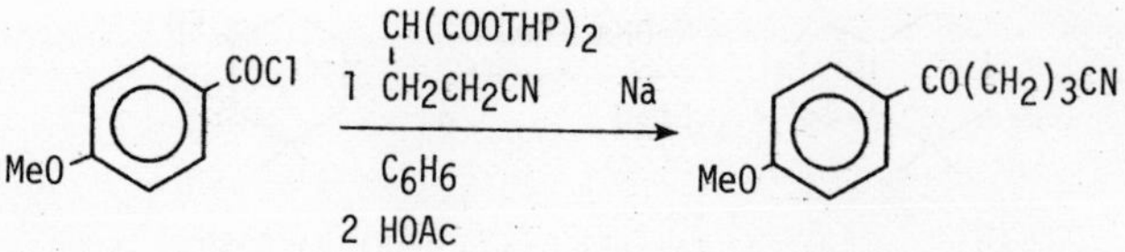

Bull Soc Chim Fr (1969) 4437

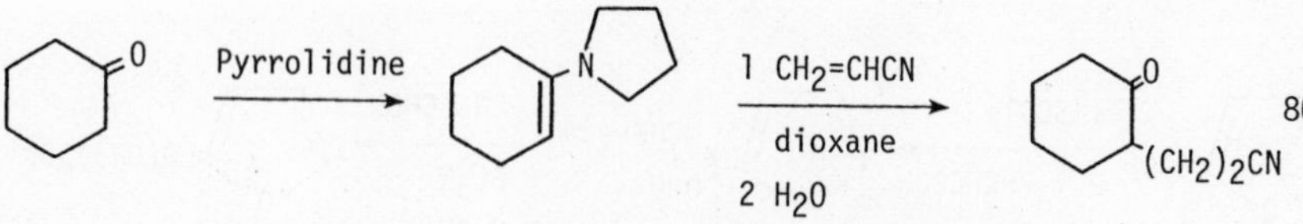

80%

JACS (1963) 85 207

80%

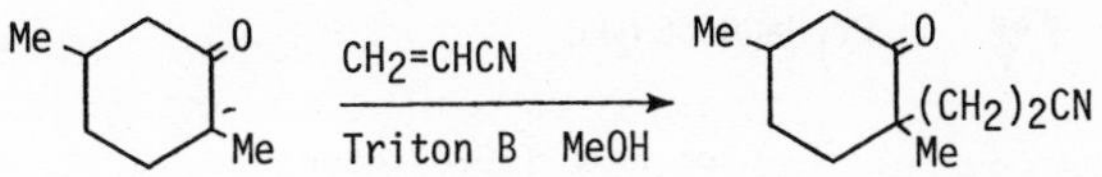

JACS (1964) 86 465
Org React (1949) 5 79

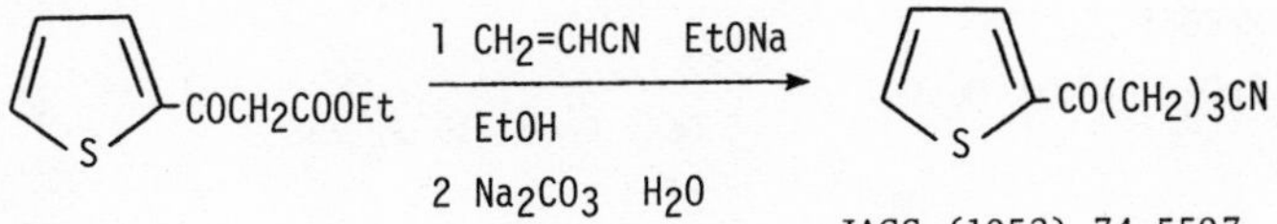

JACS (1952) 74 5597

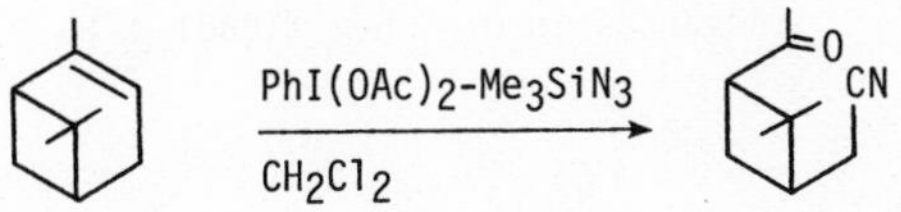

Synthesis (1972) 285

65%

Section 374   Ketone — Olefin

For the oxidation of allylic alcohols to olefinic ketones see section 168
    vol 1 (Ketones from Alcohols and Phenols)

For the oxidation of allylic methylene groups (C=C-CH₂ → C=C-CO) see
    section 170 vol 1 and 2 (Ketones from Alkyls and Methylenes)

For the alkylation of olefinic ketones see also section 177 vol 1 and 2
    (Ketones from Ketones)

Annelation reactions are not listed

$$PrCH_2C{\equiv}CBu \xrightarrow{MeCOO_2H} PrCH{=}CHCOBu$$

Ber (1954) 87 1478

PrC≡CH

            Me
         |
       i-PrCBH$_2$

1 BuC≡CCl   Me

2 MeONa   →   PrCH=CHCOCH$_2$Bu   48%

3 H$_2$O$_2$ NaOAc

                       Chem Comm (1973) 606

C$_5$H$_{11}$C≡CH

1 RLi   THF

2 (i-Pr)$_3$B   →   C$_5$H$_{11}$C=C(Pr-i)$_2$ 42%

3 MeCOCl |
                            COMe

4 CrO$_3$ H$_2$SO$_4$ Me$_2$CO

                       Tetr Lett (1973) 795

PhCOOLi

Me$_2$C=CHLi   Et$_2$O
             →   PhCOCH=CMe$_2$ 40%

                       JCS (1950) 2012
                       Advances in Org Chem (1960) 2 1

EtCHCOCl
|
Me

1 CH$_2$=CH$_2$ AlCl$_3$ PhNO$_2$
                  →   EtCHCOCH=CH$_2$
2 PhNEt$_2$ |
                             Me

                       Gazz (1967) 97 610
                       (Chem Abs 68 12401)

PhCOOH   HOCH$_2$C≡CH - - - - - - - → PhCOOCH$_2$C≡CH   660°
                                           →   PhCOCH=CH$_2$ <80%
                                      (vapor phase)

                       JACS (1972) 94 5086

PhCH$_2$OH

1 MeMgI
             →   PhCH=CHCOMe
2 Me$_2$CO

                       Tetr Lett (1964) 3481

PhCHO

CH$_3$COMe NaOH H$_2$O
               →   PhCH=CHCOMe   65-78%

                       Org Synth (1932) Coll Vol 1 77 78 81 283
                       Org React (1968) 16 1

$$\text{PrCHO} \quad \xrightarrow[\text{2 TsOH}]{\overset{\displaystyle 1 \; \overset{\text{COMe}}{|}}{\text{CH}_2\text{COOBu-t} \quad \text{piperidine} \quad \text{EtOH}}} \quad \text{PrCH=CHCOMe} \qquad 73\%$$

Acta Chem Scand (1963) <u>17</u> 2216

$$\xrightarrow[\text{Et}_2\text{O}]{\overset{\text{Me}}{|}\text{ClCHCOMe} \quad \text{Mg}}$$

Rec Trav Chim (1950) <u>69</u> 307

$$\text{Me}_2\text{CH}_2\text{CHO} \quad \xrightarrow[\text{NaH} \quad \text{MeOCH}_2\text{CH}_2\text{OMe}]{(\text{EtO})_2\text{POCH}_2\text{COMe}} \quad \text{Me}_2\text{CH}_2\text{CH=CHCOMe} \qquad 76\%$$

Bull Soc Chim Fr (1967) 2477
JACS (1969) <u>91</u> 5675
Org React (1965) <u>14</u> 270

$\text{Ph}_3\text{P=CHCOR}$ Ber (1962) <u>95</u> 1513

$$\xrightarrow[\text{acid}]{\text{Polyphosphoric}}$$

25%

Tetrahedron (1968) <u>24</u> 553

$$\overset{\text{O}}{\overset{/\backslash}{\text{PhCHCH}_2}} \quad \xrightarrow[\text{2 Hydrolysis}]{1} \quad \text{PhCH=CHCOMe} \qquad <70\%$$

Angew (1965) <u>77</u> 1134
(Internat Ed <u>4</u> 1075)

EtSCH=CH_2, RLi   JACS (1973) <u>95</u> 2694

$$\xrightarrow[\text{2 Acid}]{1 \; \text{Li} \quad \text{t-BuOH} \quad \text{RNH}_2}$$

Synthesis (1972) 391
Advances in Org Chem (1972) <u>8</u> 1

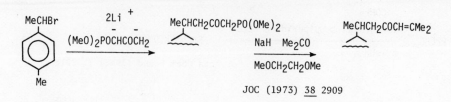

$$MeCHBr \xrightarrow[(MeO)_2POCHCOCH_2]{2Li^+} MeCHCH_2COCH_2PO(OMe)_2 \xrightarrow[MeOCH_2CH_2OMe]{NaH \quad Me_2CO} MeCHCH_2COCH=CMe_2$$

JOC (1973) 38 2909

$$PrBr \xrightarrow[2 \ Et_2NCH=CHCOMe]{1 \ Mg} PrCH=CHCOMe$$

Bull Soc Chim Fr (1960) 515
(1966) 287

ArOC=CHCOR'   Compt Rend (1964) 258 5234
|
R

MeCOCl AlCl₃

CHCl₃

Tetr Lett (1968) 5455

$$PhCOCH_2 \xrightarrow[\substack{1 \ LiN(Pr-i)_2 \\ 2 \ PhCH_2SCH_2Br \\ 3 \ NaIO_4 \\ 4 \ Toluene \ (reflux)}]{} PhCOC=CH_2$$ 52%
| |
Me   Me

Chem Comm (1974) 135

$$(CH_2)_9 \begin{matrix} CO \\ CH_2 \\ CH_2 \end{matrix} \xrightarrow[\substack{2 \ Me_2S_2 \\ 3 \ NaIO_4 \ MeOH \ H_2O \\ 4 \ 50°}]{1 \ LiN\text{–i-Pr} \quad THF} (CH_2)_9 \begin{matrix} CO \\ CH \\ \| \\ CH \end{matrix}$$

JACS (1973) 95 6840
PhSeCl, H₂O₂   JACS (1973) 95 6137
PhSeBr, NaIO₄   JACS (1973) 95 5813
Chem Comm (1973) 695

CH₂CH₂COMe $\xrightarrow[H_2O]{HIO_4 \quad dioxane}$ CH=CHCOMe

Chem Comm (1973) 746

O₂, palladium acetylacetonate   JOC (1971) 36 752

Review: The Aldol Condensation   Org React (1968) 16 1

i-PrCH$_2$COCH$_3$ $\xrightarrow[\text{2 (COOH)}_2]{\text{1 MeCHO KOH i-PrOH}}$ i-PrCH$_2$COCH=CHMe

JCS C (1970) 1469

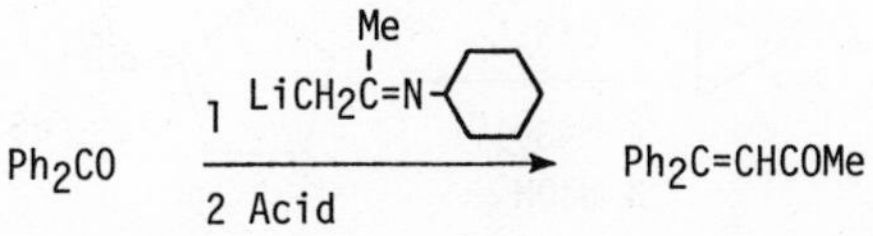

Ph$_2$CO $\xrightarrow[\text{2 Acid}]{\text{1 LiCH}_2\overset{\text{Me}}{\underset{}{C}}\text{=N-}\bigcirc}$ Ph$_2$C=CHCOMe   50%

Angew (1968) 80 8
(Internat Ed 7 7)

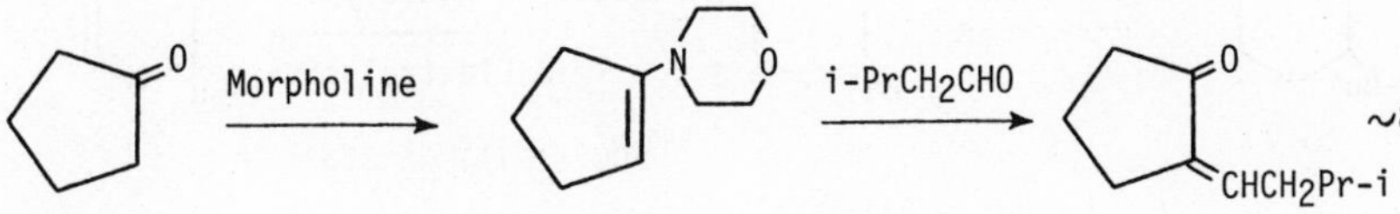

Ber (1962) 95 1493

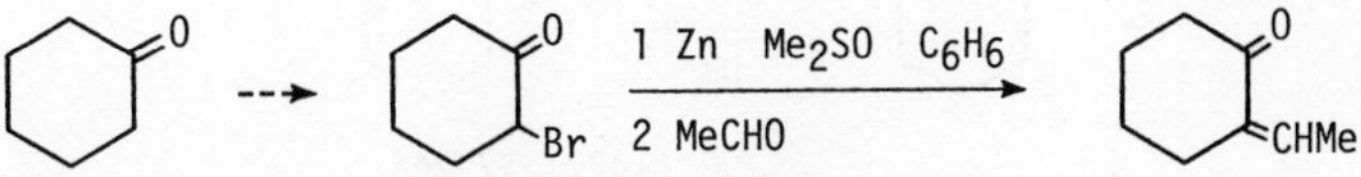

JACS (1967) 89 5727

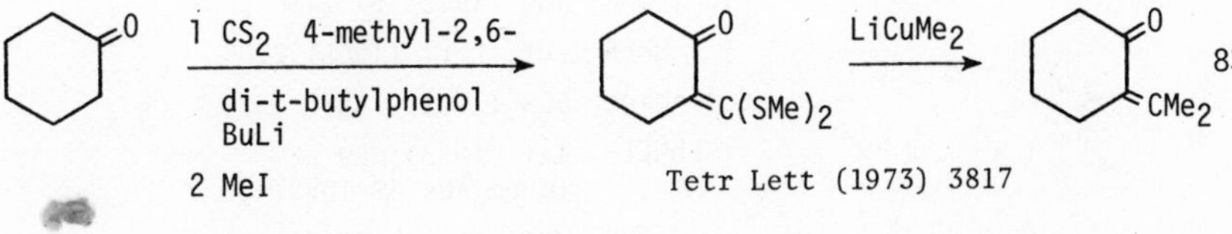

Tetr Lett (1973) 3817

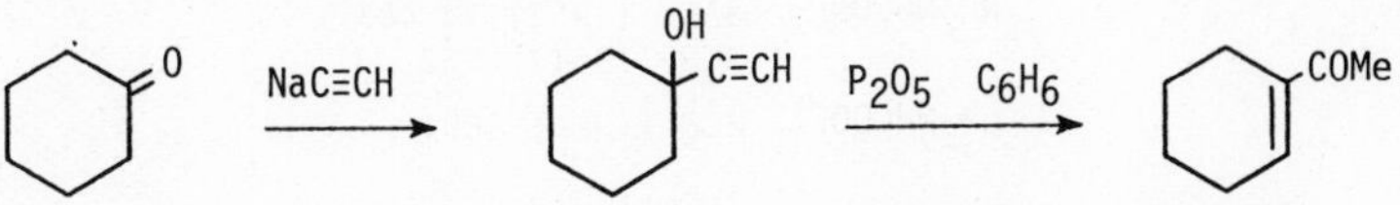

Org Synth (1955) Coll Vol 3 22
JACS (1953) 75 4740

$Ph_2CO \xrightarrow[\text{NaH THF}]{Ph_2POCH{=}C(NHBu)Ph} Ph_2C{=}CHCOPh$ 62%

Tetr Lett (1971) 1401 1419
JOC (1971) 36 2892

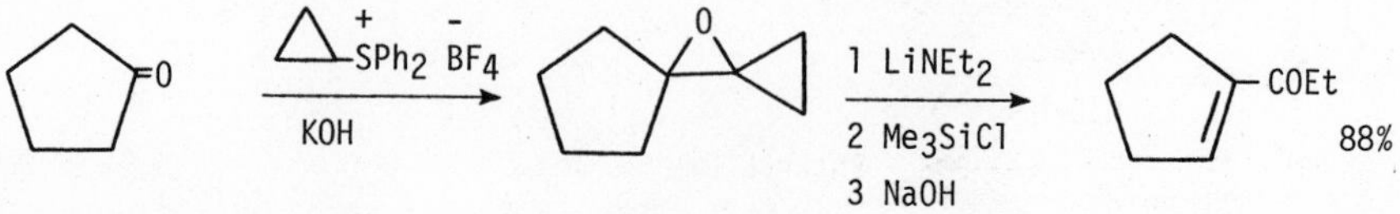

SPh2 BF4
KOH
O
1 LiNEt2
2 Me3SiCl
3 NaOH
COEt
88%

JACS (1973) 95 289

O
t-Bu
Me
BrCHCOOEt
O
COOEt
Me
t-Bu
1 Hydrolysis
2 Electrolysis
COMe
t-Bu

JOC (1971) 36 3232

O
Br2 CH2Cl2
THF
O
Br
MgO
DMF
O

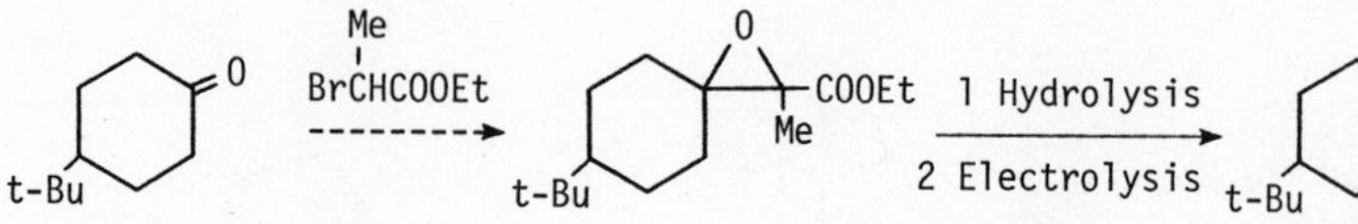

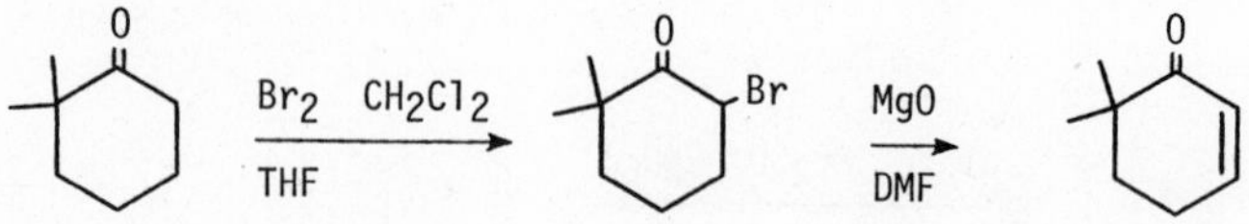

JOC (1972) 37 268

Dehydrohalogenation with: $CaCO_3$, DMA	JCS (1961) 2532
LiBr, Li_2CO_3, Me_2SO	JCS Perkin I (1972) 50
LiBr, Li_2CO_3, DMF	Tetrahedron (1973) 29 2575
LiCl, DMF	JOC (1972) 37 2436
HMPA	Tetr Lett (1968) 2105
Collidine	JCS (1961) 1583
$PhNEt_2$	Izv (1953) 889 (Chem Abs 49 1082)
Pyr	Helv (1942) 25 821
$NH_2NHCONH_2$	JACS (1952) 74 483 JCS (1961) 2532
$NH_2NHCOOEt$	JCS (1959) 1691

continued

2,4-Dinitrophenylhydrazine   JACS (1952) $\underline{74}$ 3951
                                          (1950) $\underline{72}$ 2290
$(MeO)_2PO_2^- \ NMe_4^+$ Bull Soc Chim Fr (1971) 2962 2551
            $AgNO_3$ JACS (1960) $\underline{82}$ 130

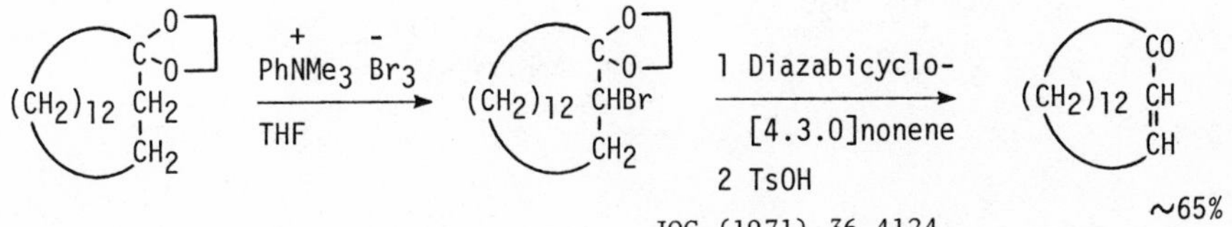

JOC (1971) $\underline{36}$ 4124   ~65%

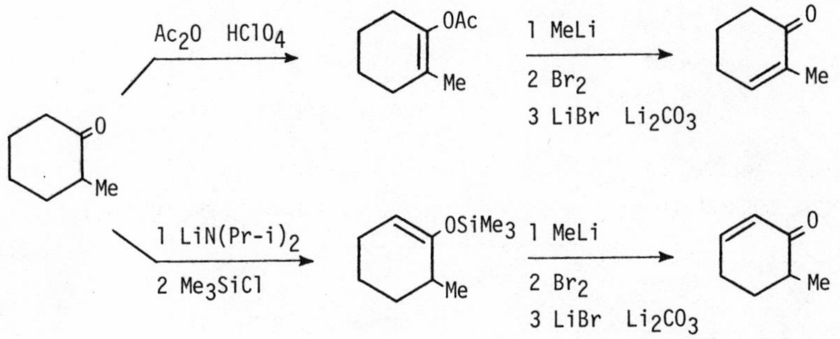

JOC (1973) $\underline{38}$ 2576

$EtCH=CH_2$ $\xrightarrow{\begin{array}{l}1\ B_2H_6\quad THF\\ 2\ HC\equiv CCOMe\\ 3\ Air\end{array}}$ $EtCH_2CH_2CH=CHCOMe$ 72%

JACS (1970) $\underline{92}$ 3503

$MeCH=CH_2$ $\xrightarrow{\begin{array}{l}1\ MeCOCl\quad AlCl_3\quad CH_2Cl_2\\ 2\ Quinoline\end{array}}$ $MeCH=CHCOMe$

Org Synth (1971) $\underline{51}$ 115
$MeCOOH, (CF_3CO)_2O$ JCS (1953) 3628

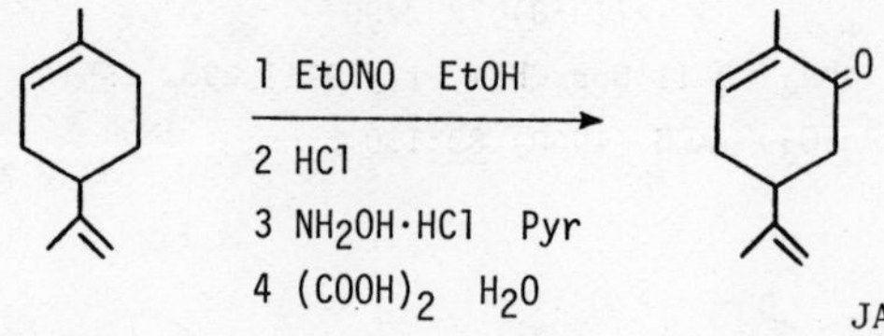
1 EtONO EtOH
2 HCl
3 NH2OH·HCl Pyr
4 (COOH)2 H2O
O

~60%

JACS (1951) 73 5856

NOF JOC (1968) 33 3699

$C_5H_{11}C{\equiv}CCH_2OMe$ —— 1 BuLi $Me_2NCH_2CH_2NMe_2$; 2 Me_2SO_4; 3 BuLi; 4 MeI; 5 Acid ——→ $C_5H_{11}C(Me){=}CHCOMe$

Tetr Lett (1973) 2585

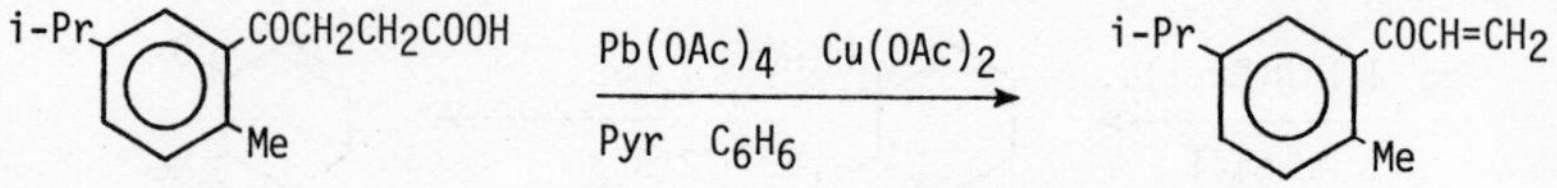
i-Pr
COCH2CH2COOH
Me
Pb(OAc)4 Cu(OAc)2
Pyr C6H6
i-Pr
COCH=CH2
Me

31%

Synthesis (1973) 541
JOC (1968) 33 2008

$C_6H_{13}COCH(Me)COOMe$ —— 1 HCHO Me_2NH, HCl dioxane; 2 MeI; 3 DMF 75° ——→ $C_6H_{13}COC(Me){=}CH_2$ 52%

Tetr Lett (1973) 5037

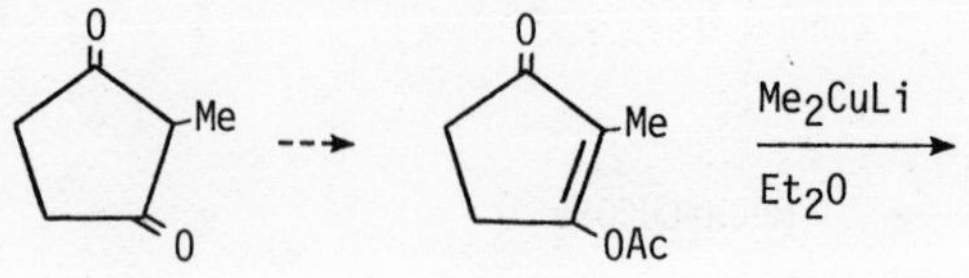
O
Me
O
O
Me
OAc
Me2CuLi
Et2O
O
Me
Me

Tetr Lett (1973) 2071

Via enol ether Chem Comm (1973) 907

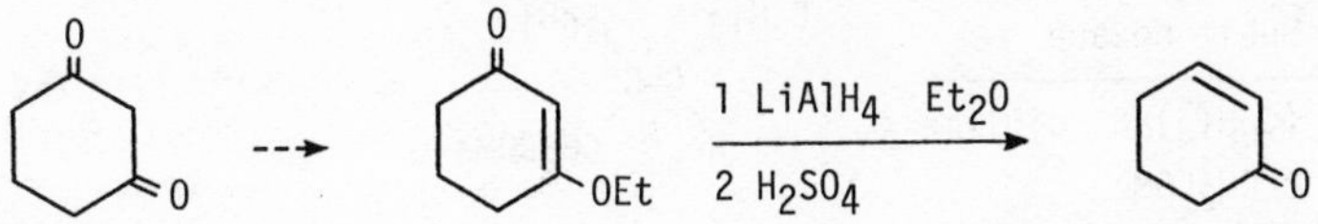

1 LiAlH$_4$ Et$_2$O

2 H$_2$SO$_4$

62-75%

Org Synth (1973) Coll Vol 5 294

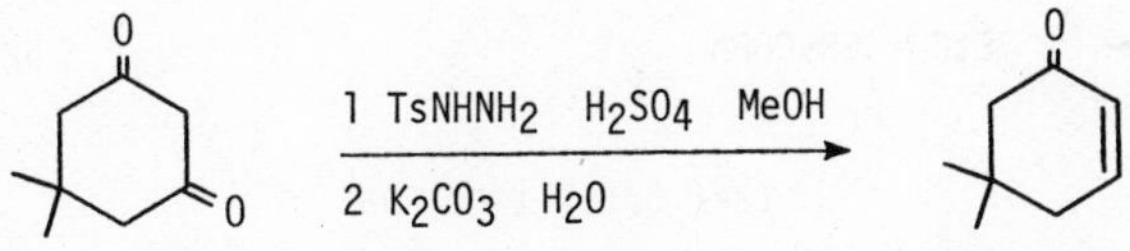

1 TsNHNH$_2$ H$_2$SO$_4$ MeOH

2 K$_2$CO$_3$ H$_2$O

52%

JOC (1973) 38 3637

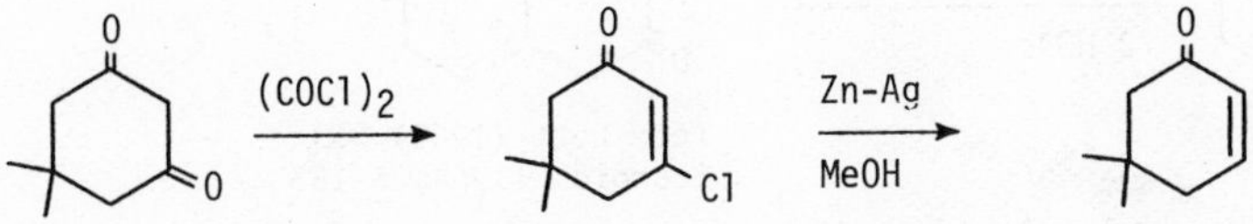

(COCl)$_2$

Zn-Ag

MeOH

75%

JOC (1973) 38 3658

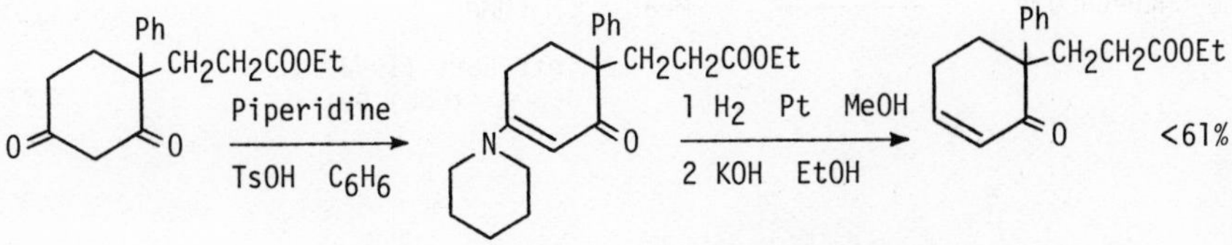

Ph

CH$_2$CH$_2$COOEt

Piperidine

TsOH C$_6$H$_6$

Ph

CH$_2$CH$_2$COOEt

1 H$_2$ Pt MeOH

2 KOH EtOH

Ph

CH$_2$CH$_2$COOEt

<61%

Ber (1964) 97 1723

1 NBS

2 PhS$^-$

3 m-Chloroperbenzoic
  acid

SOPh

Cl

Cyclohexene

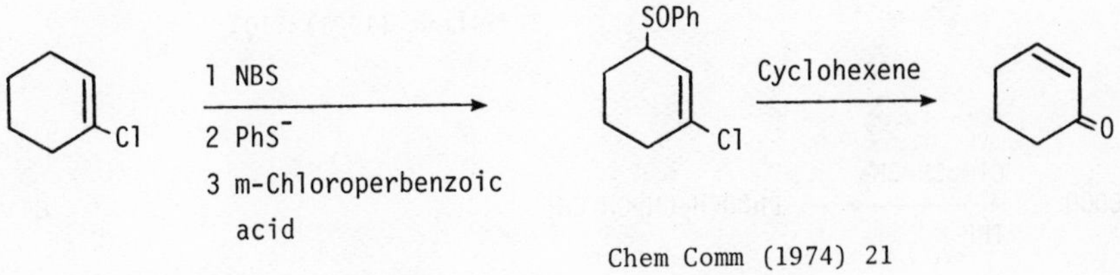

Chem Comm (1974) 21

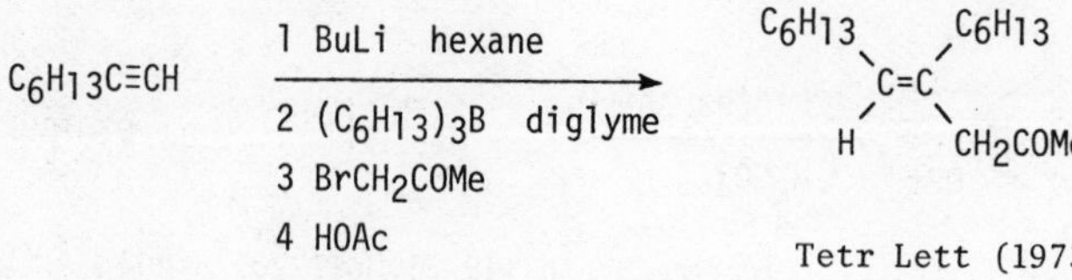

$C_6H_{13}C\equiv CH$

1 BuLi   hexane

2 $(C_6H_{13})_3B$ diglyme

3 $BrCH_2COMe$

4 HOAc

$$\begin{array}{c} C_6H_{13} \qquad C_6H_{13} \\ \diagdown \quad C=C \quad \diagup \\ H \qquad\qquad CH_2COMe \end{array}$$

75%

Tetr Lett (1973) 4491

$EtCH_2COPr$

$$\begin{array}{c} Ph \\ | \\ Ph_3P=C=COLi \end{array}$$

$$\begin{array}{c} EtCH=CCH_2COPh \\ | \\ Pr \end{array}$$

52%

Tetr Lett (1972) 933

1 t-BuOK   t-BuOH

2 HOAc

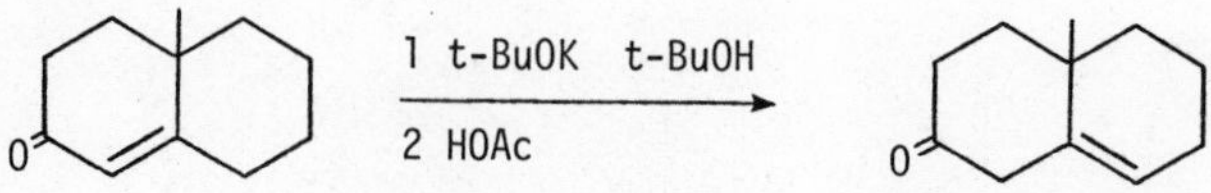

Tetr Lett (1962) 669
Steroids (1964) 3 183

Via enol trichloroacetate   Tetrahedron (1969) 25 1717

$Me_2CHCH=CHCOMe$

$h\nu$ Et_2O

$Me_2C=C \cdot CH_2COMe$

75%

Tetr Lett (1964) 1203
JCS C (1966) 571

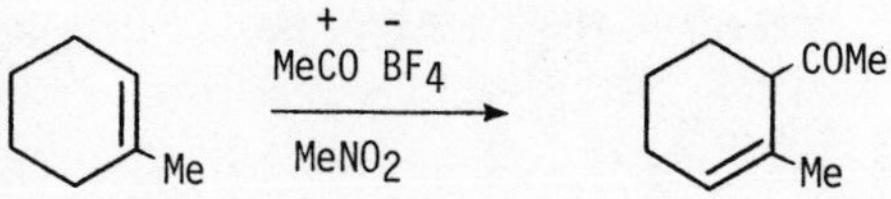

$\overset{+}{MeCO}\ \overset{-}{BF_4}$

$MeNO_2$

70%

Tetr Lett (1971) 3101

PhCOOH

$ClMgCH=CH_2$

THF

$PhCOCH_2CH_2CH=CH_2$

84%

Synthesis (1970) 189

$$\frac{PhOCH_2CH=CH_2 \quad Ph_3P}{Pd(OAc)_2 \quad C_6H_6}$$

70%

Tetr Lett (1973) 121

π-Methallylnickel

bromide　DMF

JACS (1967) $\underline{89}$ 2755

$$\frac{1 \; HCOOEt}{MeONa}$$

$$2 \; CH_2=CCH_2OH$$
$$\quad\quad\;\; | $$
$$\quad\quad\;\; Me$$

$$\frac{1 \; \triangle}{2 \; K_2CO_3}$$

62%

JOC (1970) $\underline{35}$ 570

$$\overset{OH}{\underset{}{Me_2CCH=CH_2}}$$

$$\xrightarrow[]{\overset{\overset{\textstyle Me}{|}}{CH_2=COMe} \quad H_3PO_4}$$

$$Me_2C=CHCH_2CH_2COMe$$

94%

Helv (1967) $\underline{50}$ 2091 2095
Tetr Lett (1969) 3243
JACS (1973) $\underline{95}$ 553

$$\overset{OH}{\underset{\underset{Me}{|}}{EtCCH=CH_2}}$$

$$\xrightarrow[MeONa]{Diketene}$$

$$\overset{MeCOCH_2COO}{\underset{\underset{Me}{|}}{EtCCH=CH_2}}$$

$$\xrightarrow[]{Al(OPr-i)_3}$$

$$\underset{\underset{Me}{|}}{EtC=CHCH_2CH_2COMe}$$

Tetrahedron (1969) $\underline{25}$ 1667

$$\underset{\underset{Me}{|}}{CH_2=CCH_2Cl}$$

$$\xrightarrow[K_2CO_3 \quad EtOH]{MeCOCH_2COMe}$$

$$\underset{\underset{Me}{|}}{CH_2=CCH_2CH_2COMe}$$

Org Synth (1973) Coll Vol 5 767

$$MeC{\equiv}CCH_2CH_2COMe \xrightarrow[\text{EtOAc}]{H_2 \quad Pd\text{-}CaCO_3} MeCH{=}CHCH_2CH_2COMe \qquad 78\%$$

JCS (1951) 2445
JACS (1972) 94 507

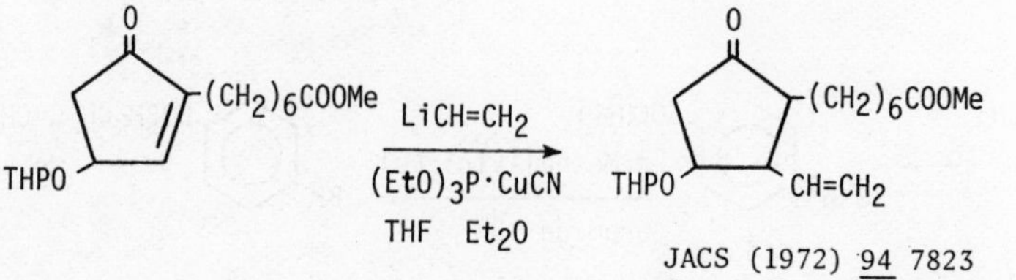

LiCH=CH₂ / (EtO)₃P·CuCN / THF Et₂O

JACS (1972) 94 7823
Can J Chem (1970) 48 1626
Helv (1971) 54 1939
Org React (1972) 19 1

(i-Bu)₂AlCH=CHR   Tetr Lett (1972) 4083

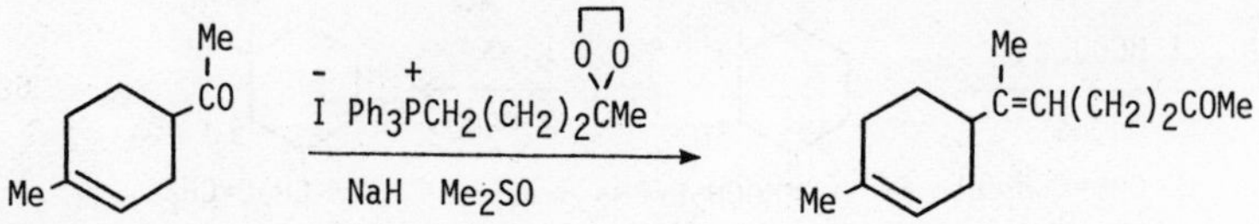

JCS Perkin I (1972) 2653

Also via: Acetylenic ketones   Section 309
                β-Hydroxyketones   330

Section 375   Nitrile — Nitrile

Dinitriles

$$PhCH_2Cl \xrightarrow[\text{Me}_2SO]{CH_2(CN)_2 \quad NaH} (PhCH_2)_2C(CN)_2 \qquad 75\%$$

JOC (1961) 26 4112

JACS (1954) <u>76</u> 1076

JACS (1958) <u>80</u> 1752

~40%

Org Synth (1963) Coll Vol 4 273 274 66
JACS (1965) <u>87</u> 4403

PhCH$_2$CN $\xrightarrow{\text{CH}_2=\text{CHCN EtONa}}$ PhCHCN   20-30%
                                                             |
                                                             CH$_2$CH$_2$CN

JACS (1943) <u>65</u> 437
Org React (1949) <u>5</u> 79

MeCH=CHCN $\xrightarrow[\text{DMF}]{\text{Electrolysis   Et}_4\overset{+}{\text{N}} \overset{-}{\text{OTs}}}$ MeCHCH$_2$CN
                                                                                        |
                                                                                    MeCHCH$_2$CN

Compt Rend <u>C</u> (1967) <u>265</u> 751

Me$_2$C(CH$_2$)$_2$COOH $\xrightarrow[\text{Na}_2\text{CO}_3 \text{ MeOH}]{\text{Electrolysis}}$ Me$_2$C(CH$_2$)$_4$CMe$_2$ 60%
| | |
CN   CN   CN

Bull Soc Chim Fr (1970) 183
Z Naturforsch (1947) <u>2b</u> 185

Section 376   Nitrile — Olefin

αβ, βγ, and higher olefinic nitriles

$$BuC{\equiv}CH \xrightarrow[\substack{2 \ MeLi \ \ Et_2O \\ 3 \ (CN)_2}]{1 \ (i\text{-}Bu)_2AlH \ \ hexane} BuCH{=}CHCN \qquad 87\%$$

JACS (1968) <u>90</u> 7139

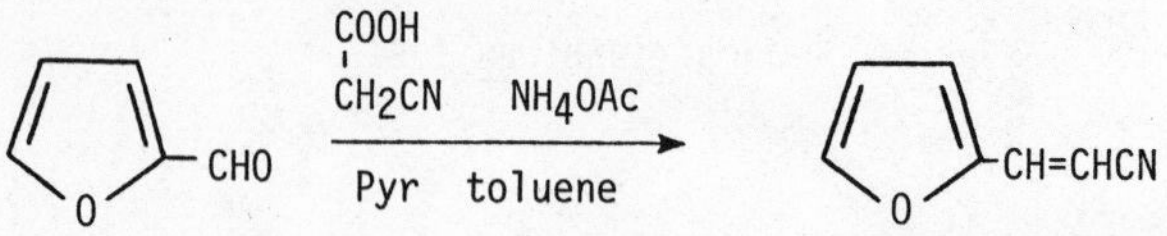

Org Synth (1973) Coll Vol 5 585
JCS Perkin I (1973) 2241
JACS (1949) <u>71</u> 3562

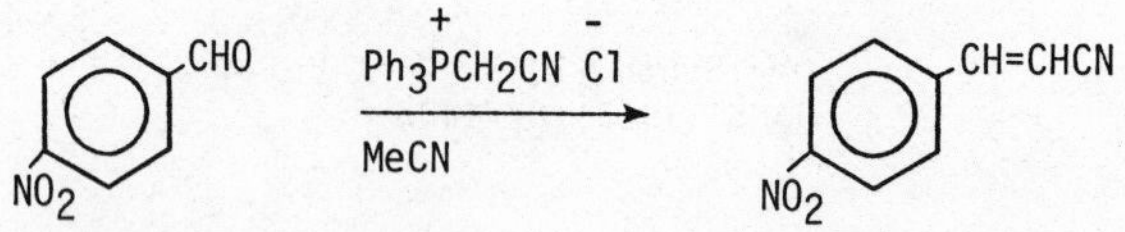

74-78%

42%

Ber (1970) <u>103</u> 2077
JOC (1971) <u>36</u> 2026

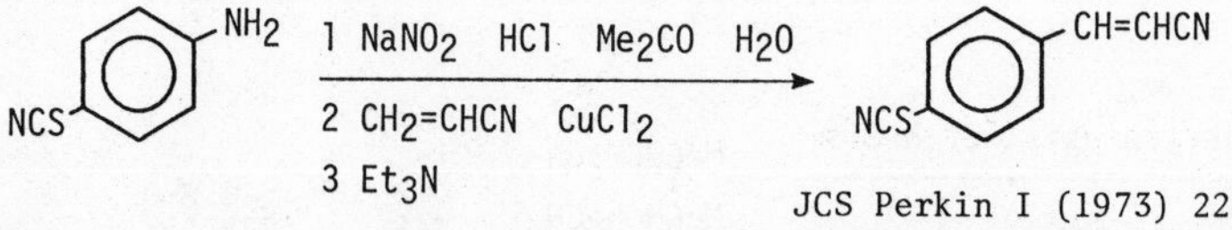

60%

JCS Perkin I (1973) 2241
Org React (1960) <u>11</u> 189

$$PhH \xrightarrow[HOAc]{CH_2{=}CHCN \ \ Pd(OAc)_2} PhCH{=}CHCN$$

Tetrahedron (1969) <u>25</u> 4819

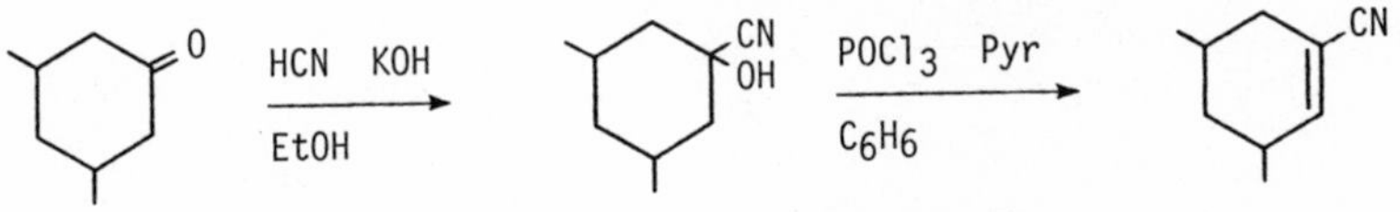
HCN KOH
EtOH
CN
OH
POCl3 Pyr
C6H6
CN

JOC (1972) 37 2201
Tetrahedron (1968) 24 3127
Can J Chem (1969) 47 3266

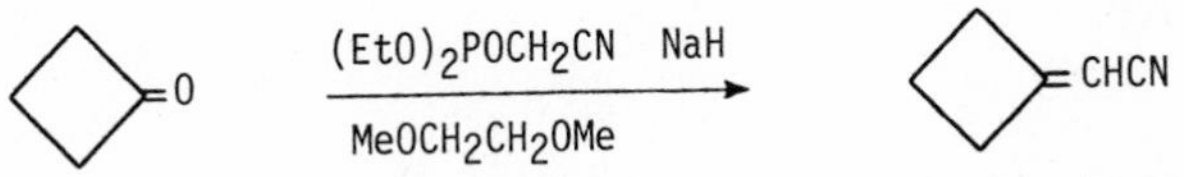
(EtO)2POCH2CN NaH
MeOCH2CH2OMe
CHCN

75%

JOC (1971) 36 1024
(1965) 30 505

Ph_2CO —(1 CH_3CN BuLi THF hexane; 2 H_3PO_4 H_2O)→ $Ph_2C{=}CHCN$ 58%

JOC (1968) 33 3402
Angew (1972) 84 767
(Internat Ed 11 722)

$PhCH_2CN$ —(HCHO $PhCH_2\overset{+}{N}Me_3$ $\overset{-}{O}H$; EtOH)→ $PhC(CN){=}CH_2$ 86%

JACS (1950) 72 5645

EtCHO —($CH_2{=}CHCN$ Ph_3P; t-BuOH)→ $EtCH{=}CHCH_2CN$ 44%

Tetr Lett (1967) 2401
(1964) 1653

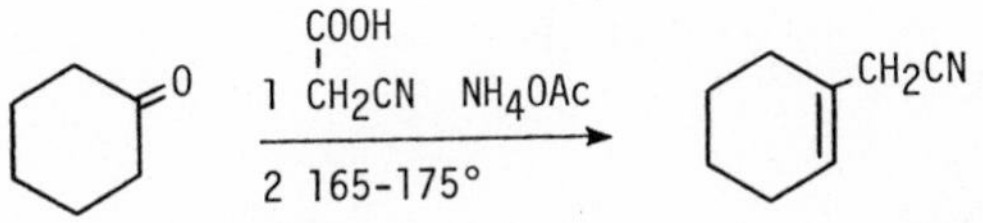
COOH
1 CH2CN NH4OAc
2 165-175°
CH2CN

76-91%

Org Synth (1963) Coll Vol 4 234
(1973) Coll Vol 5 585

Via enamine Chem Pharm Bull (1973) 21 1601

$$\underset{\underset{Ph}{|}}{EtCHCN} \xrightarrow[\text{NaOH \quad Me}_2SO]{HC\equiv CH \quad PhCH_2\overset{+}{N}Et_3 \ \overset{-}{C}l} \underset{\underset{Ph}{|}}{\overset{CH=CH_2}{\underset{|}{EtCCN}}}$$

80%

Tetr Lett (1966) 5489

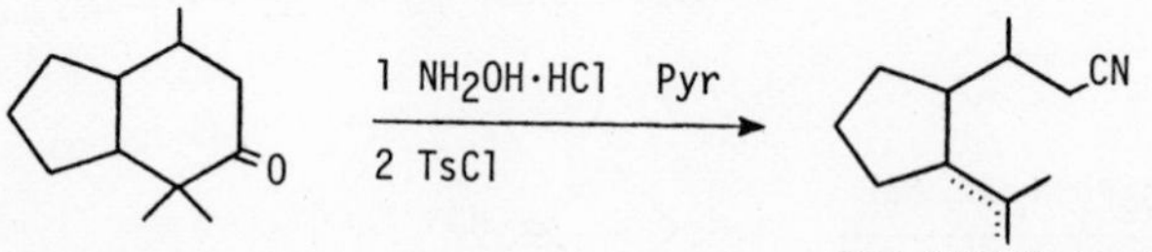

64%

JOC (1970) <u>35</u> 186
   (1962) <u>27</u> 1961 1965
Coll Czech (1962) <u>27</u> 377

Also via: Acetylenic nitriles (Section 310)

Section 377   <u>Olefin — Olefin</u>

$$EtC\equiv CEt \xrightarrow[\text{2 I}_2 \quad NaOH]{\text{1 B}_2H_6 \quad THF} \underset{\text{cis} \quad \overset{Et}{} \quad \text{trans}}{EtCH=C-\underset{\underset{Et}{|}}{\overset{\overset{Et}{|}}{C}}=CHEt}$$

68%

JACS (1968) <u>90</u> 6243
JOC (1973) <u>38</u> 1617

$$BuC\equiv CH \dashrightarrow BuC\equiv CCl \xrightarrow[\substack{\text{2 PrC}\equiv CH \\ \text{3 MeONa} \\ \text{4 i-PrCOOH}}]{\substack{\text{Me} \\ \text{i-PrCBH}_2 \\ \text{1} \quad \text{Me}}} \underset{\text{trans \ trans}}{BuCH=CHCH=CHPr}$$

45%

Chem Comm (1973) 606

$$BuC\equiv CC\equiv CBu \xrightarrow[\substack{2\ HOAc \\ 3\ H_2O_2\ \ NaOH\ \ H_2O}]{1\ Dicyclohexylborane\ \ THF} BuCH=CHCH=CHBu$$

cis   cis   79%

JACS (1970) <u>92</u> 4068

$$BuC\equiv CH \xrightarrow[\substack{2\ CuCl\ \ THF \\ 3\ H_2SO_4\ \ H_2O}]{1\ (i\text{-}Bu)_2AlH\ \ hexane} BuCH=CHCH=CHBu$$

trans trans   73%

JACS (1970) <u>92</u> 6678
Annalen (1960) <u>629</u> 222

$360°$ 42%

JACS (1953) <u>75</u> 384

$$\begin{array}{c} CH_3 \\ MeCOH \\ MeCOH \\ CH_3 \end{array} \xrightarrow{HBr\ \ H_2O} \begin{array}{c} MeC=CH_2 \\ MeC=CH_2 \end{array}$$

Org Synth (1955) Coll Vol 3 312

$Me_2SO\ \ 160°$ JOC (1964) <u>29</u> 123

$$PhCHO \xrightarrow[EtOLi\ \ EtOH]{PhCH=CHCH_2\overset{+}{P}Ph_3\ \overset{-}{Cl}} PhCH=CHCH=CHPh$$

~60%

Org Synth (1973) Coll Vol 5 499

$$\xrightarrow[\substack{2\ CH_2=CHCH=CH_2\ \ CuCl \\ NaOAc\ \ Me_2CO\ \ H_2O \\ 3\ KOH\ \ MeOH}]{1\ NaNO_2\ \ HCl\ \ H_2O}$$

57-62%

Org Synth (1963) Coll Vol 4 727
Org React (1960) <u>11</u> 189

PrBr $\xrightarrow[\text{2 CH}_2\text{=CHCH=CHCH}_2\text{Br}]{\text{1 Mg   THF}}$ PrCH$_2$CH=CHCH=CH$_2$

Bull Soc Chim Fr (1964) 2485

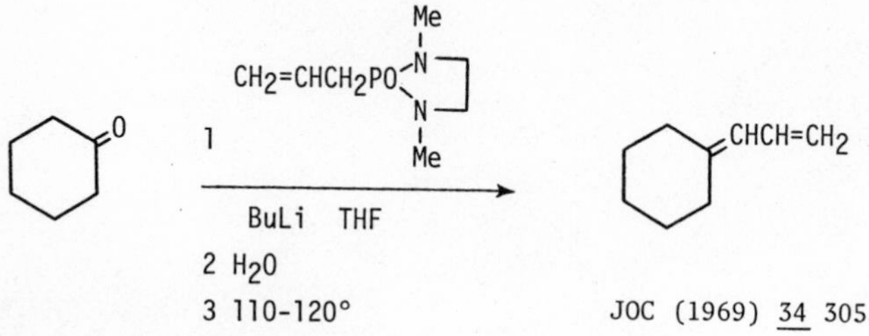

JOC (1969) 34 3053

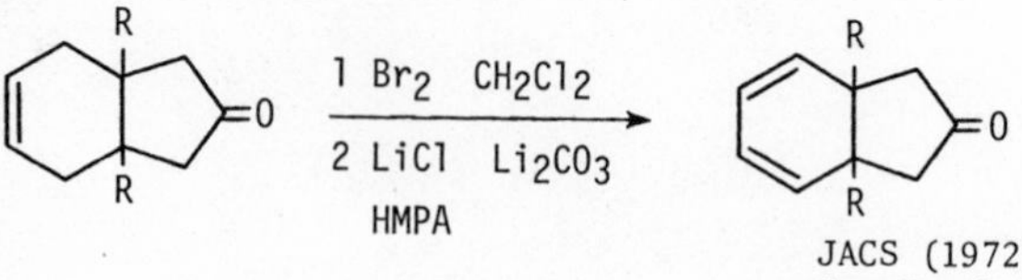

JACS (1972) 94 2155
Org Synth (1973) Coll Vol 5 285

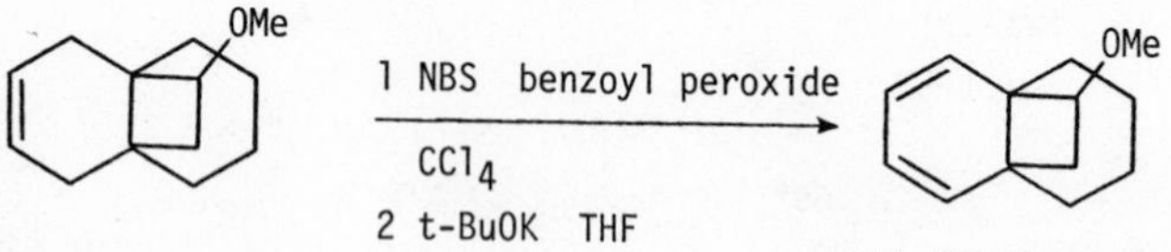

59%

JACS (1972) 94 7118

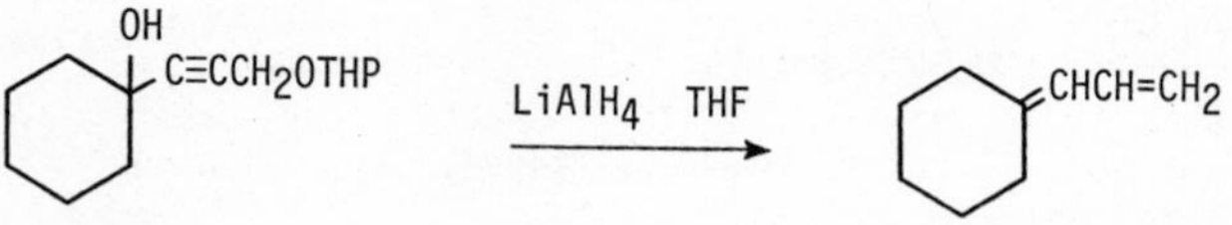

52%

Acta Chem Scand (1972) 26 2540

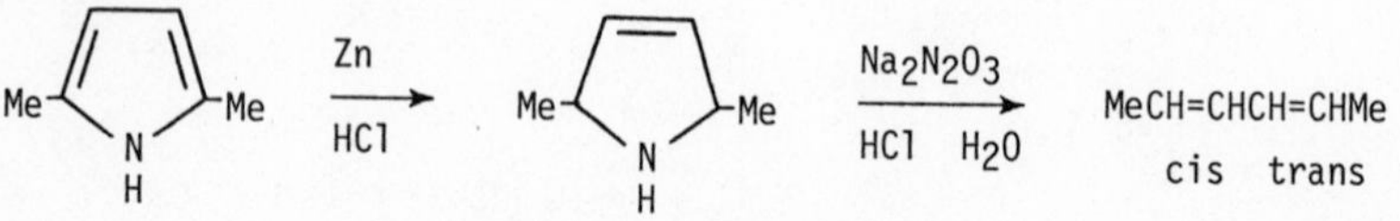

JACS (1966) 88 1335 2858

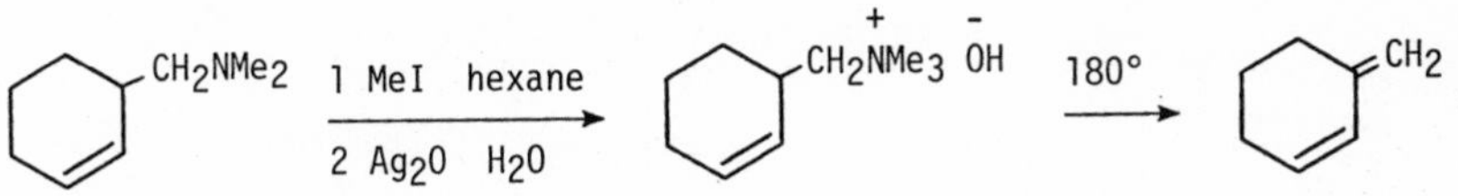

JOC (1972) <u>37</u> 2201

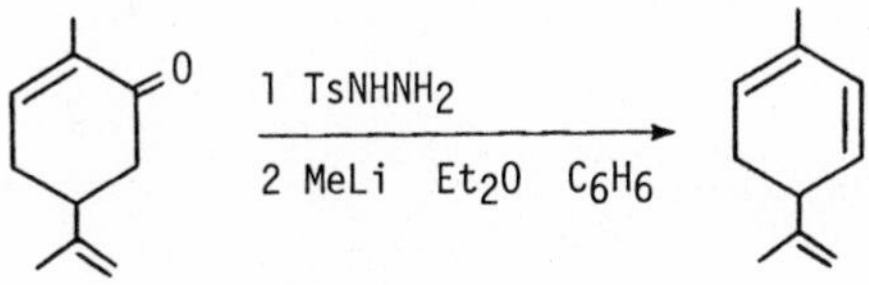

80%

JACS (1968) <u>90</u> 4762

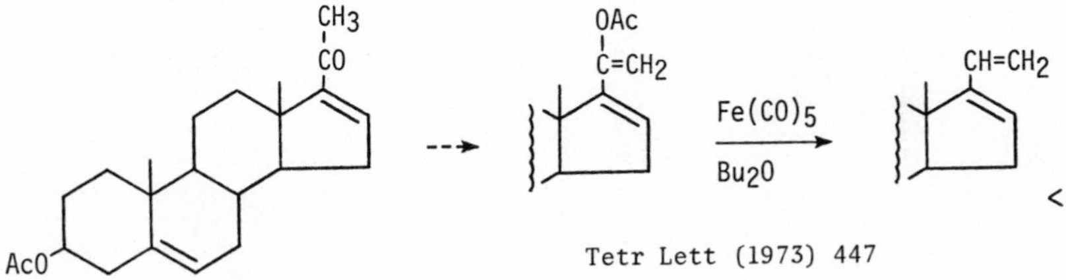

Tetr Lett (1973) 447

PhCH=CHBr →(1 Mg THF / 2 SOCl₂)→ PhCH=CHCH=CHPh

JOC (1972) <u>37</u> 3749

Mg; CuCl   Angew (1967) <u>79</u> 101
(Internat Ed <u>6</u> 85)

K₄[Ni₂(CN)₆]   Tetr Lett (1970) 4567

$C_6H_{13}C{\equiv}CCH_2C{\equiv}C(CH_2)_6Cl$ $\xrightarrow[\text{heptane}]{(i\text{-Bu})_2AlH}$ $C_6H_{13}CH{=}CHCH_2CH{=}CH(CH_2)_6Cl$ 82%

JOC (1963) <u>28</u> 1254

Reviews: Reductions by Metal-Ammonia Solutions and Related Reagents
Advances in Org Chem (1972) <u>8</u> 1

A Comparison of Methods Using Lithium/Amine and Birch
Reduction Systems   Synthesis (1972) 391

Aust J Chem (1955) <u>8</u> 512

Electrolytic reduction   JOC (1969) <u>34</u> 3970

$CH_2=CHCH_2Br$ $\xrightarrow{\begin{array}{l}1\ Mg\ \ Et_2O\\2\ BrCH_2CHOEt\\ \ \ \ \ \ \ \ \ \ \ Br\\3\ Zn\ \ ZnCl_2\\ \ \ \ BuOH\end{array}}$ $CH_2=CHCH_2CH=CH_2$ <59%

Org Synth (1963) Coll Vol 4 748

$CH_2=CHOBu$ $\xrightarrow{(CH_2=\overset{Me}{\overset{|}{C}}-CH_2)_3B}$ $CH_2=CHCH_2\overset{Me}{\overset{|}{C}}=CH_2$ 80-90%

Tetr Lett (1971) 2127

$CH_2=CHCH_2Br$ $\xrightarrow{\begin{array}{l}1\ PhSCH_2CH=CH_2\ \ BuLi\ \ THF\\ \ \ 1,4\text{-diazabicyclo}[2,2,2]octane\\2\ Li\ \ EtNH_2\end{array}}$ $CH_2=CHCH_2CH_2CH=CH_2$

Tetr Lett (1968) 5629
(1969) 3707

$(MeO)_2CHCH_2CH_2\overset{Me}{\overset{|}{C}}=CHCH_2Cl$ $\xrightarrow[THF\ \ HMPA]{CH_2=\overset{Me}{\overset{|}{C}}CH_2MgCl}$ $(MeO)_2CHCH_2CH_2\overset{Me}{\overset{|}{C}}=CHCH_2CH_2\overset{Me}{\overset{|}{C}}=CH_2$

Tetr Lett (1969) 1393
Org Synth (1955) Coll Vol 3 121
Bull Soc Chim Fr (1964) 2485
Tetr Lett (1972) 1471

PhH $\xrightarrow{\overset{COCl}{\overset{|}{(CH_2)_8COCl}}}$ $PhCO(CH_2)_8COPh$ $\xrightarrow[MeO(CH_2)_4OMe]{h\nu}$ $CH_2=CH(CH_2)_2CH=CH_2$ <53%

JOC (1971) <u>36</u> 1838

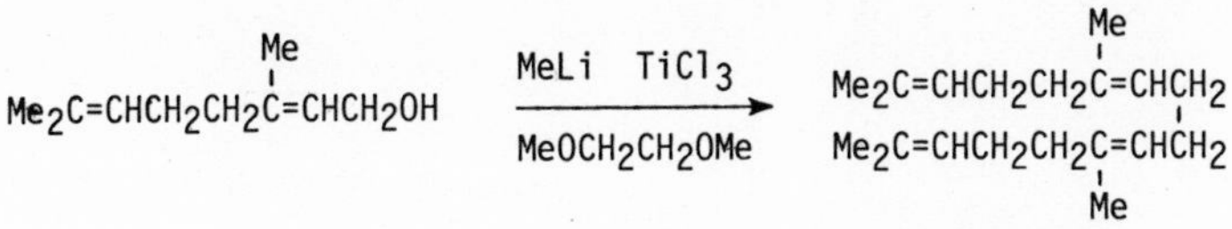

Me
$Me_2C=CHCH_2CH_2C=CHCH_2OH$
MeLi TiCl3
MeOCH2CH2OMe
Me
$Me_2C=CHCH_2CH_2C=CHCH_2$
$Me_2C=CHCH_2CH_2C=CHCH_2$
Me

JACS (1968) 90 209

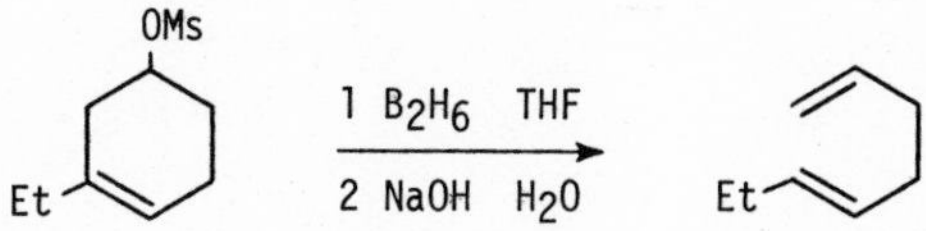

OMs
Et
1 B2H6 THF
2 NaOH H2O
Et

Synthesis (1971) 229

$$PrC{\equiv}C(CH_2)_4C{\equiv}CCONHBu\text{-}i \xrightarrow[\text{EtOAc}]{H_2 \quad Pd\text{-}CaCO_3} PrCH{=}CH(CH_2)_4CH{=}CHCONHBu\text{-}i$$

JCS (1950) 115

$$CH_2{=}CH(CH_2)_4Br \dashrightarrow CH_2{=}CH(CH_2)_4MgBr \xrightarrow[FeCl_3 \quad THF]{CH_2{=}CHBr} CH_2{=}CH(CH_2)_4CH{=}CH_2$$

Synthesis (1971) 303

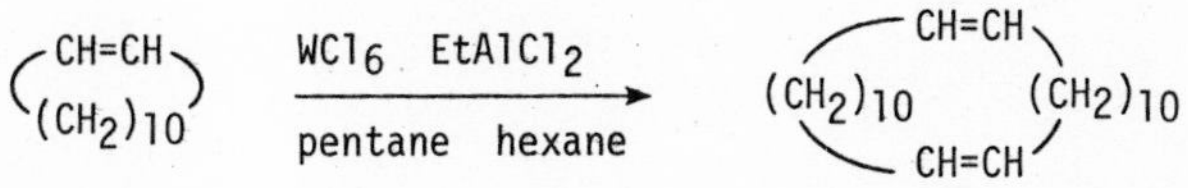

CH=CH
(CH2)10
WCl6 EtAlCl2
pentane hexane
CH=CH
(CH2)10
(CH2)10
CH=CH

~31%

Synthesis (1972) 134

$$CH_2{=}CH(CH_2)_8COONa \xrightarrow[\text{MeOH}]{\text{Electrolysis}} CH_2{=}CH(CH_2)_{16}CH{=}CH_2$$

JCS (1953) 2393